CHEMICAL ANALYSIS

Vol. 1. **The Analytical Chemistry of Industrial Poisons, Hazards, and Solvents.** *Second Edition.* By Morris B. Jacobs.
Vol. 2. **Chromatographic Adsorption Analysis.** By Harold H. Strain (*out of print*)
Vol. 3. **Colorimetric Determination of Traces of Metals.** *Third Edition.* By E. B. Sandell
Vol. 4. **Organic Reagents Used in Gravimetric and Volumetric Analysis.** By John F. Flagg (*out of print*)
Vol. 5. **Aquametry: Application of the Karl Fischer Reagent to Quantitative Analyses Involving Water.** By John Mitchell, Jr. and Donald Milton Smith (*temporarily out of print*)
Vol. 6. **Analysis of Insecticides and Acaricides.** By Francis A. Gunther and Roger C. Blinn (*out of print*)
Vol. 7. **Chemical Analysis of Industrial Solvents.** By Morris B. Jacobs and Leopold Scheflan
Vol. 8. **Colorimetric Determination of Nonmetals.** Edited by David F. Boltz
Vol. 9. **Analytical Chemistry of Titanium Metals and Compounds.** By Maurice Codell
Vol. 10. **The Chemical Analysis of Air Pollutants.** By Morris B. Jacobs
Vol. 11. **X-Ray Spectrochemical Analysis.** *Second Edition.* By L. S. Birks
Vol. 12. **Systematic Analysis of Surface-Active Agents.** *Second Edition.* By Milton J. Rosen and Henry A. Goldsmith
Vol. 13. **Alternating Current Polarography and Tensammetry.** By B. Breyer and H. H. Bauer
Vol. 14. **Flame Photometry.** By R. Herrmann and J. Alkemade
Vol. 15. **The Titration of Organic Compounds** (*in two parts*). By M. R. F. Ashworth
Vol. 16. **Complexation in Analytical Chemistry: A Guide for the Critical Selection of Analytical Methods Based on Complexation Reactions.** By Anders Ringbom
Vol. 17. **Electron Probe Microanalysis.** *Second Edition.* By L. S. Birks
Vol. 18. **Organic Complexing Reagents: Structure, Behavior, and Application to Inorganic Analysis.** By D. D. Perrin
Vol. 19. **Thermal Methods of Analysis.** By Wesley Wm. Wendlandt
Vol. 20. **Amperometric Titrations.** By John T. Stock
Vol. 21. **Reflectance Spectroscopy.** By Wesley Wm. Wendlandt and Harry G. Hecht
Vol. 22. **The Analytical Toxicology of Industrial Inorganic Poisons.** By the late Morris B. Jacobs
Vol. 23. **The Formation and Properties of Precipitates.** By Alan G. Walton
Vol. 24. **Kinetics in Analytical Chemistry.** By Harry B. Mark, Jr. and Garry A. Rechnitz
Vol. 25. **Atomic Absorption Spectroscopy.** By Walter Slavin
Vol. 26. **Characterization of Organometallic Compounds** (*in two parts*). Edited by Minoru Tsutsui
Vol. 27. **Rock and Mineral Analysis.** By John A. Maxwell
Vol. 28. **The Analytical Chemistry of Nitrogen and Its Compounds** (*in two parts*). Edited by C. A. Streuli and Philip R. Averell
Vol. 29. **The Analytical Chemistry of Sulfur and Its Compounds** (*in three parts*). By J. H. Karchmer
Vol. 30. **Ultramicro Elemental Analysis.** By Günther Tölg
Vol. 31. **Photometric Organic Analysis** (*in two parts*). By Eugene Sawicki
Vol. 32. **Determination of Organic Compounds: Methods and Procedures.** By Frederick T. Weiss
Vol. 33. **Masking and Demasking of Chemical Reactions.** By D. D. Perrin
Vol. 34. **Neutron Activation Analysis.** By D. De Soete, R. Gijbels, and J. Hoste
Vol. 35. **Laser Raman Spectroscopy.** By Marvin C. Tobin
Vol. 36. **Emission Spectrochemical Analysis.** By Morris Slavin
Vol. 37. **Analytical Chemistry of Phosphorus Compounds.** Edited by M. Halmann
Vol. 38. **Luminescence Spectrometry in Analytical Chemistry.** By J. D. Winefordner, S. G. Schulman, and T. C. O'Haver
Vol. 39. **Activation Analysis with Neutron Generators.** By Sam S. Nargolwalla and Edwin P. Przybylowicz

CHEMICAL ANALYSIS

A SERIES OF MONOGRAPHS ON
ANALYTICAL CHEMISTRY AND ITS APPLICATIONS

Editors

P. J. ELVING · I. M. KOLTHOFF

Advisory Board

J. Badoz-Lambling	J. J. Lingane	E. B. Sandell
George E. Boyd	John A. Maxwell	Eugene Sawicki
Raymond E. Dessy	Louis Meites	Carl A. Streuli
Leslie S. Ettre	John Mitchell	Donald E. Smith
Dale J. Fisher	George H. Morrison	Wesley Wendlandt
Barry L. Karger	Charles H. Reilley	James D. Winefordner
	Anders Ringbom	

VOLUME 39

A WILEY-INTERSCIENCE PUBLICATION

JOHN WILEY & SONS
New York / London / Sydney / Toronto

Activation Analysis with Neutron Generators

SAM S. NARGOLWALLA

Scintrex, Ltd.
Concord, Ontario, Canada

EDWIN P. PRZYBYLOWICZ

Research Laboratories
Eastman Kodak Company
Rochester, New York

A WILEY-INTERSCIENCE PUBLICATION

JOHN WILEY & SONS
New York / London / Sydney / Toronto

Chem
QD
606
.N37

Copyright © 1973, by John Wiley & Sons, Inc.

All rights reserved. Published simultaneously in Canada.

No part of this book may be reproduced by any means, nor transmitted, nor translated into a machine language without the written permission of the publisher.

Library of Congress Cataloging in Publication Data

Nargolwalla, Sam S.
   Activation analysis with neutron generators.

   (Chemical analysis, v. 39)
   "A Wiley-Interscience publication."
   Includes bibliographical references.
   1. Radioactivation analysis. 2. Neutron sources. I. Przybylowicz, Edwin P., joint author. II. Title. III. Series.

QD606.N37   545'.822   73-8793
ISBN 0-471-63031-4

Printed in the United States of America

10 9 8 7 6 5 4 3 2 1

FOREWORD

From the beginning of neutron activation analysis in the late 1930s, the source of neutrons has been the factor limiting its widespread use. Natural radioisotope sources of neutrons have been and, in large part, still remain laboratory curiosities. Research reactors have been available since the 1940s, but their sophisticated engineering plants and government-licensed procedures have often discouraged analysts, despite the missionary-like efforts of a number of proponents. On the other hand, the introduction of relatively inexpensive neutron generators in the late 1950s made activation analysis for certain elements available to all at a purchase price and maintenance cost comparable to many other sophisticated analytical techniques. With the neutron generator, activation analysis can be exploited for its reliability in major and minor constituent analysis rather than for its high sensitivity.

The comprehensive treatment of activation analysis with neutron generators presented in this book provides an excellent introduction to the subject for newcomers to the field, and at the same time can serve as a comprehensive guide and reference source for the seasoned analyst. It presents an exhaustive treatment of the uniqueness of the activation analysis technique as it specifically applies to the neutron generator source. It can be particularly recommended to industrial analysts who are searching for new ways of solving specific problems with instruments they can control in their own laboratories.

The collaboration of the authors in this area, and indeed the generation of this book, were the outgrowth of a leave of absence Dr. Przybylowicz took from Eastman Kodak Company. His time was spent at the National Bureau of Standards as an Industrial Research Associate in a study of the potential application of activation analysis to problems at Eastman Kodak. This volume attests to the success of his studies and this collaboration, and indeed to the widespread applicability of "neutron generator activation analysis" to the industrial laboratory.

W. WAYNE MEINKE, Chief
Analytical Chemistry Division
Institute for Materials Research
National Bureau of Standards
Washington, D.C.

PREFACE

This book had its genesis in 1968 while one of the authors (E.P.P.) was on assignment as Research Associate in the Activation Analysis Section of the National Bureau of Standards, Gaithersburg, Maryland. Shortly before that time the Bureau's 14-MeV neutron activation facility had become operational and had demonstrated that activation analysis with neutron generators could be a very precise and accurate technique in the analytical laboratory. The development of such a facility, however, required a substantial amount of engineering know-how, as well as a background in activation analysis. Several members of the Activation Analysis Section of the Bureau were qualified in this area; thus it was decided that this information should be made generally available. Since there was no comprehensive text on this subject available it seemed appropriate to document all that had been learned in setting up this facility so that others might have the benefit of this experience.

We believe we have accurately represented the state of the art in the field of activation analysis with neutron generators. It is difficult to keep from being myopic in writing about a technique with which one has had close familiarity. We hope we have neither oversold nor underrated the technique and its applications but have presented a fair assessment of its capabilities and limitations.

Inevitably the writing of such a book becomes a series of compromises. The scope and treatment of certain topics must be limited to be concise, yet for the person interested in detail, this may appear to be an incomplete treatment. We have tried to strike a balance between these two extremes, but only the reader can determine whether we have succeeded. Specifically, we originally planned to include a chapter on statistical treatment of data but found that the volume had already exceeded original guidelines. Consequently, this discussion was eliminated in the belief that such information could be obtained from other readily available sources.

A book like this would not be possible without the help of many people. First, we are indebted to Dr. W. Wayne Meinke of the National Bureau of Standards for his encouragement of this project at the outset and to the tremendous library services provided by the Bureau in the field of activation analysis. We would like to thank Miss Sheryl Birkhead of the Bureau staff who did much in the early planning of this book to provide us with bibliographic information. Many workers in the field of activation analysis were generous with their technical information. A special note of appreciation goes to Dr. Hugh Dibbs of the Department of Energy, Mines and Resources, Mines Branch, Government of Canada, for his thorough review of the final manuscript and the many suggestions he made which improved it significantly. We are solely responsible for errors which may appear. Thanks are due to the staff of the Eastman Kodak Research Laboratories for their help in putting this manuscript together, specifically for the editorial services of Mrs. Ardelle Kocher, the art work of Loren Jenné, and the typing of Mrs. Mary Mickelson. We are grateful to Mrs. Betty Lou Willard, who provided yeoman service in the typing, assembly, and correction of the manuscript.

Finally, a word of appreciation to our families who, in addition to providing encouragement, also knew many lonely hours while we pursued our goal of putting this manuscript together.

S. S. NARGOLWALLA
Concord, Ontario, Canada

E. P. PRZYBYLOWICZ
Rochester, New York

February 1973

ACKNOWLEDGMENTS

The authors are most grateful for permission from publishers and authors to reproduce the following figures and tables.

J. Prud'homme, Texas Nuclear Corporation Neutron Generators 1062, Texas Nuclear Division, Nuclear Chicago, Austin, Texas	Figs. 2.13, 3.2–3.4, 3.9, 3.10, 3.12–3.16, 3.21, 4.6, and 4.7
P. L. Jessen, "Design Considerations for Low Voltage Accelerators," Kaman Nuclear Report, KN-68-459(R), 1968. Published by Kaman Sciences Corporation, Colorado Springs, Colorado	Figs. 2.14 and 2.15
H. P. Dibbs, "Activation Analysis with a Neutron Generator," Research Report R155, 1965. Mines Branch, Department of Energy, Mines and Resources, Ottawa, Canada. Reproduced by permission of Information Canada	Figs. 2.16, 2.17, 4.12, and 4.24 Tables A-2 and A-3
H. J. Price, *Graphical Analysis of Theoretical Neutron Flux*, 1964. Published by Kaman Sciences Corporation, Colorado Springs, Colorado	Figs. 2.19–2.29
B. T. Kenna and F. J. Conrad, *Health Phys.*, **12**, 566 (1966). By permission of Microforms International Marketing Corporation, Elmsford, New York	Figs. 2.30, 2.31, and 2.32
J. Csikai, M. Buczkó, Z. Bödy, and A. Demény, *At. Energy Rev.*,**VII**, No. 4, (1969), International Atomic Energy Agency (IAEA), Vienna, Austria	Figs. 2.35, 6.3 A.1a–A.1e, A.2a–A.2e, A.3a–A.3d Tables A-1, 6.1, 6.2, 6.3, and 6.5
D. E. Wood, in *Advances in Activation Analysis*, Vol. 2, p. 277 (1972). By permission of the Academic Press, New York, New York	Fig. 3.11
L. D. Hall, *A Little Bit About Almost Nothing*, Second Edition, 1963, Ultek Division, Perkin-Elmer Corporation, Palo Alto, California	Figs. 3.19, 3.20, and 3.24

R. Steinberg and D. L. Alger, "A Solution to Tritium Pump Contamination for Small Accelerators," National Aeronautics and Space Administration Technical Memorandum TMX-52578, 1969. Courtesy of NASA, Lewis Research Center, Cleveland, Ohio

Figs. 3.22 and 3.33

J. H. Coon, *Fast Neutron Physics*, Part 1, Vol. IV, J. B. Marion and J. L. Fowler, Eds., *Interscience Monographs and Texts in Physics and Astronomy*, Interscience, New York, 1960, p. 677. By permission of John Wiley and Sons Inc., New York, New York

Figs. 3.25, 3.28, and 3.30

W. W. Meinke and R. W. Shideler, *Michigan Memorial Phoenix Project Report* MMPP-191-1, 1961, University of Michigan, Ann Arbor, Michigan

Figs. 3.31 and 4.18

O. Reifenschweiler, "Sealed-off Neutron Tube: The Underlying Research Work," Philips Research Report 16, 401 (1961). © N. V. Philips' Gloeilampenfabrieken, Eindhoven, Netherlands

Fig. 3.33

H. I. Oshry, "Fast Neutron Activation Analysis with a Portable Source," *Proceedings 1961 International Conference, Modern Trends in Activation Analysis*, p. 28, 1961. By permission of The Texas A and M University, College Station, Texas

Fig. 3.34

Courtesy of Kaman Sciences Corporation, Colorado Springs, Colorado. A-711 Accelerator Tube, manufactured by Kaman Sciences Corporation, Colorado Springs, Colorado

Fig. 3.35

J. E. Bounden, P. D. Lomer, and J. D. L. H. Wood "High-Output Neutron Tubes," *Proceedings 1965 International Conference, Modern Trends in Activation Analysis*, p. 182, 1965. By permission of The Texas A and M University, College Station, Texas

Fig. 3.36

N. G. Goussev, *Engineering Compendium on Radiation Shielding*, Vol. 1, p. 12, 1968. Springer-Verlag New York Inc., New York, New York

Fig. 4.1

E. E. Watson, R. G. Cloutier, and J. D. Berger, *Health Phys.*, **17**, 739 (1969). By permission of Microforms International Marketing Corporation, Elmsford, New York

Fig. 4.3

J. J. Broerse and F. J. Van Werven, *Health Phys.*, **12**, 83 (1966). By permission of Microforms International Marketing Corporation, Elmsford, New York

Fig. 4.4

L. R. Day and M. L. Mullender AERE (Gt. Brit.) Rept. AWRE NR-1/63, 1963. UKAEA copyright

Figs. 4.5, 4.16, and 4.17

E. P. Przybylowicz, private communication, 1972, Eastman Kodak Company, Rochester, New York

Fig. 4.13

B. L. Twitty, "Neutron Activation Facility at the National Lead Company of Ohio," Report NLCO-955, Summary Technical Report, April 1, 1965 to June 3, 1965, issued August 11, 1965. The work depicted herein was performed for the USAEC under contract No. AT-(30-1)-1156 by National Lead Company of Ohio, Cincinnati, Ohio

Figs. 4.19 and 4.20

J. D. L. H. Wood, D. W. Downton, and J. M. Bakes, "A Fast-Neutron Activation Analysis System with Industrial Applications," *Proceedings 1965 International Conference, Modern Trends in Activation Analysis*, p. 175, 1965. By permission of The Texas A and M University, College Station, Texas

Fig. 4.21

A-711 Neutron Generator, manufactured by Kaman Sciences Corporation, Colorado Springs, Colorado

Figs. 4.22 and 4.23

E. Cerrai and F. Gadda, *Energ. Nucl.*, **9**, 317 (1962)

Fig. 5.3

O. U. Anders and D. W. Briden, *Anal. Chem.*, **36**, 287 (1964). Reprinted from copyright by The American Chemical Society by permission of the copyright owner — Figs. 5.4 and 6.6

H. P. Dibbs, "The Determination of Oxygen by Fast Neutron Activation Analysis," Technical Bulletin TB 55, 1964. Mines Branch, Department of Energy, Mines and Resources, Ottawa, Canada. Reproduced by permission of Information Canada — Fig. 5.5

W. E. Mott and J. M. Orange, *Anal. Chem.*, **7**, 1338 (1965). Reprinted from copyright by The American Chemical Society by permission of the copyright owner — Fig. 5.6

F. F. Dyer, L. C. Bate, and J. E. Strain, *Anal. Chem.*, **39**, 1907 (1967). Reprinted from copyright by The American Chemical Society by permission of the copyright owner — Fig. 5.7

S. C. Mathur and G. Oldham, *Nucl. Energy*, September/October, 1967. By permission of Microforms International Marketing Corporation, Elmsford, New York — Table 6.4

E. Ricci, *J. Inorg. Nucl. Chem.*, **27**, 41 (1965). By permission of Microforms International Marketing Corporation, Elmsford, New York — Figs. 6.4 and 6.5; Table 6.7

J. Hoste et al., *Instrumental and Radiochemical Activation Analysis*, p. 74, 1971. © The Chemical Rubber Co. Used by permission of The Chemical Rubber Co. — Table 6.8

S. S. Nargolwalla, J. Niewodniczanski, and J. E. Suddueth, *J. Radioanal. Chem.*, **5**, 403 (1970) — Appendix IV edited from text

CONTENTS

1. Introduction   1

    I. General Comments   1
    II. Principle of Technique   2
    III. Advantages   3
    IV. Limitations   6
    V. Cost   6
    VI. Applications   7
    VII. Summary   7
         References   7–8

2. Production and Interaction of Fast Neutrons   9

    I. Introduction   9
    II. Basic Concepts   9
         A. Nuclear Reaction Notation   10
         B. Q Value or Interaction Energy   10
         C. Coulomb Barrier   14
    III. Neutron Production by Small Accelerators   15
         A. Neutron Yield Calculations   16
         B. Neutron-Producing Reactions   19
             1. D–D Reaction   19
             2. D–T Reaction   23
         C. Thermalization of Fast Neutrons   29
         D. Experimental Fast Neutron Flux Distributions   32
    IV. Fast-Neutron-Induced Reactions   42
         A. Classification of Nuclear Reactions   43
         B. Rare Reactions   56
         References   60

3. The Neutron Generator   63

    I. Introduction   63
    II. Pumped Neutron Generator   67
         A. Ion Sources   68
             1. Radiofrequency Ion Source (RF)   69

	a. Theory and Operation of the RF Ion Source	70
	2. Penning Ion Source (PIG)	72
	a. Theory and Operation of the PIG Source	73
	3. Duoplasmatron Ion Source (DP)	74
	a. Operation of the DP Source	76
	4. Anomalies in the Comparison of Ion Sources	76
B.	Gas Supply Regulation	78
C.	Accelerating Tube	80
	1. Van de Graaff Accelerating Tube	81
	2. Einsel-Lens Accelerating Tube	82
D.	Pulsing Systems	83
	1. Postacceleration Pulsing Unit	85
	2. Preacceleration Pulsing Unit	86
	3. Dual-Pulsing System	88
E.	Vacuum Systems	89
	1. Vacuum Pumps	90
	a. Oil Diffusion Pumping System	90
	b. Ion Pump	92
	c. Turbomolecular Vacuum Pump	97
	2. Vacuum-Measuring Devices Used in Neutron Generators	97
	a. Thermocouple Gauge	98
	b. Penning (Philips) Gauge	98
	3. Materials for High-Vacuum Systems	99
	4. Design Criteria for Neutron Generator Vacuum Systems	100
F.	Targets for Neutron Generators	101
	1. Target Considerations	102
	2. Deuterium and Tritium Targets for Use with Neutron Generators	104
	a. Frozen-Deuterium Targets (D_2O Ice)	104
	b. Absorbed-Hydrogen-Isotope Solid Target	105
	c. Preparation of Gas-in-Metal Targets	105
	d. Cooling of Gas-in-Metal Targets	106
	e. Characteristics of Gas-in-Metal Targets under Bombardment	107
	f. Target Configurations for "Long-Term" Irradiations	112
G.	High-Voltage Power Supply	113
H.	Safety Considerations for Neutron Generator Operation	114
I.	Operational Characteristics of the Pumped Neutron Generator	116
III. Sealed-Tube Neutron Source		118
A.	Introducton	118

B. Design Considerations of a Portable Sealed-Tube Neutron
   Source 119
   1. Ion Sources for Sealed-Tube Operation 120
   2. Replenisher System 125
   3. Targets for Sealed-Tube Neutron Source 126
   4. Performance of Sealed Tubes 126
IV. Evaluation and Impact of Low-Voltage Accelerator Neutron
   Sources 128
   References 129

4. Radiation Hazards and Shielding Considerations for Neutron Generator Facilities 132

I. Introduction 132
II. Radiation Exposure 133
   A. Radiation Dose-Flux Equivalence 134
      1. X- and Gamma Rays 135
      2. Monoenergetic Electrons and Beta Particles 136
      3. Protons and Charged Particles 138
      4. Neutrons 139
III. Hazards in Neutron Generator Operation 142
   A. Tritium—Internal Hazard And Contamination 143
      1. Physical and Biological Characteristics 143
      2. Release Mechanisms for Tritium in Neutron
         Generators 144
      3. Tritium Contamination 145
      4. Handling and Replacement of Tritium Targets 150
      5. Sampling and Measurement of Tritium 153
   B. External Hazards from Neutron Generator Operation 156
      1. Neutrons 156
         a. Attenuation of Fast Neutrons 156
         b. Shielding Calculations 162
         c. "Sky Shine" 170
         d. Neutron Detectors 171
      2. Bremsstrahlung from Target and Target
         Bombardment 173
         a. Radiation Levels 174
         b. Shielding and Control 174
      3. Induced Radioactivity in Materials of
         Construction 175
      4. Other Radiation Hazards from Beam Operation 176
         a. Prompt Gamma Radiation 176

b. Radiation Hazard from "Blank" Target Experiments	177
5. Nonnuclear Hazards	178
IV. Design and Construction of Biological Shields	178
A. Design Considerations	179
B. Construction of the Biological Shield	179
1. General Considerations	179
2. Typical Biological Shield Structures	180
V. Shielding of Detector Systems	196
References	198

5. Preparation and Transportation of Samples — 202

I. Introduction	202
II. Sample Preparation	203
A. Sample Container	203
1. Selection of Container Material	203
2. Geometrical Configuration of Container	204
B. Sample Encapsulation	207
1. Powder Samples	207
2. Liquid Samples	208
3. Metallic Rods and Disks	209
4. Reactive Metals	209
C. Sample Handling and Cleaning	210
III. Sample Transport	211
A. Sample Transport Systems	212
B. Auxiliary Sample Transport Equipment	221
References	222

6. Sources and Reduction of Systematic Error — 225

I. Introduction	225
II. Existence and Evaluation of Systematic Error	226
A. Nuclear Constants	227
B. Interfering Nuclear Reactions	229
C. Nuclear Reaction Recoil Effect	231
D. Error from Sample Blank	241
E. Errors from Neutron Flux Normalization Methods	243
F. Errors from Sample Attenuation	252
G. Instrumental Errors	269
References	271

7. Applications	**274**
I. Introduction	274
II. Prediction of Analysis Conditions	277
A. Activation Curves	278
B. Decay Curves	279
C. Counting Integral Factor	279
D. Use of Working Curves	280
III. Activation Analysis Conditions for the Elements	282
Introduction to Appendices	625
Appendix I. Activation Cross Sections for D–T Neutrons	626
Appendix II. Calculated Sensitivities for $\simeq$ 15-MeV and Thermal Neutron Activation Analysis with a Neutron Generator	634
Appendix III. Experimental Sensitivities for 14.7-MeV and Thermal Activation Analysis with a Neutron Generator	638
Appendix IV. Experimental Sensitivities for $\simeq$ 3-MeV Neutron Activation Analysis	643
Index	659

CHAPTER

1

INTRODUCTION

I. GENERAL COMMENTS

Neutron activation analysis has gained wide acceptance in recent years as a very sensitive analytical method for the qualitative and quantitative determination of a large number of elements. Although first reported in 1936 by Hevesy and Levi,[1] the technique was not widely used until the mid-1950s when sources for neutron irradiation became more readily available. In early activation analyses isotopic sources were used for the production of neutrons; these, however, were low in flux and consequently of limited usefulness. The increased availability of research reactors and the development of small, relatively inexpensive neutron generators were instrumental in accelerating the growth of this field. Great strides have been made during the past few years in developing activation analysis with neutron generators to the level of a quantitative analytical technique. For certain elements, this technique is comparable in precision, accuracy, and selectivity to such techniques as x-ray fluorescence and atomic absorption and, where useful, offers certain advantages in speed and uniqueness.

Several monographs on activation analysis are available.[2-7] These deal primarily with activation using reactor neutrons. While the theory of activation analysis is the same regardless of the irradiation source, the technology of utilizing neutron generators for activation analysis is sufficiently different from that involved in reactor activation that it requires separate discussion. The successful use of neutron generators for analysis can be attributed in large measure to technological advances in equipment design which have had as their basis a thorough understanding of the facts that control the precision, accuracy, sensitivity, and selectivity of this technique.

The need for a comprehensive treatment of activation analysis with neutron generators has prompted the writing of this treatise. The commercial availability and moderate cost of neutron generators have placed this technique within reach of every well-equipped industrial and academic analytical laboratory. The investment in time and effort in setting up an activation analysis laboratory utilizing a neutron generator is no greater than that involved for many contemporary instrumental techniques. Our intent is to describe comprehensively the technology involved in carrying out quantitative analysis with a neutron generator and to discuss reported applications of this technique so as to provide a substantive "state of the art" summary. We intend the book to serve both as an introduction to the subject for those just entering the field and as a guide and reference source for veteran practitioners. We hope that by bringing together the significant information on the use of neutron generators for activation analysis, this treatise will stimulate more general usage of the technique.

Information on the data processing and statistical treatment of gamma-ray spectra is not given in this text since this subject has been extensively treated in other books on general activation analysis procedures. The reader is referred to two recent texts which give excellent coverage of this topic.[6,7]

II. PRINCIPLE OF TECHNIQUE

The principles of activation analysis are well known and have been described in detail.[2,7] In general, the procedure involves the bombardment of a sample with a flux of particles which, upon interaction, converts the elements in the sample to various radionuclides. These nuclides decay with the emission of characteristic radiation at a rate that is well defined for each nuclide. These two parameters, namely, the energy emission spectrum and the half-life, permit the unambiguous identification of the elements originally present in the sample.

The activity of a radionuclide produced by a nuclear reaction from nucleus B can be expressed as a function of the operational parameters of the technique in the following manner:

$$A = DCN_B\sigma_B\phi\left[1 - \exp\left(\frac{-0.693t_i}{T_{1/2}}\right)\right]\left[\exp\left(\frac{-0.693t_d}{T_{1/2}}\right)\right] \quad (1.1)$$

where

A = the measured count rate at time t_d, in counts/sec
D = the branching ratio for the emission being measured
C = the detection efficiency of the counter
N_B = the number of target nuclei of element B exposed to the incident flux
σ_B = the activation cross section of the particular reaction for element B at a given incident particle energy, cm^2
ϕ = the particle flux of constant intensity and energy, particles/(cm^2 sec)
t_i = irradiation time
t_d = decay time from the termination of the irradiation to the start of counting
$T_{1/2}$ = the half-life of the product nuclide.

The units of time should be the same throughout the equation.

In carrying out an analysis, the factors, C, σ_B, and ϕ are not specifically evaluated because they can vary considerably with sample size, time, irradiation conditions, and counting arrangements. Instead, to provide higher accuracy, a comparative method in which a standard is irradiated simultaneously with the unknown is used. Thus the amount of element B in an unknown is given by the following equation:

$$\text{element } B \text{ (in unknown)} = \text{element } B \text{ (in standard)} \times \frac{A \text{ (of unknown)}}{A \text{ (of standard)}} \tag{1.2}$$

Equation 1.1, however, is useful in arriving at optimum conditions for a given analysis.

Activation analysis differs from other analytical methods in that it is based on a nuclear interaction rather than on an interaction with the electrons of the atom. The method is thus independent of the chemical or physical state of the target element and provides a true elemental analysis technique. This fact has an important consequence. The technique is sensitive only to nuclear properties such as neutron activation cross sections, half-lives of the product isotopes, and energy emission of the product isotopes, and is unrelated to the chemical properties of the elements. Thus elements that are chemically similar and difficult to differentiate by methods that depend on the behavior of the outer shell of electrons may often be readily discriminated by activation methods. This nuclear basis makes activation analysis unique among analytical methods and ensures its role in the analytical sciences.

III. ADVANTAGES

To appreciate fully the significance of neutron activation as a technique for compositional analysis, one must compare its operational parameters and capabilities with those of other techniques designed to perform similar functions. In this sense it should be compared with such techniques as the microchemical combustion methods, titrimetry, solution and flame photometric methods, x-ray absorption and emission, mass spectrometry, optical emission, and electrochemical methods.

Four primary factors govern the choice of a particular analytical method for a problem in compositional analysis: selectivity, accuracy, precision, and sensitivity. The selection will depend on the weight given each of these factors based on the objectives of the analysis and the nature of the matrix.

In addition, considerations of speed and economy often control the final choice of a method. Analytical chemists are continually seeking to develop new methods which will complement existing methods of analysis and thereby provide a full spectrum of approaches for the compositional analysis of all types of matrices. The development of the neutron generator has made it possible for every analytical laboratory which needs this kind of versatility to set up its own irradiation facility.

Activation analysis with neutron generators has been used to provide a totally automated, nondestructive analytical procedure in a number of applications. These attributes make neutron activation one of the favored techniques for use in extraterrestrial exploration.[8-12]

The potential for automation makes this technique very attractive for many applications, especially in industrial process control. Pierce et al.[13] recently discussed plans to interface a PDP-8 computer with a high-output, sealed-tube neutron generator. The computer will control the generator operation, sample transfer, counting sequence, gamma-ray spectrometry, and data reduction. Because generator activation involves short-lived isotopes, it is also possible to use the technique for on-stream analyses. A further result of this fact is that the analysis time is usually short. The best example of rapid analysis is the determination of oxygen, which, in most cases, can be completed in less than 2 min, including printout of final results. The sample size range that can be used in activation analysis with neutron generators is very large. Samples ranging from a few milligrams to decagrams may be run, depending on the analysis and the desired sensitivity. Since dissolution of the sample is not necessary, the size of the sample, *per se*, does not present the usual problems encountered in conventional "wet chemistry."

The technique is also highly selective, owing to several parameters that are under the control of the analyst. Nuclear properties such as half-life, reaction cross section, and decay scheme permit the analyst to differentiate between elements in a matrix through a choice of irradiation, decay, and counting times. Through a choice of target materials and moderating media, neutron energies of thermal (0.025 eV), 2.8-, or 14.7-MeV energy can be obtained. Laboratories which find it necessary to use this technique for a variety of problems employ a "turret" type target holder, which permits the rapid interchange of targets. Cosgrove[14] has described a system in which three targets are held in such a turret assembly, permitting the rapid interchange of a low- and a high-flux tritium target, as well as a deuterium target. A discussion of the various neutron-producing reactions that can be induced with a neutron generator and the types of nuclear reactions that can be obtained with neutrons of different energies is given in Chapter 2. Another

III. ADVANTAGES

parameter than can be readily controlled to attain the desired level of selectivity for gamma-ray counting is the detector system. Spectra are recorded with the use of sodium iodide crystal detectors coupled to a multichannel analyzer. In some cases, high-resolution, lithium-drifted germanium detectors can be used; however, the modest flux obtainable from a neutron generator together with the low efficiency of these detectors makes their use with this technique limited at present. In cases where sodium iodide detector systems prove inadequate for the desired degree of resolution, subsequent data-handling procedures using computer routines can be applied.

Perhaps the most significant advances made recently with this technique have been in improved accuracy and precision. The standard deviation of an activation analysis result is controlled by the total accumulated counts, which, in turn, are a function of the variables in Eq. 1.1. Two of the major sources of imprecision are caused by the anisotropic and time-varying neutron output from the generator target and the size of the target relative to the sample being irradiated. These factors result in a nonhomogeneous flux at the sample-irradiation position. Various schemes have been studied to "homogenize" the integrated flux experienced by the sample; techniques for doing this successfully have been achieved in a number of laboratories. The details of this source of imprecision as well as other instrumental factors which affect precision are discussed in Chapter 6.

It is notable that each source of potential imprecision has been thoroughly studied and brought under control to the extent that under optimum conditions, a relative standard deviation of 0.25% can be achieved in an analysis. Such precision compares favorably with that obtained from other quantitative analytical methods.

A significant improvement has also been achieved in the accuracy of this technique. Activation analysis is usually carried out on a comparative basis because factors such as the irradiation flux cannot be precisely controlled from one experiment to the next. Thus the accuracy achieved will depend first on the accuracy of the compositional analysis of the standard and, second, on matrix differences between the sample and standard which may result in their receiving a different flux dosage. This can be caused by the different attenuation of the neutron flux between sample and standard. Similar attenuation can take place with the emitted gamma radiation. These two factors, if left uncontrolled, can cause a severe error in the measured activity, particularly in activation analysis using neutron generators where very large sample sizes are used.

Methods have been developed and applied to handle these attenuation problems in a precise, quantitative manner. Results of such procedures have indicated that the accuracy of neutron generator activation analyses can be very high. Deviations of less than 0.5% relative have been reported in the literature.

IV. LIMITATIONS

There are, of course, certain limitations to the technique when it is compared to others in the analytical "arsenal." The major deficiency at this time appears to be the lack of sufficient flux from neutron generators to provide the desired sensitivity for certain elements. Cross sections for reactions with fast neutrons are usually one to three orders of magnitude smaller than for neutron capture reactions with thermal neutrons. This fact, coupled with an output of 10^{10} to 10^{11} neutrons/sec from a typical tritium target, results in a lower measurable limit for most elements of approximately 0.1 to 10 mg. For many applications, this sensitivity is sufficient; however, there could be a larger number of potential applications for neutron generators if a factor of 10 to 100 increase in flux were available. Some of the problems in the achievement of higher neutron fluxes are discussed in Chapter 3. At present, activation analysis with neutron generators has about the same sensitivity range as the x-ray fluorescence technique.

One of the most important limitations of the neutron generator is target lifetime. A discussion of the factors affecting the useful life of targets is given in Chapter 3 along with a discussion of some of the new developments in generator design which claim constant neutron output for prolonged periods.

V. COST

The cost of setting up an activation facility employing a neutron generator, while modest by some standards, is not inconsequential for a small laboratory. In the United States, the cost of a neutron generator with associated sample-handling devices and shielding is approximately $35,000. A versatile multichannel analyzer system with a shielded detector adds an additional $15,000, to bring the total cost of equipment to $50,000. Obviously, more auxiliary facilities can be added for specific purposes or for increased versatility; the figures given here will cover a minimum system having a reasonable degree of flexibility. Systems designed for more limited use, such as single-element determinations, can be purchased for a fraction of this cost. A system for the determination of oxygen at the milligram level could be set up for less than $20,000. The cost of maintaining a facility of this kind is nominal, the major expenditure being the cost of target replacement. Targets cost in the neighborhood of $100 and have a useful life of 4 to 6 mA-hr of integrated beam current.

VI. APPLICATIONS

An attempt has been made in Chapter 7 to summarize all of the reported activation analysis procedures utilizing neutron generators. This work has been largely aided by the recent availability of the activation analysis bibliography[15] issued by the National Bureau of Standards. This section has been organized with the view of providing a quick reference to conditions for the determination of any given element, with a summary of possible interferences.

VII. SUMMARY

It is clear that the technique of neutron activation analysis with neutron generators provides a number of distinct advantages for the analytical chemist. The method is based on a unique principle that provides a totally different access to the compositional nature of the sample than other techniques, and it is particularly powerful for the analysis of light elements where other instrumental techniques often fail. It is capable of high precision and accuracy and, in fact, is superior in these respects to some instrumental analytical methods. The technique has certain parameters which can be manipulated to obtain the desired degree of selectivity and, for certain elements, can provide a sensitivity capable of measuring parts per million in certain matrices. Many elements can be determined at the milligram level with no difficulty. The commercial availability of neutron generators has placed this technique within reach of every well-equipped analytical laboratory. Although activation analysis with a neutron generator is at present limited in sensitivity, continuing efforts in research and development will undoubtedly result in improvement in the near future.

REFERENCES

1. G. Hevesey and H. Levi, *Math.-Fys. Medd.*, **14** (5), 3 (1936), Institute of Theoretical Physics, University of Copenhagen.
2. R. C. Koch, *Activation Analysis Handbook*, Academic Press, New York, 1960.
3. W. S. Lyon, Jr., Ed., *Guide to Activation Analysis*, Van Nostrand, Princeton, New Jersey, 1964.
4. D. Taylor, *Neutron Irradiation and Activation Analysis*, George Newnes, London, 1964.
5. H. J. M. Bowen and D. Gibbons, *Radioactivation Analysis*, Oxford University Press, London, 1963.
6. D. DeSoete, R. Gijbels, and J. Hoste, *Neutron Activation Analysis*, Wiley, New York, in press.
7. J. M. A. Lenihan and S. J. Thomson, Eds., *Activation Analysis, Principles and Applications*, Academic Press, New York, 1965.

8. L. E. Fite, E. L. Steele, R. E. Wainerdi, E. Ibert, and W. Wilkins, USAEC Report TID-18257, 1963, pp. 33–60.
9. J. I. Trombka and A. E. Metzger, "Neutron Methods for Lunar and Planetary Surface Compositional Studies," in *Analysis Instrumentation—1963*, Plenum Press, New York, 1963, pp. 237–250.
10. L. E. Fite, E. L. Steele, R. E. Wainerdi, E. Ibert, P. Jimenez, C. Samson, and M. To-on, USAEC Report TID-19999, 1963.
11. J. S. Hislop and R. E. Wainerdi, *Anal. Chem.*, **39**, 28A (1967).
12. J. A. Waggoner and R. J. Knox, "Elemental Analysis Using Neutron Inelastic Scattering," in *Radioisotopes for Aerospace*, Part 2, J. C. Dempsey and P. Polishuk (Eds.), Plenum Press, New York, 1966, pp. 270–291.
13. T. B. Pierce, R. K. Webster, R. Hallet, and D. Mapper, "Developments in the Use of Small Digital Computers in Activation Analysis Systems," Proceedings, 1968 International Conference on Modern Trends in Activation Analysis, National Bureau of Standards, Special Publication 312, Vol. II, 1969, p. 1116.
14. J. F. Cosgrove, "Routine Determination of Major Components by Activation Analysis," Proceedings, 1968 International Conference on Modern Trends in Activation Analysis, National Bureau of Standards, Special Publication 312, Vol. I, 1969, p. 457.
15. G. J. Lutz, R. J. Boreni, R. S. Maddock, and W. W. Meinke, National Bureau of Standards, Technical Note 467, 1968.

CHAPTER

2

PRODUCTION AND INTERACTION OF FAST NEUTRONS

I. INTRODUCTION

In activation analysis, information describing the nature, energy, intensity, and spatial distribution of the bombarding beam of particles is of vital concern to the analyst. More often than not, serious discrepancies between the analytical result and the "true" value can be attributed to the lack of (or limited knowledge of) the incident beam characteristics. The beam picture is further confused if secondary particles capable of activating the sample are produced near the sample or within the sample matrix itself. Degradation of a supposedly pure primary beam by scattering within the sample, or by its attenuation by the hardware associated with sample mounts, can conceivably invalidate expected activity computations based on fundamental nuclear constants. The anisotropic and polyenergetic nature of the usable beam precludes an accurate *a priori* assessment of possible interferences from flux and cross-section information. Despite this complex situation, the neutron generator user is capable of making accurate and precise analyses based on techniques which have been developed. These techniques take into consideration the beam characteristics described above which are inherent in neutron production with small, low-voltage, charged-particle accelerators.

The subject matter considered in this chapter is concerned primarily with basic methods and nuclear reactions used for the production of high-energy neutrons by the utilization of neutron generators capable of accelerating deuterons up to about 0.5 MeV. Considerable emphasis is placed on those aspects which govern characteristics of the usable neutron flux. In the concluding section of this chapter, analytically significant nuclear reactions induced by fast neutrons are discussed. As a preface to these discussions, a short treatment of some of the relevant nuclear physics definitions and basic concepts is presented to establish a general background pertinent to the solution of problems related to activation analysis with neutron generators.

II. BASIC CONCEPTS

Neutron generators produce neutrons by accelerating a beam of deuterons up to 500-keV kinetic energy and permitting this beam to impinge upon a suitable target. A working knowledge of neutron production systematics is of fundamental importance both to the accelerator design engineer and to the analyst for optimum utilization of this equipment.

A. NUCLEAR REACTION NOTATION

When energetic nuclear particles impinge upon a target, there is a certain probability that a nuclear transformation may occur. The science of nuclear physics attempts to describe appropriate mechanisms by which interactions between bombarding particles and target nuclei take place. Since many of the concepts in this area of particle physics are very complex, only the simplest description of the interaction processes is given here. The nuclear reaction will be characterized in terms of the identity of the incident particle, the target nucleus, and the reaction product or products. A typical reaction can be written as

$$X + a \to Y + b \tag{2.1}$$

or in shorthand notation,

$$X(a, b)Y$$

where a is the projectile, X the target nucleus, b the emitted light particle, and Y the residual heavy nucleus. For purposes of this discussion, neutron production with generators, other than the sealed-tube type, is accomplished by accelerating ionized deuterium; the target is either deuterium or tritium embedded in a suitable metal matrix, and the emergent light particles are high-energy neutrons.

B. Q VALUE OR INTERACTION ENERGY

In a manner analogous to that used with chemical reactions, a nuclear reaction can be written to indicate the energy Q released or consumed, as

$$X + a \to Y + b + Q \tag{2.2}$$

The Q value for a particular reaction can be calculated using the Einstein equation $E = mc^2$ where E is the energy associated with the mass m, and c is the velocity of light. Hence it can be shown that 1 g of matter is equivalent to an energy, $E = 1 \times (3 \times 10^{10})^2$, or 9×10^{20} ergs. The nuclear interaction energy Q is that associated with the difference between the mass of the products and reactants. Thus if m_Y and m_b represent the individual masses of the products, and m_a and m_X the masses of the reactants, the difference Δm is given by

$$\Delta m = (m_a + m_X) - (m_Y + m_b) \tag{2.3}$$

and

$$Q = \Delta mc^2 \tag{2.4}$$

II. BASIC CONCEPTS

If Q is positive, the reaction is termed exoergic (energy releasing); if negative, the reaction is endoergic (energy consuming). In nuclear reactions, masses are expressed in atomic mass units (amu), and a conversion is necessary to express the energy in MeV units. The energy associated with nuclear mass is $E = mc^2 = 1.66 \times 10^{-24} \times 9 \times 10^{20}$ or 1.49×10^{-3} erg. The electronic charge of an electron is 1.6×10^{-19} C and the work W corresponding to 1 eV is $W = q \times V$, or $(1.6 \times 10^{-19} \text{ C} \times 1 \text{ V})$, or 1.6×10^{-19} J, or 1.6×10^{-12} erg. Therefore, 1 amu has an energy equivalent of $1.49 \times 10^{-3}/1.6 \times 10^{-12}$ or 0.931×10^9 eV or 931 MeV.

Consider the general case of neutron production,

$$X + d \rightarrow Y + n + Q \tag{2.5}$$

where n is the emitted neutron and d is the bombarding deuteron. If we apply the laws of conservation of energy and momentum for nonrelativistic particles, the kinematics of this reaction can be evaluated. Information relating the neutron energy angular distribution to the incident deuteron bombarding energy can be obtained. In considering the dynamics of such two-body interactions, two frames of references are used: the laboratory frame or (L) system, and the center-of-mass frame or the (C) system. In the former, the target nucleus is considered to be at rest before collision, whereas in the latter, the center-of-mass of the incident deuteron plus the target nucleus is at rest. The (L) system may be visualized as that which exists from the viewpoint of an external observer, and the (C) system as that which exists if the observer were located at the center-of-mass of the incident particle plus the target nucleus. The conditions before and after collision for the two frames of references are illustrated in Fig. 2.1. In the (L) system the target

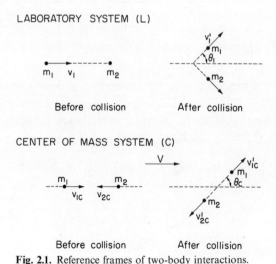

Fig. 2.1. Reference frames of two-body interactions.

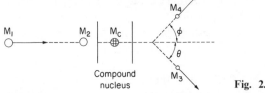

Fig. 2.2. Dynamics of two-body interactions.

nucleus of mass m_2 is stationary before collision and the bombarding particle of mass m_1 has a velocity v_1. After collision, m_1 is deflected with a velocity v_1' at the scattering angle θ_1 measured from its initial line of motion, and m_2 recoils in some direction with some velocity. In the (C) system, the frame of reference itself is moving with a velocity V with respect to the (L) system. A more detailed treatment of these systems is available in texts on modern physics and other handbooks[1-4] and will not be included here. For purposes of theoretical discussion the (C) system is simpler; however, experiments are generally carried out in the (L) frame of reference.

For nonrelativistic particles, a typical two-body interaction in the (L) system can be represented as shown in Fig. 2.2. The systematics for this reaction are shown in Table 2.1.

TABLE 2.1. Notation of Two-Body Interaction in the Laboratory (L) System

Property of Particle	Notation in Laboratory Coordinates				
Rest mass	m_1	m_2	m_c	m_3	m_4
Velocity	V_1	0	V_c	V_3	V_4
Kinetic energy	E_1	0	E_c	E_3	E_4
Momentum	$P_1 = \sqrt{2m_1 E_1}$	0	$P_c = P_1$	$P_3 = \sqrt{2m_3 E_3}$	$P_4 = \sqrt{2m_4 E_4}$
Angle	0	—	0	θ	ϕ

It has been shown[4] that, in the case of nonrelativistic particles, the energy of interaction (or Q value) can be expressed as follows:

$$Q = E_3\left(1 + \frac{m_3}{m_4}\right) - E_1\left(1 - \frac{m_1}{m_4}\right) - \frac{2\sqrt{m_1 E_1 m_3 E_3}}{m_4}\cos\theta \quad (2.6)$$

where

$$Q = (m_1 + m_2 - m_3 - m_4)c^2$$

In nuclear reactions with a positive Q, involving charged particles, only the coulomb barrier need be considered. In those reactions for which Q is negative, a certain amount of energy must be added to the reaction system before it can proceed. The obvious way to achieve this goal is to accelerate the incident particle to the desired threshold energy until the reaction just begins to occur. This excess energy, therefore, depends on the Q value and on coulomb barrier considerations. The coulomb barrier is discussed in the next section. For the moment, however, the Q relationship (Eq. 2.6) can be reexamined, and two important observations may be made. First, this equation is independent of the reaction mechanism, and second, relationships permitting the computation of the kinetic energy of the expelled neutron as a function of the laboratory angle θ can be derived. It has been shown[4] that for a constant Q

$$\sqrt{E} = a \pm \sqrt{a^2 + b} \qquad (2.7)*$$

where

$$a = \frac{\sqrt{m_1 m_3 E_1}}{m_3 + m_4} \cos \theta$$

$$b = \frac{m_4 Q + E_1(m_4 - m_1)}{m_3 + m_4}$$

The question then arises as to whether the negative reaction energy Q associated with endoergic reactions is adequate to promote the nuclear transformation process. If we recall that the incoming projectile, possessing a kinetic energy equivalent to the negative Q value, must share some fraction of this energy with the target nucleus so that momentum be conserved, then the threshold energy the incident particle must have can be approximated[4] by

$$E_T = -Q\left(\frac{m_1 + m_2}{m_2}\right) \qquad \text{for} \quad m_2 \gg \frac{Q}{c^2} \qquad (2.8)$$

and the energy carried away by the emergent neutron for $\theta = 0°$ is

$$E_3 = E_T\left[\frac{m_1 m_3}{(m_3 + m_4)^2}\right] \qquad (2.9)$$

A further question is whether E_T is indeed sufficient to propagate the nuclear reactions. The answer is "yes" if the incident and expelled particles are both uncharged (neutrons), and "no" if either of them is charged. The concept of

---

* For finite bombarding energies, E_3 is single valued; and for $Q > 0$, $m_4 > m_1$
$$\sqrt{E_3} = a + \sqrt{a^2 + b}$$

an effective threshold energy can best be understood with the help of the following short discussion on the coulomb barrier. As will be noted later, barrier considerations apply in cases of both neutron production and the interaction of neutrons where the emitted particle is charged.

C. COULOMB BARRIER

When charged-particle interactions take place, the positively charged incident particle approaching the positively charged target nucleus experiences electrostatic repulsion. The degree of repulsion increases with decreasing distance between them, until the incident particle comes within the range of the nuclear forces. This coulomb repulsion gives rise to a potential barrier, which has been described in detail elsewhere.[2, 5] The height of this barrier around a nucleus of charge $Z_1 e$ and radius r_1 for a positively charged particle of charge $Z_2 e$ and radius r_2 can be estimated as the energy of coulomb repulsion when the particles are just in contact from the equation

$$V = \frac{Z_1 Z_2 e^2}{r_1 + r_2} \tag{2.10}$$

and the nuclei radii r of the particles can be approximated from $r \simeq 1.5 \times 10^{-13} A^{1/3}$, where A is the mass number of the nucleus in question. Suppose that the calculated coulomb potential for a particular nuclear reaction is V. Then according to the classical theory of momentum conservation, the incident charged particle must possess at least $[V(M_1 + M_2)]/M_2$ of energy to surmount the barrier and initiate the nuclear reaction; the nuclear masses M_1 and M_2 are defined as shown in Fig. 2.2. In the quantum mechanical treatment, there always exists a small but finite probability that the charged particle will penetrate or tunnel through the barrier even if its energy is less than that of the barrier itself. Further discussion of the theoretical probability of barrier penetration, or the tunneling effect, as it is called, will not be pursued here; interested readers are referred to useful derivations and formulations[5] for the computation of this penetrability. The discussion thus far can be clarified by giving a simple example which illustrates the effect of the different constraints dictating the propagation of nuclear reactions.

Consider the specific nuclear reaction,

$$^{14}\text{N}(\alpha, p)^{17}\text{O} \tag{2.11}$$

which has a Q value of -1.20 MeV. After consideration is given to momentum conservation, E_T is calculated to be 1.54 MeV. The coulomb barrier for this reaction, computed from Eq. 2.10, is 3.4 MeV. In order that momentum be conserved, the energy required to overcome this barrier is calculated to be 4.4 MeV. It is clear that the initial energy requirement (1.20 MeV) based on the Q value is quite insufficient to promote this reaction. Lord Rutherford actually used α-particles of over 7 MeV in his experiments with this particular reaction. To go a step further, it has been assumed thus far that the incident particle penetrating the barrier will undergo a head-on collision with the target nucleus, or an s-type interaction.* A short but excellent treatment of collisions which are not head-on is given in reference 5. These collisions are termed "p" and "d" interactions for centrifugal barrier calculations. For our example, it can be shown that the centrifugal barrier is yet another additional quantity of energy; it is calculated to be approximately 0.3 MeV for p collisions and 0.87 MeV for d-type interaction. The total barrier height is the sum of the coulomb and centrifugal barriers. However, it is most important to remember that only the coulomb barrier contributes to the minimum energy requirement for promoting nuclear reactions, except in some very special cases where s-type interactions do not lead to the reaction of interest. It should also be emphasized that barrier considerations apply both to the incident and the expelled charged particle.†

III. NEUTRON PRODUCTION BY SMALL ACCELERATORS

The production of neutrons by bombarding suitable targets with the isotopes of hydrogen is particularly attractive since the net energy gain by the reaction system is quite large. For instance, a proton entering a nucleus adds about 8 MeV of energy to the system. In the case of an α-particle, the net gain is not quite as much since the difference between approximately 28 MeV of binding energy required to break up an α-particle into two neutrons and two protons, and the 32 MeV gained by the reaction system from the addition of four extra nucleons, is only 4 MeV. When compared with the entry of a deuteron, however, even the energies brought in by protons and α-particles are considered small. The deuteron requires only 2 MeV to split into a neutron and a proton but adds 16 MeV into the reaction system, a net gain of 14 MeV.

Another consideration in favor of the deuteron as a projectile was proposed by J. R. Oppenheimer and M. Phillips. According to their theory the deuteron, because of its low binding energy of 2 MeV, behaves as a relatively loose combination of a proton and a neutron. In its approach

---

* When the original direction of motion of the incident particle does not pass through the target nucleus center, the projectile-target system possesses angular momentum l which can be quantized in integral multiples of $h/2\pi$. The interactions s, p, d, and so on, correspond to $l = 0, 1, 2$, respectively.

† Although texts[6-12] on activation analysis deal primarily with (n, γ) interactions where barrier considerations do not apply, the effect of the fast flux component of reactor systems can be properly evaluated only by estimating threshold reaction interferences, which necessarily include coulomb barrier calculations.

toward a nucleus, electrostatic repulsion tends to deflect the proton, while the neutron remains unaffected. Also, if the incident energy of the deuteron exceeds the neutron-proton binding energy, the deuteron can split, and the proton will be repelled. However, the neutron will enter the target nucleus since it experiences no potential barrier by virtue of its electrical neutrality. This two-stage process, followed by a nuclear reaction at low incident deuteron energies, is generally not energetic enough to expel a charged particle. But since the nucleus is in an excited state, it will lose its excess energy by emitting gamma radiation. Nuclear reactions of the (d, p) type are extremely common with almost all elements, and at low deuteron energies, the Oppenheimer-Phillips theory applies.

At higher incident energies, the nucleus can adsorb the deuteron as one unit to form a compound nucleus. The end result from both mechanisms of interaction is, of course, the same. Charged-particle-induced reactions of this type compete with each other, the degree of competition depending on the incident particle energy. One such competing reaction is the (d, n) reaction. A large number of reactions in this category have been reported, especially with targets of low mass numbers. For example, the interaction of approximately zero-energy deuterons with deuterium nuclei is accompanied by the emission of moderate outputs of neutrons with kinetic energies of the order of 2.5 MeV by the reaction,

$$^2H + {}^2H \rightarrow {}^3H + {}^1n \tag{2.12}$$

Similarly, if the target consists of tritium atoms, neutrons of about 14 MeV are produced by the reaction,

$$^2H + {}^3H \rightarrow {}^4He + {}^1n \tag{2.13}$$

In view of the Oppenheimer-Phillips mechanism and the low binding energy of the deuteron, coulomb barrier hindrance is negligible and this makes the deuteron ideal as an accelerating particle for low-voltage accelerators. Some useful neutron-producing reactions and their systematics are summarized in Table 2.2.

A. NEUTRON YIELD CALCULATIONS

The total yield of neutrons produced in a nuclear reaction is proportional to the flux of incident particles and the number of target nuclei exposed to the incident beam, or

$$F = N\sigma\phi \tag{2.14}$$

where F is the neutron yield (neutron/sec), N is the number of target nuclei per square centimeter of target, σ is the reaction cross section (in cm^2), and ϕ is the incident particle rate (expressed as particles/sec). The relationships

III. NEUTRON PRODUCTION BY SMALL ACCELERATORS

TABLE 2.2. Neutron Production from Charged-Particle-Induced Reactions

Reaction	Threshold Energy of Incident Particle (MeV)	Q Value (MeV)	Emitted Neutron Energy at Threshold Bombarding Energy (MeV)
^{2}H(d, n)^{3}He	—	+3.266	2.448
^{3}H(p, n)^{3}He	1.019	−0.764	0.0639
^{3}H(d, n)^{4}He	—	+17.586	14.064
^{9}Be(α, n)^{12}C	—	+5.708	5.266
^{12}C(d, n)^{13}N	0.328	−0.281	0.0034
^{13}C(α, n)^{16}O	—	+2.201	2.07
^{7}Li(p, n)^{7}Be	1.882	−1.646	0.0299

between cross section and the energy of the incident particles for various reactions are described in a later section of this chapter. This dependence is called the excitation function of the reaction. Equation 2.14 assumes that the target is extremely thin compared to the range of the incident particles in the target matrix. If the target thickness X is large compared to this range, the number of neutrons produced per second is given by the expression,

$$dF = \sigma\phi N \quad \text{or} \quad dF = \sigma\phi\left(\frac{\rho N_a}{A}\right) dx \quad (2.15)$$

where ρ is the density of the target matrix material, N_a is Avogadro's number, and A is the gram-atomic weight of the target material. The cross section σ is given in units of 10^{-24} cm^2, which is a barn. Equation 2.15 can be reduced to

$$dF = \sigma\phi\left(\frac{\rho N_a}{A}\right)\left(\frac{dx}{dE}\right) dE$$

$$= \sigma\phi\left(\frac{\rho N_a}{A}\right)\left(\frac{dE}{dx}\right)^{-1} dE \quad (2.16)$$

The term dE/dx is called the "stopping power" and represents the loss of energy by the projectile in its passage through the target material. Therefore the total production rate F in neutrons/sec is

$$F = \phi\left(\frac{N_a}{A}\right) \int_{E_T}^{E_{\max}} \frac{\sigma(E)\, dE}{[(1/\rho)(dE/dx)]} \quad (2.17)$$

where E_T, the threshold energy, is given by Eq. 2.5, and $E_{\max}$ is the energy of the incident particle. The beam currents can be related to particles/sec by

$$\phi = 6.25 \times \frac{I \times 10^{18}}{Z} \quad (2.18)$$

18   PRODUCTION AND INTERACTION OF FAST NEUTRONS

where ϕ is the particle output per second, I is the beam current (in A), and Z is the charge on the incident particle (in C). For example, the charge on a single deuteron is 4.8×10^{-10} esu or 1.6×10^{-19} C. Therefore, 1 A of deuteron current is equivalent to 6.25×10^{18} deuterons/sec.

It is appropriate at this point to introduce briefly the concepts of atomic and molecular beams. By definition, an ionized deuterium atom when accelerated through a potential difference of 200 kV will acquire an energy of 200 keV. Similarly, a singly charged deuterium molecule will pick up the same amount of total energy, but will share it equally between the two deuterium nuclei, each possessing 100 keV of kinetic energy. It will be shown later, from cross-section and thick-target penetration considerations, how differently these particles affect the neutron yield. The neutron yields from most of the reactions given in Table 2.2 are shown graphically in Figs. 2.3 and 2.4, the data being taken from references 13 and 8, respectively. The principal conclusion that can be drawn from a study of these graphs is that

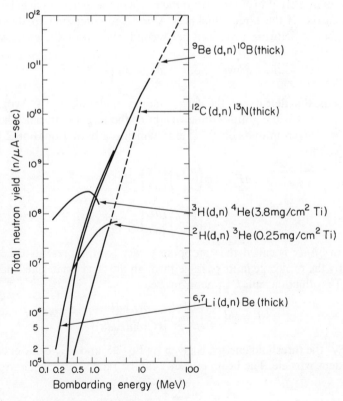

Fig. 2.3. Neutron yields: (d, n) reactions.

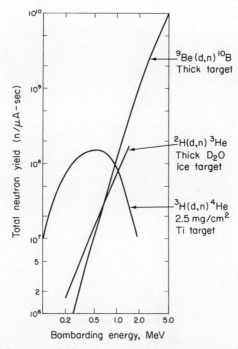

Fig. 2.4. Experimental neutron yields under normal laboratory conditions.

the deuteron energy required to initiate these reactions is very low. This conclusion is by far the most important for the design of economical high-current, low-voltage neutron generators.

B. NEUTRON-PRODUCING REACTIONS

The two nuclear reactions used for the production of fast neutrons by low-voltage accelerators can be expressed as follows:

$$^2H + {}^2H \rightarrow {}^3He + {}^1n + 3.266 \text{ MeV}$$
$$E_n \simeq 2.5 \text{ MeV (laboratory angle } \theta = 0°) \tag{2.19}$$

and

$$^2H + {}^3H \rightarrow {}^4He + {}^1n + 17.586 \text{ MeV}$$
$$E_n \simeq 14 \text{ MeV (laboratory angle } \theta = 0°) \tag{2.20}$$

1. D–D REACTION

The dependence of the cross section on the energy of the incident deuteron beam, or the excitation function, for this reaction is depicted in Fig. 2.5,

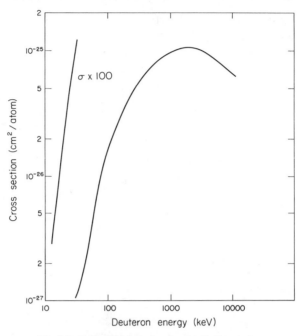

Fig. 2.5. Excitation function for the D–D reaction.

which has been taken from reference 14. It is important to bear in mind that this curve represents the probability of interaction per single atom and not for a target of finite thickness. It may be assumed that the neutron yield from very thin targets, as a function of incident deuteron energy, will follow the shape illustrated in Fig. 2.5. For this reaction, a cross-section value of 3×10^{-29} cm^2/atom can be approximated from the curve at about 13-keV incident deuteron energy. The broad resonance peak has a maximum value of about 10^{-25} cm^2/atom, or 100 mb, at about 2-MeV bombarding deuteron energy. At acceleration voltages normally used in neutron generators of about 150 to 200 kV, the cross section for the D–D reaction is about 3×10^{-26} cm^2/atom.

In the earlier treatment of the dynamics of two-body interactions, the nonrelativistic case for the calculation of the angular distribution of neutron energies as a function of incident deuteron energy was considered. With the high energy stability of modern-day accelerators, the relativistic treatment should in fact be considered. These calculations are complex; however, computed tables are available. Fowler and Brolley[15] have presented a series of tables, from which Table 2.3 has been abstracted, for reactions with isotopes of hydrogen. The reader is strongly urged to study this work, which

III. NEUTRON PRODUCTION BY SMALL ACCELERATORS

discusses in great detail charged-particle-induced reactions producing high-energy neutrons. Transformations from the center-of-mass to the laboratory system are also presented in a useful and lucid manner. The principal observation to be made from an examination of Table 2.3 is that the energy spread of the expelled neutrons as a function of laboratory angle is significant. For example, at 200-keV incident deuteron energy, the energy of the emitted neutrons at 0° is 3.05 MeV compared to the energy 2.0997 MeV at 150°. The extent of this energy anisotropy increases with increasing deuteron energy, as is shown by Burrill[13] and plotted in Fig. 2.6. The energy peaks in the forward direction with increasing deuteron energy until at about 4.45 MeV, the deuteron disintegrates with the production of discrete neutron groups which can be discriminated by time-of-flight experiments.

The D–D neutron output not only varies in energy with the angle of emission, but is also anisotropic with respect to its intensity. The differential neutron yield for a very thin target at laboratory angles of 0° and 90° is illustrated in Fig. 2.7. The degree of anisotropy in intensity is illustrated in

TABLE 2.3. Neutron Energy as a Function of Laboratory Angle θ and Incident Deuteron Energy; D–D Reaction (Relativistic Case)

	Deuteron Energy (MeV)				
θ	0.1000	0.2000	0.3000	0.4000	0.5000
0	2.8504	3.0500	3.2177	3.3690	3.5099
5	2.8489	3.0477	3.2148	3.3654	3.5057
10	2.8443	3.0408	3.2059	3.3543	3.4934
15	2.8367	3.0294	3.1912	3.3372	3.4731
20	2.8261	3.0135	3.1710	3.3129	3.4451
30	2.7966	2.9695	3.1148	3.2456	3.3675
40	2.7571	2.9108	3.0400	3.1564	3.2648
50	2.7092	2.8400	2.9501	3.0495	3.1422
60	2.6546	2.7600	2.8492	2.9301	3.0057
70	2.5955	2.6741	2.7415	2.8032	2.8613
80	2.5340	2.5854	2.6311	2.6739	2.7149
90	2.4721	2.4970	2.5218	2.5467	2.5716
100	2.4116	2.4116	2.4171	2.4256	2.4360
110	2.3545	2.3316	2.3198	2.3137	2.3113
120	2.3021	2.2590	2.2321	2.2135	2.2003
130	2.2557	2.1954	2.1558	2.1269	2.1048
140	2.2165	2.1420	2.0921	2.0549	2.0258
150	2.1852	2.0997	2.0419	1.9984	1.9641
160	2.1624	2.0690	2.0057	1.9579	1.9199
170	2.1486	2.0505	1.9839	1.9334	1.8933
180	2.1440	2.0443	1.9766	1.9253	1.8845

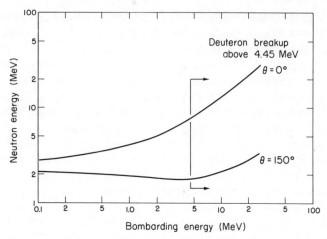

Fig. 2.6. Neutron energy as a function of incident deuteron energy for the D–D reaction.

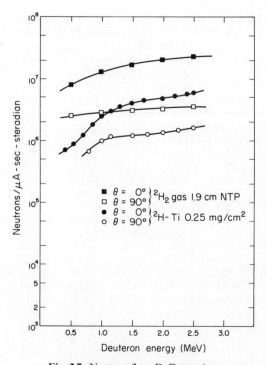

Fig. 2.7. Neutron flux: D–D reaction.

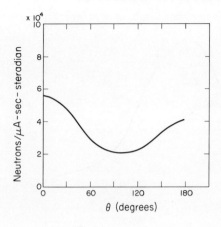

Fig. 2.8. Differential neutron yield as a function of angle θ for the D–D reaction; deuteron energy, 150 keV; thick D–Ti target.

Fig. 2.8. Therefore, Fig. 2.7 cannot be used for the estimation of the 4π yield. The difference between thick- and thin-target neutron yields may also be noted. As was mentioned before, the angular distribution of D–D neutrons is not isotropic in energy or in intensity; however, the differential yield can be calculated for a thick target of deuterium absorbed in titanium. The yield as a function of laboratory angle[16] is shown in Fig. 2.8. A good estimate for the 4π thick-target yield at 150- to 200-keV incident deuteron energy is of the order of 10^6 neutrons/sec-μA. An important factor in activation analysis is the usable neutron flux, as seen by the sample. In considering that the neutrons emitted must traverse the target coolant and sample-holder mechanism, the maximum D–D neutron flux that can be obtained in this configuration (at $\simeq 0.5$ cm from the target) is of the order of 5×10^7 neutrons/(cm^2 sec). This generalization is a good estimate for a 2.5-mA beam of 200-keV deuterons bombarding a thick titanium target on which about 6 cm^3 of deuterium gas has been absorbed per square inch of target surface.

2. D–T REACTION

This reaction is a prolific source of high-energy neutrons, as can be seen from the excitation function[14] illustrated in Fig. 2.9 which can be compared with the D–D cross-section curve (Fig. 2.5). For thin targets, the D–T cross section at about 150-keV incident deuteron energy is about 4.55×10^{-24} cm^2/atom. The shape of the curve is a varying function of the incident particle energy, rising from a value of about 2×10^{-28} cm^2/atom at about 7-keV deuteron energy to a peak of about 5×10^{-24} cm^2/atom or 5 barns at about 110-keV incident deuteron bombarding energy. From cross-section considerations, it would be expected that at 150-keV incident deuteron

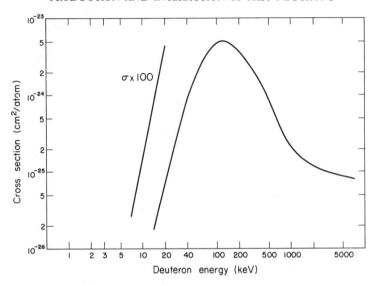

Fig. 2.9. Excitation function for the D–T reaction.

energy, the output from the D–T reaction would be about 300 times greater than that from the D–D reaction. Experimentally, it has been observed that this improvement is indeed approximated for thick-target neutron production. The prolific yield from the D–T reaction makes it particularly useful for analytical applications requiring a high degree of sensitivity; however, the D–D reaction is sometimes preferred in special cases because of minimization of the number of interferences.

For the relativistic case, the dependence of the emergent neutron energy as a function of laboratory angle is shown in Table 2.4, which is condensed from Fowler and Brolley.[15] The relatively small spread of the neutron energy through large changes in the laboratory angle may be noted. For example, the difference between the neutron energy at 150° and that at 0° is only about 1.7 MeV for incident deuterons of 200 keV. Analogous to the D–D reaction, the energy anisotropy increases with increasing deuteron energy. An interesting observation to be made from an examination of Table 2.4 is that at an angle of about 100°, the neutron energy remains essentially constant at a value of about 14 MeV and is independent of acceleration voltages to about 500 kV. This behavior is shown in Fig. 2.10, constructed from data given by Benveniste and Zenger[17] in their excellent dissertation on the D–T reaction. Therefore, a sample placed at an angle of 100° would experience a neutron flux whose energy is independent of the bombarding deuteron energy within the accelerating limits of neutron generators. The

TABLE 2.4. Neutron Energy as a Function of Laboratory Angle θ and Incident Deuteron Energy: D–T Reaction (Relativistic Case)

θ	Deuteron Energy (MeV)				
	0.1000	0.2000	0.3000	0.4000	0.5000
0	14.7569	15.0937	15.3671	15.6064	15.8254
5	14.7542	15.0898	15.3623	15.6007	15.8190
10	14.7462	15.0782	15.3478	15.5838	15.7998
15	14.7329	15.0589	15.3238	15.5557	15.7681
20	14.7143	15.0322	15.2906	15.5168	15.7240
30	14.6625	14.9574	15.1975	15.4031	15.6010
40	14.5923	14.8565	15.0722	15.2617	15.4356
50	14.5064	14.7331	14.9191	15.0331	15.2340
60	14.4074	14.5914	14.7436	14.8758	15.0038
70	14.2987	14.4362	14.5519	14.6560	14.7530
80	14.1837	14.2725	14.3503	14.4221	14.4904
90	14.0661	14.1058	14.1454	14.1849	14.2245
100	13.9495	13.9409	13.9434	13.9517	13.9635
110	13.8374	13.7830	13.7503	13.7292	13.7151
120	13.7330	13.6364	13.5716	13.5237	13.4860
130	13.6394	13.5053	13.4121	13.3407	13.2824
140	13.5591	13.3932	13.2760	13.1847	13.1091
150	13.4942	13.3029	13.1666	13.0596	12.9703
160	13.4467	13.2368	13.0866	12.9682	12.8690
170	13.4177	13.1965	13.0378	12.9125	12.8073
180	13.4079	13.1829	13.0214	12.8938	12.7866

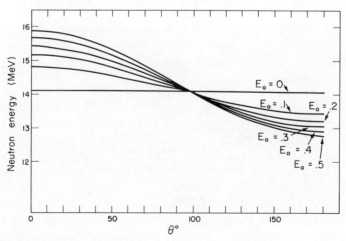

Fig. 2.10. Neutron energy as a function of laboratory angle θ; E_0 is the energy of the incident deuteron in MeV (D–T reaction).

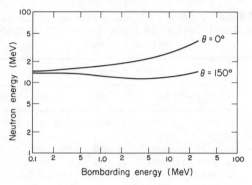

Fig. 2.11. Neutron energy as a function of incident deuteron energy for the D–T reaction.

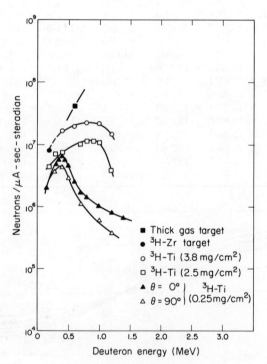

Fig. 2.12. Neutron flux: D–T reaction.

degree of anisotropy in energy of the emitted neutrons at bombarding deuteron energies up to a few megaelectron volts[13] is illustrated in Fig. 2.11.

Although the neutron flux obtained from the D–T reaction is not quite isotropic, the degree of anisotropy is considerably less than that from the D–D reaction. For thick targets, Burrill[13] gives the neutron yields as illustrated in Fig. 2.12. The effect of target thickness on the neutron intensity is clearly depicted. Experiments have proved that thick-target fluxes closely approximate the isotropic case. Also, thick-target yield curves (Fig. 2.12) can be compared with the thin-target yield curve[18] shown in Fig. 2.13. The latter

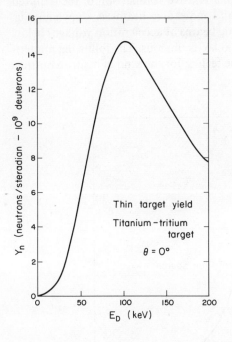

Fig. 2.13. Differential neutron yield at $\theta = 0°$ from a "thin" target for the D–T reaction.

curve closely approximates the cross-section curve previously shown in Fig. 2.9. A thorough study[19] has shown that when using thick targets, the analytical sensitivity is improved as the incident deuteron energy is increased. For deuterons of energy in the range of 300 to 400 keV, the neutron intensity reaches a peak,* and a further increase in the deuteron energy does not serve any useful purpose.

---

*Mainly because practical tritiated targets cannot be made thick enough to absorb the entire beam.

For fundamental nuclear physics studies, a thin target is desirable so that the degradation of the incident deuteron energy is minimized. The 4π yield from the D–T reaction can be approximated by multiplying the differential yield shown in Fig. 2.12 by 4π. Estimated usable neutron fluxes of the order of 10^{10} neutrons/(cm² sec) can be obtained at $\simeq 0.5$ cm from a thick target containing 3 to 5 Ci/in.² of tritium absorbed in a titanium matrix. This flux has been estimated for a 2.5-mA deuteron beam of 150- to 200-keV energy and is approximately 200 times as large as that estimated from the D–D reaction under equivalent experimental conditions.

In Section III.A, a brief reference was made to atomic and molecular beam effects. For a constant neutron yield, a relative comparison of the required power for predominantly atomic and molecular beams is shown in Fig. 2.14.[20] The advantage of using atomic beams at acceleration voltages below 200 kV and molecular ions above 200 kV is obvious. The following numerical example will give the reader some feeling for the kind of neutron outputs

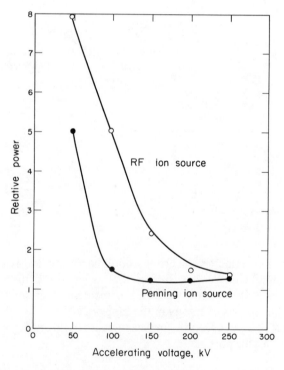

Fig. 2.14. Relative power versus accelerating voltage at a given neutron yield for the D–T reaction.

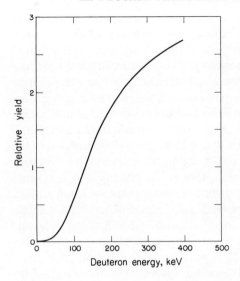

Fig. 2.15. Thick-target relative neutron yield versus deuteron energy for the D–T reaction.

obtained from both atomic and molecular beams. The thick-target yield curve given by Jessen[20] and shown in Fig. 2.15 shows that the neutron yield for 150-keV deuterons is about five times greater than that for 75-keV deuterons. Since each molecule is composed of two deuteron atoms, the yield from an atomic beam is 5/2 or 2.5 times greater than that for a molecular beam. However, if the situation at 400-kV acceleration voltage is considered, the yield (at 400 kV) is about 1.5 times that for 200-keV deuterons and therefore the neutron yield for atomic ions is about 1.5/2 or 0.75 times that from a molecular beam. Because ion sources do not produce either 100% pure atomic or molecular ions, an accurate comparison of neutron outputs as a function of acceleration voltage cannot be theoretically predicted. The atomic-to-molecular ratio of the ion beam must be considered for the particular ion source being used. The above-mentioned example gives only an approximate measure of the relationship between the neutron output and the acceleration voltage for a given beam identity.

C. THERMALIZATION OF FAST NEUTRONS

High-energy neutrons, such as those generated from the D–D and D–T reactions, can be reduced to thermal energies[o] (0.025 eV) by multiple collisions with nuclei of low atomic number. The most effective slowing-down

---

[o] Actually the neutron spectrum shows a Maxwellian spread. For convenience, 0.025 eV is taken as "representative" energy.

media are those rich in hydrogen. Experiments[21] show that about 10 to 16 cm of water or paraffin is adequate to thermalize D–D and D–T neutrons.

Because the fast flux can be considered to be essentially isotropic, the thermal flux increases with the distance from the target to the point of measurement in the moderator. A broad maximum is reached at about 5 cm into the moderator and then the thermal flux drops off with distance from the target. Although it would be advantageous to perform irradiations in the region of maximum thermal flux, the fast flux at this location is significant and serious interferences can be encountered in the determination of many elements having large fast neutron capture cross sections. For a 2.5-mA beam of 200-keV deuterons bombarding a thick 3- to 5-Ci/in.^{2} tritium target, thermal fluxes of the order of 5×10^8 neutrons/(cm^2 sec) can be easily obtained at the point of maximum production in the moderator. Dibbs[22] has reported distributions as shown in Figs. 2.16 and 2.17. In this comprehensive report, details of the flux measurement method are given using the foil activation technique. Fast flux measurements were made by

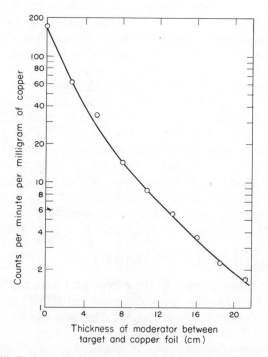

Fig. 2.16. Fast neutron flux distribution with distance in water moderator.

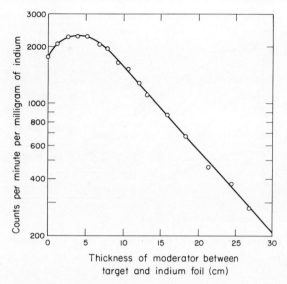

Fig. 2.17. Thermal neutron flux distribution with distance in water moderator.

counting the positron activity from the ^{63}Cu(n, 2n)^{62}Cu reaction induced by 14-MeV neutrons. The thermal flux distribution was measured using indium foils and counting the β-emission from the product of the reaction ^{115}In(n, γ)^{116m}In. The cadmium ratio, that is, the ratio of the activity induced in a bare foil to that covered with cadmium (0.04 in. thick), at the maximum thermal flux position was about 7.5. Using a thick target and a 1-mA deuteron beam at 150 keV, an absolute fast flux of 6×10^8 neutrons/(cm^2 sec) and a thermal flux at the optimum position of 1×10^8 neutrons/(cm^2 sec) were measured. Similar measurements are reported by Burrill[13] and are shown in Fig. 2.18 for three neutron energies. Other workers[23] have reported comparable data.

Insofar as the D–D reaction is concerned, thermal fluxes of the order of 10^5 neutrons/(cm^2 sec) have been measured[24] using suitable moderators. This reaction has some unique but limited applications which are described in Chapter 7. Aside from activation analysis, the reaction is useful for nuclear physics experiments at the undergraduate and graduate levels. The nominal shielding requirements permit experiments to be performed in a teaching laboratory.

In concluding this section, the availability of other particle and electromagnetic fluxes from neutron generators should be recognized. Useful fluxes of high-energy protons (15 MeV) and gamma rays (20 MeV) can be obtained with minor modifications to the generator. Bremsstrahlung are also

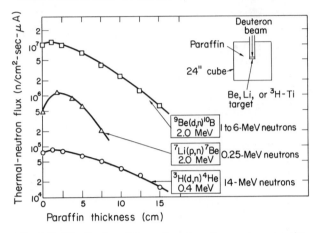

Fig. 2.18. Distribution of thermal neutron flux; various energies.

produced by simply reversing the accelerating voltage polarity and allowing an electron beam to bombard a suitable high Z target. Many useful studies concerning the measurement of biological damage by protons and the attenuation of high-energy gamma radiation can be carried out. Additional information regarding this topic has been provided by Prud'homme.[18] In this treatise, only the neutron-producing capability of the generator will be discussed.

D. EXPERIMENTAL FAST NEUTRON FLUX DISTRIBUTIONS

Thus far, the energy and intensity characteristics of the neutrons produced from D–D and D–T reactions have been approached from basic nuclear considerations, and the illustrations given describe the effect of thick targets on the neutron yield. Thermal flux measurements from the slowing down of fast neutrons have also been described. However, the experimenter is most concerned with the fast flux reaching the sample boundaries. The usable flux suffers considerable degradation, owing to the substantial amount of hardware between the target and the sample. This flux distribution is extremely complex and varies considerably from one installation to another, depending on such factors as the materials used for target cooling jackets, coolants, and sample holders. As pointed out in the introduction to this chapter, characteristics of the usable flux can account for many of the systematic errors in an analysis.

One of the principal assumptions made in the treatment of neutron production was that the neutrons emanate from a point source. In the practical case, the neutron source is, at best, a nonuniform disk, highly irreproducible because of the uneven nature of tritium deposition on the target and the instability of the deuteron beam. A number of theoretical and experimental studies of neutron distribution around a target will be described. For simplicity, the flat circular target will be considered as a homogeneous neutron source. Then the neutron flux at a point external to the target can be calculated by integrating over the active surface of the disk. Results from such calculations agree reasonably well with those from experimental flux measurements even at distances close to the target. The knowledge of the variation in the deuteron-beam diameter or in the position of the beam on the target is helpful in the estimation of the flux incident on a sample. Comprehensive computations of the flux as a function of position in front of a target have been carried out by Price.[25]

Figure 2.19 describes the geometry and definition of terms used for the interpretation of Figs. 2.20 to 2.29. From a study of these flux curves, it is possible to estimate the degree of flux variation for a particular sample shape or size. A recent mathematical treatment[26] permits the estimation of flux gradients around a disk-shaped neutron source. When the theoretical calculations are compared with experimental results, good agreement is noted. The conclusion drawn in this work is that the observed, sharp flux gradients are inherent in the system and not due to nonideal behavior of the experimental parameters.

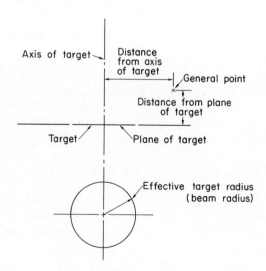

Fig. 2.19. Geometry and explanation of terms used in flux graphs (Figs. 2.20–2.29).

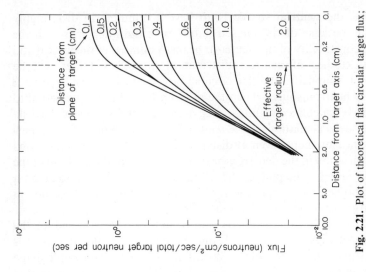

Fig. 2.21. Plot of theoretical flat circular target flux; effective target radius is 0.3 cm.

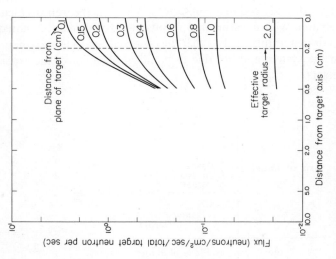

Fig. 2.20. Plot of theoretical flat circular target flux; effective target radius is 0.2 cm.

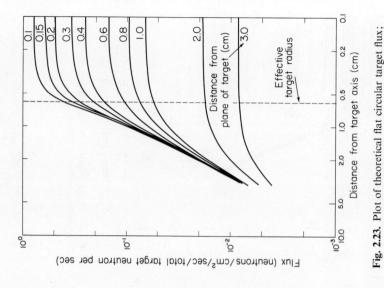

Fig. 2.23. Plot of theoretical flat circular target flux; effective target radius is 0.6 cm.

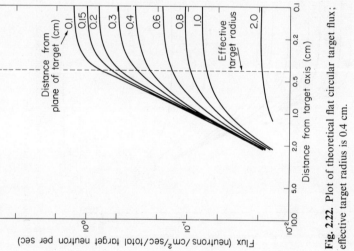

Fig. 2.22. Plot of theoretical flat circular target flux; effective target radius is 0.4 cm.

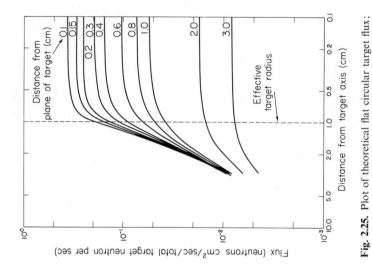

Fig. 2.25. Plot of theoretical flat circular target flux; effective target radius is 1.0 cm.

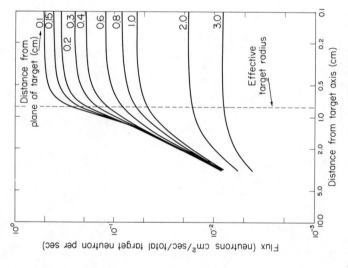

Fig. 2.24. Plot of theoretical flat circular target flux; effective target radius is 0.8 cm.

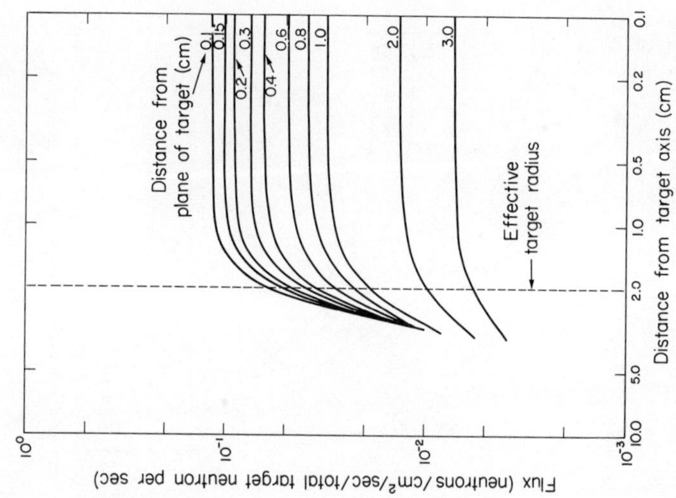

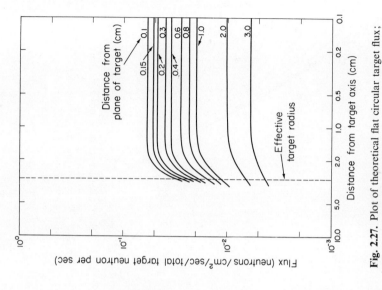

Fig. 2.27. Plot of theoretical flat circular target flux; effective target radius is 3.0 cm.

Fig. 2.26. Plot of theoretical flat circular target flux; effective target radius is 2.0 cm.

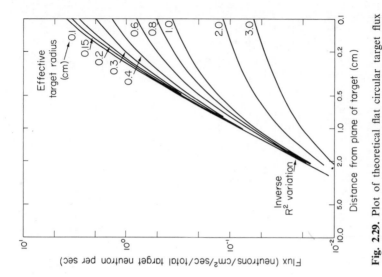

Fig. 2.29. Plot of theoretical flat circular target flux versus distance from the plane of the target along the axis of the target for various effective target radii.

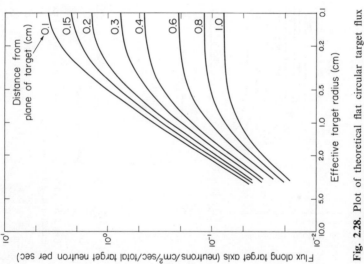

Fig. 2.28. Plot of theoretical flat circular target flux versus effective target radius for various distances from the plane of the target along the axis of the target.

Op de Beeck[27] has recently studied the effect of source dimensions on the neutron flux, and has found that targets with large diameters do not provide the expected improvements in flux. In fact, computations show that flux gradients are increased with increasing source diameter with the possibility also of poor neutron economy. Darrall and Oldham[28] have investigated neutron flux variations on the axis of a 14.7-MeV neutron generator. Experimental data taken by them were in good agreement with the theoretical derivations described in detail in their report. A range of elements having threshold energies varying from 1.5 eV to 13.1 MeV were used and their results indicated considerable deviation from the inverse square law at close distances from the target, to about 2% at a distance of 10 cm. A further extension of this work, showing neutron flux variations directly above and below the target, subsequently was reported by Oldham and Bibby.[29] If target edge and cooling-jacket effects were disregarded, the flux pattern was reasonably symmetrical around the tritium target. A detailed examination of this report is recommended. Gibbons[30] has recently summarized data on neutron flux variations in the vicinity of tritium targets.

Although an accurate value of the flux would be most useful for the determining of activation cross sections, the assumptions made in the theoretical approach are only approximate. For example, the total emission rate of neutrons is seldom accurately known. The diameter of the beam is difficult to assess, and the maintenance of a homogeneous and isotropic neutron source is subject to many, and sometimes uncontrollable, parameters, for example, homogeneous deposition of tritium on the target and electrical stability of the beam. In view of these difficulties, the nuclear analytical chemist has approached this problem experimentally.

The neutron spectrum is a vital factor in the interpretation of results. A simple and less approximate method has been described by Ricci.[31] Errors in analysis, due to neutron scattering before activation of the sample, are treated. This scattering causes an energy spread and the assumption of monoenergetic neutrons can no longer be used. Ricci's calculations yield results which permit the determination of the neutron spectrum impinging upon the sample. This spectrum is divided into four main energy groups; their energy ranges are obtained from scattering calculations. The discussion given[31] is illustrated by a practical example in which Ricci shows that errors as large as 15% may be encountered if due consideration is not given to the perturbed, incident neutron flux. This detailed paper is recommended for further study since it describes why the agreement in the literature of values for fast neutron reaction cross sections is at best mediocre. A conclusion that may be drawn from this work is that for high accuracy in analysis, flux normalization or standardization of neutron generators by well-designed foil activation techniques leaves much to be desired, in spite of their general acceptance today. The principal reason for this is the nonuniformity of structural configurations between various installations. If, however, the experimenter is prepared to accept a margin of error of the magnitude described in Ricci's work, flux normalization by foil activation techniques is appropriate. But for many precise analytical applications involving characterization of standard materials, the analyst must consider all aspects of the perturbed or usable flux experienced by the sample. In addition, perturbations caused by the sample matrix itself must be considered. Some of the methods describing these techniques are discussed in Chapter 6. At this point, however, some pertinent studies regarding the experimental measurement of flux distributions are worth mentioning. Flux patterns from 14.7-MeV neutron generators are of importance to quantitative analysis, health physics, and radiation damage studies. Using radiographic techniques, Kenna and Conrad[32] obtained the flux density map for a uniform target (Fig. 2.30). Experiments with a fresh target yielded results (Fig. 2.31) which show very little variation in flux from one side of the target to the other. The effect of target deterioration resulting in nonuniformity of the target (10–15-hr operation) is shown in Fig. 2.32. They observed significant flux asymmetry of up to 30% at a distance of 1.25 cm from the target, the extent of the variation falling off with increasing distance from the neutron source.

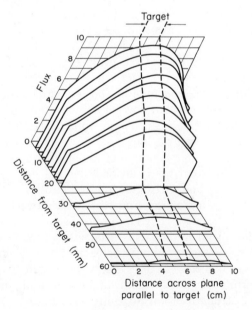

Fig. 2.30. Relative flux density map by radiography; middle of uniform target.

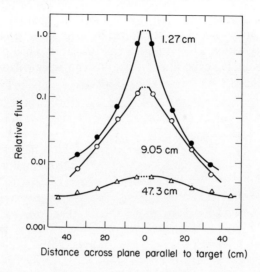

Fig. 2.31. Relative flux as a function of distance; uniform target.

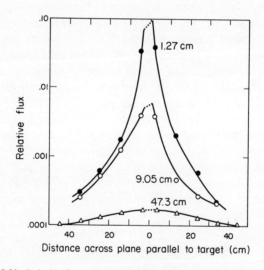

Fig. 2.32. Relative flux as a function of distance; nonuniform target.

A more recent study by Priest et al.[33] illustrates that the flux patterns are far from isotropic. Flux maps developed by them show regions of maximum reproducibility within which samples can be placed. Their data also indicate that beyond 8 cm the experimental fast flux can be calculated by the inverse square law (Fig. 2.33).

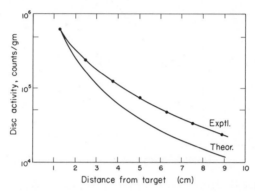

Fig. 2.33. Decrease in neutron intensity with distance from the target.

Information relating to the nature and characteristics of the usable flux has brought out some of the salient points an experimenter must consider in the interpretation of activation data. It is emphasized that each neutron generator installation is sufficiently unique, so that direct comparison of activation data obtained from several neutron generators could be in error and result in ambiguous analyses. If a particular application demands a high degree of accuracy and precision, then each analyst must evaluate the flux characteristics for his individual system, or find some method whereby the flux pattern can be related to a sample and a comparative standard in a reproducible manner. Various techniques, to be described later, deal with such problems of equivalent standard-sample response. As is shown in Chapter 6, these techniques have the capability of providing highly precise data regardless of the inherent instability and heterogeneous nature of the usable neutron flux.

IV. FAST-NEUTRON-INDUCED REACTIONS

In this section, nuclear reactions produced by fast neutrons are described. In addition, a brief treatment of some of the secondary reactions of importance to the activation analyst is given. Finally, the possibilities for the promotion of some rare nuclear reactions are shown insofar as they could result in interference in some applications. During this discussion, only information directly useful to the analyst is given in detail. However, brief

IV. FAST-NEUTRON-INDUCED REACTIONS

mention is made of topics such as cross-section energy dependencies in order to provide the reader with a more complete picture of variables affecting the yield from fast-neutron-induced reactions. This approach has been selected since it is the purpose of this section to assist in better evaluating the experiment prior to the first irradiation and thus to provide the maximum amount of information from each experiment. Nuclear physics texts contain much of the theoretical background on the mechanisms of nuclear reactions. The interested reader is directed to some dissertations[34–40] for details of the many models suggested for nuclear reactions. The physics of excitation functions have been elaborated by other authors[41–44] engaged in cross-section measurements. They have drawn comparisons between calculated cross-section values as a function of energy and those determined experimentally. The literature is also rich in tabulations[22, 44–60] which give both calculated and experimental data on excitation functions.

From the activation analyst's viewpoint, however, only the application of fast-neutron-induced reactions is considered. In the following subsections, the various types of reactions possible from bombardment of samples are described from which working estimates of expected interferences can be made.

A. CLASSIFICATION OF NUCLEAR REACTIONS

The principal nuclear reactions of interest in activation analysis with neutron generators are listed here in the approximate order of increasing threshold energies.

$$(n, \gamma): {}^{A}Z + {}^{1}n \rightarrow {}^{(A+1)}Z + \gamma \tag{2.21}$$

$$(n, n'\gamma): {}^{A}Z + {}^{1}n \rightarrow {}^{A}Z + {}^{1}n + \gamma \tag{2.22}$$

$$(n, p): {}^{A}Z + {}^{1}n \rightarrow {}^{A}(Z-1) + p \tag{2.23}$$

$$(n, \alpha): {}^{A}Z + {}^{1}n \rightarrow {}^{(A-3)}(Z-2) + \alpha \tag{2.24}$$

$$(n, 2n): {}^{A}Z + {}^{1}n \rightarrow {}^{(A-1)}Z + 2n \tag{2.25}$$

In order to provide some insight into the probabilities of occurrence of these reactions, a brief qualitative description of the interaction energetics involved is presented.

In the center-of-mass system, the capture of a neutron by a target nucleus results in the product nucleus acquiring an excitation energy equivalent to the sum of the incident neutron kinetic energy and the binding energy of the neutron in the product nucleus. A variety of reactions can take place, depending on the particles emitted from the product nucleus. The preferred reaction will depend on the energy required for the particle, or particles, to escape from the compound nucleus. If this energy is less than that imparted to the product nucleus, the reaction is promoted. The specific reaction that takes place depends not only on the characteristics and properties of the compound nucleus but also on the competition between various modes of deexcitation of the compound nucleus. For incident neutron energies much less than the nuclear binding energy, the observed reactions will depend strongly on the magnitude of the binding energy. For nuclei of mass number $A < 20$, the neutron binding energy exhibits large, periodic fluctuations from nucleus to nucleus. For mass numbers between 20 and 130, the neutron binding energy shows, on the average, a very slow increase from about 8 to 8.5 MeV, followed by a slow decrease to 7.5 MeV for the heaviest elements. This general trend in the binding energy, however, includes anomalous variations. For example, neutron capture by nuclei of odd mass and neutron number results in excitation energies about 1 to 2 MeV higher than that for even nuclei adjacent to them. These variations are also observed in the region of the so-called neutron magic numbers; that is, the neutron binding energy in a nucleus having one neutron less than a closed shell will be considerably higher than the average, whereas that for an already closed shell will be significantly lower. The three principal energies of interest to the user of neutron generators are ~ 0.025 eV (thermal), ~ 2.8 MeV, and ~ 14 MeV. The cross-section dependency on energy for the reactions included in the earlier classification is discussed briefly below.

The (n, γ) reaction is in all cases exoergic and the cross section in most cases decreases with increasing neutron energy. Some nuclides, however, display high resonance absorption at certain neutron energies. Many tabulations are available in the literature which present this dependency of cross section on energy; the Brookhaven series[56] is recommended. Isotopic thermal neutron cross sections vary by as much as 10^{10} barns and do not follow any definite relationship as a function of Z or A. For 14-MeV neutrons, the (n, γ) cross sections may usually be neglected as they are of the order of a few millibarns, except for some special cases when they can be significant. In general, for neutron energies between 1 and 15 MeV, the (n, γ) cross sections increase as the third power of the mass number A up to a mass number $A \simeq 100$, when a saturation value is reached. For D–D neutrons, certain elements exhibit relatively high (n, γ) cross sections which can be useful for interference-free analytical applications. Perkin et al.[58] measured the (n, γ) cross sections for about 30 elements. Using chemical separations, they were able to identify the residual radioactive products and measure the absolute activity induced from the irradiation of selected targets with 14.7-MeV neutrons. The errors resulting from capture reactions from neutrons of lower energy were allowed for in their final values. Table 2.5 gives the results obtained by this group, together with a limited number of measurements

TABLE 2.5. List of Radiative Capture (n, γ) Cross Sections for 14.5-MeV Neutrons[a]

Target Nuclide	Neutron Number	$\sigma(n, \gamma)$[b] (mb)	Standard Deviation (%)	Target Nuclide	Neutron Number	$\sigma(n, \gamma)$[c] (mb)
^{23}Na	12	0.33	10	—	—	—
^{26}Mg	14	0.2	25	—	—	—
^{27}Al	14	0.53	25	—	—	—
^{30}Si	16	0.49	10	—	—	—
^{41}K	22	3.5	20	—	—	—
^{50}Ti	28	3.5	30	—	—	—
^{55}Mn	30	0.76	10	—	—	—
^{63}Cu	34	2.56	15	—	—	—
^{65}Cu	36	6.3	30	—	—	—
^{71}Ga	40	1.9	10	—	—	—
^{82}Se	48	0.65	30	—	—	—
^{81}Br	46	3.5	25	—	—	—
^{89}Y	50	2.9	10	—	—	—
^{96}Zr	56	≤ 4	—	—	—	—
^{104}Ru	60	13.6	20	—	—	—
^{110}Pd	64	2.0	20	—	—	—
^{127}I	74	2.5	20	—	—	—
^{138}Ba	82	1.3	30	—	—	—
^{139}La	82	1.48	10	^{139}La	82	1.1 ± 0.2
^{142}Ce	84	≤7.5	—	—	—	—
^{141}Pr	82	3.33	10	^{141}Pr	82	2.1 ± 1.0
^{160}Gd	96	18.5	30	^{160}Gd	96	3.0 ± 1.0
^{165}Ho	98	≥9.45	—	^{164}Dy[d]	98	1.5 ± 0.5
^{186}W	112	4.0	20	—	—	—
^{198}Pt	120	1.7	20	^{164}Dy[e]	98	8 ± 3
^{205}Tl	124	2.0	20	^{175}Lu	104	2 ± 1
^{208}Pb	126	3.05	15	—	—	—
^{209}Bi	126	1.45	12	—	—	—
^{232}Th	142	5.2	15	—	—	—
^{238}U	146	3.3	15	—	—	—

[a] Note: Values from reference 58 have been corrected for the low-energy neutron error.
[b] Reference 58.
[c] Reference 59.
[d] Isomeric production.
[e] Ground-state production.

made by Wille and Fink.[59] It can be noted that the cross sections are of the order of 5 mb except for the light elements. A plot of these cross sections (Fig. 2.34) with respect to the neutron number shows slight evidence for lower values in regions of magic numbers. For D–D neutrons, the Brookhaven compilations indicate (n, γ) cross sections for a few of the elements of the order of 100 mb. A recent and excellent compilation by Csikai et al.[60] of (n, γ) cross sections for 1-, 3-, and 14-MeV neutrons as a function of target

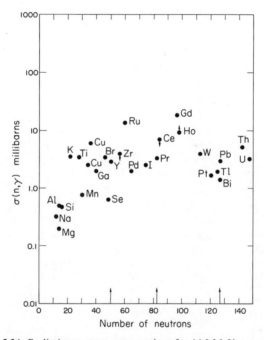

Fig. 2.34. Radiative capture cross sections for 14.5-MeV neutrons.

neutron number is shown in Fig. 2.35. From these graphs, maximum (n, γ) cross-section values of ~ 100 mb for 1-MeV, ~ 30 mb for 3-MeV, and ~ 15 mb for 15-MeV neutrons can be observed.

All (n, n'γ) reactions are slightly endoergic and have no coulomb barrier. About 60 stable isotopes are known to possess an isomeric state of a half-life greater than 1 μsec. From the limited data available, the cross-section values lie between 100 and 600 mb. The only useful reactions of this type, induced by D–D and D–T neutron activation, are those where the metastable state for the excited nucleus is sufficiently long that the activity could be measured

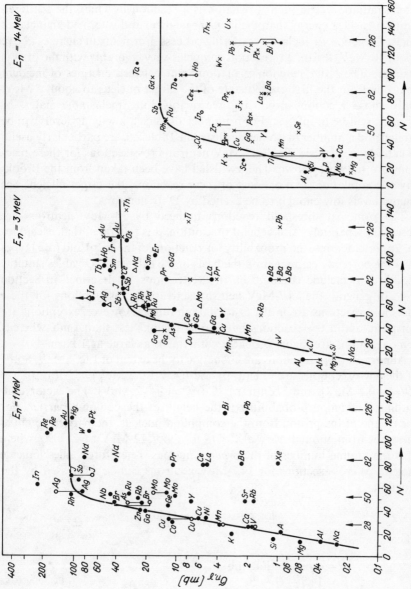

Fig. 2.35. Radiative capture cross sections for fast neutrons versus target neutron number, at 1, 3, and 14 MeV.

after irradiation. The (n, n'γ) reaction is a reaction in which the scattered neutron has less energy than the incident neutron and is termed an inelastic nuclear reaction or inelastic scattering process. For neutron energies above about 10 MeV, the (n, 2n) reaction competes very favorably with the (n, n'γ) reaction. The latter reaction is predominant between energies of incident neutrons from the first excited state of the target nucleus to about 5 MeV. Only those reactions producing reasonably long metastable states, or isomers, will be of interest. For a number of nuclei, low-lying levels can be excited and a number of isomers, as listed in Table 2.6, are particularly useful in activation analysis. The 14.7-MeV neutron cross sections for these reactions are not well known. The few listed have been taken from the Brookhaven compilations.[57] For most of these reactions, the target nucleus has energy levels low enough to be excited by D–D neutrons.

The principal threshold reactions induced by 14-MeV neutrons are described separately. As a general precaution, it is reiterated that, except for (n, 2n) interactions, the probability for reactions of the (n, p) and (n, α) types must be correctly estimated on the basis of the calculated Q values suitably modified to include coulomb barrier restrictions, as described in Section II.C. In general, the 14.7-MeV neutron activation cross sections for these threshold reactions are in the 5- to 500-mb range. However, exceptions are apparent, and a few reactions have cross sections less than 1 mb, whereas some reactions are highly probable with values as large as 1 barn.

Almost all (n, p) reactions are endoergic to the extent of 1 to 4 MeV. Some of the few exceptions are ^{47}Ti(n, p)^{47}Sc ($Q = +0.2$ MeV), ^{58}Ni(n, p)^{58}Co ($Q = +0.4$ MeV), and ^{64}Zn(n, p)^{64}Cu ($Q = +0.2$ MeV). The factor controlling interaction probabilities is the height of the coulomb barrier. For the escape of the proton from the compound nucleus, the barrier potential increases from about 1 MeV at $Z = 3$ to about 12 MeV at $Z = 90$, where Z is the atomic number of the *product* nucleus. Literature values indicate that (n, p) cross sections for 14.7-MeV neutrons can be expressed by the function

$$\log \sigma = 0.308F - 0.461 \qquad (2.26)$$

where

σ = cross section, mb
$F = 14.8 - Q - B$

$$B = \frac{Z_1 Z_2 e^2}{R_1 + R_2} \times 6.242 \times 10^5 \text{ MeV} \ (e = 4.8 \times 10^{-10})$$

Z_1 = atomic number of the product nucleus
Z_2 = atomic number of the proton

TABLE 2.6. Useful (n, n'γ) Reactions Induced by 14.7-MeV Neutrons

Reaction	Cross Section (mb)	Half-life	Principal Radiation Emitted (MeV)
^{73}Ge(n, n'γ)^{73m}Ge	—	0.53 sec	IT 0.054, γ 0.0135
^{75}As(n, n'γ)^{75m}As	13	0.0168 sec	IT 0.0243, 0.304, γ 0.279
^{77}Se(n, n'γ)^{77m}Se	—	17.5 sec	IT 0.162
^{79}Br(n, n'γ)^{79m}Br	—	4.8 sec	IT 0.21
^{81}Br(n, n'γ)^{81m}Br	—	37 μsec	IT 0.27, γ 0.28
^{87}Sr(n, n'γ)^{87m}Sr	—	2.83 hr	IT 0.388
^{89}Y(n, n'γ)^{89m}Y	400 ± 47	16 sec	IT 0.908
^{90}Zr(n, n'γ)^{90m}Zr	—	0.81 sec	IT 2.32, 0.1327, γ 2.18
^{93}Nb(n, n'γ)^{93m}Nb	—	13.6 years	IT 0.0304
^{105}Pd(n, n'γ)^{105m}Pd	—	37 μsec	IT 0.18278, γ 0.3190
^{107}Ag(n, n'γ)^{107m}Ag	—	44.3 sec	IT 0.0931
^{109}Ag(n, n'γ)^{109m}Ag	—	40 sec	IT 0.0877
^{111}Cd(n, n'γ)^{111m}Cd	—	48.6 min	IT 0.15, γ 0.247
^{113}Cd(n, n'γ)^{113m}Cd	—	14 years	IT 0.27
^{113}In(n, n'γ)^{113m}In	—	100 min	IT 0.393
^{115}In(n, n'γ)^{115m}In	81 ± 9	4.45 hr	IT 0.335
^{115}Sn(n, n'γ)^{115m}Sn	—	159 μsec	IT 0.107, γ 0.12, 0.499
^{135}Ba(n, n'γ)^{135m}Ba	—	28.7 hr	IT 0.268
^{137}Ba(n, n'γ)^{137m}Ba	—	2.55 min	IT 0.6616
^{151}Eu(n, n'γ)^{151m}Eu	—	58 μsec	IT 0.175, γ 0.0216
^{167}Er(n, n'γ)^{167m}Er	—	2.3 sec	IT 0.2078
^{176}Lu(n, n'γ)^{176m}Lu	—	3.7 hr	β^- 1.31, 1.22 γ 0.08836
^{178}Hf(n, n'γ)^{178m}Hf	—	4.3 sec	IT 0.0888, γ 0.0932, 0.4268, 0.3257, 0.2136
^{179}Hf(n, n'γ)^{179m}Hf	—	18.6 sec	IT 0.161, γ 0.217
^{180}Hf(n, n'γ)^{180m}Hf	12.4 ± 0.5	5.5 hr	IT 0.058, 0.5, γ 0.0933, 0.4436
^{180}Ta(n, n'γ)^{180m}Ta	—	8.1 hr	β^- 0.61, 0.71, γ 0.0933, 0.103
^{180}W(n, n'γ)^{180m}W	—	0.0052 sec	IT 0.22, 0.5, γ 0.10
^{183}W(n, n'γ)^{183m}W	—	5.3 sec	IT 0.11, γ 0.16, 0.21
^{191}Ir(n, n'γ)^{191m}Ir	—	4.9 sec	IT 0.0418, γ 0.047, 0.082, 0.1294
^{193}Ir(n, n'γ)^{193m}Ir	—	12 days	IT 0.08
^{197}Au(n, n'γ)^{197m}Au	200 ± 50	7.2	IT 0.13, 0.4095, γ 0.2793
^{199}Hg(n, n'γ)^{199m}Hg	—	43 min	IT 0.375, γ 0.158
^{201}Hg(n, n'γ)^{201m}Hg	—	9.3 μsec	IT 0.52
^{204}Pb(n, n'γ)^{204m}Pb	—	66.9 min	IT 0.91, γ 0.2893, 0.3747, 0.90
^{207}Pb(n, n'γ)^{207m}Pb	200	0.8 sec	IT 1.064, γ 0.5697

R_1 = radius of the product nucleus, cm
R_2 = radius of the proton, cm
Q = interaction energy of the reaction, MeV.

R_1 and R_2 can be calculated from the approximate expression $R \simeq 1.5 \times 10^{-13} A^{1/3}$ cm. Equation 2.27, however, is approximate and is based on (n, p) cross sections of 37 stable isotopes producing radioactive gamma emitters of half-lives between 1 sec and 24 hr. Deviations from Eq. 2.26 can be as great as a factor of 2 to 4. Some typical excitation functions for (n, p) reactions taken from recent compilations[49, 61] are shown in Figs. 2.36 to 2.39. The

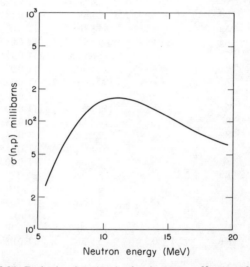

Fig. 2.36. Excitation function for (n, p) reaction: ^{60}Ni(n, p)^{60}Co.

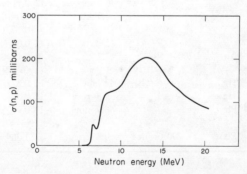

Fig. 2.37. Excitation function for (n, p) reaction: ^{24}Mg(n, p)^{24}Na.

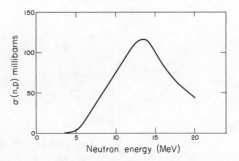

Fig. 2.38. Excitation function for (n, p) reaction: ^{56}Fe(n, p)^{56}Mn.

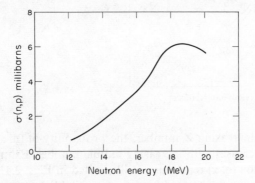

Fig. 2.39. Excitation function for (n, p) reaction: ^{197}Au(n, p)^{197}Pt.

isotope effect is demonstrated in Fig. 2.40, also taken from reference 61. A general analysis of (n, p) cross sections reveals an increasing trend with proton number Z up to $Z = 16$, the cross sections being ≤ 10 to $\simeq 300$ mb. In the region $16 \leq Z \leq 24$, a broad maximum is observed. With increasing Z, the (n, p) cross section gradually decreases to a value of 2 to 3 mb for the heaviest elements.

The (n, α) reactions are invariably endoergic to about the same extent as (n, p). However, with increasing Z number, the (n, α) Q-value gradients become less negative, then slightly positive, and finally, for very high Z number elements, the reaction becomes exoergic to the extent of 5 to 7 MeV. This favorable situation is partially negated because the coulomb barrier for the emission of an α-particle from the compound nucleus is approximately twice as high as that for proton emission. The barrier increases from about 1 MeV for $Z = 2$ to about 22 MeV for $Z = 90$. For about 26 stable isotopes the (n, α) cross section for the production of gamma emitters with half-lives from 1 sec to 24 hr ranges from 0.5 to 150 mb. Since the coulomb barrier

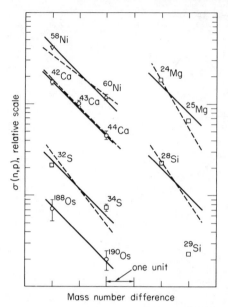

Fig. 2.40. Isotope effect for (n, p) cross sections.

increases with increasing Z number, the probability of (n, α) reactions, on the average, decreases with increasing atomic number. Examples of excitation functions for (n, α) reactions are illustrated in Figs. 2.41 to 2.44. These figures have been reconstructed from those available in references 49 and 61. In general the (n, α) cross-section behavior exhibits a decrease from about 100 mb down to about 1 mb from the lightest to the heaviest element. Two definite maxima are observed within this range. One is located in the region between the elements sodium and chlorine ($\gtrsim$ 100 mb) and the other in the region of rare earths ($\simeq$ 10 mb). At the neutron magic number 50, the (n, α) cross sections are considerably higher than those for the neighboring nuclides. Gardner and Poularikas,[41] have given a relationship describing the isotopic dependency of (n, α) cross sections.

The (n, 2n) reactions are predominantly endoergic. The negative Q values are of the order of 7 to 13 MeV. However, no coulomb barrier exists and the threshold energies are controlled by the interaction energy. For about 26 stable nuclides leading to positron and gamma emitters with half-lives between 1 sec and 24 hr after activation, the (n, 2n) cross sections vary between 18 and 1200 mb. The dependence of the cross section on energy for (n, 2n) reactions can be summarized as follows: For mass numbers below 40, the threshold energies are generally in excess of 14 MeV and the reaction is highly improbable ($\sigma^{(n, 2n)}$ = 5 to 50 mb) using neutrons from the D–T reaction and 150-kV acceleration. Slightly above A = 40, the threshold

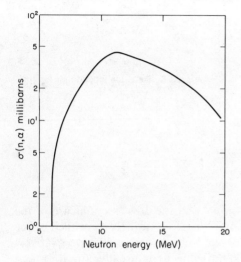

Fig. 2.41. Excitation function for (n, α) reaction: ^{63}Cu(n, α)^{60}Co.

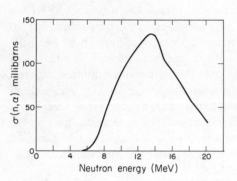

Fig. 2.42. Excitation function for (n, α) reaction: ^{27}Al(n, α)^{24}Na.

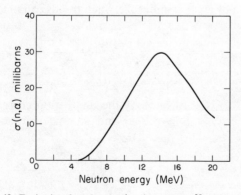

Fig. 2.43. Excitation function for (n, α) reaction: ^{59}Co(n, α)^{56}Mn.

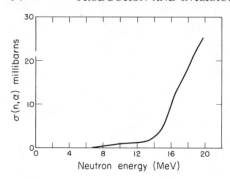

Fig. 2.44. Excitation function for (n, α) reaction: ^{197}Au(n, α)^{194}Ir.

energies are only a little below 14 MeV and the cross sections are of the order of a few millibarns. The reaction is therefore strongly controlled by the Q value. However, a few nuclides around mass number 50 have relatively high (n, 2n) cross sections ($\simeq$ 1000 mb) due to unusually low threshold energies in this mass region. The threshold energy keeps decreasing with a corresponding increase in the (n, 2n) cross section until mass number 90 when the cross-section values go through another minimum, because of the rising threshold. The minima at neutron numbers 28 and 50 can be attributed to the threshold or Q effect, and those at neutron numbers 82 and 126, to the relative competition with the (n, n'γ) reaction. The nuclear level density controlling the inelastic scattering reaction decreases to a much larger extent than that for the (n, 2n) reaction. The composite effect on the (n, 2n) cross section is illustrated in Fig. 2.45, which has been constructed from graphical data given in reference 61. Typical excitation functions taken from reference 49 are shown in Figs. 2.46 to 2.48.

The systematic trends observed in the behavior of (n, p), (n, α), and (n, 2n) cross sections are of great significance. Such trends can assist the analyst in

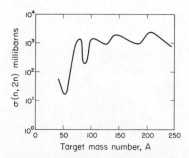

Fig. 2.45. Behavior of (n, 2n) cross sections at neutron magic numbers.

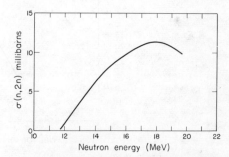

Fig. 2.46. Excitation function for (n, 2n) reaction: ^{14}N(n, 2n)^{13}N.

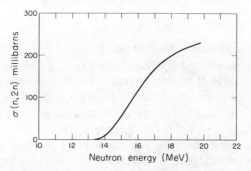

Fig. 2.47. Excitation function for (n, 2n) reaction: ^{46}Ti(n, 2n)^{45}Ti.

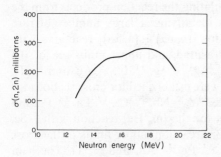

Fig. 2.48. Excitation function for (n, 2n) reaction: ^{88}Sr(n, 2n)^{87m}Sr.

estimating previously unmeasured cross sections. The measurement of cross sections of the above-mentioned reactions is not discussed in this text. Suffice it to mention that a host of techniques have been used for their determination. These have been fully described in the literature.[34, 36, 62] The poor communication that generally exists between the pure physicist, who is most qualified and experienced in cross-section measurement technology, and the analytical chemist, who invariably demands highly accurate data, has been appropriately expressed by Cox[61] in a summary on fast neutron cross sections.

... The user is interested primarily in obtaining cross section data for his research and the cross-section measurer is interested in physics....

... Experimental physicists would do the data users and data evaluators a great service by including all pertinent details of the experimental procedure so that intercomparison between data from various sources could be made on an objective basis.

In addition to the predominant reactions discussed thus far, there are some relatively rare reactions which are also induced with high-energy neutrons. Some of these reactions, which are not often experienced by the activation analyst, are described in the following section.

B. RARE REACTIONS

The principal reason why the activation analyst has not considered some of the fast-neutron-induced reactions noted here, is that their existence has been reported mainly in physics journals. Cross-section compilations give many reactions that are energetically possible with 14-MeV neutrons; in fact, most of these measurements were made using neutron generators. Although the majority of the rare reactions are less probable than those reported earlier, some do have significant cross sections and can present problems in certain analytical applications. Bramlitt and Fink[63] have made a comprehensive study involving cross-section measurements of these reactions. Using a 400-kV Cockcroft-Walton generator and radiochemical separation procedures, they succeeded in isolating the reaction products from the irradiation of thick samples and in identifying a large number of rare reactions. Activation cross sections for [(n, nα) + (n, αn)] reactions with ^{65}Cu, ^{70}Zn, ^{71}Ga, and ^{93}Nb were measured, and upper limits set for the isotopes for ^{51}V, ^{76}Ge, ^{81}Br, ^{87}Rb, ^{107}Ag, ^{109}Ag, ^{115}In, ^{197}Au, and ^{203}Tl. For the (n, 2p) reaction, upper limits were established for cross sections for ^{29}Si, ^{41}K, ^{45}Sc, ^{50}Ti, ^{51}V, ^{55}Mn, ^{75}As, ^{89}Y, ^{93}Nb, ^{133}Cs, ^{139}La, ^{141}Pr, and ^{159}Tb, and upper limits were also set for the (n, ^{3}He) reaction with ^{45}Sc, ^{93}Nb, ^{197}Au, and ^{205}Tl. For the (n, 3n) reaction, upper limits of the cross section were set for target isotopes ^{141}Pr, ^{197}Au, and ^{203}Tl. The sum [(n, np) + (n, pn) + (n, d)] cross-section value for ^{58}Ni was 520 ± 120 mb, and an upper limit for the same reaction with the target isotope ^{92}Mo was established. In addition to the above, they studied the probability of some (n, p), (n, 2n), (n, α), and (n, γ) reactions; some data from their work are given in Table 2.7. A comprehensive table, which includes cross-section measurements by other experimenters, is also given in their report. The (n, t) cross sections have been studied by Poularikas[64], and cross sections[50] and excitation functions[48] for this reaction have also been reported. In spite of the low probability for the promotion of most of these rare reactions, high accuracy in analyses requires a knowledge of their expected interferences. In a limited number of cases, some of these reactions, for example, ^{90}Zr(n, n'p)^{89m}Y and ^{93}Nb(n, n'α)^{89}Y, have been reported[65] as useful for analytical purposes.

Summarizing information as reported by the limited number of measurements, we can generalize that (n, n'p) cross sections are comparable to (n, p) values. The (n, t) cross sections are of the order of 1 to 2 mb except for the lightest elements. Further, the (n, n'α) cross sections are of the order of 1 to 3 mb, and the (n, ^{3}He) cross sections are about 0.1 mb.

It is apparent from a review of cross-section studies related to neutron generator activation analysis that intercomparison of data is difficult, if not impossible, because of the lack of detail about the experimental conditions.

IV. FAST-NEUTRON-INDUCED REACTIONS

In addition, the wide variety of experimental arrangements used by the analyst adds to the confusion when attempting to correlate the data. As mentioned before, a large quantity of cross-section data is available at present, in tabular form. In view of the need for accurate and intercomparable cross-section data, the graphical cross-section survey performed by Csikai et al.[60] has been selected for inclusion in this treatise and is given in Appendix I.

Some interesting interactions resulting from the bombardment of certain matrix materials can introduce interferences as a result of induced secondary reactions. A reaction of this type, induced by recoil protons in hydrogenous materials, was encountered by Gilmore and Hull[66] in their efforts to analyze nitrogen in hydrocarbons. Although the method was successful for nitrogen determinations in the percent range, the technique was limited in the parts per million range. Since the recoiling protons possess, on an average, an energy high enough to induce such reactions as $^{13}C(p, n)^{13}N$, nitrogen determinations based on the measurement of ^{13}N from the reaction $^{14}N(n, 2n)^{13}N$ in a hydrocarbon matrix are subject to this systematic error. Gilmore and Hull have reported an equivalent nitrogen content of 900 ppm, resulting from the recoil secondary reaction, for the specific conditions of their experiments. An analogous situation also exists in hydrogen-oxygen systems, such as water. In this case the secondary reaction $^{16}O(p, \alpha)^{13}N$ is induced. Nitrogen determinations in aqueous solutions can therefore be subject to error. Both Braier et al.[67] and Nargolwalla[68] have estimated interferences to be of the order of 5000 ppm of equivalent nitrogen in an aqueous system. Their results compare favorably with those reported by Gilmore and Hull.[66] It should be pointed out, however, that proper discriminatory beta- and gamma-ray spectroscopic techniques can be used in other cases to separate the ^{13}N activity from similar half-life species such as ^{62}Cu and ^{27}Mg.

The reactor activation analyst often places great emphasis on compilations of gamma-ray spectra for the identification of radioactive nuclides, and the neutron generator user is no exception. To date, a number of such compendiums are available[65, 69, 70] which describe the conditions for data acquisition and also indicate calculated sensitivities based on a fixed activity detection limit. Activation analysis facilities employing neutron generators, more often than not, are quite different from one another, particularly in the irradiation and counting configurations used. In these cases the reference spectra can only provide qualitative data and rough quantitative estimates of expected activities. Recently a comprehensive survey of the elements irradiated with D–D neutrons was also reported.[71]

The presentation of the subject matter in this chapter has been designed to assist the analyst in the proper assessment of analytical conditions for his purpose. For many applications, requiring only 10 to 20% accuracy, highly

TABLE 2.7. Rare Nuclear Reactions with 14.7-MeV Neutrons

Upper Limits for (n, 2p) Cross Sections at 14.7 MeV

Reaction	Monitor Reaction and Cross Section (mb)	Upper Limit (n, 2p) Cross Section mb
^{29}Si(n, 2p)^{28}Mg	^{27}Al(n, α)^{24}Na (114)	0.50
^{41}K(n, 2p)^{40}Cl	^{41}K(n, α)^{38}Cl (30)	0.13
^{45}Sc(n, 2p)^{44}K	^{45}Sc(n, α)^{42}K (63)	0.21
^{50}Ti(n, 2p)^{49}Ca	^{50}Ti(n, p)^{50}Sc (28)	0.28
^{51}V(n, 2p)^{50}Sc	^{51}V(n, p)^{51}Ti (55)	0.030
^{55}Mn(n, 2p)^{54}V	^{55}Mn(n, α)^{52}V (33)	0.30
^{75}As(n, 2p)^{74}Ga	^{75}As(n, α)^{72}Ga (93)	0.50
^{89}Y(n, 2p)^{88}Rb	^{27}Al(n, p)^{27}Mg (82)	0.030
^{93}Nb(n, 2p)^{92}Y	^{93}Nb(n, α)^{90m}Y (5.9)	0.50
^{133}Cs(n, 2p)^{132}I	^{133}Cs(n, α)^{130}I (1.0)	0.005
^{139}La(n, 2p)^{138}Cs	^{139}La(n, α)^{136}Cs (1.87)	0.046
^{141}Pr(n, 2p)^{140}La	^{141}Pr(n, γ)^{142}Pr (2.3)	0.84
^{159}Tb(n, 2p)^{158}Eu	^{159}Tb(n, p)^{159}Gd (2.2)	0.080

Upper Limits on (n, ^{3}He) Reaction Cross Sections at 14.7 MeV

Reaction	Monitor Reaction and Cross Section (mb)	Upper Limit (n, ^{3}He) Cross Section mb
^{45}Sc(n, ^{3}He)^{43}K	^{45}Sc(n, α)^{42}K (53)	0.30
^{93}Nb(n, ^{3}He)^{91m}Y	^{93}Nb(n, α)^{90m}Y (5.3)	0.060
^{197}Au(n, ^{3}He)^{195}Ir	^{197}Au(n, α)^{194}Ir (0.43)	0.020
^{205}Tl(n, ^{3}He)^{203}Au	^{205}Tl(n, α)^{202}Au (0.75)	0.070

Cross Sections and Upper Limits for $[(n, n\alpha) + (n, \alpha n)]$
Reaction Sums at 14.7 MeV

Reaction[a]	Monitor Reaction and Cross Section (mb)	$(n, n'\alpha)$ Cross Section mb[a]
^{51}V(n, n'α)^{47}Sc	^{51}V(n, α)^{48}Sc (23)	<5
^{65}Cu(n, n'α)^{61}Co	^{65}Cu(n, 2n)^{64}Cu (954)	2.9 ± 0.8
^{70}Zn(n, n'α)^{66}Ni	^{68}Zn(n, α)^{65}Ni (18)	0.89 ± 0.40
^{71}Ga(n, n'α)^{67}Cu	^{69}Ga(n, p)^{69m}Zn (24)	2.1 ± 1.8
^{76}Ge(n, n'α)^{72}Zn	^{27}Al(n, α)^{24}Na (114)	<1.0
^{81}Br(n, n'α)^{77}As	^{79}Br(n, α)^{76}As (9.2)	<6.5
^{87}Rb(n, n'α)^{83}Br	^{87}Rb(n, α)^{84}Br (39)	<1.5
^{93}Nb(n, n'α)^{89m}Y	^{63}Cu(n, 2n)^{62}Cu (507)	2.5 ± 1.1
^{107}Ag(n, n'α)^{103m}Rh	^{109}Ag(n, α)^{106}Rh (10.5)	<2.0
^{109}Ag(n, n'α)^{105m}Rh	^{109}Ag(n, α)^{106}Rh (10.5)	<0.60
^{115}In(n, n'α)^{111g}Ag	^{115}In(n, α)^{112}Ag (2.7)	<0.055
^{197}Au(n, n'α)^{193}Ir	^{197}Au(n, α)^{194}Ir (0.43)	<0.040
^{203}Tl(n, n'α)^{199}Au	^{203}Tl(n, α)^{200}Au (0.37)	<0.012

[a] For brevity, the $[(n, n'\alpha) + (n, \alpha n')]$ reaction sum is designated as $(n, n'\alpha)$.

Cross Sections for $[(n, n'p) + (n, pn') + (n, d)]$, $(n, 3n)$, and (n, γ)
Reactions at 14.7 MeV

Reaction[a]	Monitor Reaction and Cross Section (mb)	Cross Section mb
^{58}Ni(n, n'p)^{57}Co	^{27}Al(n, α)^{24}Na (114)	520 ± 120
^{92}Mo(n, n'p)^{91m}Nb	^{92}Mo(n, p)^{92g}Nb (60)	<50
^{141}Pr(n, 3n)^{139}Pr	^{141}Pr(n, γ)^{142}Pr (2.3)	<10
^{197}Au(n, 3n)^{196}Au	^{27}Al(n, α)^{24}Na (114)	<0.1
^{203}Tl(n, 3n)^{201g}Tl	^{203}Tl(n, 2n)^{202}Tl (1300)	<10
^{89}Y(n, γ)^{90m}Y	^{27}Al(n, α)^{24}Na (114)	1.1 ± 0.6
^{89}Y(n, γ)^{90g}Y	^{89}Y(n, γ)$^{90m+g}$Y (2.9)	1.8 ± 0.6
^{93}Nb(n, γ)^{94m}Nb	^{27}Al(n, α)^{24}Na (114)	0.44 ± 0.26
^{141}Pr(n, γ)^{142}Pr	^{27}Al(n, α)^{24}Na (114)	2.3 ± 1.1

[a] For brevity, the $[(n, n'p) + (n, pn') + (n, d)]$ reaction sum is designated as $(n, n'p)$.

detailed studies concerning flux perturbations are not necessary. However, with increasing demands for more accurate and precise analytical techniques, more is expected from the neutron generator facility. Under these circumstances, the analyst must be fully aware of all the parameters influencing flux characteristics so that account can be taken of anticipated interferences in a given situation.

REFERENCES

1. S. Glasstone, *Principles of Nuclear Reactor Engineering*, Van Nostrand, Princeton, New Jersey, 1955.
2. H. Etherington, Ed., *Nuclear Engineering Handbook*, McGraw-Hill, New York, 1958.
3. R. M. Eisberg, *Fundamentals of Modern Physics*, Wiley, New York, 1964.
4. R. Evans, *The Atomic Nucleus*, McGraw-Hill, New York, 1955.
5. G. Friedlander, J. W. Kennedy, and J. M. Miller, *Nuclear and Radiochemistry*, Wiley, New York, 1964.
6. G. B. Cook and J. F. Duncan, *Modern Radiochemical Practice*, Oxford University Press, London, 1951.
7. R. C. Koch, *Activation Analysis Handbook*, Academic Press, New York, 1960.
8. W. S. Lyon, Jr., Ed., *Guide to Activation Analysis*, Van Nostrand, Princeton, New Jersey, 1964.
9. J. M. A. Lenihan and S. J. Thomson, Eds., *Activation Analysis, Principles and Applications*, Academic Press, New York, 1965.
10. D. DeSoete, R. Gijbels, and J. Hoste, *Neutron Activation Analysis*, Wiley-Interscience, New York, 1972.
11. D. Taylor, *Neutron Irradiation and Activation Analysis*, George Newnes, London, 1964.
12. H. J. M. Bowen and D. Gibbons, *Radioactivation Analysis*, Oxford University Press, London, 1963.
13. E. A. Burrill, "Neutron Production and Protection," High Voltage Engineering Corporation, Burlington, Massachusetts, 1963.
14. C. F. Barnett, W. B. Gauster, and J. A. Ray, Oak Ridge National Laboratory, Report ORNL-3113, 1961.
15. J. L. Fowler and J. E. Brolley, Jr., *Rev. Mod. Phys.*, **28**, 103 (1956).
16. J. D. Seagrave, E. R. Graves, S. J. Hipwood, and C. J. McDole, Los Alamos Scientific Laboratory, Report LAMS-2162, 1958.
17. J. Benveniste and J. Zenger, University of California Radiation Laboratory, Report UCRL-4266, 1954.
18. J. T. Prud'homme, "Texas Nuclear Corporation Neutron Generators," Texas Nuclear Corporation, Austin, Texas, 1962.
19. K. Perry, G. Aude, and J. Laverlochere, Preprints, "Symposium on Trace Characterization—Chemical and Physical," Paper No. 54, National Bureau of Standards, Gaithersburg, Maryland, October 3–7, 1966.

IV. FAST-NEUTRON-INDUCED REACTIONS

20. P. L. Jessen, "Design Considerations for Low Voltage Accelerators," Kaman Nuclear Report KN-68-459(R), Kaman Nuclear Corporation, Colorado Springs, Colorado, 1968.
21. S. Nagy, J. Csikai, and I. Angeli, *Nucl. Instrum. Methods*, **91** (3), 345 (1971).
22. H. P. Dibbs, "Activation Analysis with a Neutron Generator," Department of Energy, Mines and Resources, Ottawa, Canada, Mines Branch Research Report R155, 1965.
23. W. W. Meinke and R. W. Shideler, Michigan Memorial Phoenix Project, Report MMPP-191-1, 1961.
24. S. S. Nargolwalla, J. Niewodniczanski, and J. E. Suddueth, "Activation Analysis with the Cockcroft-Walton Neutron Generator," in *Activation Analysis Section: Summary of Activities July 1968 to June 1969*, P. D. LaFleur (Ed.), NBS Technical Note 508, issued July 1970.
25. H. J. Price, "Graphical Analysis of Theoretical Neutron Flux Produced by Fast-Neutron Generators," Kaman Nuclear Corporation, Colorado Springs, Colorado, 1964.
26. J. Op de Beeck, *J. Radioanal. Chem.*, **1**, 313 (1968).
27. J. Op de Beeck, *Radiochem. Radioanal. Lett.*, **1** (4), 281 (1969).
28. K. G. Darrall and G. Oldham, *Nucl. Energy*, July–August 1967, p. 104.
29. G. Oldham and D. M. Bibby, *Nucl. Energy*, November–December 1968, p. 167.
30. D. Gibbons, Proceedings, IAEA Conference on Utilization of Neutron Generators in Physics and Chemistry, Budapest, Hungary, May 1969, in press.
31. E. Ricci, *J. Inorg. Nucl. Chem.*, **27**, 41 (1965).
32. B. T. Kenna and F. J. Conrad, *Health Phys.*, **12**, 566 (1966).
33. H. F. Priest, F. C. Burns, and G. L. Priest, *Nucl. Instrum. Methods*, **50**, 141 (1967).
34. J. B. Marion and J. L. Fowler, Eds., *Fast Neutron Physics*, Part II, Wiley, New York, 1963.
35. H. Feshbach, C. E. Porter, and V. F. Weisskopf, *Phys. Rev.*, **96**, 448 (1954).
36. E. Segré, Ed., *Experimental Nuclear Physics*, Vols. I–III, Wiley, New York, 1953.
37. H. Feshbach, D. C. Peaslee, and V. F. Weisskopf, *Phys. Rev.*, **71**, 145 (1947).
38. J. M. Blatt and V. F. Weisskopf, *Theoretical Nuclear Physics*, Wiley, New York, 1952.
39. A. Chatterjee, *Phys. Rev.*, **134**, B374 (1964).
40. D. G. Gardner, *Nucl. Phys.*, **29**, 373 (1962).
41. D. G. Gardner and A. D. Poularikas, *Nucl. Phys.*, **35**, 303 (1962).
42. D. G. Gardner and Yu-wen Yu, *Nucl. Phys.*, **60**, 49 (1964).
43. M. Bormann, *Nucl. Phys.*, **65**, 257 (1965).
44. J. Picard and C. F. Williamson, *Nucl. Phys.*, **63**, 673 (1965).
45. B. T. Kenna and F. J. Conrad, "Tabulation of Cross Sections, Q-Values, and Sensitivities for Nuclear Reactions of Nuclides with 14 MeV Neutrons," Sandia Corporation Report SC-RR-66-229, Sandia Corporation, Albuquerque, New Mexico, 1966.
46. S. C. Mathur, J. B. Ashe, and I. L. Morgan, "Compilation of Neutron Reaction and Total Cross Section at 14 MeV," Contract AF-33(616)-8160, Texas Nuclear Corporation, Austin, Texas, 1964.
47. E. B. Paul and R. L. Clarke, *Can. J. Phys.*, **31**, 267 (1953).

48. S. Pearlstein, Brookhaven National Laboratory, Report BNL-897(T-365), 1964.
49. P. Jessen, M. Bormann, F. Dreyer, and H. Neuert, *Nucl. Data, Sect. A*, **1**, 103 (1966).
50. W. Nagel and A. H. W. Aten, Jr., *Physica*, **31**, 1091 (1965).
51. A. Chatterjee, *Nucleonics*, **23** (8), 112 (1965).
52. S. Pearlstein, *Nucl. Data, Sect. A*, **3**, 327 (1967).
53. A. Kjelberg, A. C. Pappas, and E. Steinnes, *Radiochim. Acta*, **5**, No. 1, 28 (1966).
54. M. P. Menon and M. Y. Cuypers, *Phys. Rev.*, **156**, 1340 (1967).
55. J. Kantele and D. G. Gardner, *Nucl. Phys.*, **35**, 353 (1962).
56. B. P. Bayhurst and R. J. Prestwood, Los Alamos Scientific Laboratory, Report LA-2493, 1961.
57. Brookhaven National Laboratory, "Neutron Cross Section," Report BNL-325, 2nd ed., Supplement 1 (1960) and Supplement 2, five volumes, 1964–1966.
58. J. L. Perkin, L. P. O'Connor, and R. F. Coleman, AERE (G.B.), Report AWRE 0-59/57, 1957.
59. R. G. Wille and R. W. Fink, *Phys. Rev.*, **118**, 242 (1960).
60. J. Csikai, M. Buczkó, Z. Bödy, and A. Demény, Atomic Energy Review, Vol. VII, No. 4, IAEA, Vienna, 1969.
61. S. A. Cox, "Fast Neutron Cross Sections," in Proceedings, Second Conference on Neutron Cross Sections and Technology, National Bureau of Standards Special Publication 299, Vol. II, 1968.
62. D. J. Hughes, *Pile Neutron Research*, Addison-Welsey, Cambridge, Massachusetts, 1953.
63. E. T. Bramlitt and R. W. Fink, *Phys. Rev.*, **131**, 2649 (1963).
64. A. Poularikas, M. S. Thesis, University of Arkansas, Fayetteville, Arkansas, 1962.
65. M. Cuypers and J. Cuypers, "Gamma-Ray Spectra and Sensitivities for 14-MeV Neutron Activation Analysis," Activation Analysis Research Laboratory, Texas A and M University, College Station, Texas, 1966.
66. J. T. Gilmore and D. E. Hull, *Anal. Chem.*, **34**, 187 (1962).
67. H. A. Braier, W. E. Mott, D. F. Rhodes, and R. A. Stallwood, "Applications of Neutron Activation Analysis at Gulf Research," Report 651, Gulf Research and Development Company, Pittsburgh, Pennsylvania, 1963.
68. S. S. Nargolwalla, Ph.D. Thesis, University of Toronto, Toronto, Canada, 1965.
69. G. Aude and J. Laverlochere, "Spectres Gamma de Radioelements Formés par Irradiation sous Neutrons de 14 MeV, Section d'Application des Radioelements, SAR-G-63-38, C.E.N., Grenoble, France, 1963.
70. J. T. Strain and W. T. Ross, Oak Ridge National Laboratory, Report ORNL-3672, 1965.
71. S. S. Nargolwalla, J. Niewodniczanski, and J. E. Suddueth, *Trans. Am. Nucl. Soc.*, **12** (2), 510 (1969).

CHAPTER

3

THE NEUTRON GENERATOR

I. INTRODUCTION

At the turn of the century naturally occurring α-emitters were the only sources of high-energy particles available for nuclear structure research. At that time, scientists had expressed the need for particles of higher intensity and energy than those obtained from the decay of a limited number of radioisotopes. In addition, the need to vary both the energy and intensity of particles to suit experimental conditions was envisaged.

In 1932 the first device for accelerating charged particles was constructed. In the history of nuclear science, the pioneering work of J. D. Cockcroft and E. T. S. Walton[1] stands out as the first successful experiment in the acceleration of atomic ions with kinetic energy sufficient to induce nuclear reactions. Although Rutherford had previously demonstrated nuclear transformation effects with high-energy α-particles, the Cockcroft-Walton experiments[1] yielded protons which were *artificially* produced by the ionization of hydrogen gas in a discharge tube and by the subsequent acceleration of the gaseous ions by means of a high voltage. By permitting the protons to impinge upon lithium, Cockcroft and Walton were able to induce the disintegration of ^{7}Li through the ^{7}Li(p, α)^{4}He reaction. Significantly, the reaction was detectable at a proton energy of 0.125 MeV, a value considerably less than the calculated potential barrier experienced by a proton entering the ^{7}Li nucleus.

The voltage-multiplier or cascade-rectifier principle was used by Cockcroft and Walton for the generation of the high voltage required for the ion-acceleration process. The voltage-multiplier circuit is illustrated in Fig. 3.1. The two capacitors, C_1 and C_2, are charged on alternate half-cycles; then, with vacuum tube rectifiers S_1 and S_2 acting as switches, twice the input voltage appears across the capacitors when they are discharged in series. The transformer T is the source of the initial AC potential. A unit of two capacitors and two rectifiers is called a voltage doubler. Several doublers in series constitute a voltage multiplier. By starting with 100,000 V across the secondary of a transformer, Cockcroft and Walton were able to obtain an output voltage of nearly 800,000 V. The main advantages of a voltage-doubler system are the lack of moving parts and the ability to carry

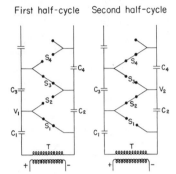

Fig. 3.1. Cockcroft-Walton voltage multiplier.

reasonably large currents at moderate accelerating voltages, advantages ideal for the operation of *low-cost* charged-particle accelerators.

Particle accelerators are generally distinguished by the method of applying the electrical field which provides the accelerating force to the particle. Applying the electric field from a voltage multiplier between two terminals provides the simplest approach for charged-particle acceleration. The majority of accelerators are of this direct potential type, although the circuitry used today is considerably more sophisticated in order to satisfy the increasing needs of the experimenter. Two basic methods are used for the production of high voltage for direct potential accelerators: (1) voltage multipliers, and (2) Van de Graaff voltage generators where a terminal is charged electrostatically. Both methods are used for the acceleration of monoenergetic particles with a minimum of energy spread. Voltage-multiplier systems can produce potentials of up to 350 kV and, with appropriate ion sources, yield beam currents of up to 10 mA. Electrostatic high-voltage generators of the Van de Graaff type have the capability of providing 50-mA beam currents at 1000 kV; however, average beam currents used are considerably lower.

The great demand for relatively inexpensive neutron sources was felt mostly by small laboratories and academic institutions and initiated considerable activity in the accelerator manufacturing industry. By 1959, this requirement was partially satisfied by the introduction of low-beam-current machines from several manufacturers. Each machine was priced in the region of $20,000 and was capable of an output of 10^{10} neutrons/sec using the D–T reaction. In the last 13 years, great strides have been taken to improve the earlier designs, and today low-voltage accelerators, providing neutron outputs of about 3×10^{11} neutrons/sec, are commercially available in the United States for approximately the same price.

Since all positive-ion accelerators are basically similar in design, they can be represented by the diagram shown in Fig. 3.2. Essentially, the major

I. INTRODUCTION

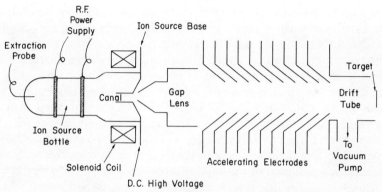

Fig. 3.2. Schematic of pumped neutron generator.

components are (1) an ion source, (2) an evacuated accelerator tube, and (3) a target upon which the accelerated ions impinge to produce neutrons. The overall performance of a neutron generator is dependent on a complex combination of interactions between its individual components. The intention here is to discuss the effect of these interactions on the overall performance of the system critically. To provide a basis for these detailed discussions, a short but general description of each subsystem of the neutron generator is given.

The neutron generator is maintained at a low pressure and deuterium gas is admitted at a controlled rate into the ion source (Fig. 3.2) where it is ionized. The ions produced are directed, by the application of a suitable electric field, into the accelerating section, where they are accelerated up to an energy equivalent to the potential drop across the accelerating section for singly charged positive ions. After passage through a potential-free section, called the drift tube, the accelerated ions strike a suitable target, resulting in the generation of high-energy neutrons. By operating a vacuum pump continuously during beam production, the excess gas admitted into the generator for optimum ion-source operation is reduced to a minimum and collisional losses from the ion beam are minimized. In general, the ion source operates at gas pressures about 100 to 1000 times greater than the normal static pressure of the entire accelerator system, which is typically in the 10^{-8}-torr range. The ion-source assembly is connected with a focusing system to optimize the beam current. Pulsing of the ion beam may be accomplished by the inclusion of electrostatic deflector plates in the beam path, or by the use of suitable switching circuits in the ion source or extraction potential power supplies.

In Chapter 2 (Section III.B.2) the effect of atomic and molecular beams was discussed. It may be recalled that for the D–T reaction, at acceleration voltages less than 200 kV, atomic beams are more efficient neutron producers than molecular beams; therefore, some accelerators employ magnetic beam separators so that only the atomic component of the beam is permitted to impinge upon the target. All ion sources produce a mixture of both atomic and molecular ions. This is achieved in the ion source by the bombardment of an atom or a molecule with electrons of sufficient energy to knock out an orbital electron. The ejected electron in a high-vacuum system can travel a large distance before striking another gas atom or molecule. In order to restrict the physical size of an ion source, the free electron must be contained in a limited volume and yet be kept away from the walls of the source. This is achieved by the imposition of magnetic and electrical fields in directions optimum for this function. Ionization is performed by three principal methods, each related to the type of ion source used. These ion sources are (1) the radiofrequency source (RF), (2) the Penning ion source (PIG), and (3) the duoplasmatron source (DP). These sources can be characterized by the method employed to ionize the gas. The RF source performs this function by the application of an RF field; the PIG and DP sources employ magnetic mirrors.

An essential component of neutron generator systems is the production of high voltage for the acceleration of ions. The high-voltage power supplies fall into two main categories: (1) electrostatic, using the Van de Graaff principle, and (2) the Cockcroft-Walton system of voltage-doubling circuits. Modern-day accelerators operate at 100- to 400-kV acceleration potentials obtained from power supplies having a current capacity of 5 to 10 mA.

Evacuation of the accelerator is achieved by means of several different types of pumping units. The trend today is toward the use, almost exclusively, of ion-getter pumps or ion pumps as they are called, although turbomolecular and diffusion pumps are also occasionally used. Static vacuums of 10^{-8} torr and dynamic vacuums of 10^{-6} torr can be maintained with these pumps. Generally, small mechanical pumps are employed to "rough out" the system prior to starting the main pump.

Neutron-producing targets and the development of suitable cooling devices for the target have been the subject of much debate and research. In general, gas-on-metal occluded targets are used, and both tritium and deuterium gas can be readily absorbed onto a metal film deposited on a metal backing plate. Generally, target loadings for 200-kV accelerators are either 5 Ci/in.^{2} of tritium or 6 cm^3/in.^{2} of deuterium. However, numerous experiments have been performed with wide varieties of gas loadings and target thicknesses. Frozen heavy water has also been used and makes an excellent deuterium target; gas targets which contain tritium gas at several atmospheres pressure are also used. The poor operating life of tritium targets has provided the incentive for much of the target research reported in the literature. Ingenious devices have been designed that employ multiple-disk target holders or rotating tritium-loaded metal ribbons. Target lifetimes of tritium-loaded rare earths have been the subject of considerable investigation. Despite the development of higher output accelerators, target design today is considerably behind the basic ion-source and accelerator design. Some unique cooling devices have been invented to reduce the rate of loss of tritium from the target under bombardment. Methods for target reloading during generator operation have also been successfully practiced.

Although the accent today is on higher and higher flux machines, there is also a considerable demand for moderate flux accelerators which can be operated on a continuous basis for relatively long times. The sealed-tube source fulfills this requirement adequately. Recent developments in sealed-tube sources indicate the feasibility of 10^{11}-neutrons/sec output for over 100 hr of continuous operation. Some prototypes presently being tested have operated for over 1000 hr. In pulsed operation particularly, the sealed tube is capable of giving 10^9 neutrons/10 μsec pulse.

The reliability of neutron generator subsystems depends strongly on the level of beam operation. Compromises are often necessary to maintain these subsystems in a high state of operational reliability with a minimum of downtime. In this chapter, significant research contributing to the development of reliable neutron generators is described. The interaction of each component on the total generator performance is critically examined and assessed. We hope that these discussions will assist the reader in the selection and operation of neutron generators, in a manner most useful and economical for the particular application.

II. PUMPED NEUTRON GENERATOR

In this section the operational characteristics and design considerations affecting the performance and reliability of pumped neutron generators are discussed. The major components, briefly discussed earlier, and some of the essential supporting units are also described. Although neutron generator facilities vary sufficiently from one another to warrant an exhaustive discussion of each system, a full evaluation of their individual performance is obviously not within the scope of this book. However, the literature abounds in descriptions of this nature, and workers entering the field for the first time and wishing to reproduce a system in their own laboratories are referred to the many references cited in the individual sections of this chapter. Here we shall describe a few selected systems in detail in order to illustrate major design and performance differences. Attention is directed toward the intrinsic behavior of components of the generator which, when integrated, serve as limiting factors for determining the ultimate reliability of the complete facility. The ultimate performance of each system depends strongly on the complex interaction of each individual component in the system as a whole. It is this overall behavior which must be understood and which determines whether a particular combination of independent components is suitable for trouble-free operation of the complete system.

There are probably about 500 neutron generators in worldwide operation today that can be classified as low-voltage, positive-ion accelerators. Of these, about 30% have been or are now being used for compositional analysis using fast neutrons. In this area of interest, there appears to be a continuous influx of performance claims which leave a would-be investigator in some confusion in the selection of a system best suited to his particular needs.

The significant aspects of accelerator design is heavily stressed so that a proper and fair assessment of a unit system can be made. Performance comparison of the individual components is made based on experiences of research workers in this field. We shall also provide operational data to assist the uninitiated in putting together an optimum system for his needs. Finally, a status report on accelerators today is given with some "predictions" of future development.

A. ION SOURCES

Before 1950, nearly all accelerators were equipped with conventional filament-type ion sources. Although hydrogen-discharge experiments[2] carried out as early as 1920 had proved that the ionization of hydrogen in long glass tubes exhibited a pure Balmer spectrum demonstrating the existence of atomic hydrogen ions, possibilities for extracting these atomic ions were not investigated. However, certain key observations made during these experiments prompted the development of the RF ion source. It was noted that the intense plasma in the center of the tube was almost all atomic, while the plasma in the vicinity of the metal electrodes exhibited a molecular spectrum, indicating that metals were good catalysts for the recombination of atomic ions. Further work,[3] in which the plasma was formed by an induction coil wrapped around an electrodeless gas discharge tube, showed that glass was a suitable container because it promoted very little ion recombination. Smith[4] classified a number of materials on the basis of this recombination effect on a relative scale: Pyrex, 2×10^{-5}; quartz, 7×10^{-4}; alumina, 0.33; and platinum, 1.0. For most metals studied, the recombination coefficient was about unity. Moak et al.,[5] from the Oak Ridge National Laboratory, point out the requirements of a good ion source for accelerators, to which we have added some functional recommendations based on the maintenance and servicing of such equipment. The combined list is as follows:

1. Stability and long life.
2. Large beam currents.
3. Low gas consumption.
4. High proton percentage.
5. Simplicity of construction.
6. Low power consumption.
7. Compactness.
8. Easy access for replacement or maintenance.
9. Ability to turn off and on instantaneously.
10. Negligible contribution to the total downtime of the accelerator.
11. Design to prevent life-shortening of other accelerator components not directly related to ion production.
12. Rugged construction.

Since the RF ion source of today is basically similar both in construction and operation to the original Oak Ridge design, the reader is referred to the original paper[5] for further details pertaining to actual construction and performance.

1. RADIOFREQUENCY ION SOURCE (RF)

The basic RF ion source described here is a modification of the original design by Thoneman and co-workers[6] and is shown in Fig. 3.3. At about

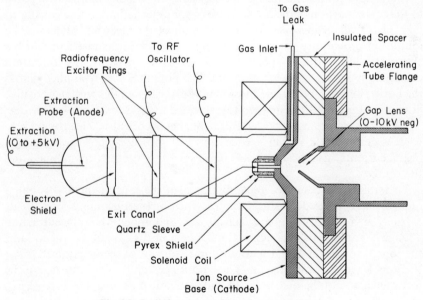

Fig. 3.3. Radiofrequency ion source and gap lens.

10^{-2}-torr pressure, a typical source can produce beam currents in excess of 1 mA with an atomic-to-molecular ratio of about 9. The gas efficiency of this source is low, compared with other sources, with the result that pumping requirements are significantly greater than those with other sources in order to keep the dynamic vacuum of the accelerator at reasonable pressures. The source requires a 30- to 100-MHz oscillator with an output of about 100 W, although RF source oscillators producing up to 450 MHz have been used. The ions produced are extracted by supplying a positive voltage of several thousand volts to the extraction electrode (Fig. 3.3), which draws from 5 to 10 mA. The source has the capability for pulsing only at very low frequencies, since plasma stabilization takes a significant fraction of the pulsed time.

a. *Theory and Operation of the RF Ion Source.* In Fig. 3.3, hydrogen or deuterium is fed at a controlled rate into the glass envelope via a gas leak. If an RF voltage is applied across the excitor rings, an electrical field is generated between the two ring electrodes located outside and in contact with the glass envelope. Since glass is an insulator, the electric field lines traverse the glass and pass into the gas. The power supplied to the RF oscillator is adequate to create and maintain intense ionization of the gas. This gas discharge glows with a bright pink color characteristic of the atomic hydrogen spectrum. The plasma color (a bright pink) is also one of the principal guides to the performance of the source. For high ion currents, a magnetic field, whose lines of force are in the direction of the long axis of the bottle, is applied to concentrate the ion beam. Restricting the electron paths to the center portion of the bottle causes a spiraling action of the electrons to take place, which increases the path length of the electrons and thus the ionization probability per unit path length. Owing to the restriction of the electrons toward the center section of the glass envelope, the ionization density is highest at this location. If a DC potential is now applied across the bottle, ions can be directed toward the exit canal of the bottle. A positive potential of up to 5 kV is applied to a shielded tungsten extraction electrode. The glass shield protects the electrode from backaccelerated electrons. These sources can operate within a wide range of pressures; however, for high beam currents, gas pressures of 10^{-3} to 10^{-1} torr are maintained in the source.

To minimize the surface recombination of the ions at the exit canal, a quartz sleeve is used to shield the canal from the oncoming ions. In its capacity as a shield, the quartz acts as a virtual anode when the extraction voltage is applied. Its focusing action, shown in Fig. 3.4, is such as to force the ions through the canal without bombardment of the canal itself. After the ions leave the exit canal they are no longer under the influence of the focusing action of the quartz sleeve and would diverge unless focused. A gap or electrostatic focusing lens is used to prevent this divergence and directs

II. PUMPED NEUTRON GENERATOR

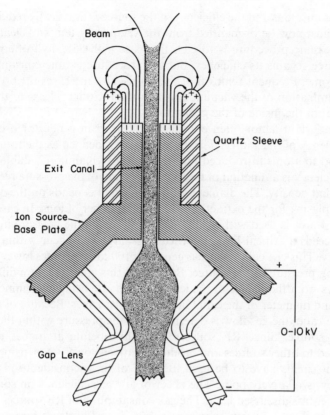

Fig. 3.4. Focusing action of the ion-source quartz sleeve and gap lens.

the ions into the accelerating section of the generator. A variable DC voltage of 0 to 10 kV is generally applied to a cone-shaped aluminum electrode and the focal length of the beam is determined by the focus-voltage setting. The focused beam then enters the accelerating section (to be described later), which also acts as a focusing device.

Ion sources operated on the RF principle possess some distinct advantages, as well as some shortcomings. Because of the high atomic-to-molecular ratio (about 9 : 1) for an optimally operating source, a maximum theoretical efficiency for neutron production is obtained in the region up to about 200 kV. The high atomic-to-molecular ratio is due to the almost complete glass construction of the source. However, because of the strong reducing property of hydrogen, some reduction of the quartz components, as well as sputtering of any aluminum in the system, will occur. Once this happens, the walls of the glass envelope act as a catalyst for the recombination of atomic ions and the efficiency of the source is drastically reduced. The source must now be dismantled from the vacuum system and cleaned. The usual cleaning procedure is to etch the glass with 25% hydrofluoric acid. The source regains its efficiency, but the procedure is time-consuming and troublesome. Frequent removal of the source also increases the probability of contamination of the vacuum system and of tritium release to the atmosphere from the inside of the generator.

An important factor often considered by neutron generator users is the gas efficiency of the source. The gas efficiency is defined as the ratio of ions produced to atoms introduced into the source to maintain a stable plasma. This efficiency is a function of the ion density and of the pressure required to obtain that density. The diameter of the ion canal depends on the density of the gas plasma for the extraction of a given number of ions. In cases where the ion source is used with a suitable lens system, if the pressure in the lens zone exceeds a critical level, electrical discharge will occur within the lens geometry. This critical pressure is generally 100 to 1000 times lower than the operating pressure of the source. Because of this great pressure differential, gas flows from the source to the lens area, the rate being determined by the length and diameter of the canal linking the two zones. Because of this high pressure ratio, the gas flow is determined by the pressure within the operating ion source. Since RF sources generally operate at higher pressures compared to other sources and produce relatively lower ion densities, rather long exit canals have to be used. Because of these considerations the gas efficiency is relatively poor. The effect of low gas efficiency on some other components is discussed later. The gas consumption of RF sources drawing about 1 to 2 mA of beam current is about 20 cm^3 of deuterium per hour.

2. PENNING ION SOURCE (PIG)

This source[7] employs the Philips ion gauge principle for gas discharge and is designed to operate on a continuous basis. A cold-cathode, high-voltage arc minimizes input power, cooling, and inlet gas-flow requirements. An important consideration in the PIG source is the lack of complex electronics; only a simple, stable power supply is necessary for operation. The cathode life is long because it does not use a heated cathode. A cross-section view of a typical PIG source is illustrated in Fig. 3.5. The two cathodes are connected electrically so that the anode is positive with respect to both. A variable potential difference from 0 to 10 kV is applied between the electrodes for the initiation and maintenance of an intense plasma in the source. A steady-state magnetic field of about 400 to 1000 G is essential for proper functioning of the source. This field is supplied either by a permanent magnet or by a solenoid. Typical operating gas pressures of 10^{-3} to

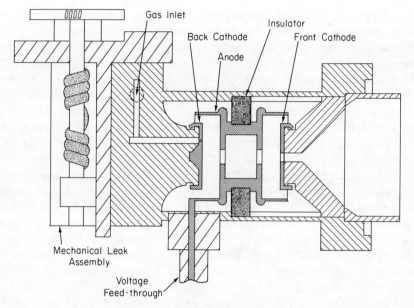

Fig. 3.5. Schematic of Penning ion source.

2×10^{-2} torr at gas-flow rates in the range of 25 cm^3/hr must be maintained for optimum operation.

The atomic-to-molecular ion ratios reported by various research groups are conflicting. Values of 0.03 and 0.5 have been reported by Jessen[8] and Barnett et al.,[9] respectively. However, for an average source, a value of 0.4 is considered normal. The cathodes are generally constructed from aluminum or aluminum alloys because of their high sputtering thresholds.[10] The ion source can supply more than 3 mA of usable beam current. Other than hydrogen isotopes, helium, argon, nitrogen, and oxygen can also be ionized without significant design changes to the source. If heat transfer conditions have been carefully considered, a typical PIG source will provide over 200 hr of continuous operation at beam currents of about 2 mA, before cathode or anode replacement becomes necessary.

a. *Theory and Operation of the PIG Source.* The operating principle of the PIG source is based on the magnetic mirror effect which can be described as follows. Free electrons released from the cathode, or present in the vicinity of the cathode, are accelerated toward a tubular-shaped anode (Fig. 3.5) located in the arc chamber of the source. The radial motion of the electrons toward the inside surface of the ion-source housing is restricted by the imposition of an axial magnetic field. This field causes the electrons to extend their path lengths considerably by causing them to travel spirally across the ion chamber. In this volume these electrons create an intense gas plasma by further collisions. In order to maintain an effective magnetic field the aluminum caps are mounted on iron cathodes. In addition, to maintain an unrestricted flow of electrons to the anode, all parts of the ion source must be constructed from metals, which results in a low atomic-to-molecular ion ratio in the extracted beam. Because of the high ion density characteristics of the PIG source, very small diameter exit canals can be used. Consequently, the gas efficiency of this source is superior to that of the RF source.

If permanent magnets are utilized, the only other electronics required for the operation of the PIG source is a well-regulated DC power supply. For a high degree of reliability over a long period the source has some distinct advantages over the RF source. The absence of insulator surfaces eliminates the periodic cleaning process necessary for RF ion sources. Any such cleaning procedure necessarily involves removal of the source from the main vacuum system of the generator, a process during which contamination from tritium can occur.

Probably the only disadvantage of the PIG source in terms of maintenance is the need for replacement of cathodes and anodes. (These components deteriorate rapidly if not installed properly.) Some workers also find the lower atomic-to-molecular ratio a disadvantage. The RF ion source possesses a distinct advantage in this one respect. When deterioration of a PIG source occurs, this ratio is also affected as with the RF source. Overheating of the source due to improper heat transfer between the cathode cap and the cathode mount accelerates breakdown. In Fig. 3.6, an exploded view of a dismantled PIG source illustrates this type of damage; the deterioration of the aluminum anode may be clearly seen. In spite of extensive electrode damage, PIG sources that have operated over several hundred hours are known to continue to provide useful beam currents up to 60% of their maximum rating. If long life were the only consideration in the selection of an ion source, it would appear that the Penning source offers better sustained performance than the RF ion source.

3. DUOPLASMATRON ION SOURCE (DP)

A high-intensity ion source recently introduced in small-accelerator technology is a modified version of the Von Ardenne source. This source, called the duoplasmatron (DP) source, contains three electrodes and generates electrons from a heated cathode. The electrons are accelerated through an intermediate electrode which provides both electrostatic and magnetic focusing. During startup, the filament heater draws 50 to 100 W; once the

Fig. 3.6. Exploded view of Penning ion source; from left to right, mechanical gas leak connected to source housing, anode, retaining sleeve, front cathode, and assembly holder.

arc is struck, ion bombardment keeps the cathode hot so that further filament power is not needed. This ion source can generate ion currents of the order of 1 A; however, for small-accelerator use, ion currents of 1 to 10 mA are sufficient. The extracted beam contains 60% of atomic ions. The gas efficiency of these sources can be as high as 95%, although 60% is more typical. The magnetic field strength is about 5000 to 16,000 G in a small gap between the magnet pole pieces. Owing to the lack of sustained high-current capacity of the cathode, burnout occurs with some frequency.

A schematic of a duoplasmatron source is shown in Fig. 3.7. In comparison to the RF and PIG sources its construction is considerably more complicated. As in the PIG source, a magnetic field is used to restrict electron movement into a spiral path in order to increase the probability of electron collisions with other gas molecules and neutral atoms. Since the plasma intensity is extremely high, small exit canals can be used, allowing the source to function at the relatively high pressure of 5×10^{-1} to 5 torrs. Although the source has some advantages in terms of high beam currents and good efficiency for gas utilization, its design parameters are very critical. Furthermore, it suffers from short cathode life, particularly when operated at high beam currents.

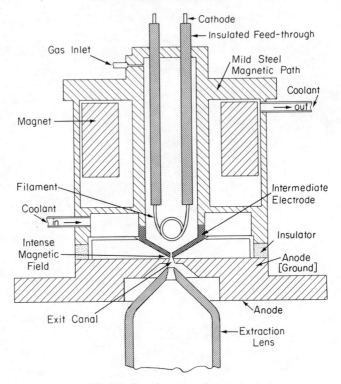

Fig. 3.7. Duoplasmatron ion source.

a. *Operation of the DP Source.* During startup the filament is raised to a high temperature, thus producing copious quantities of free electrons in a very small volume, which causes intense ionization of the gas. The large number of ions formed are permitted to exit through a canal whose opening varies from about 0.025 to 0.075 cm in diameter. The alignment problem in this source is most critical for optimum operation. Any masking of the exit canal, due to imprecision in the design of manufacture, leads to a rapid deterioration of the source. The intense proton or deuteron beams produced can be suitably directed by quadrupole magnets into beam tubes for use as neutron producers for activation analysis or for nuclear physics and ion implantation research. Neutron generators using this source are often quite elaborate in construction.

4. ANOMALIES IN THE COMPARISON OF ION SOURCES

Descriptions given thus far, relevant to the theory and operation of the three types of ion sources, permit definite comparisons to be made among the sources. Although the advantages and disadvantages can be deduced from these descriptions, experience gained by many workers in the field shows that comparisons based on the operating characteristics of a particular source can lead to much misunderstanding in source selection. The source must be considered in conjunction with the rest of the accelerating system and must be judged according to its performance as an integral part of the small accelerator. The end use of an accelerator is one of the prime factors influencing the selection of a particular ion source. Operating beam load and the freedom from contamination of the accelerator vacuum significantly affect the operation of a source. All ion sources will function reasonably well if the generator is routinely operated at about 25% of its rated beam capacity. Total operating times based on reduced operational parameters cannot be extrapolated to equivalent integrated beam times at maximum beam currents. This is due to many other considerations, such as beam optics, ability of the accelerating tube to maintain a "tight" beam configuration, and electron bombardment damage to key ion-source components. The selection of an ion source must therefore be made on the basis of a particular application. The specific application will, in general, define the limits within which the accelerator system as a whole must perform. The downtime that can be tolerated in the particular application is also an important factor. Since the cleanliness of a vacuum system improves ion-source life, factors affecting vacuum contamination must also be considered. The ease and frequency of source reconditioning must of course also be taken into account.

Thus, in view of the many factors affecting ion-source operation, comparisons made only from design considerations or operating experience with a source can be misleading. The analyst is primarily interested in obtaining steady beams for a reasonable time and in conducting a minimum of maintenance of the ion source, with the added requirement of being able to maintain high beam currents if and when necessary. Factors such as gas economy and component costs can also influence the selection of a source. Generally, however, the investigator is restricted in ion-source selection because a particular type of source is generally associated with a given make of accelerator. Nevertheless, there are sufficient alternatives offered by neutron generator manufacturers so that the user should decide which options will best suit his purpose.

The development of ion sources has, however, continued. The use of better materials in the construction of ion-source components has led to a considerable lengthening in ion-source life. Similarly, improvements in electrostatic optics have resulted in more efficient gas utilization. It can be concluded that research on ion-source development is by no means lagging behind research into the production of improved neutron generators. Experimenters are also demanding more rapid ion-source response for pulsed operation because the stability of ion sources today for pulsed operation is still unsatisfactory. The need for long-term irradiations of the order of days is increasing as more applications of low-voltage accelerators are made. The ability of ion sources to produce ions other than those of hydrogen isotopes has revolutionized basic design thinking, especially for ion implantation work.

B. GAS SUPPLY REGULATION

Probably the most important single component responsible for the efficient working of a neutron generator is the device which controls the gas supply to the ion source in a steady and reproducible manner. It is also often necessary for the gas regulator or "leak" to ensure the purity of the gas entering the ion source. Gas purity is ensured by leaks which have the ability to discriminate against contaminant gases, or by devices constructed with great care to prevent contamination from the structural material of the gas leak. It is generally preferable if the gas leak can be operated from the operating console of the neutron generator, that is, outside the biological shield surrounding the generator.

Some of the early designs for such regulatory devices were very simple in their construction and principle of operation. In these devices a small orifice, located in the ion-source housing and directly connected to the main gas bottle, was sufficient to control the entry of the gas to the ion source. The pressure was regulated by a manually controlled gas regulator attached to the bottle. Although the simplicity of these devices is a great advantage, major difficulties were experienced in obtaining steady-state operations of the ionized beam. These difficulties are due to the fact that normal gas regulators are only efficient in their regulatory function when operating at above atmospheric pressure. Since ion sources require a small flow of gas into a high vacuum, stability of flow cannot be guaranteed. In addition, a small amount of particulate contamination can render the gas leak ineffective. Furthermore, orifice "leaks" do not possess the ability to discriminate against contaminant gases. To ensure even a moderate degree of gas purity, these devices must be cleaned regularly using rigorous procedures. Some of these gaseous contaminants can, of course, be handled by the installation of suitable cold traps; however, the traps have to be very close to the ion source and present a design problem. Frequent flushing of the gas leak is also necessary, particularly after the accelerator has not been used for an extended time. In spite of these deficiencies, orifice-type gas leaks are still used.

An adaptation of the early fixed-orifice "leak" is a gas leak with an adjustable orifice. This type of regulation is performed by merely adjusting a small screw located across the orifice. A system of gears is generally employed to turn the screw either manually, or remotely by a high-torque, low-rev/min electric motor. Figure 3.8 shows a schematic of an adjustable gas leak. Provided that precautions are taken to maintain the "leak" free from undesirable contaminants, the mechanical "leak" is extremely reliable, is almost free from maintenance problems, and provides extremely sensitive gas control and fast response. The response of the "leak" can be adjusted to the experimenter's requirements by the choice of an electric motor and gear train.

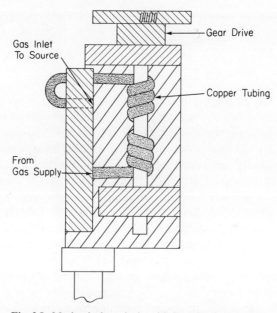

Fig. 3.8. Mechanical gas leak, with Penning ion source.

One of the most popular gas "leaks" is the permeable-membrane gas regulator. Metallic membranes which selectively permit the passage of hydrogen isotopes are commonly used. It is well known that palladium and palladium alloys exhibit temperature-dependent permeability to the passage of hydrogen isotopes. Since the lattice structure of these metals only permits the passage of hydrogen and its isotopes, the purity of the gas entering the ion source is ensured. Palladium-silver alloys are generally preferred because of their low deterioration compared with pure palladium; these are known as "palladium leaks" (Fig. 3.9). The palladium envelope separates the vacuum from the high-pressure side of the leak. Hydrogen or deuterium gas

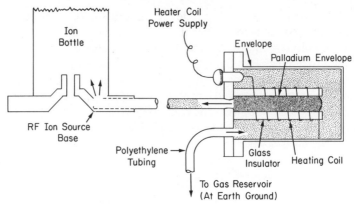

Fig. 3.9. Palladium leak.

is introduced to the high-pressure side and, depending on the pressure differential and the rate of heating provided by the resistance coil, a steady and accurate flow of high-purity gas is introduced into the ion source. Although this type of leak is very efficient, great care must be exercised in its use with high-purity gas because gaseous contamination or moisture can readily clog the palladium envelope. Blockage is generally manifested as a reduced gas output when the same heating current is used. The leak then needs to be cleaned by a bakeout procedure simultaneously with a reversed hydrogen flow, that is, from the low-pressure side of the leak as installed. If this bakeout procedure is ineffective, the leak must be dismantled and thoroughly cleaned in solvent and dried before reinstallation. One of the most useful spare parts that neutron generator users can stock is an additional gas leak.

Although the gas leak component may appear to be a seemingly minor part of a neutron generator, its neglect may seriously hamper the performance of other major components such as the vacuum pump and the ion source due to contamination. It is always good practice to check the leak performance at frequent intervals, and to operate the neutron generator at a pressure closer to the "starving" gas pressure rather than at excess gas or "rich" gas pressures.

C. ACCELERATING TUBE

The intense ion plasma produced inside the ion source is directed through the exit canal and into the focusing and accelerating sections of the neutron generator. Generally, these functions take place in a single tube composed of electrodes and insulators, which permit the dual action of focusing and acceleration. The focusing electrodes (or electrode) carry a negative DC potential with respect to the terminal high voltage (100 to 500 kV) applied to the first accelerating stage. By utilizing a system of resistors, a potential difference is maintained between each accelerating stage. This division of high voltage can be evenly distributed between the individual stages so that the ion beam receives a boost in energy equivalent to the potential drop at each stage. At the end of the accelerating tube the ions have acquired an energy equivalent to the total potential drop between the high-voltage terminal and ground. As described in Section II.A.1.a and illustrated in Fig. 3.4, the electric field lines between the first accelerating stage and the focus electrode have a "pinching" effect on the ion beam which ensures that the majority of the ions are forced toward the center of the beam tube. Since each successive accelerating stage is electrically negative with respect to the stage preceding it, focusing is maintained throughout the length of the accelerating tube. Two basic types of accelerating tubes are most commonly in use today, the Van de Graaff and the Einsel-lens tubes. Although the general principle of acceleration remains essentially the same, design features show sufficient dissimilarities that a separate description of each type is warranted.

1. VAN DE GRAAFF ACCELERATING TUBE

This tube consists of a number of identical electrodes separated by insulators, the latter forming the vacuum wall of the tube itself. The Van de Graaff lens, or tube, is schematically depicted in Fig. 3.10. The potential at each electrode is supplied by a voltage-dividing circuit consisting of a bank of resistors. Approximately equal potential differences exist between the elec-

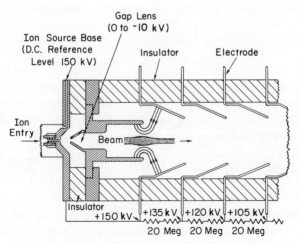

Fig. 3.10. Van de Graaff accelerating tube and gap lens

trodes so that the ion beam gains energy in constant increments as it passes by each lens system. For example, the acceleration of ions up to 150 keV can be achieved by dividing the voltage equally along a tube consisting of 10 electrodes with a bank of ten 20-MΩ resistors connected as shown in Fig. 3.10. Since many stages or lenses are used, the voltage difference between each lens is not too high (15 kV in the example above); this type of accelerating system is useful for air-insulated systems. To simplify construction it has often been questioned whether lengthening of the insulators, and thus reducing the number of electrodes, would not provide a better tube design. In this situation the voltage drop across each stage would be substantially larger. The distribution of the electric field depends strictly on the uniformity of the insulator resistance along its outside surface; any contamination of this surface would severely affect the uniformity of the electrical charge along the length of the insulator. In addition, to limit the power dissipation due to secondary electron bombardment of the insulator surfaces would require the design of the electrodes to be such as to shield the insulators from a direct view of the bombarding electrons. The insulators, constructed from silica, glass, or alumina, may act as deposition sites for sputtered metal. In addition, they themselves may be reduced by the deuterium gas. Both processes reduce their insulating effectiveness.

The focusing ability of a lens system is determined by the ratio of the potential difference across the lens to the energy of the ions passing through the lens. This ratio implies, therefore, that the lens located at the beginning of the accelerator tube possesses stronger focusing action than the lens situated at the end of the tube. Sharp focusing can thus be obtained by varying the voltage across the first one or two lens systems without affecting the overall optics of the accelerating tube appreciably. Lens systems such as these are assembled by bonding together insulators and aluminum electrodes using vacuum-type "O"-rings or epoxy bonding to affect a good seal.

2. EINSEL-LENS ACCELERATING TUBE

A typical version of the Einsel-lens accelerating tube is illustrated in Fig. 3.11. One of the important features of this system is the construction of the electrodes. These tubular-shaped elements have insulating shields to protect both the lens gap and the main insulators separating the electrodes from scattered radiation. In this system, two lenses are generally employed and the focusing range is extremely wide. It is necessary that this lens system be kept very clean since any variation in the insulating characteristics would greatly affect the focusing and accelerating properties of the entire tube. Owing to the limited number of lenses, the operation of the tube is very sensitive to surface contamination. Generally, accelerating tubes of this type

II. PUMPED NEUTRON GENERATOR

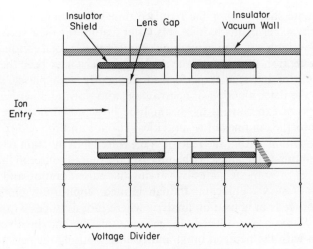

Fig. 3.11. Einsel-lens accelerating tube.

are externally insulated by oil or by gas under pressure. With a controlled atmosphere in contact with the outside of this tube, the accelerating and focusing properties can be kept constant over long periods.

In the assembly of a typical Einsel-lens accelerating tube, great care is taken to align each section or lens system properly. A jig is necessary for the proper assembly of the tube. Irreparable damage can be done to the tube, and to some critical sections of the generator itself, if the accelerating section is carelessly assembled. Frequently, the insulators closest to the gap lens need attention. Sputtered metal often gets deposited on these early stages, resulting in occasional arcing and loss of focus. If the tube is not cleaned frequently, severe arcing can result with possible loss of vacuum in the system and damage to the tube.

D. PULSING SYSTEMS

Not much information is available in the literature regarding pulsing systems, and in attempting to describe such devices, we have relied heavily on the Texas Nuclear report by Prud'homme[11] which gives a discussion appropriate for the purpose of this book.

For many applications, particularly those associated with basic research in nuclear structure, a pulsed neutron source is necessary. An acceptable pulsing system must have certain essential characteristics. The neutron pulse duration and repetition rate must be variable over a wide range. The resolution of the neutron pulse must be sharp; that is, the pulse must rise and terminate sharply after the required time with a minimum of residual beam. Many experiments require this latter feature for their success.

In the development of pulsing units as integral parts of a continuous-duty neutron generator, the needs of the nuclear physicist have been the dictating factors. One of the studies performed frequently with pulsed sources is related to the determination of neutron age and slowing-down parameters. In this type of experiment, the beam tube of a pulsed neutron source is inserted into the confines of a subcritical reactor and the neutron yield at different locations within the reactor is measured as a function of time after the neutron burst. The nuclear chemist uses the pulsed source in the study of the half-lives of isomeric states by varying the pulse duration and the delay between pulses. A significant, though limited, application of the pulsed neutron source is in activation analysis where radioisotopes with half-lives less than 11 sec are produced. In general, this method involves a short irradiation with the neutron burst and counting during the period between pulses. One of the major difficulties with this type of experiment is the unreliability of the detection equipment when it is located near the neutron source. Most detection systems are notoriously unstable or susceptible to drift in the high-radiation field generated during the neutron burst or activation cycle. Often the detector system retains a memory from a previous neutron burst, thus affecting the reliability of counting from subsequent activation cycles. In spite of these difficulties, the elements listed in Table 3.1 offer possible analytical applications using "in-place" counting techniques and a pulsed neutron source. This table reveals some interesting possibilities in the region of atomic numbers less than 6. These elements present major analytical problems with most other nuclear activation methods since no long-lived nuclides are available for analytical purposes. By appropriate selection of the irradiation time and the repetition rate, it is possible to

TABLE 3.1. Useful Nuclear Reactions for Pulsed Neutron Activation Analysis

Element	Reaction	Half-life (msec)	Bombarding Neutron Energy
Lithium	^{6}Li(n, p)^{6}He	820	14 MeV
	^{7}Li(n, γ)^{8}Li	890	Thermal
Beryllium	^{9}Be(n, α)^{6}He	820	14 MeV
Boron	^{11}B(n, γ)^{12}B	25	Thermal
	^{11}B(n, α)^{8}Li	890	14 MeV
Carbon	^{12}C(n, p)^{12}B	25	14 MeV
Potassium	^{39}K(n, 2n)^{38}K	950	14 MeV

analyze for the light elements listed in this table, often with a minimum of matrix interference.

Most standard-model neutron generators are constructed so that a pulsing unit can be incorporated later without undue modifications to the basic system. Several types of pulsed units utilizing different principles are used at present. The more common types will be described in detail. Basically, pulsers use the principle of pre- and postacceleration electrostatic deflector plates to perform the pulsing action. The methods of charging these plates are essentially similar. Dual-pulsing units utilizing both pre- and postaccelerating plates are also used. In this case, the charging of the plates is done synchronously. Limited use has also been found for pulsers that sinusoidally vary the beam on the target. Some typical systems are described below.

1. POSTACCELERATION PULSING UNIT

As the name implies, this unit is located between the target and the accelerating tube. The essential components are a set of deflection plates and an electronic pulser. Applying a voltage across the deflection plates causes the beam to be deflected off the target and onto a beam-catcher assembly cooled by water or Freon. The voltage pulse is applied by the electronic pulser to the deflection plates, which cancels the potential across the plates and permits the beam to continue undeflected onto the target. The frequency and width of the beam pulse on the target depend on the width and repetition rate of the pulse that is applied to the plates by the electronic pulser. A schematic of a typical postacceleration pulsing system is shown in Fig. 3.12. A block diagram showing the electronic units for this pulsing system is shown in Fig. 3.13. In Fig. 3.12, the deflector plates are about 8 in.

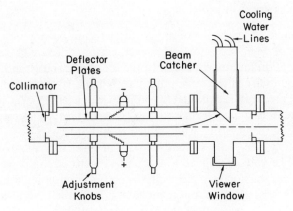

Fig. 3.12. Postacceleration pulsing assembly.

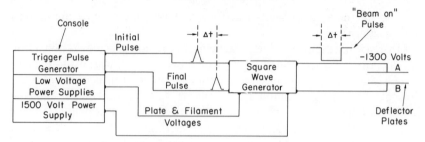

Fig. 3.13. Block diagram of electronic components for postacceleration pulsing unit.

long by 1 in. wide and are insulated from the main body of the assembly. The lateral deflection of the beam can be adjusted by varying the distance between the plates; mechanical controls are used for this purpose.

The initial actuating pulse is supplied by a system composed of a trigger generator, a square-wave generator, and the necessary power supplies (Fig. 3.13). The controls for these units are located on the operating console of the neutron generator. In the operating mode, two pulses are supplied by the trigger generator to the square-wave generator, the second pulse being delayed by a preset time after initiation of the first pulse. The frequency of these pulses can be varied. The typical frequency range is from 10 to 10^5 Hz. The time delay between the pulses is also variable, the typical range being 1 to 10^4 μsec. For convenience, the trigger generator is equipped with prompt and delay outputs which can be interfaced to suitable counting equipment. An easy and safe way to align the beam is to experiment with a hydrogen beam on an aluminum or tantalum target.

2. PREACCELERATION PULSING UNIT

The principle of operation of the preacceleration pulsing unit is basically similar to that described above. However, certain differences warrant additional description and explanation. In the schematic shown in Fig. 3.14, the deflection plates are at the same potential. When a voltage is applied across these plates, the beam is deflected onto the base of a lens system composed of an Einsel lens and an electrostatic gap lens. This is necessary for the convergence of the entering beam, which would otherwise possess considerable divergence. The lens system consists of a set of insulators and three electrodes. The voltage applied to the center electrode can be varied from 0 to +5 kV, and the adjacent electrodes are at ground potential. In this lens system the focusing effect on the beam is extremely sensitive to small changes in the voltage applied to the center electrode. The electronic com-

II. PUMPED NEUTRON GENERATOR

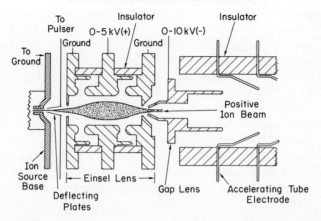

Fig. 3.14. Schematic of preacceleration pulsing unit; pulse-on-target condition.

ponents required to operate this type of pulsing unit are shown in Fig. 3.15. A similarity between this circuit and that used for the postacceleration pulsing unit may be noted. For neutron generators where the high-voltage terminal is at the full accelerating potential, the signals from the trigger generator are transmitted by a light telemetry circuit to the high-voltage terminal. This transmitting system generally consists of two neon glow tubes mounted on some part of the accelerator which is at ground potential. The initial and final trigger pulses drive the neon tubes, which are optically coupled to individual photomultiplier tubes via light pipes. The neon tubes conduct at maximum current when the trigger pulse arrives, and glow brightly throughout the duration of the pulse. The light signal is then converted to an electrical pulse by the photomultipliers, which are located in

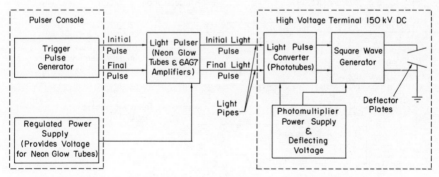

Fig. 3.15. Block diagram of electronic components for preacceleration pulsing unit.

the high-voltage terminal. The output of the photomultipliers then actuate the multivibrator in the square-wave generator. The remainder of the operation sequence is identical with that for the postacceleration unit.

From the descriptions given of the two systems it can be seen that there is little choice between them in terms of their operation. A subtle difference is the location of the beam-catcher plates which, when hit by the deuteron beam, produce D–D neutrons. This production results from the interaction of the primary deuteron beam with deuterons embedded in the catcher plate from previous bombardments. However, this problem is not significant if the beam catcher is suitably shielded from the target by moderating material and if the location of the beam-catcher mechanism in the drift tube is far from the target end.

3. DUAL-PULSING SYSTEM

In using a neutron generator in a pulsed mode where counting is carried out in close proximity to the target, it is desirable that the neutron background between pulses be at a minimum. The two systems described earlier suffer from the fact that a residual beam exists. Although the acceptance level of this residual beam depends on the experiment, a residual beam is to be avoided if possible because it can complicate counting information stored during the "off" stage of the pulsing operation. This problem can be avoided by using a dual pulser. This system, when operating at maximum efficiency, can provide a reduction in the residual beam from 0.3 to 0.0005% of the maximum pulse strength. Such a reduction clearly establishes the superiority of this unit. On the other hand, the dual pulser requires complicated electronic circuitry and must be tuned very carefully for maximum efficiency. The operation of this system is shown in Fig. 3.16. The initial and final

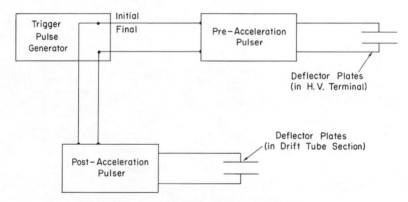

Fig. 3.16. Block diagram of a dual-pulsing system.

trigger pulses are fed to the pre- and postaccelerating stages of the unit. The current pulse produced by the preacceleration stage is accelerated past the postacceleration stage deflector plates unchanged. The residual beam normally present when ion-source pulsing is used is almost completely eliminated. Pulses of width as small as 1 μsec are not distorted because of the short transit time of the ions from the source to the deflector plates of the postpulser. In this system, the two pulsing stages operate synchronously.

In addition to the common types of pulsers discussed above a few special types have also been used with neutron generators, including pulsers which vary the beam on the target in a sinusoidal manner, systems capable of providing neutron bursts in the range of 10 to 100 msec, and systems possessing a wide range of pulse widths and frequency ranges.

E. VACUUM SYSTEMS

The selection of the appropriate pumping system is perhaps the most important consideration in the operation of a reliable, stable, and versatile neutron generator. In selecting a vacuum system, one must recognize certain factors which relate to the ultimate use of the generator. These factors include the nature of the gas used, the desired static pressure in the accelerator, the dynamic pressure at which the accelerator will ultimately operate, and finally, the degree and nature of the contaminants which can be safely vented in the laboratory. It is also very important that the vacuum system be simple to operate and easily accessible for maintenance or replacement. In this section some of the major vacuum systems which have been and are now being used to maintain high-vacuum conditions in neutron generators are described. Some preventive maintenance steps and signs of impending failure in some of these units are stressed in the discussion. Methods of maintaining clean vacuums over a long period are also described. Finally, some comparisons between the performance of the various vacuum pumps as components of neutron generators are made.

The two parameters of importance in a vacuum system are the ultimate vacuum attainable under given laboratory conditions and the pumping speed. The pumping speed is generally expressed in liters per second. A pumping speed of 1 liter/sec implies that a vessel of 1-liter capacity will have its pressure reduced to 0.37 of its initial pressure every second, provided wall effects are neglected. At high vacuums, the effects of outgassing or desorption from the walls of the system and other internal surfaces become critically important. If the system has not been "baked out" even at a moderate vacuum, the presence of water can extend the pumpdown time significantly. In any calculation of ultimate pressure that can be obtained in a given system, the effect of outgassing must be taken into account. For a typical unbaked system constructed from stainless steel with a total surface area of 2000 in.² and using a pump with a pumping speed of 200 liters/sec, Hall[12] has shown that a pressure of 4×10^{-8} torr is obtained after 10 hr. Under a given gas load, as is the case with neutron generators, the load factor must also be considered in the estimation of the ultimate vacuum for a given system using a given pumping speed. In a well baked-out system, at ultrahigh vacuum (less than 10^{-9} torr), the pressure in the system is only slightly higher than that in the pump itself.

An essential requirement for all high-vacuum pumping systems is a "roughing" pump which performs the initial pumpdown from atmosphere to as low a vacuum as necessary for the main pumping unit to start up. The mechanical pump is almost universally used for this initial "roughing-out" phase of the pumping sequence. Exceptions are aspirators and sorption pumps, which are employed when the complete elimination of oil vapors is desired. The operation of the typical oil-sealed mechanical pump is found in most introductory physics texts. It is sufficient to mention here that the pumping action is performed by compression of the gas by an eccentrically located vane and subsequent exhaustion of the compressed gas to the atmosphere. The pump operates in an oil bath. Because all oils have a vapor pressure, hydrocarbon vapors can diffuse back from the pump and into the vacuum system. Foreline traps equipped with molecular sieves, activated alumina, or liquid nitrogen traps are generally used to eliminate this source of contamination from the main vacuum system. In a typical neutron generator vacuum system, the forepump or "roughing" pump will pump down the entire system to a pressure of about 10^{-3} to 10^{-2} torr in 15 to 30 min. Once this pressure is reached, the main pumping system is turned on to reduce the pressure to 10^{-6} to 10^{-8} torr.

1. VACUUM PUMPS

To achieve the necessary vacuum for optimum operation of a neutron generator, three types of pumps are used. Some of the major differences between these units are discussed here, and those that are significant are pointed out. The auxiliary equipment necessary to complete the pumping unit as used with the neutron generator is described briefly. Recent and significant developments relating to tritium control are presented to illustrate the effort needed to minimize this hazard.

a. *Oil Diffusion Pumping System.* This pumping system has been in use for many years for a wide variety of applications. The device operates as shown in Fig. 3.17. Oil vapors generated in the heater section of the pump pass through a concentric chimney and are forced out through a small orifice.

II. PUMPED NEUTRON GENERATOR

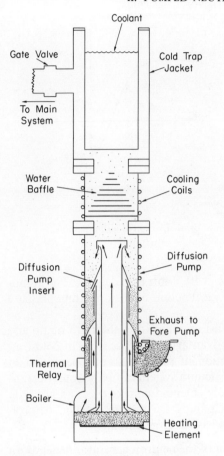

Fig. 3.17. Typical oil diffusion pump assembly with cold trap connected.

The expanding vapor leaving the orifice is forced downward at high velocity and carries with it gas molecules from the vacuum manifold in the jet stream. The gas molecules collecting at the bottom of the pump are then removed by a "roughing" pump. The walls of the pump are cooled by cooling coils. Upon contact with the walls, the oil vapors condense and return to the boiler. A thermal relay acts as a safety device to prevent contamination of the main vacuum system if the cooling-water supply fails or is not turned on before the oil heater. Diffusion pumps are connected via a vacuum manifold to the main vacuum system and are equipped with a cold trap constructed from stainless steel. The trap is filled with a Dry Ice-acetone mixture or with liquid nitrogen, and a water baffle is also used to assist in condensation of the vapors. The cold trap condenses water vapor so that contamination of the oil is prevented. For neutron generators, diffusion pumps having a

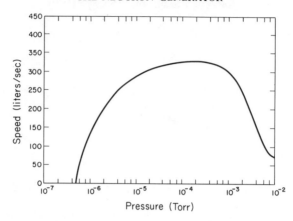

Fig. 3.18. Pumping speed of air for a 320-liter/sec oil diffusion pump.

pumping speed of about 300 liters/sec at 10^{-5} torr are adequate. The pumping speed of an oil diffusion pump is a function of the operating pressure (Fig. 3.18).

The main advantage of an oil diffusion pump is its ability to pump any kind of gas for long periods. A disadvantage is the necessity of cold-trap recharging during continuous operation. If the cold trap is permitted to attain room temperature, the entire vacuum system can be seriously contaminated. An isolation valve, preferably solenoid controlled, is an absolute necessity in this case. Oil diffusion pumps are seldom used in neutron generators today, except in machines having large volumes.

b. *Ion Pump.* Optimum operation of the neutron generator depends on minimizing oil vapor or other contamination of the accelerator vacuum. For this reason, oil diffusion pumps are undesirable. An alternative pump which can provide an excellent vacuum with no possibility of vapor contamination is the ion pump. In addition to cleanliness and high-vacuum capability, the ion pump is a more compact unit than the diffusion pump and can be installed directly in line with the accelerator tube of the generator. The absence of moving parts and relatively trouble-free operation make this pump a very attractive choice for generator users. The pump also measures its own pressure and is capable of pumping for long periods at high speeds. The pumping speed for a particular gas is almost independent of the operating pressure. Some of the major disadvantages of this pump are (1) it can store tritium if tritium targets are used, (2) the pumping of inert gases is inefficient and extremely slow, and (3) because these pumps employ a magnetic field for their operation, the pump has to be magnetically shielded from the ion beam to prevent undesirable perturbations of the latter, which can result in serious damage to accelerator components from stray beam and loss of focus. This magnetic effect is particularly serious when low atomic number ions like hydrogen or deuterium are accelerated. The life of the ion pump is directly proportional to the pressure in the system being pumped. It is not uncommon to find these pumps working for as long as three years at an average pressure of 10^{-6} torr. If the system is leak-tight, the pump can be shut down for long periods and then reactivated when desired without a preliminary "rough-out" cycle.

An ion pump operates by using a cold-cathode discharge in a magnetic field to sputter a reactive metal such as titanium. This sputtering produces a "gettering effect" on the gas, trapping it within a metallic film formed from the sputtering process. The process of using a metal film as a gas collector was first described in a patent by Penning.[13] He showed that by using a device like that illustrated in Fig. 3.19, a discharge could be maintained at

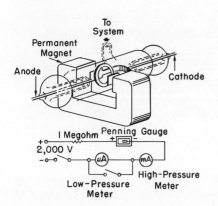

Fig. 3.19. Penning gauge and circuit.

high vacuums. However, the pumping capability of Penning's invention was not applied until many years later. The basic design shown in Fig. 3.19 forms the basis of modern-day ion pumps.

For the ion pump to operate, the pressure in the system must be reduced to below 10^{-2} torr by means of a mechanical oil pump or some similar device. When this pressure is reached, voltage is applied to the ion pump for startup. The initial voltage is only about 400 to 500 V. However, the current drawn by the pump is over 1 A. The action of the pump is depicted in Fig. 3.20. In a typical ionization process, electrons ejected from the cathode are caused to spiral by the imposition of a suitable magnetic field and ultimately find their way to the anode. In the collisions with gas molecules enroute to

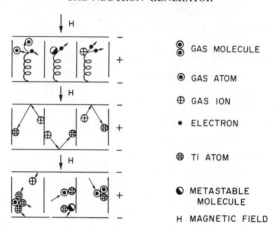

Fig. 3.20. Pumping mechanism of the vacuum ion-getter pump.

the anode, additional electrons are ejected and a self-sustaining discharge is obtained. The positive ions that are produced bombard the cathode and liberate atoms. These atoms, being uncharged, are sputtered mainly onto the anode surface. By this action the reactive metal film is continuously replenished and is available for trapping additional gas molecules. In addition to this mechanism for the removal of gas molecules by ionization, collisions between electrons and gas molecules produce various types of neutral species such as dissociated molecules (atoms) and metastable atoms and molecules. Since such neutral species possess high potential energies, they are readily trapped by the sputtering film which can attain saturation almost immediately. Since the current drawn by the pump elements is directly proportional to the rate of electron-molecule collisions, that is, the density of the gas, the current meter in the voltage supply to the pump can be calibrated to read directly in pressure, and thus the pump acts as its own pressure gauge.

Normally the operating range of ion pumps is from about 10^{-3} to 10^{-10} torr, and the operating life of the pump in a clean system is inversely proportional to the pressure at which it operates. Manufacturers of ion pumps generally rate the life as high as 40,000 hr at 10^{-6} torr; however, the customer pump-life guarantee is generally of the order of 400 hr at 10^{-5} torr. In neutron generators, however, the chance for accidental contamination of the ion pump is relatively high, particularly during target-changing operations and during the switching period from the "roughing" pump to the main ion pump. Since the sputtering rate of the titanium is a direct function of the pressure, the pump is self-regulating and the pumping speed is constant over a wide range of operating pressures. The pumping of gases such as nitrogen, oxygen, and water vapor occurs by the trapping of these molecules in the anode film. The pumping of hydrogen, the isotopes of hydrogen, and the noble gases occurs mainly at the cathode. The speed of pumping hydrogen is considerably greater than for other gases. In Fig. 3.21, the pumping speeds of a 140-liter/sec pump for a number of gases are illustrated. In considering the life of these pumps, we should mention the effects of startup on pump life since the previous remarks regarding the effects of pressure on pump life apply only to continuous operation. During startup, it must be remembered that owing to the low voltage, the sputtering rate is also low and the life of the pump is unaffected.

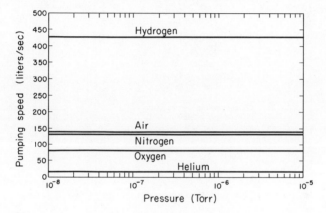

Fig. 3.21. Pumping speed of a 140-liter/sec vacuum ion pump for several gases.

As mentioned earlier, the ion pump is a storage device and as such will accumulate large quantities of tritium in the pump elements. This is particularly hazardous since pump maintenance requires its occasional removal from the vacuum system of the neutron generator. A solution to the tritium problem has been suggested by Steinberg and Alger.[14] In their experiments, reduction of tritium contamination was achieved by separating the pumping function during static and dynamic beam operation. A bulk-titanium sublimator was incorporated into the vacuum system. It was estimated that the pump life was increased by 10 years by this technique. Essentially, the bulk sublimator is only switched on during beam operation at pressures of about 10^{-5} torr. Under the increased gas load during this period of operation the regular ion pump is valved off. During the standby period a bank of molecular sieve sorption pumps and an ion-getter pump are used to maintain the normal static vacuum. The bulk sublimator is a device which contains a rod of titanium, heated to 1500°F, which is allowed to sublimate onto the surface of a water-cooled chamber. The depositing titanium either forms stable compounds with the gas molecules or physically entrains them within the sublimed layer. The life of a sublimator of this type is rated at 20,000 hr at a hydrogen load of 10 atm cm^3/hr. The life of an ion-getter pump in such a differentially pumped system is estimated to be about 50,000 hr at 10^{-7} torr. Since the tritium is permanently buried under each layer of deposited titanium, no outgassing of this hazardous isotope occurs. A typical pumpdown cycle, as reported by Steinberg and Alger,[14] is shown

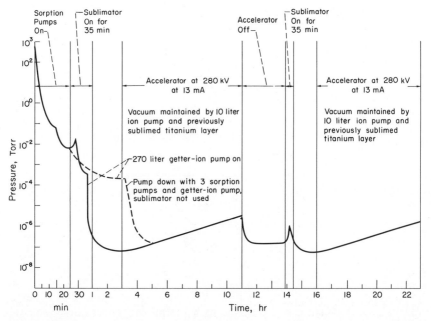

Fig. 3.22. Typical pumpdown and operating cycle for a 300-kV accelerator (volume of system, 300 liters) showing effect of bulk sublimator.

in Fig. 3.22. This cycle was run on a 300-kV accelerator of about 300-liter volume. A schematic of this pumping system in a regular pump-type neutron generator is shown in Fig. 3.23. Although the Steinberg-Alger system offers a good solution for the elimination of the tritium problem, no cost estimate for a commercial unit is available yet.

II. PUMPED NEUTRON GENERATOR

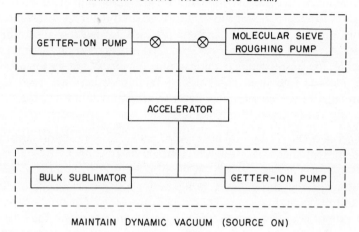

Fig. 3.23. Block diagram of a pumping system with bulk sublimator for a small accelerator.

c. *Turbomolecular Vacuum Pump.* A third type of pump potentially applicable to neutron generators is the turbomolecular pump. The operation of this pump is similar to the workings of a turbine air blower. The clearance between the rotor and the stator is kept extremely small and the rotor is electrically driven at extremely high velocities. These pumps can maintain compression ratios as high as 10^5 to 10^6 between the outlet and the inlet. A mechanical pump is connected to the outlet which is held at about 10^{-4} torr, so that the inlet vacuum is of the order of 10^{-10} torr. Since the compression ratio varies with the mass of the particles being pumped, hydrogen is pumped less efficiently than the heavier gases like hydrocarbons. The pump is therefore extremely effective in limiting gaseous hydrocarbon contamination entering the main vacuum system from the mechanical pump. The major disadvantage thus far experienced with this type of pump is the high rate of wear and tear of the high-speed rotor assembly. A cooling-water system has to be used to maintain the integrity of the rotor pump bearings. With improvement in design, these pumps could compete favorably with the commonly used ion-getter pumps.

2. VACUUM-MEASURING DEVICES USED IN NEUTRON GENERATORS

Thus far the discussion of the vacuum system for neutron generators has been limited to a description of the principal pumping systems and their operation. A short description of some of the methods employed in the measurement of the pressure in a neutron generator is in order. Exhaustive treatment of vacuum-measuring devices is beyond the scope of this treatise; therefore, only a few devices commonly used in neutron generators are discussed briefly. The reader is referred to the treatment of vacuum gauges by Leck[15] for more detailed information regarding the various types of gauges in use. Two principal types of vacuum gauges are used for the measurement of the pressure in a neutron generator. These are the thermocouple and the Penning (Philips) gauges.

a. *Thermocouple Gauge.* This type of gauge is used almost universally for the measurement of "rough" vacuum, that is, up to 10^{-3} torr. The operation of the device is based on the heat transfer from a metal to a gas. Current is passed through the hot junction of a thermocouple element located in the vacuum system. The temperature of this junction is dependent on the rate of heat conduction from the junction element to the impinging gas molecules. This rate, in turn, is dependent on the existing gas pressure in the vacuum system. The gas pressure of the system, therefore, is related to the filament temperature, which generates some voltage in the thermocouple. Thus a voltage versus pressure calibration can be obtained. The workings of a typical thermocouple gauge can be understood from the illustration shown in Fig. 3.24.

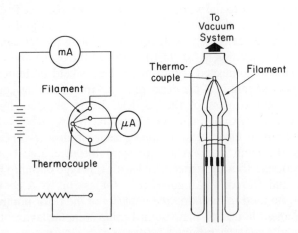

Fig. 3.24. Thermocouple gauge and basic circuit.

b. *Penning (Philips) Gauge.* This gauge is used to measure pressures below 10^{-3} torr. The gauge consists of a ring-shaped anode and two symmetrically located cathodes. An example of this gauge is shown schematically in Fig. 3.19.

3. MATERIALS FOR HIGH-VACUUM SYSTEMS

The materials used in the construction of high-vacuum systems such as neutron generators are of extreme importance. The ultimate vacuum possible in a typical accelerator system depends to a large measure on the type of materials in contact with the active volume. A general materials guide is given in Table 3.2. The materials listed are by no means the only ones available; however, they are representative of the types which can be used at the operating pressures indicated.

TABLE 3.2. Materials Evaluation for High-Vacuum Systems

Pressure Range (torr)	Suggested Materials	Joining Techniques	Cleaning Techniques	Bakeout
Atmospherics to 10^{-3} torr	Metals, glass, ceramics, organics	Solder, braze, weld, organic adhesives, organic gaskets	Superficial	Not required
10^{-3}–10^{-6} torr	Metals, glass, ceramics, organics	Solder, braze, weld, organic adhesives, organic gaskets	Detergent cleaning, remove fluxes	Not required
10^{-6}–10^{-10} torr	Copper, Monel, stainless steel, ceramics, glass, Viton A	Heli-arc weld, fluxless braze, Viton A gaskets, all-metal joints	Acid, alkali, or both; detergent cleaning	May be required for 10^{-10}-torr range
Below 10^{-10} torr	Copper, Monel, stainless steel, ceramics, glass	Heli-arc weld, fluxless braze	Acid, alkali, or both; detergent cleaning	Required

Some of the general, but important, properties which must be considered in the selection of materials for use in all-metal neutron generator systems are that they should be nonmagnetic, strong, resistant to corrosion, easily machined and joinable, and that they also possess good mechanical properties at elevated temperatures, and have low outgassing characteristics. Aluminum, stainless steel, and OFHC (oxygen-free high conductivity) copper satisfy most of these needs for the construction of neutron generator vacuum systems. Viton "O"-rings are well suited for use in sections of the system in which O-rings have to be used, for example, accelerator sections and target mounts. It has been shown that below 200°C gas evolution from Viton is negligible. Small amounts of water vapor can be removed easily by baking out in vacuum for two days at 200°C. Viton O-rings do not require a high-vacuum grease in order to seal; however, because Viton is a fluorinated rubber, and therefore of substantial hardness, some pressure is necessary to effect a good vacuum seal.

4. DESIGN CRITERIA FOR NEUTRON GENERATOR VACUUM SYSTEMS

Optimal integration of all components for a neutron generator vacuum system calls for consideration of certain factors. With reference to the "roughing" pump system, the rubber tubing connecting the roughing pump to the main vacuum system should be as short as possible and as large as possible in diameter. All oil-sealed roughing pumps should be equipped with an isolation valve, preferably electrically operated, located in the vacuum line to the pump, so that in the event of electrical power failure, the vacuum side of the roughing pump will be opened to the atmosphere to prevent atmospheric pressure from forcing the oil from the pump back into the main vacuum system. Most neutron generator vacuum systems are equipped with an isolation valve which is only opened during the "roughing" cycle. A molecular sieve trap is generally located between the roughing pump and the main vacuum system. This trap minimizes backstreaming of oil vapors, thus improving the roughing pressure.

Valves should also be included in the main drift-tube section of the vacuum system in a manner such that leak checks can be performed by suitable isolation of suspected subsections. The main vacuum pump should be installed with suitably located isolation valves so that rapid removal for maintenance can be effected. A solenoid-operated gate valve situated between the tritium target holder and the rest of the vacuum system is a useful addition for the protection of the entire neutron generator system. Should the ion beam accidentally burn through the target, the influx of coolant into the vacuum system can be stopped effectively by this valve, which is actuated at some threshold pressure.

An important factor in generator maintenance is the procedure by which the neutron generator is taken up to atmospheric pressure for maintenance or for a target change. It is advantageous to bleed in dry nitrogen gas to the system, a technique that can reduce startup times considerably. Finally, in the selection of a pumping unit for a specific accelerator, the speed and capacity of the pump must be matched to handle the total volume of the system adequately under the gas loads normally used. For most neutron generators, pumping systems capable of removing 200 liters/sec are adequate, although for long-drift-tube generators, a pump speed of 300 liters/sec is preferable.

F. TARGETS FOR NEUTRON GENERATORS

The neutron output and the angular distribution of neutrons obtained from deuterium and tritium targets have already been discussed in considerable detail in Chapter 2. In this section, attention is drawn to the preparation of neutron-producing targets commonly used in neutron generators with a view to increasing the tritium target life. Several special target designs are described here. In investigating significant aspects of neutron generator development, it can be concluded that research in the area of target development has been neglected compared with research on other design features of the low-voltage accelerator. It is surprising that although many of the physicochemical properties of adsorbed hydrogen isotopes have been studied in considerable depth, very little progress has been made in the preparation of satisfactory deuterium and tritium targets. In the meantime, the basic accelerator has experienced great improvements in stability, high beam currents, and sophisticated vacuum systems. The need for target improvement, however, has been considered important enough to devote entire international conferences [16, 17] to a discussion of the development of more suitable targets. An outcome of these discussions is the recognition of specific properties desired by experimenters. These properties can be summarized as follows:

1. The target must have a long lifetime to permit activation experiments at constant neutron fluxes for long irradiation periods.
2. The target must have a high neutron yield per unit area of active surface; thus the target nuclei must be deposited in saturated quantities over the smallest area possible.
3. The contamination of targets from other neutron-producing nuclei by auxiliary nuclear reactions with the bombarding beam should be minimized.
4. The target nuclei of interest must be deposited in a suitable matrix onto a thin backing surface both to reduce the attenuation of the neutron beam and to facilitate cooling of the target.
5. The target design should be such as to permit rapid replacement without exposure to tritium.

Deuterium or tritium targets are almost exclusively used for neutron production in activation analysis with low-voltage particle accelerators. These targets can generally be classified into two categories, gaseous and solid. In a few applications, liquid targets are also used. In the case of gaseous targets, the target holder itself has to be isolated from the main vacuum system by an extremely thin window or by the incorporation of a differential pumping system.

It is obvious that for maximum neutron yield per unit of incident beam current, the target must be composed of the pure isotope rather than a compound of the isotope. Consider the general case of a beam of particles impinging upon a target material. If N is the number of target nuclei per square centimeter, and X is the stopping cross section of the target material in eV-cm^2/target atom, then for thin targets, the target thickness in terms of beam energy loss can be expressed as $\Delta E = NX$. Now if M bombarding particles of energy E are incident on a target with a neutron production cross section equal to $\sigma(\theta, E)$ barns/sr, the yield is $NM\sigma(\theta, E)$ neutrons/sr. Therefore, the yield Y per steradian per incident beam particle is

$$Y(\theta, E) = \frac{\Delta E}{X} \sigma(\theta, E) \qquad (3.1)$$

Similarly, for targets thick enough to stop the beam,

$$Y(\theta, E) = \int_0^E \frac{\sigma(\theta, E)}{X(E)} dE \qquad (3.2)$$

and the neutron yield varies inversely as the stopping cross section per target atom. Thus the neutron yield from a gas target of tritium or deuterium gives a tenfold improvement over those targets in which tritium or deuterium is deposited in a metal matrix. In practice, gas targets are severely limited because the high beam current on thin windows can create window failure. Therefore, where large beam currents are desired, deuterated or tritiated metal targets are always used.

1. TARGET CONSIDERATIONS

In evaluating the capability of a target, it is imperative to consider certain factors such as target preparation, behavior of the target under bombardment, energy loss of the impinging beam within the thickness of the target, and the mode by which the active nuclei on the target are depleted with a corresponding loss of neutron flux.

In general, thin films of deuterium or tritium can be prepared by the vacuum evaporation-condensation process. Several methods have been employed for the evaporation of the target nuclei and for the deposition of the matrix metal in which the active nuclei are incorporated. Of specific interest to neutron generator users is the tritium target. Tritium gas is sorbed into a condensed film of a metal which forms hydrides of sufficient thermal stability to withstand deuteron bombardment at energies between 150 keV and 3 MeV.

Several materials are suitable matrices for impregnation by tritium. Zirconium and titanium are satisfactory for the formation of stable hydrides, although some rare earths, namely erbium and yttrium, and other metals such as tungsten, silver, and molybdenum have also found limited use. Before proceeding to a detailed examination of the manufacture and characteristics of neutron generator targets, the concept of stopping cross sections, as it applies to targets under bombardment by the isotopes of hydrogen, should be discussed.

It may be recalled that the stopping cross section is inversely proportional to the neutron yield and directly proportional to the energy degradation of the impinging charged-particle beam as it penetrates the target. Considering these dependencies, it is obvious that the stopping cross section of different matrix materials contributes directly to the ultimate target performance, in terms of the decreasing neutron yield with increasing bombardment time. For instance, in the case of gas targets, the ultimate performance will depend strongly on the stopping cross section of the thin window separating the main accelerator vacuum system from the target chamber. The stopping cross section per deuteron for a number of different types of deuterium (or tritium) targets is given in Fig. 3.25 as a function of proton energy. The data shown were taken from a review article by Whaling[18]; the solid lines are experimental data whereas the dashed lines were obtained using a semitheoretical relationship.[18] The dotted sections of the curves were obtained by extrapolation between the data for neighboring elements. Since the dependence of the stopping cross section on the atomic number is not well known in the region below 1 MeV, the reliability of the dotted curves below this

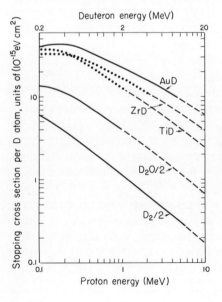

Fig. 3.25. "Molecular" stopping cross sections per deuteron for various types of deuterium (or tritium) targets.

energy is questionable. Using Fig. 3.25, one can compute the energy degradation of the impinging beam in units of keV/mg cm^2 by multiplying the ordinate by the factor 598 (for H), 12.57 (for Ti), 6.60 (for Zr), and 3.05 (for Au). Two abscissa scales are also provided so that the stopping cross sections of deuterons can be interrelated. For example, the stopping cross section X_d for deuterons at energy $2E$, is equivalent to X_p, the stopping cross section for protons at energy E. The conversion factors which permit energy degradation calculations have been abstracted from a comprehensive table given by Coon.[19]

2. DEUTERIUM AND TRITIUM TARGETS FOR USE WITH NEUTRON GENERATORS

In the discussion to follow, emphasis is placed on the description, preparation, and characteristics of solid targets for use with neutron generators. Because of their rare use for activation analysis, only a brief mention of gas targets is made. Most of the treatment of solid targets is limited to those in which hydrogen isotopes are absorbed into a titanium or a zirconium film. Some attention, however, is given to specialized target studies relating to the use of rare earths.

a. *Frozen-Deuterium Targets* (D_2O *Ice*). Deuterium-ice targets are very effective at deuteron energies above 1 MeV. Using 500 μA of a 1-MeV deuteron beam, Van Dorsten[20] has measured a neutron yield of 10^{10} neutrons/sec. This yield corresponds closely to the yield calculated from the known D–D reaction cross section for 1-MeV energy deuterons. A thick target (ca. 20 μm) was maintained at a temperature of $-100°C$. By depositing the D_2O ice on a rotating copper disk, Van Dorsten was able to reduce the energy density deposited in the target substantially and thus maintain its integrity. In addition, the deuteron beam was continuously and rapidly deflected through small angles by electrostatic fields so as to trace Lissajou figures on the ice surface. With respect to ice targets, some consideration has also been given to the use of T_2O ice.[21] However, contamination of the ice from condensable vapors in the vacuum system and the chemical decomposition of T_2O from its own beta activity make the use of tritium-ice targets unattractive.

Typical systems in which frozen D_2O targets deposited on metal plates are refrigerated by liquid nitrogen have been described by Coon[19] and by Manley et al.[22] These studies indicate that rapid vaporization of D_2O occurs if the target temperature is permitted to exceed $-100°C$. Significant reduction in the quantity of vaporization can be obtained by defocusing the deuteron beam.

b. *Absorbed-Hydrogen-Isotope Solid Target.* In their search for a matrix suitable for tritium impregnation and a high neutron output, Graves et al.[21] prepared both tritium and deuterium targets by impregnating various materials with these isotopes of hydrogen. The preparation of such targets was reported as early as 1937 by Penning and Moubis.[23] In their work, Graves et al. employed both tantalum and zirconium as matrix materials, and also suggested that the use of titanium should be considered because of its low stopping cross section and consequently greater penetrability for bombarding deuterons.

The extent of impregnation determines much of the behavior of a gas-impregnated metal target. Hydrogen-isotope loadings up to a ratio of two atoms per Zr or Ti atom have been reported[24, 25]; however, a 1 : 1 ratio is typical and may be considered satisfactory. The principal advantage for the use of gas-in-metal targets lies in their good stability and high thermal conductivity. They are capable of handling large beam currents, which would normally destroy the thin windows of gas targets and cause significant vaporization of the $D_2 O$ ice target.

c. *Preparation of Gas-in-Metal Targets.* Numerous techniques have been employed[21, 26-30] for the preparation of gas-in-metal targets by the impregnation of the hydrogen isotopes into a metal matrix like zirconium or titanium. In a typical procedure, a backing plate of tungsten, gold, copper, or platinum about 0.01 in. thick is selected. About 10 μg- to 10 mg/cm^2 of zirconium or titanium is deposited on the backing plate. The plate serves to form the vacuum seal and acts as a good thermal conductor for cooling purposes. In the deposition of the thinner layers of zirconium or titanium, a vacuum evaporation technique is used; the thicker layers are deposited by melting the matrix material directly onto the backing plate. The melting process often results in a nonuniform deposit. The zirconium or titanium is heated to a temperature of about 400°C in an atmosphere of the desired hydrogen isotope. On cooling, the metal matrix becomes impregnated with the isotope gas. Hydrogen-isotope gas pressures from about 2 to 100 torr are employed. Within this pressure range, the degree of impregnation is relatively constant.

To ensure uniform and optimum loading of the target, the metal matrix is first thoroughly outgassed in a high vacuum, which also prevents absorption of possible contaminants which can adversely influence subsequent impregnation of the active hydrogen-gas isotope. During the outgassing procedure, a pressure of about 10^{-6} torr is maintained and the outgassing process is carefully controlled. It is also important that the backing plate itself be outgassed before zirconium or titanium is deposited. The outgassing temperature in this case depends on the material of the backing plate. If tungsten is used, a temperature of 2000°C is necessary to outgas it adequately. Similarly, the zirconium or titanium used should be outgassed prior to the deposition step.

In order to obtain good adhesion between the evaporated metal deposit and the backing plate, some treatment of the backing-plate surface is often necessary. A method commonly used consists in heating the backing plate to about 400°C by electron bombardment or by thermal radiation. If a copper backing material is used, the presence of a thin oxide layer significantly improves the adhesion of the metal to the backing plate. In general, smooth and polished surfaces on backing plates are not conducive to good adhesion.

  d. *Cooling of Gas-in-Metal Targets.* A neutron generator accelerating a deuteron beam of 200 keV at a current of 2.5 mA results in the dissipation of 500 W of heat in the gas-in-metal target. Since the hydrogen isotope will outgas from the target at temperatures above 250°C the cooling side of the backing-plate surface must be cooled efficiently.

Experiments performed with 50 to 100 W of beam power deposited over a 0.5-in.-diameter target surface have shown the feasibility of simple air-jet cooling. Further improvements in air-jet cooling are found if the target is rotated during the air spray and if a water-vapor spray is employed in the air jet. For activation analysis, however, beam powers from 500 to 750 W are generally desired for maximum sensitivity, and liquid cooling of the target-backing surface is absolutely necessary. Numerous types and designs of cooling jackets are employed which use either water or Freon as the coolant. The design of water jackets varies from the simple device shown in Fig. 3.26 to a more sophisticated system in which both the front and the back surfaces of the target are subjected to peripheral cooling.[31] The efficiency of target cooling is reflected in the degree of degradation of the target. Target-life data

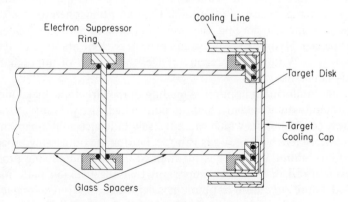

Fig. 3.26. Water jacket for cooling gas-in-metal targets.

recently reported[31] certainly suggest the superiority of peripheral cooling. This cooling system is shown schematically in Fig. 3.27. A distinct advantage in using the peripheral cooling system is the ability of the experimenter to locate a sample for irradiation directly in contact with the backing surface of the target, provided it is properly insulated, so that the Faraday-cup beam-monitoring principle, almost universally used for beam current measurement, is not affected. An increase by at least a factor of 2 in target life is obtained using peripheral cooling rather than the more commonly used water-jacket-type cooling device. Some of the desirable features of efficient target cooling systems for beam powers of the order of 500 to 750 W can be summarized as follows: (1) high cooling capacity to prevent the outgassing of hydrogen isotope from the metal matrix; (2) minimum amount of extraneous construction material in order to reduce neutron beam degradation and scatter; and (3) efficient insulation of the target from the neutron generator so that beam currents can be measured accurately.

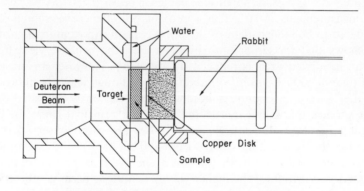

Fig. 3.27. Peripheral cooling of gas-in-metal targets.

e. *Characteristics of Gas-in-Metal Targets under Bombardment.* Examples of the neutron yield from the D–D and D–T reactions using zirconium or titanium have been given in Chapter 2. Comparisons between the neutron yields from gas-in-metal targets and those from gaseous and ice targets were also given. In most of these cases, the assumption of a 1 : 1 gas-to-metal ratio was made. It may also be recalled that the D–T neutron yield of gas-in-metal targets falls rapidly with increase in deuteron beam time on the target and that the D–D yield from T–Ti targets increases with integrated beam currents. The latter effect is due to the displacement of tritium atoms by the bombarding deuterons. The behavior of zirconium-type targets impregnated with tritium under deuteron bombardment is

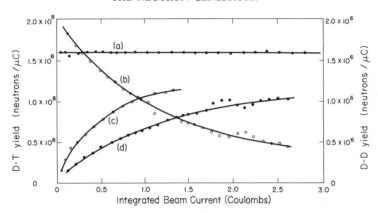

Fig. 3.28. Neutron yield as a function of integrated deuteron beam current for thick gas-in-metal targets. (a) Zr-D target, D-D yield. (b) Zr-T target, D-T yield. (c) Zr target, D-D yield. (d) Zr-T target, D-D yield.

illustrated in Fig. 3.28. A monatomic deuteron beam at 250 keV was used for these measurements.[19] To a first approximation, the curves shown can be considered to be independent of the beam current. An examination of Fig. 3.28 shows that for zirconium-tritium targets, the neutron yield from a 0.5-cm^2 irradiated area is reduced by about one-half after an integrated beam time of 600 μA-hr/cm^2 of target area. Although it has been observed that after a certain amount of bombardment, carbon is deposited on the target surface, this is not considered to be the principal cause for target depletion in terms of neutron yield. It has also been observed that in increasing the accelerating voltage from 100 to 400 kV, the target lifetime increases proportionally with the beam energy. It is presumed that the increased penetration of the higher energy deuterons permits utilization of tritium atoms in the deeper layers of the thick target. It is therefore advantageous for thick targets to accelerate deuterons considerably above the 110-keV energy at which the D-T cross section for thin-target neutron production is greatest.

In activation analysis using D-T neutrons, the buildup of neutrons from the D-D reaction can introduce systematic errors. Several investigators have attempted to determine quantitatively the degree of D-D interference in D-T analyses; comparisons of the data available are confusing. In Fig. 3.28 the D-D buildup, as reported by Coon,[19] for a new zirconium metal target can be used to determine the D-D flux in any hypothetical sample position. In comparison, the D-D buildup for a zirconium-tritium target is more gradual, although ultimately, after an integrated beam current of about 3.0 C, the total D-D neutron buildup in both cases is about the same. At this integrated beam current, it is interesting that the D-T yield from the zirconium-tritium target is only 50 times higher than the D-D output. Nargolwalla et al.[32] have shown that for some elements D-D activation is significant, and the presence in a sample of elements such as cadmium, indium, barium, gold, and some rare earths can produce interfering activities significant enough to introduce systematic errors in the analyses for most elements by D-T neutron activation. In the careful cross-section measurement experiments carried out by Wille and Fink,[33] the estimate of D-D neutron buildup in a used target was as high as 25% of the D-T neutron flux. Recent work at the National Bureau of Standards[34] illustrates quantitatively the D-D flux buildup as a function of integrated beam current from titanium-hydrogen targets of various thicknesses. The results, presented in Fig. 3.29, were obtained using a deuteron beam (ca. 50% atomic) impinging

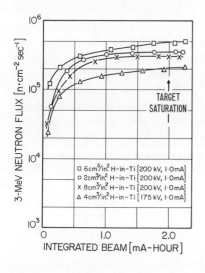

Fig. 3.29. D-D neutron yield versus integrated deuteron beam current for hydrogen-in-titanium targets for various target loadings.

on the target over an area of about 5 cm^2 (diffused beam). Hydrogen-impregnated targets were used to simulate as closely as possible the normal titanium-tritium target used for D-T neutron activation analysis. Experiments with blank copper targets indicated a saturated D-D neutron builup of a factor of 10 less than in the case of hydrogen-titanium targets.

For the experiment mentioned above, the beam current was stabilized at about 1.0 mA. The D-D flux measurements were made at the normal sample position of a dual sample-biaxial rotating assembly using the threshold reaction ^{27}Al(n, p)^{27}Mg; $\sigma = 1.5$ mb for 3-MeV neutrons. The D-T flux for a standard 5-Ci/in.^{2} tritium target at an equivalent sample position and beam current was of the order of 10^8 neutrons/(cm^2 sec). At the point

(2 mA-hr) where the D–D reaction reaches saturation, the ratio of D–T/D–D neutron flux was calculated to be about 100 on the basis of a D–T target half-life of 1 mA-hr. Using normal deuterium targets and bombarding at identical beam currents, a 3-MeV neutron flux of 5×10^5 neutrons/(cm^2 sec) was measured. This flux closely approximates that measured at saturation in the D–D buildup experiments using H–Ti targets. Knowledge of the response of elements for D–D neutron activation[32] and of the integrated beam on the D–T target, permits prediction of interferences due to the D–D buildup process in any D–T analysis. In Fig. 3.29, the effect of reducing the acceleration voltage on the saturation flux level is clearly indicated.

In an attempt to relate the theoretical buildup of deuterons in a tritium target from their bombardment history, several targets, similar to those referred to in Fig. 3.28, were analyzed after prolonged bombardment and found to contain about 10 at. % of deuterons, a quantity predicted from beam current considerations. An analysis of the tritium remaining in the thick targets showed that no more than 10% of the tritium was lost, although much greater proportions of tritium were lost from the thin layer penetrated by the deuteron beam. The behavior described above appears to agree with results obtained from most experiments done with thick targets.

Experiments involving the use of thin targets, that is, where the range of impinging deuterons was greater than the target thickness, have also been reported. For deuteron energies in the kiloelectron-volt range, experiments by Conner, as reported by Coon,[19] show target behavior as depicted in Fig. 3.30.

In these experiments, tritium targets (tritium-in-titanium) with tungsten backing plates were bombarded by 450-keV deuterons. A beam current of about 80 µA was permitted to illuminate a $\frac{3}{16}$-in.-diameter circle. Target thicknesses were such as to degrade the incident deuteron beam energy by about 80 keV in the tritium-titanium layer. The tritium-to-titanium ratio of the targets was reported as only 0.07. Higher ratio targets behaved in approximately the same way in the incident deuteron energy range 150 to 450 keV. It can be observed from Fig. 3.30 that the D–T neutron yield was reduced to one-half of its original value after 0.25 C of integrated beam current. At this point, the D–D neutron yield rose to four times that of the D–T neutron output. This behavior is quite different from the thick-target behavior described earlier. It appears that for a thin target the deuterons saturate the surface, thus offering an efficient means for the production of a good D–D neutron target.

In comparing the large number of data on tritium target lifetimes, sufficient disagreement exists to prevent the formulation of any general conclusions. This may be well understood in light of the many, and often

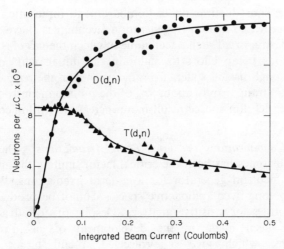

Fig. 3.30. D–T and D–D neutron yields versus integrated beam current for "thin" tritium-in-titanium targets.

unreported, conditions which radically affect target lifetimes. It should be recognized that conditions such as nonuniformity of the tritium deposit, bombardment area, heat dissipation in the target, average temperature of the target, and therefore the cooling characteristics, and target thicknesses used are the principal factors governing the target life. Target lifetime data seldom provide this information, and therefore it is not surprising that for a standard 5-Ci/in.^{2} tritium-in-titanium target bombarded by 150- to 200-keV deuterons, values for target half-life (neutron production) from 0.25 to 8 mA-hr have been reported in the literature. In a recent conference[17] Morgan described an excellent study of the behavior of tritium-titanium targets under bombardment. In this work, the half-life of various targets was measured as a function of beam spot size, beam current, and tritium loadings. The proceedings of the conferences on neutron-producing targets[16,17] are worthy of detailed study.

A considerable amount of research has been performed in the search for other materials for impregnation with tritium. Efforts by an Oak Ridge National Laboratory group, as reported by Strain,[35] indicated that Er–T and Y–T showed erratic behavior with various loadings and were susceptible to rapid hydration when exposed to the atmosphere.

Although these rare earth targets are thermally more stable than Ti–T targets, their half-lives were unexpectedly shorter under deuteron bombardment. Work performed at the Centre d'Etudes Nucléaires de Grenoble, France,[36] confirms the higher thermal stability of the Er–T and Y–T targets as well as the short lifetime under bombardment. In an attempt to avoid target replacement, tritium regeneration techniques have been used. Hollister[37] has reported such a technique in which the target is bombarded by tritium ions in an effort to maintain the initial neutron yield. The technique, though useful, suffers from inefficiency in the regeneration step. According to Strain,[35] only about 9% of the initial ions introduced into a high-efficiency RF ion source result in an increase in tritium concentration in the target.

f. *Target Configurations for "Long-Term" Irradiations.* The problem of short target lifetimes has been the critical factor limiting the application of the pumped neutron generator for long-term irradiations. With current targets many long-lived radioactive species cannot be produced in large enough quantities for sensitive analyses. Despite this disadvantage, great strides have been made in the development of gas-in-metal target configurations, which give a prolonged, useful target life. Some configurations available today are said to allow half-lives for neutron production of 20 mA-hr. One of the early concepts was built by Meinke and Shideler[38] and is shown in Fig. 3.31. Essentially, the target consisted of a 10.5- by 2.0-in. Ti–T ribbon rotated in front of the deuteron beam. The active

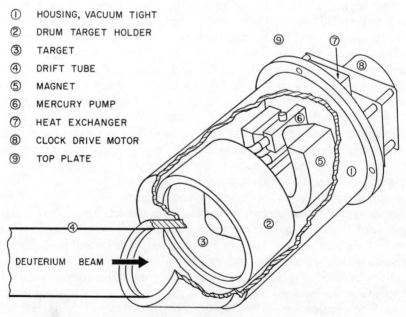

Fig. 3.31. Long-lived rotating target assembly (Meinke-Shideler design).

target strip was clamped to the inside of a stainless steel drum. The back surface of the target was cooled by mercury circulating through a channel provided for this purpose. The target drum was rotated at a rate of 0.00725 in./min, corresponding to about 200 min for 95% target utilization in a 1.5-in.-wide beam at 1 mA. This target was expected to give a constant 10^9-neutron/(cm^2 sec) fast flux for a period of 22 hr. Rotating targets have since appeared on the market. Morgan[17] has reported target half-lives of from 10 to 80 hr using a commercially available rotating target. A recent development by Booth[39] has also proved successful.

No mention has been made so far of the radiological hazards associated with both the replacement and use of tritium targets. This subject is treated in detail in Chapter 4. The subject of the targets used in sealed-tube generators has been included in the last section of this chapter, where the sealed-tube neutron source is described. Although a literature survey on the subject of targets reveals a voluminous quantity of data, improvement in gas-in-metal targets has not been significant when compared to other advances in neutron generator technology. Whereas beam currents obtainable have more than tripled compared to those that could be obtained a decade ago, tritium target lifetimes have remained relatively unchanged. Target research has been directed toward devices which permit low-density energy depositions, thus increasing the effective target life. There does not appear to be further interest in searching for thermally stable hydrides of alloyed materials. Perhaps metal alloys with better characteristics than titanium or zirconium would be a fruitful area for target research.

G. HIGH-VOLTAGE POWER SUPPLY

The development of reliable high-voltage power supplies for use in conjunction with neutron generators has been relatively rapid and satisfactory. In the use of neutron generators for activation analysis, a mention of power supplies is generally omitted since few problems due to improper function of high-voltage units are experienced. The principal factor from the cost point of view, however, is the high-voltage power supply output.

A review of the specifications of neutron generators manufactured today reveals that two basic types of power sources for the acceleration voltage are used. These are the electrostatic (the Van de Graaff or Felici design) and the Cockcroft-Walton voltage-doubler or cascade-rectifier arrangement discussed earlier. Although there is little choice between these two designs, some of their important differences are worthy of consideration. The electrostatic high-voltage supply gives near-ripple-free outputs and excellent regulation when compared to the cascade-rectifier system. However, the latter is more compact, free of moving parts, and less expensive. Another important difference between them is their response to sudden changes in the load, for example, beam-switching operations. The electrostatic supply appears to respond to this situation more rapidly.

Although power supply breakdowns are relatively rare, when they do occur the repair is time-consuming and often expensive. The oil-insulated, transformer-type supplies used with the Cockcroft-Walton machines require careful oil-conditioning procedures prior to initial startup, and the quality of the oil must be checked periodically for carbonaceous deposits. Insulation standards must be maintained and standard spark-resistance tests are helpful in determining the general condition of the oil. Some of the desirable features of a high-voltage power supply for optimum operation of a neutron generator are as follows:

1. Good voltage regulation, to minimize the effect of rapid voltage fluctuations.
2. Low temperature coefficient, 0.1%/°C, to reduce drift problems after initial equilibrium temperature has been reached.
3. Fast recovery from sudden load changes.

Interested readers are referred to more exhaustive descriptions[40, 41] of methods for particle acceleration. Some of the effects of acceleration-voltage variations on the beam characteristics of a neutron generator are most important and are discussed in Section II.I of this chapter.

H. SAFETY CONSIDERATIONS FOR NEUTRON GENERATOR OPERATION

In this section, methods for ensuring the safety of the neutron generator system itself are considered. Personnel hazards are discussed in Chapter 4. In principle, two major sources of hazards exist, namely, uncontrolled heat generation in the target-cooling system and secondary effects on the system from primary breakdown of vacuum components.

It may be recalled that 200-keV deuterons at a beam current of 2.5 mA deposit 500 W of beam power on a small area (ca. 3 cm^2). If this beam were permitted to impinge upon a blank tantalum plate 0.01 in. thick and suitably cooled by circulating water, the beam side of the blank plate would glow at red heat almost as soon as the beam was turned on. Should the cooling be interrupted for a few seconds, the plate would fail, with drastic results. This kind of a mishap has been known to destroy a $20,000 installation. It is obvious that proper attention must be given to the design of a dependable cooling system coupled to an interlock system for the protection of all key components against target failure. A simple approach to this problem is to include in the downstream side of the cooling lines a flow switch that is electrically coupled both to the main power input to the generator and also to suitably located solenoid-actuated isolation valves in the vacuum system of the accelerator. If, for any reason, the coolant flow is interrupted, completion of an electrical circuit through the flow switch instantly shuts down the power input to the machine and actuates the solenoid valves which isolate components such as the ion source, vacuum pump, accelerator tube, and the generator front end from the incoming coolant. The operation of the flow-switch interlock should be periodically checked, and the installation of thermocouples in the inlet and outlet sides of the target cooling lines is recommended. Readouts located at the console often warn the experimenter of impending disaster. An undesirable heat situation due to coolant interruption or beam displacement can be detected ahead of the failure point and shutdown immediately initiated. As a rule of thumb, the inlet-to-outlet temperature difference for a 500-W heat dissipation load on a 3-cm^2 area of a tritium-titanium target should not be greater than 30°F, with the outlet temperature less than 110°F. The coolant flow rate should be adjusted to maintain the temperature difference below this threshold. If this is not possible, a small refrigeration unit should be installed in the inlet side of the target cooling. For closed-loop cooling systems and beam powers greater than 100 W, a refrigeration unit is necessary. Coolant flow rates vary from 1 to 5 liters/min, depending on the beam current used, water-jacket design, and refrigeration. Typical differential temperatures between the inlet and outlet ports on the target cooling assembly with a beam current of 2.5 mA are approximately 20°F. Freon has been used as a coolant in certain commercial cooling systems. However, degradation of circulating pump seals and Freon replenishing problems have discouraged the widespread use of Freon.

The electrical interlock system is particularly useful for the protection of the accelerator from vacuum failures other than breakdown of the cooling system. If ion pumps are used, the pump itself is protected against burnout by a thermally activated switch which turns off the pump power at some safe pressure. However, this protective feature, generally included in the manufacturer's specifications, does not protect the pump against gross contamination resulting from vacuum failure. Pump-isolating solenoid-actuated valves are most useful for this purpose. They not only protect the pump from contamination, but also isolate the pump from the main system for maintenance and repair. A well-protected neutron generator system may cost up to 10% more than a basic unprotected system; however, this additional cost is more than offset in terms of ease of maintenance and protection of key components of the entire neutron generator facility.

I. OPERATIONAL CHARACTERISTICS OF THE PUMPED NEUTRON GENERATOR

For the purpose of familiarizing the operator with the behavior of a neutron generator, certain characteristics of the basic accelerator must initially be investigated. In view of the neutron hazard with tritium or deuterium targets, all such studies relating to the beam behavior are conducted using hydrogen and a blank target of copper or tantalum. It should be noted that once the generator has been used with tritium-loaded targets and deuterium gas, the use of hydrogen and blank targets does not eliminate neutron production. Enough residual deuterium and tritium remain in the system to produce appreciable neutron outputs. As long as visual beam observations are made from a reasonable distance (minimum of 6 ft) from the high-voltage terminal end of the accelerator, the x-ray hazard from "backstreaming" electrons is negligible. Beam performance data, as a function of key control settings of the gas leak, focus voltage, high voltage, and suppressor voltage, are obviously important for optimization of the beam current.

The beam emerging from the focus lens diverges as it leaves the lens opening. The multistage accelerating tube enables the establishment of a uniform field gradient and must provide moderate focal properties to control and contain the beam within the confines of the accelerator tube. If the focusing properties of the accelerator tube are too strong, beam crossover can occur early along the drift tube or even in the accelerator tube itself. The quality and size of the beam spot appearing on the target depend in a large measure on the beam optics and beam current. The beam current, as measured at the target, decreases rapidly with reduction in the accelerating voltage because of degradation of the focal properties of the focus lens-accelerating structure combination.

The effect of high-voltage variation on the beam spot size for a Cockcroft-Walton accelerator with optimized beam current is shown in Fig. 3.32. These data were taken from a generator with an RF ion source. Performance data taken from a recent study[42] on a 200-keV accelerator equipped with a Penning ion source are shown in Table 3.3. Another significant effect on the beam size is the focus control. For activation analysis a diffused beam is desirable for economical use of the target, minimum beam variation, and more uniform sample irradiation.

Some of the fixed and nonvariable features included in the assembly of the neutron generator also affect the beam characteristics. Items in this group include collimators, beam-handling magnet systems, and magnetic fields from the use of ion pumps.

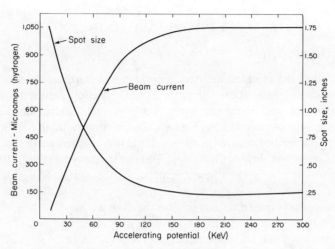

Fig. 3.32. Beam current and beam spot size as a function of accelerating voltage (RF source; beam current optimized).

TABLE 3.3. Effect of Acceleration Voltage on the Diameter of the Beam Spot (Penning Source, Beam Current Optimized)

Acceleration Voltage (kV)	Peaked Beam Current (mA) Approximate Diameter of Beam Spot (in.)				
	0.5	1.0	1.5	2.0	2.5
200	0.5	0.625	0.75	1.25	1.5
195	0.5	0.75	1.0	1.5	—
190	0.625	0.75	1.25	—	—
185	0.75	1.0	1.5	—	—
180	0.75	1.25	—	—	—
175	0.75	1.5	—	—	—
170	1.0	—	—	—	—
160	1.5	—	—	—	—

III. SEALED-TUBE NEUTRON SOURCE

A. INTRODUCTION

Ever since the pumped neutron generator was introduced as a source for neutron irradiations about 20 years ago, the search for an even simpler, more efficient, and more readily portable neutron source has continued. Although isotopic sources such as Cf, Ra–Be, or Po–Be fulfill the requirement of portability and ease of handling, the disadvantages of a mixed neutron spectrum, low neutron flux, and the inability to turn off the source of neutrons at will, gave sufficient impetus for the continued development of compact monoenergetic sources of high-energy neutrons. The pumped neutron generator system does not possess the drawbacks of the isotopic source; however, the limited target life and its lack of portability have stimulated the development of the sealed-tube neutron source. As early as 1937, Penning and Moubis[23] designed a sealed tube requiring no external gas supply or pumping system. The Penning tube produced about 10^5 neutrons/sec and had a limited life because of the rate at which the sealed-in gas was consumed in the discharge. It was not until about 20 years later that tubes with neutron outputs between 10^7 and 10^8 neutrons/sec and a half-life of several hundred hours became available. The work of Reifenschweiler and Van Dorsten[43] contributed significantly to this development. A summary of their efforts at the Philips Research Laboratories in the Netherlands is given by Reifenschweiler.[44]

The design of suitable components and the improvement of existing designs were considerably accelerated because of the value of a sealed-tube generator as a tool for oil-well logging.[45–47] In this operation, a small sealed-tube neutron source is lowered into a bore hole together with a detector system. Measurement of the scattered neutrons, or the excited gamma radiation, provides an indication as to the presence of oil-bearing strata. One of the more useful applications of the sealed tube has been as an educational tool. Numerous nuclear physics and radiochemical studies are brought into the reach of a normal university laboratory with this device. As will be seen later in this section, the availability of recently developed high-output tubes has had a great impact on activation analysis.

Many comparisons have been made regarding the utility of sealed-tube and pumped neutron generators. We hope that the detailed discussions in this section will point out the advantages and drawbacks of both types of neutron sources. The controversy that surrounds one or the other of these neutron sources exists primarily between the manufacturers. It is therefore often difficult and confusing for new workers in this field to appreciate fully the value of each particular source for activation analysis. To clarify this point, the two types of generators have been deliberately segregated in this chapter. We hope that this will help the reader to arrive at the appropriate conclusions for the maximum use, both economically and technically, of the source best suited for his purpose.

B. DESIGN CONSIDERATIONS OF A PORTABLE SEALED-TUBE NEUTRON SOURCE

The problems encountered in the development of a sealed neutron source are summarized well by Reifenschweiler.[48] The outline of these problems can probably be appreciated from the detailed descriptions of the development and operation of the pumped neutron generators discussed in Sections I and II of this chapter. These problems arise principally from differences between the two types of accelerators. It may be recalled that in a pumped accelerator the acceleration of the positive ions is conducted from a relatively high-pressure region of the ion source (10^{-2} torr) through a region at a relatively low pressure (10^{-5} torr). The equilibrium between these regions is maintained by continuously evacuating the gas in the low-pressure region with a high-speed pumping system. Most of the problems associated with the early development of the sealed-tube generator involved the maintenance of this equilibrium without resorting to an external gas supply or pumping system. Because the tube is entirely self-contained, the design of the acceleration stage, or stages, and of the ion source necessarily demands satisfactory operation of both these units at the same gas pressure. Consequently, the design criteria involved in maintaining a relatively low voltage discharge in the ion source and at the same time permitting ion acceleration without breakdown in the high voltage between the electrodes are stringent.

If the sealed-tube generator is to possess some distinct advantages over the pumped version, it must be superior in terms of high neutron production over prolonged periods. Since the target in the sealed tube cannot be replaced, the design of appropriate target-replenishing systems operating inside the tube must be considered. Finally, since there is no possibility of an external gas supply, and since gas absorption in the discharge process is significant, some type of pressure-regulating device, also operating internally, is essential for the proper performance of the tube.

In the description of the sealed-tube neutron source a few major designs currently in use will be discussed. Although the functions of the individual components are, of course, similar, for example, ionization, acceleration, replenishment, and pressure regulation, the performance of each integral tube as evaluated by the individual manufacturer and through customer case histories varies sufficiently to warrant independent discussion.

In general, the sealed-tube generator consists of an ion source, accelerating structure, target section, and a replenisher sealed inside a glass tube which is insulated with oil or pressurized gas. For mechanical protection the glass tube is encapsulated in a steel housing through which appropriate cables are drawn via feedthroughs. The average tube is designed to permit continuous and/or pulsed operation. The accelerating voltage varies from 100 to 200 kV and neutron outputs up to 10^{11} neutrons/sec for a period of several hundred hours are now available. The cost of these high-flux tubes is perhaps slightly less than the cost of an equivalent flux, pumped model. The development and performance of the principal components of the sealed-tube neutron source are outlined below.

1. ION SOURCES FOR SEALED-TUBE OPERATION

Ion sources operating on the RF or the Penning principle are currently used in sealed tubes. The fact that the tube must operate at a constant gas pressure suggests the use of an ion source which will perform efficiently at relatively low gas pressures, and an accelerating structure which will operate at relatively high pressures. The Penning source certainly satisfies the requirement for stable operation at low pressures. The source, as used in sealed-tube generators, is an adaptation of the original Penning design. One of the earlier designs of a sealed-tube generator, reported by Reifenschweiler,[48] shows the location of a Penning source inside the tube; a schematic diagram is shown in Fig. 3.33. The source requires a positive high voltage from 1 to 4 kV which is applied to a cylindrical anode via a ceramic insulator in the ion-source wall. The magnetic field necessary for efficient operation of the source is supplied by a small permanent magnet which produces a magnetic field of about 600 G in the center of the discharge. In the range of pressures employed, the source produces primarily molecular ions. The tube, including the steel jacket, measures 42 cm in length and is 5 cm in diameter. The high voltage in this earlier model was supplied via a glass cone; a silicon-rubber plug is used for the negative high-voltage connection.

A unique variation on conventional sealed-tube designs was offered by Oshry.[49] A modified cold-cathode ion source was used except that no auxiliary power supply was needed to drive it. This was achieved by connecting it in series with the accelerating electrode, such that all the current from the high-voltage supply passed through the source itself. The sealed-neutron source and its associated, miniaturized Van de Graaff power supply are shown in Fig. 3.34. Typical operating conditions included a positive 132-kV accelerating voltage and an ion-source current of 12 μA. Beam currents up to 6 μA were reported. Under these conditions, a total neutron output of 10^8 neutrons/sec and a maximum usable neutron flux of 5×10^6 neutrons/(cm^2 sec) were quoted. The electron trap used in the Oshry design consisted of a pair of rings located symmetrically around the target and maintained at

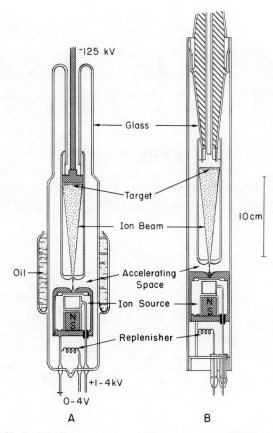

Fig. 3.33. Sealed-off neutron tube with Penning source (Philips design).

−3000 V. A static pressure of 10^{-3} torr was maintained in the tube by the use of a deuterium-impregnated titanium element, the deuterium being released by varying the temperature of the titanium element.

The Penning-type ion source has been used successfully[50] for the operation of the Kaman sealed tube (Kaman Nuclear, Colorado Springs, Colorado). This tube is guaranteed for continuous operation with an initial output in excess of 10^{11} neutrons/sec, and more than 5×10^{10} neutrons/sec after 100 hr of operation. A photograph of the Kaman high-output tube is shown in Fig. 3.35. The tube employs a high-voltage power supply which uses pressurized sulfur hexafluoride for insulation. The sealed tube is encased in a stainless steel dome which is also insulated with sulfur hexafluoride under pressure.

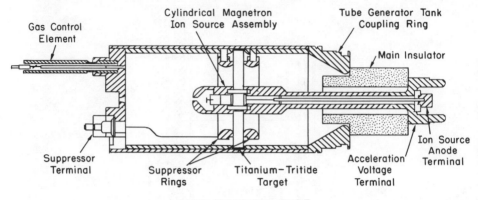

ACCELERATION TUBE

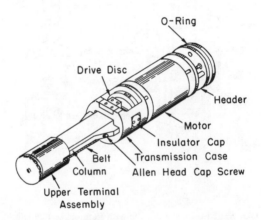

ELECTROSTATIC HIGH VOLTAGE MACHINE

Fig. 3.34. Ion accelerating tube and high-voltage supply (Oshry design).

An example of the use of an RF ion source for sealed tubes is given by Wood et al.[51] Typical operating parameters of the L-tube (Fig. 3.36) given by them are quoted below:

Gas pressure	1.5×10^{-2} torr
ion-source power	500 W at 20 MHz
accelerating voltage	110 kV
suppression voltage	375 V
total tube current	1.5 mA
neutron output	10^{10} neutrons/sec
tube life	100 hr

III. SEALED-TUBE NEUTRON SOURCE

Fig. 3.35. High-flux sealed tube (Kaman design).

The L-tube was designed at the Services Electronic Research Laboratories, Baldock, England (S.E.R.L.), and is about 56 cm long and 5.6 cm in diameter. Under normal operating conditions the target is at high voltage and the ion source at ground potential. A titanium replenisher is used to control the pressure inside the tube. In the equipment shown in Fig. 3.36, the target is held at 400 V positive with respect to the target shield to suppress

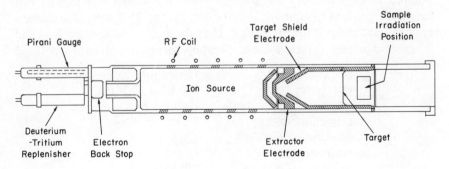

Fig. 3.36. L-tube (S.E.R.L. design).

secondary electrons. Success with the L-tube led to the development of the high-output S.E.R.L. P-tube. Bounden et al.[52] and Downton and Wood[53] have discussed the performance of this tube. The P-tube specifications include a rated neutron output of 10^{11} neutrons/sec for a life of 100 hr. The size of the tube is ideally suited for laboratory operation. In the design of this tube, shown schematically in Fig. 3.37, an RF ion source is used;

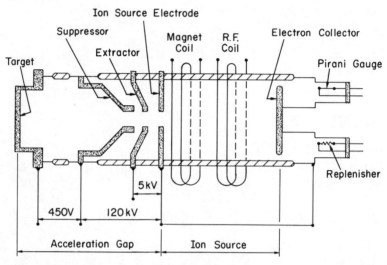

Fig. 3.37. P-tube schematic (S.E.R.L. design).

however, its extraction system is an adaptation of the Thoneman and Harrison design,[54] and operates at −5 kV with respect to the ion source. In contradistinction to the L-tube, the P-tube utilizes an axial magnetic field to increase the total ion current (note the similarity with the solenoid used in pumped models utilizing RF sources where the solenoid is located just before the gap lens to improve ionization inside the RF bottle). The accelerating voltage in the P-tube is 120 kV. The tube itself is about 40 cm long and 5.5 cm in diameter. Typical P-tube operating parameters given are as listed

gas pressure	15 millitorr
ion-source power	500 W at 14 MHz
magnetic field strength	100 G
extraction voltage	5 kV
accelerating voltage	120 kV
beam current	12 mA
suppression voltage	450 kV

III. SEALED-TUBE NEUTRON SOURCE

neutron output 10^{11} neutrons/sec
useful target diameter   28 mm

The Schlumberger tube is specially designed to withstand temperatures up to 150°C for well-logging operations and is described by Frentrop and Sherman.[55] This tube is only about 11 in. long and 1.2 in. in diameter, and utilizes a Penning ion source. The essential components of the tube are shown schematically in Fig. 3.38. The quoted neutron yield is of the order of 10^8 neutrons/sec, and the average life of the tube is given as over 200 hr.

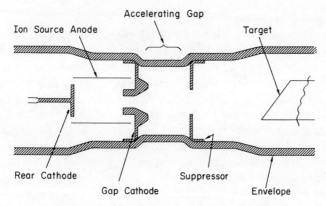

Fig. 3.38. Internal view of the Schlumberger tube.

The designs of ion sources and of accelerating structures for the tubes described thus far are by no means the only designs available today. It is known that inexpensive and reliable tubes rated at about 10^7- to 10^{10}-neutrons/sec output are manufactured in the U.S.S.R., Poland, Hungary, and other countries where they are used routinely in well-logging operations. Technical specifications of these tubes, however, are not available.

2. REPLENISHER SYSTEM

It is obvious that once a tube is filled with deuterium, or a mixture of deuterium and tritium, at some operating pressure, the gas discharge would soon remove the gas and reduce the pressure below acceptable limits. To maintain a suitable gas pressure throughout the life of the tube, the early Philips tubes included a hydrogen replenisher patented by Nienhuis.[56] Essentially, this device consists of a zirconium spiral around a tungsten wire situated behind the ion source. Depending on the temperature, zirconium will absorb or release hydrogen. A voltage is applied across the spiral to adjust the pressure level in the tube to the desired value. Since zirconium absorbs hydrogen-like gases rather slowly, it was soon replaced by the titanium-soot replenisher system. In this device the titanium is vaporized in an inert gas atmosphere and then condensed in the form of a sootlike deposit of particle diameter from one to several hundred angströms. The deposition surface is the inside wall of a nickel cylinder which can be heated. Release of the impregnated hydrogen isotope from the titanium soot is achieved by varying the temperature of the nickel cylinder.

3. TARGETS FOR SEALED-TUBE NEUTRON SOURCE

In the early stages of sealed-tube development, the normal gas-in-metal disk targets deposited on gold or silver backings were used. The disadvantages of these targets with no provision for replenishing have already been discussed in detail under pumped neutron generators. Research into the feasibility of using self-loading targets ("drive-in" targets) showed that, for a tube filled with a mixture of equal amounts of deuterium and tritium, self-loading of the original target could be obtained. Upon target saturation, the stability and constancy of the neutron output could be maintained over long periods, provided a replenisher system was included to allow for the gas cleanup from the discharge.

Experiments[48] with a gold target, a replenisher, and a tube filling of 50% deuterium and 50% tritium showed that after 15 hr of operation, the target reached the saturation point. With this arrangement, and using an accelerating voltage of 125 kV, a neutron output of 8×10^7 neutrons/sec per 100 μA of beam current was obtained. A considerable improvement in the neutron yield was gained by replacing the all-gold target by a titanium-plated drive-in target. Under identical conditions, a neutron output of 2.4×10^8 neutrons/sec per 100 μA was obtained for the titanium target, a value three times greater than that for the all-gold target.

4. PERFORMANCE OF SEALED TUBES

The performance of high-output sealed neutron tubes was discussed recently.[57] Life test data obtained by Reifenschweiler[57] are shown in Fig. 3.39. Neutron outputs between 2 and 3×10^{10} neutrons/sec were maintained for over 1000 hr. In the operation of this tube the average pressure was about 10^{-4} torr. In this report,[57] a brief mention was also made of a 5-mA tube operating at 200-kV accelerating voltage and a maximum neutron output of 3.5×10^{11} neutrons/sec. To date, no tube-life data for this tube have been made available. On the basis of such recent advances in tube design, the possibility in the near future of a tube producing

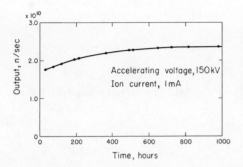

Fig. 3.39. Life test data of a high-yield neutron tube (Philips design).

10^{12} neutrons/sec and having a life in excess of several hundred hours does not seem to be an unreasonable extrapolation. Long-term performance of the Kaman tube has been reported by Jessen,[58] and life tests conducted in the laboratory, together with customer case histories, are summarized in his paper. The performance of the Kaman tube is illustrated in Fig. 3.40. Typical performance data (Fig. 3.41) for the S.E.R.L. tubes have also been reported by Downton and Wood.[53]

Although design details have not been made available, some sealed-tube manufacturers offer tubes capable of pulsed operation. The reader is referred to an excellent report by Elenga and Reifenschweiler[59] for details on this topic. Tubes with neutron outputs between 1 and 5×10^8 neutrons/sec and capable of providing 10^9-neutron/sec pulses have been developed. Pulse duration times vary from 5 μsec to 10 sec. With a special tube having a

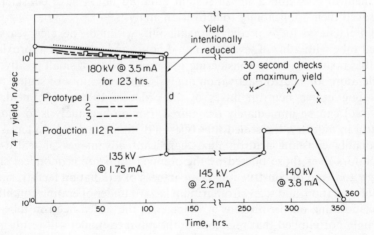

Fig. 3.40. Neutron yield versus time for three high-output tubes (Kaman design).

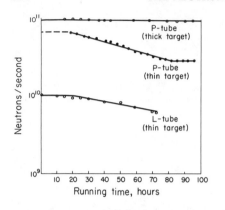

Fig. 3.41. Tube-life performance (S.E.R.L. design).

neutron output between 1 and 2.5×10^8 neutrons/sec, pulsed outputs as high as 5×10^{10} neutrons/sec were obtained which had pulse duration times between 3 and 500 μsec.

IV. EVALUATION AND IMPACT OF LOW-VOLTAGE ACCELERATOR NEUTRON SOURCES

Until recently, the high neutron outputs from pumped neutron generators far outweighed the long-life advantage offered by the sealed-tube accelerator. The analytical sensitivity obtained from the latter was too low for many applications. Today, however, sealed tubes with guaranteed outputs of 10^{11} neutrons/sec and a life of several hundred hours have considerably improved their applicability to activation analysis.

Manufacturers today predict the availability within the next few years of a sealed tube with a life of several hundred hours and an output approximating 10^{12} neutrons/sec. In considering the applicability of small positive-ion accelerators for neutron activation analysis, the versatility of sealed tubes providing usable neutron fluxes of the order of 10^{10} to 10^{11} neutrons/(cm^2 sec) can be immediately recognized. For on-line analysis for quality control in industry, the sealed tube offers a distinct advantage in terms of a reasonably constant neutron flux. Significant advantages of sealed-tube generators over those offered by the pumped units are economical shield design and ease of construction, compactness of irradiation facility, simplicity of operation, and safety from tritium hazard under operating conditions. By elaborating on the obvious advantages of the sealed neutron tube, it is certainly not implied that phaseout of the pumped model will result in the near future. If anything, the tremendous analytical applicability of small low-voltage accelerators should spur the manufacturers of pumped accelerators to introduce innovations in target design, solve cooling problems associated with very high flux machines (3-mA beam current), and alleviate the biological hazard from tritium handling. Progress in the last instance has already been made. If the application of pumped models is to increase, the needs of industry must be taken into account during the design and incorporation of improvements in accelerator systems. As viewed by the user, the neutron generator is but one component in a sophisticated analytical system. Therefore the goal in generator design must be to produce a "blackbox" neutron source that will perform with a minimum of servicing. A sealed-tube generator capable of high neutron fluxes appears to fulfill this goal very closely at the present time.

REFERENCES

1. J. D. Cockcroft and E. T. S. Walton, *Proc. Roy. Soc. (London)*, **A136**, 619 (1932).
2. R. W. Wood, *Proc. Roy. Soc. (London)*, **A97**, 455 (1920).
3. J. J. Thompson and G. P. Thompson, *Conduction of Electricity through Gases*, Vol. II, Cambridge University Press, London, 1928.
4. W. V. Smith, *J. Chem. Phys.*, **11**, 110 (1943).
5. C. D. Moak, H. Reese, Jr., and W. M. Good, *Nucleonics*, **9** (3), 18 (1951).
6. P. Thoneman, J. Moffat, O. Roaf, and J. Sanders, *Proc. Phys. Soc. (London)*, **61**, 483 (1948).
7. E. M. Penning, *Physica*, **4**, 71 (1937).
8. P. L. Jessen, "Design Considerations for Low Voltage Accelerators," Kaman Nuclear Report KN-68-459(R), Kaman Nuclear Corporation, Colorado Springs, Colorado, 1968.
9. C. F. Barnett, P. M. Stier, and G. E. Evans, *Rev. Sci. Instrum.* **24**, 394 (1953).
10. G. Wehner, *Phys. Rev.*, **93**, 663 (1954).
11. J. T. Prud'homme, "Texas Nuclear Corporation Neutron Generators," Texas Nuclear Corporation, Austin, Texas, 1962.
12. L. D. Hall, "A Little Bit About Almost Nothing," 2nd ed., Ultek, Palo Alto, California, February 1963.
13. F. M. Penning, "Coating by Cathode Disintegration," U.S. Patent 2146025, February 7, 1939.
14. R. Steinberg and D. L. Alger, "A Solution to Tritium Pump Contamination for Small Accelerators," NASA Technical Memorandum TMX-52578, Lewis Research Center, Cleveland, Ohio, April 1969.
15. J. H. Leck, *Pressure Measurement in Vacuum Systems*, Reinhold, New York, 1957.
16. "Accelerator Targets for the Production of Neutrons," Euratom Report EUR 1815, e, revised English edition, 1964.
17. "Accelerator Targets for the Production of Neutrons," Euratom Report EUR 2641, d,f,e, 1966.

18. W. Whaling, *Handbuch der Physik*, Vol. XXXIV, Springer-Verlag, Berlin, 1958, p. 193.
19. J. H. Coon, "Targets for the Production of Neutrons," in *Fast Neutron Physics*, Part I, Vol. IV (Interscience Monographs and Texts in Physics and Astronomy), J. B. Marion and J. L. Fowler (Eds.), Interscience, New York, 1960, pp. 677–720.
20. A. C. van Dorsten, *Philips Tech. Rev.*, **17**, 109 (1955).
21. E. R. Graves, A. A. Rodrigues, M. Goldblat, and D. I. Meyer, *Rev. Sci. Instrum.*, **20**, 579 (1949).
22. J. H. Manley, L. J. Hayworth, and E. A. Luebke, *Rev. Sci. Instrum.*, **12**, 587 (1941).
23. F. M. Penning and J. A. H. Moubis, *Physica*, **4**, 1190 (1937).
24. A. D. McQuillan and M. K. McQuillan, *Titanium*, Academic Press, New York, 1956.
25. G. L. Miller, *Zirconium*, Academic Press, New York, 1954.
26. A. B. Lillie and J. P. Conner, *Rev. Sci. Instrum.*, **22**, 210 (1951).
27. J. P. Conner, T. W. Bonner, and J. R. Smith, *Phys. Rev.*, **88**, 468 (1952).
28. R. S. Rochlin, *Rev. Sci. Instrum.*, **23**, 100 (1952).
29. W. J. Arrol, E. J. Wilson, and C. Evans, AERE (G.B.), Report I/R 1135, 1953.
30. B. J. Massey, Oak Ridge National Laboratory, Report ORNL-2237, 1957.
31. K. Perry, G. Aude, and J. Laverlochere, Preprints, "Symposium on Trace Characterization—Chemical and Physical," paper No. 54, National Bureau of Standards, Gaithersburg, Maryland, October 3–7, 1966.
32. S. S. Nargolwalla, J. Niewodniczanski, and J. E. Suddueth, *Trans. Am. Nucl. Soc.*, **12** (2), 510 (1969).
33. R. G. Wille and R. W. Fink, *Phys. Rev.*, **118**, 242 (1960).
34. S. S. Nargolwalla and J. E. Suddueth, "Activation Analysis with the Cockcroft-Walton Neutron Generator," in *Activation Analysis Section: Summary of Activities, July 1969 to June 1970*, P. D. LaFleur and D. A. Becker (Eds.), NBS Technical Note 548, issued December 1970.
35. J. E. Strain, "Use of Neutron Generators in Activation Analysis," *Prog. Nucl. Energy, Series IX*, **4**, 137–157 (1965).
36. G. Breynat, M. Dubus, R. Gerbier, J. F. Maurin, Bull. Inform. Sci. Tech. (Paris), **98**, 81–92 (Nov. 1965).
37. H. Hollister, *Nucleonics*, **22** (6), 68 (1964).
38. W. W. Meinke and R. W. Shideler, Michigan Memorial Phoenix Project, Report MMPP-191-1, 1961.
39. R. Booth, University of California Radiation Laboratory, Report UCRL 70183, 1967.
40. E. A. Burrill, "Advances in DC Methods of Particle Acceleration," *IEEE Trans. Nucl. Sci.*, **NS-10**, No. 3, 69 (July 1963).
41. N. J. Felici, "Cylindrical Electrostatic Generators," in *Radiation Sources*, A. Charlesby (Ed.), Pergamon Press, London, 1966.
42. J. E. Suddueth and S. S. Nargolwalla, "Activation Analysis with the Cockcroft-Walton Neutron Generator," in *Activation Analysis Section: Summary of Activities July 1968 to June 1969*, P. D. LaFleur (Ed.), NBS Technical Note 508, issued July 1970.
43. O. Reifenschweiler and A. C. van Dorsten, *Phys. Verh. Mosbach*, **8**, 163, (1957).

IV. EVALUATION OF LOW-VOLTAGE ACCELERATOR NEUTRON SOURCES

44. O. Reifenschweiler, *Nucleonics*, **18**, 69 (1960).
45. R. L. Caldwell and R. F. Sippel, *Bull. Am. Assoc. Pet. Geol.*, **42**, 159 (1958).
46. R. L. Caldwell, *Nucleonics*, **16**, 58 (1958).
47. J. Tittman and W. B. Helligan, "Laboratory Studies of a Pulsed Neutron Source Technique in Well Logging," Society of Petroleum Engineers, AIME, Caspar, Wyoming, Paper No. 1227-G, April 1959.
48. O. Reifenschweiler, "Sealed-Off Neutron Tube: The Underlying Research Work," N. V. Philips, Eindhoven, Netherlands; *Philips Res. Rep.*, **16**, 401 (1961).
49. H. I. Oshry, Proceedings, International Conference on Modern Trends in Activation Analysis, College Station, Texas, December 28–31, 1961, p. 28.
50. P. L. Jessen, "Long-Term Operating Experience with High-Yield Sealed Tube Neutron Generators," Proceedings, 1968 International Conference on Modern Trends in Activation Analysis, National Bureau of Standards Special Publication 312, Vol. II, 1969, p. 895.
51. J. D. L. H. Wood, P. D. Lomer, and J. E. Bounden, "High Output Neutron Tube for Activation Analysis," in *L'Analyse par Radioactivation et ses Applications aux Sciences Biologiques*, Presses Universitaires de France, 1964, pp. 119–122, Paris.
52. J. E. Bounden, P. D. Lomer, and J. D. L. H. Wood "High-Output Neutron Tube," in Proceedings, 1965 International Conference on Modern Trends in Activation Analysis, College Station, Texas, April 19–22, 1965, pp. 182–185.
53. D. W. Downton and J. D. L. H. Wood, "A 10^{11} Neutrons per Second Tube for Activation Analysis," in Proceedings, 1968 International Conference on Modern Trends in Activation Analysis, National Bureau of Standards Special Publication 312, Vol. II, 1969, p. 900.
54. P. C. Thoneman and E. R. Harrison, AERE (G.B.), Report GP/R 1190, 1958.
55. A. H. Frentrop and H. Sherman, "Schlumberger Tube: For Oil-Well Logging," in "Using Accelerator Neutrons," E. A. Burrill and M. H. MacGregor, *Nucleonics*, **18** (12), 72 (1960).
56. K. Nienhuis, U.S. Patent 2,766,397 (1956).
57. O. Reifenschweiler, "A High Output Sealed-Off Neutron Tube with High Reliability and Long Life," in Proceedings, 1968 International Conference on Modern Trends in Activation Analysis, National Bureau of Standards Special Publication 312, Vol. II, 1969, p. 905.
58. P. L. Jessen "Long Term Operating Experience with High Yield Sealed Tube Neutron Generators" in Proceedings, 1968 International Conference on Modern Trends in Activation Analysis, National Bureau of Standards Special Publication 312, Vol. II, 1969, p. 895.
59. C. W. Elenga and O. Reifenschweiler, "The Generation of Neutron Pulses and Modulated Neutron Fluxes with Sealed-Off Neutron Tubes," in Proceedings of an IAEA Symposium on Pulsed Neutron Research, Karlsruhe, Vol. 2, 1965, pp. 609–622.

CHAPTER

4

RADIATION HAZARDS AND SHIELDING CONSIDERATIONS FOR NEUTRON GENERATOR FACILITIES

I. INTRODUCTION

Radiation hazards involved with the operation of a neutron generator arise from the equipment itself and from induced activity in the surroundings. The operation hazards from particle accelerators differ, depending on individual design; however, the radiation hazard is common to all. This hazard is reduced to levels acceptable for operating personnel by shielding and ensuring a sufficient working distance from the neutron source. The selection of shielding materials is based on their compositional characteristics. These materials must possess the ability to attenuate radiation with negligible induced activation. Many neutron generator facilities employ an arrangement of "shadow shields" for personnel on the same level as the equipment. In such cases, the problem of "skyshine" or reflection can be significant and must be included as a necessary parameter in shielding considerations.

Next in importance to the cost of the neutron generator itself is the cost of the biological shield. Proper design and construction of the shield must be very carefully weighed before installation of the neutron source. The judicious choice of materials, together with the recognition of appropriate geometric considerations, is essential from the economic and safety viewpoints. The ultimate design is controlled by the maximum neutron output of the generator, available space, anticipated duty cycle of the neutron generator, and cost.

Since the biological tolerance to high-energy neutrons is very low, it is advisable to compute shielding requirements on the conservative side. A useful rule of thumb is to design a shield to reduce the neutron and gamma dose by a factor of 10 less than the maximum permissible limit. In many facilities this extra shielding provision has accommodated future installation of higher flux machines without further addition to the existing shield. The awareness of maximum permissible limits of absorbed radiation for acceptable shield design is most important. In this chapter therefore, concepts of dose-flux equivalence for the radiations of interest are introduced.

Perhaps the biological hazard from tritium-in-metal targets has been overemphasized; however, contamination from this isotope of hydrogen is extremely serious owing to its low tolerance level. Therefore, the entire problem of tritium target handling, installation, and storage will be discussed, and the chemical and physical behavior of tritium is outlined in some detail. Tritium-monitoring techniques, as well as recent efforts to reduce or eliminate this hazard from pumped-type neutron generators, are also described.

In addition to the hazards associated with induced activation, bremsstrahlung production in the vicinity of the high-voltage terminal, nonnuclear hazards from fire, inadequate ventilation, and irradiation experiments employing so-called blank targets are brought to the attention of the reader. Besides the personnel hazards, the effect of secondary or degraded radiation on shielded detector systems located short distances from the primary neutron source is discussed.

The reader is introduced to some basic concepts and computational methods used for shield-design calculations. Shielding of sealed neutron tubes is also included in discussions on the design and construction of biological shields. The attenuation of high-energy neutrons and gamma photons as relating to useful computational methods and existing experimental data is outlined. Suggestions for shield-design calculations and basic assumptions generally made are also offered. Pictorial representations of some typical designs are given together with radiation monitoring data. Adaption of these structures may be used for construction of new structures.

In the constructional aspects of biological shields, duty-cycle considerations, tritium target decay, and shield effectiveness are elaborated. Allowances for the inclusion of access ports for peripheral systems such as pneumatic tubes, remote beam-viewing apparatus, ventilation, and humidity control are also treated. The neutron generator itself accounts for much of the downtime experienced in such facilities, and it is hoped that the concepts and methods outlined here will be of sufficient use to eliminate difficulties arising from improper or inadequate shield design and construction.

II. RADIATION EXPOSURE

The purpose of radiation shields is to protect objects, animate or inanimate, from radiation damage where the radiation source during operation cannot be reduced in intensity or where the object cannot be conveniently removed to a distant and safe location. Limitations governing man's exposure to radiation depend not only on the effect on the individual (somatic effect) but also on the population as a whole (genetic effect). Man's response to radiation exposure is controlled by the following principal factors:

1. Specific area of the person being irradiated.
2. Quantity of radiation exposure.
3. Type of radiation exposure.

134 RADIATION HAZARDS AND SHIELDING CONSIDERATIONS

Shielding requirements necessary to maintain radiation levels within acceptable limits involve a detailed study of radiation interactions with matter. The fundamental treatment of fast neutron interactions dealing with dose calculations and biological potency is at the present time quite approximate. Therefore the discussion of fast neutron attenuation will follow a generalized format substantiated with available shielding data taken for existing shield structures. The problem of fast neutron attenuation can perhaps best be appreciated when it is realized that a 2.5×10^{11}-neutron/sec output of 14-MeV neutrons, if unshielded, will deliver a maximum permissible dose at a distance of about 450 m from the source. Furthermore, the attenuating ability of shield materials falls off drastically with an increase in neutron energy. For example, 30 cm of water will attenuate 1-MeV neutrons by a factor of 10^5, while the same thickness of water will reduce the number of 14-MeV neutrons by only a factor of 10.

For the protection of radiation users, the International Commission on Radiological Protection (ICRP) has prescribed recommendations for acceptable radiation levels, and the treatment of absorbed dose-flux equivalence to follow is based on these recommendations.

A. RADIATION DOSE-FLUX EQUIVALENCE

Since maximum permissible doses are expressed in units of dose equivalent (rem: roentgen-equivalent-man) and the design of shields is generally based on quantities such as particle flux density [number/(cm^2 sec)] or energy flux density [MeV/(cm^2 sec)], it is necessary to use conversion formulas which relate these quantities.

The relative biological effectiveness* (RBE), or quality factor, provides a normalizing quantity which accounts for differences in the biological potency of different types of radiation. Basically the RBE relates the absorbed dose (rad: roentgen-absorbed-dose) to the dose equivalent (rem). The ICRP recommendations for this quality factor for various types and energies of particles are summarized in Table 4.1. The RBE values for neutrons have been taken from Handbook 63 issued by the National Bureau of Standards.[1] Values for other particles are as suggested by the ICRP, and are tabulated by Goussev.[2]

The maximum permissible dose (MPD) from all types of radiation as recommended[3] is 100 mrem/40-hr week, or 2.5 mrem/hr. Relationships expressing the particle flux density and energy flux density necessary to deliver one MPD for various particles and particle energies are found in a recent review by Goussev,[2] from which a brief summary of useful formulations, tables, and graphs is presented here.

---

* The preferred term for radiation protection is "quality factor." The term RBE is used for controlled experiments.

II. RADIATION EXPOSURE

TABLE 4.1. Relative Biological Effectiveness of Nuclear Radiation (rem = RBE × rad)

Type of Radiation	Relative Biological Effectiveness (RBE)
Gamma rays	1
X-rays	1
Electrons	1
Positrons	1
Beta rays	1
Alpha particles	10
Protons	10
Heavy recoil atoms	20
Neutrons, thermal	3.0
0.0001 MeV	2.0
0.005	2.5
0.02	5.0
0.1	8.0
0.5	10.0
1.0	10.5
2.5	8
5.0	7
10.0	6.5
10 to 30	6

1. X- AND GAMMA RAYS

Under conditions of electron equilibrium, the absorbed dose in air can be related to the particle flux density or energy flux density by the expression

$$P = \frac{\phi \cdot E \cdot \bar{\mu}_a \cdot (1.6 \times 10^{-6})}{100} \text{ rad/sec}$$

or (4.1)

$$P = I \cdot \bar{\mu}_a \cdot (1.6 \times 10^{-8}) \text{ rad/sec}$$

where

P = absorbed dose rate in air, rads/sec
ϕ = particle flux density, particles/(cm² sec)
I = energy flux density, MeV/(cm² sec)
E = energy of monoenergetic photons, MeV/photon
$\bar{\mu}_a$ = mass energy absorption coefficient for gamma energy, E, in air, cm²/g.

Note: 1 MeV = 1.6×10^{-6} erg
1 rad = 100 ergs/g

136   RADIATION HAZARDS AND SHIELDING CONSIDERATIONS

Equation 4.1 can be used only for photon energies less than 3 MeV. Above 3 MeV the assumption of electron equilibrium does not hold. In Table 4.2 the particle flux density and the energy flux density equivalents of the maximum permissible dose of 2.5 mrad/hr are given. A more comprehensive table is given by Goussev.[2]

TABLE 4.2. Energy and Particle Flux Densities-MPD Equivalence for Various Gamma Energies

Gamma Energy (MeV)	$\bar{\mu}_a$ (cm²/g)	Flux Equivalence of 2.5 mrad/hour	
		Energy Flux [MeV/(cm² sec)]	Gamma Flux [photons/(cm² sec)]
0.01	4.54	9.56	956
0.015	1.25	34.7	2310
0.05	0.0376	1150	23100
0.1	0.0233	1860	18600
0.5	0.0297	1460	2920
1.0	0.0280	1550	1550
1.50	0.0256	1700	1130
2.0	0.0238	1820	912
3.0	0.0211	2060	686

2. MONOENERGETIC ELECTRONS AND BETA PARTICLES

For purposes of dose rate-flux calculations, the composition (wt. %) of biological tissue is assumed to be $H = 10\%$, $C = 12.3\%$, $N = 3.5\%$, and $O = 72.9\%$. The dose rate at the surface of biological tissue can be related to the flux density of monoenergetic electrons by the expression

$$P = \frac{\phi(dE/dx) \cdot (1.6 \times 10^{-6})}{100} \text{ rad/sec} \tag{4.2}$$

where

P = dose rate at the surface of biological tissue, rad/sec

ϕ = particle flux density, electrons/(cm² sec)

$\dfrac{dE}{dx}$ = ionization loss of electron energy in biological tissue, MeV/(cm² g).

Numerical values of dE/dx for the various elements are available in reference 4.

II. RADIATION EXPOSURE

In the case of a continuous beta spectrum where the energy varies continuously from 0 to E_{max}, the dose in rad per beta particle can be calculated from

$$\frac{\int_0^{E_{max}} (dE/dx)N(E) \cdot dE}{\int_0^{E_{max}} N(E) \, dE} = \frac{d\bar{E}}{dx} \text{ MeV/(cm}^2 \text{ particle)} \quad (4.3)$$

where $N(E) \, dE$ is the number of electrons of energy between E and $E + dE$, and dE/dx is the average dose produced by a single electron averaged over the entire energy spectrum from ionization loss. By investigating the actual beta spectrum of simple beta emitters in the energy range from 0.2 to 3 MeV, one can determine the dose D in Eq. 4.3. Values for dE/dx are given in reference 4, and the dose per beta particle can be calculated from a knowledge of $N(E) \, dE$ from scintillation spectrometric measurements using an organic scintillator. The dose per beta particle for different values of E_{max} is as follows:

E_{max}(MeV)	0.2	0.4	0.6	0.8	1.0	1.5	2.0	3.0
Dose [(rads-cm^2/β-particle) $\times 10^8$]	11.3	7.4	5.6	4.2	3.7	3.2	3.1	2.9

and the absorbed dose rate can be related to the flux density of beta particles having a continuous spectrum by the equation

$$P_\beta = \phi_\beta \times D \times 40 \text{ (hr/wk)} \times 3600 \text{ (sec/hr)} = 0.1 \text{ rad/wk or 2.5 mrad/hr} \quad (4.4)$$

The calculated values for dose rate-particle flux equivalence for monoenergetic electrons and beta particles delivering one MPD are tabulated in Table 4.3. It can be noted from a more extensive table[2] that in the energy range 0.3 to 10 MeV, the particle flux is almost independent of energy in both cases, and the approximate expression,

$$\phi \text{ electrons or } \beta \cong 24 \text{ particles/(cm}^2 \text{ sec)} = 2.5 \text{ mrad/hr}$$

can be used with an accuracy of ± 10 to 15%. The method of calculating the dose rate-flux equivalence is subject to errors because the contribution from < 10-keV electrons to the absorbed dose has been neglected, the dependence of dE/dx on the atomic number has not been taken into account, and shape factors of beta-ray spectra have not been considered.

Since the biological effects of these radiations are known with considerably less certainty than the magnitude of these errors, no attempt has been made to treat these additional factors in a more quantitative manner.

TABLE 4.3. Particle Flux Density-MPD Equivalence for Monoenergetic Electrons and Beta Rays

E_{max}	Monoenergetic Electrons [electrons/(cm² sec)][a]	Beta Rays [electrons/(cm² sec)][a]
0.2	16	6
0.4	21	9
0.6	22	12
0.8	23	16
1.0	24	18
1.5	24	21
2.0	24	22
3.0	23	23

[a] Particle flux equivalence to deliver a dose rate of 2.5 mrad/hr.

3. PROTONS AND CHARGED PARTICLES

If ionization is assumed to be the principal mode for energy transfer to biological tissue, the relationship between the absorbed dose rate and the particle flux density of monoenergetic charged particles can be expressed as

$$P = \frac{\phi \cdot (dE/dx) \cdot 1.6 \times 10^{-6}}{100} \text{ rad/sec} \quad (4.5)$$

where

$\dfrac{dE}{dx} = $ particle energy loss through ionization in biological tissue, MeV-cm²/g

$P = $ absorbed dose rate, rads/sec

$\phi = $ particle flux density, particles/(cm² sec).

Substituting p_i, the ratio of the elements in biological tissue, Eq. 4.5 can be rewritten as

$$P = 1.6 \times 10^{-8} \cdot \phi \sum_{i=1}^{n} \left(\frac{dE}{dx}\right)_i \cdot \left(\frac{N}{A_i}\right) \cdot p_i \cdot 10^{-2} \text{ rads/sec}$$

$$\cong 10^{14} \cdot \phi^n \cdot \sum_{i=1}^{n} \left(\frac{dE}{dx}\right)_i \cdot \left(\frac{p_i}{A_i}\right) \text{ rad/sec} \quad (4.6)$$

where p_i is the percentage ratio of the elements in biological tissue, A_i is the atomic weight of the element in question, and N is Avogadro's number.

Assuming the elemental content of biological tissue to be that given in Section II.A.2 and using available data for the proton energy loss,[5] we can use Eqs. 4.5 and 4.6 to determine the relationship between absorbed dose and particle density by rewriting Eq. 4.5,

$$P = 1.6 \times 10^{-8} \cdot \phi\left(\frac{dE}{dx}\right)(\text{RBE}) \cdot t \cdot 3600 \text{ rem/wk}$$

from which the relationship

$$\phi = \frac{P}{1.6 \times 10^{-8}(dE/dx) \cdot \text{RBE} \cdot t \cdot 3600} \text{ particles/(cm}^2 \text{ sec)} \quad (4.7)$$

can be obtained. The RBE values are taken from Table 4.1 and t is the exposure time in hours per week. The particle flux equivalent to a dose rate of 2.5 mrad/hr for various particle energies is given in Table 4.4. Note that the absorbed dose rate is given in millirads per hour and can be converted into millirems by using the appropriate RBE value.

TABLE 4.4. Particle Flux Density-MPD Equivalence for Monoenergetic Protons

Proton Energy (MeV)	Particle Flux Density, [protons/(cm² sec) equivalent to a dose rate of 2.5 mrad/hr]
0.01	0.079
0.05	0.040
0.1	0.040
0.5	0.087
1.0	0.142
5	0.552
8	0.790
10	1.00

4. NEUTRONS

Thermal neutrons incident on biological tissue cause ionization in two different ways: first, by secondary electrons from gamma radiation emitted from the nuclear reaction $^1H(n, \gamma)^2H$; and second, by recoil protons from the reaction $^{14}N(n, p)^{14}C$. The overall dose calculation for thermal neutrons therefore includes the contribution from both these processes. For the first reaction the number of gamma quanta of energy $E = 2.2$ MeV at a depth of x cm in the tissue per incident neutron is given by $N \times A_H \times \sigma_H \times e^{-x/L}$ where N is the number of neutrons, A_H is the number of hydrogen atoms per cubic centimeter of tissue, σ_H is the capture cross section for the reaction (cm^2), and L is the diffusion length of neutrons in centimeters. Neglecting the albedo (reflection), it can be shown[2] that the dose produced by the gamma rays can be expressed as

$$P_H = \frac{\phi \cdot A_H \cdot \sigma_H \cdot E \cdot \bar{\mu}_a \cdot 1.6 \times 10^{-8} \cdot \ln[(1 + \mu L)/\mu L]}{2 \times 100} \text{ rad/sec} \quad (4.8)$$

where μ and $\bar{\mu}_a$ are the attenuation and mass energy absorption coefficients for the gamma rays in biological tissue, respectively. Substituting for $A_H = 6 \times 10^{22}$ atoms/cm^3 (H = 10%), $\sigma_H = 0.33 \times 10^{-24}$ cm^2/atom, $\mu = 0.045$ cm^{-1}, $\bar{\mu}_a = 0.025$ cm^2/g, $E = 2.2$ MeV, and $L = 2.8$ cm allows Eq. 4.8 to be reduced to

$$P_H = 5.2 \times 10^{-10} \phi \text{ rad/sec}$$
$$= 5.2 \times 10^{-10} \phi \text{ rem/sec}$$

Similar calculations for the ^{31}P(n, γ)^{32}P reaction yield dose rates that are about 9% of the above. Thus the total dose rate from the (n, γ) reactions is given by

$$P_{H+P} = 5.7 \times 10^{-10} \phi \text{ rem/sec} \quad (4.9)$$

The contribution to the total dose rate from the ^{14}N(n, p)^{14}C reaction in which protons of energy $E_p = 0.62$ MeV are emitted is given by

$$P_N = \frac{\phi \cdot \text{RBE} \cdot A_N \cdot \sigma_N \cdot E_p \cdot 1.6 \times 10^{-6}}{100} \text{ rad/sec} \quad (4.10)$$

Substituting RBE = 10, $A_N = 1.5 \times 10^{21}$ atoms/cm^3 (N = 3.5%), $\sigma_N = 1.7 \times 10^{-24}$ cm^2/atom, and $E_p = 0.62$ MeV, simplifies Eq. 4.10 to

$$P_N = 2.5 \times 10^{-10} \phi \text{ rem/sec} \quad (4.11)$$

The total dose rate from thermal neutrons is therefore

$$P_{\text{total}} = 8.2 \times 10^{-10} \phi \text{ rem/sec} \quad (4.12)$$

II. RADIATION EXPOSURE

The flux density of thermal neutrons to deliver a dose rate of one MPD is given by

$$P_{\text{total}} = 8.2 \times 10^{-10} \cdot 40 \cdot 3600 \cdot \phi$$
$$= 0.1 \text{ rem/40-hr week}$$

from which

$$\phi = 840 \text{ neutrons/(cm}^2 \text{ sec)} = 2.5 \text{ mrem/hr} \quad (4.13)$$

From the dose-flux equivalence values given in Table 4.5, it may be noted that a lower value of 670 neutrons/(cm² sec) has been recommended. This value is a result of more precise calculations which include the albedo and consider the fact that the maximum ionization occurs at some depth ($L = 0.38$ cm) below the surface of the biological tissue.

TABLE 4.5. Particle Flux Density-MPD Equivalence for Thermal, Epithermal, and Fast Neutrons

Neutron Energy (MeV)	Particle Flux [neutrons/(cm² sec) equivalent to 2.5 mrem/hr]
Thermal	670
0.0001	500
0.005	570
0.02	280
0.1	80
0.5	30
1.0	18
2.5	20
5.0	18
7.5	17
10.0	17
10 to 30	10

As the energy of the neutrons is increased, the calculations become more approximate. In general, with increase in neutron energy, elastic collisions between the neutrons and the light elements play an important role in the determination of the dose rate. In the energy region from 0.5 to 10 MeV, more than 90% of the total dose is produced through the ionization process caused by the light recoiling elements.

An overall picture of the dose-flux equivalence examples given thus far is represented in Fig. 4.1 according to Goussev.[2] As more data on the interaction processes of radiation with matter become available, these given relationships will no doubt be modified to yield greater accuracy.

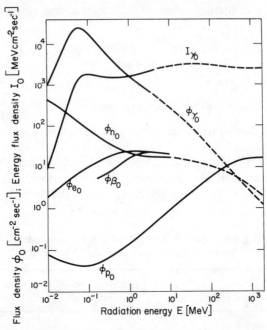

Fig. 4.1. Energy flux- and particle flux-MPD equivalence for nuclear radiation as a function of energy: ϕ_{γ_0}, I_{γ_0} gamma radiation; ϕ_{n_0} neutrons; ϕ_{β_0} beta particles; ϕ_{e_0} monoenergetic electrons; ϕ_{p_0} protons.

III. HAZARDS IN NEUTRON GENERATOR OPERATION

Similar to most apparatus of this nature, a neutron generator presents a source of potential hazard when used carelessly or operated by the uninformed. It is important, therefore, that all personnel directly or indirectly associated with the installation, operation, or maintenance of these accelerators be thoroughly familiar with the hazards. In this section, significant aspects of potential dangers are examined in detail.

The sources of potential hazard related to the operation of neutron generators can be broadly classified into two general groups: sources which contribute to contamination and possible internal assimilation of tritium by personnel, and sources such as external radiation and other nonnuclear hazards which relate to the operation of the generator. No attempt is made to establish an order of importance or severity. It is emphasized that the safe operation of a neutron generator depends on the recognition of all potential sources without difference, although means of prevention and control will vary in type and complexity depending on the hazard being considered.

III. HAZARDS IN NEUTRON GENERATOR OPERATION

A. TRITIUM—INTERNAL HAZARD AND CONTAMINATION

Tritium in man's environment results from both natural and artificial sources. In the atmosphere, cosmic ray interactions, such as (p, T) or (n, T), with oxygen and nitrogen liberate tritium. Tritium production also occurs from the interaction of the heavy-nuclei component of cosmic rays with the major elements in the atmosphere and from fusion reactions in the sun and other stars. The reactions mentioned above account for the tritium generated from natural causes.

Tritium contamination of the environment can also occur from several artificial sources. In accelerators, such as neutron generators, the use of D–T and p–T reactions represents potential sources for tritium contamination. Controlled thermonuclear experiments and the detonation of fusion devices add to the total tritium content in the atmosphere. In nuclear reactors, tritium is produced by the neutron bombardment of ^{6}Li for the manufacture of accelerator targets. Other processes such as fission, neutron capture, and deuterium activation all contribute to tritium production. Recent estimates of tritium buildup from natural sources indicate that tritium is being produced at a rate of about 4.2×10^6 C/yr and that at the present time the total world tritium content is of the order of 10^8 C. It is well known that other complex mechanisms for tritium production exist, and that the simple reactions previously mentioned are not the sole causes for the generation of tritium.

1. PHYSICAL AND BIOLOGICAL CHARACTERISTICS

Tritium, the only radioactive isotope of hydrogen, can exist as a gas, or in combination with oxygen as a liquid (T_2O). Since it is a weak beta emitter ($E_{max} = 18$ keV; $E_{average} = 5.7$ keV), the penetration of these radiations in tissue is limited to a depth of approximately 1.2×10^{-2} mm. The half-life of tritium is given as 12.36 ± 0.03 yr. Tritium decays by 100% beta emission to produce ^{3}He nuclei. In general, the tritium concentration of surface waters is of the order of one tritium atom per 10^{18} atoms of hydrogen. The specific activity is calculated to be 9600 Ci/g of tritium or 2.6 Ci/cm^3 of tritium at standard temperature and pressure. For T_2O, the specific activity is 2700 Ci/g of T_2O or 1500 Ci/g of HTO. The physical, nuclear, and chemical properties of tritium have been outlined by Jacobs.[6] For tritium hazard evaluation in neutron generator operation, only the relevant physical and biological characteristics will be emphasized.

Experiments have shown that both forms of tritium (gas and oxide vapor) are readily dispersed in air and take part in chemical reactions in the same manner as the lighter isotope, hydrogen, and H_2O vapor. After absorption in the human body, tritium is rapidly and uniformly distributed in about 90 min, and all body tissue is subject to its radiation. Tritium in the elemental gaseous form is absorbed to the extent of about 0.1% by the lungs. However, in the form of T_2O, the absorption of tritium is almost 100%. It should be noted that the gaseous T_2–H_2O vapor atmospheric exchange rate is appreciable, being estimated to be in the order of 1% per day.[6] Absorption through the lungs or skin is equivalent and takes place at the same rate. Exposure to T_2O therefore is considered to be 1000 times more hazardous than to tritium gas in the elemental form. The National Council on Radiation Protection has assigned a RBE value of 1.7 for tritium absorbed-dose calculations. The biological half-life of tritium, fortunately, is relatively short ($\simeq$ 12 days). Authorities are generally agreed that the maximum permissible tritium breathing concentration for a 40-hr-week industrial worker be no greater than 5×10^{-6} μCi/cm^3.[7,8] The ICRP has recommended a maximum air concentration for tritium of 2×10^{-6} μCi/cm^3 for continuous occupational exposure over a 168-hr week. For unrestricted areas, the USAEC[8] has established that a concentration of 2×10^{-7} μCi/cm^3 be the maximum permissible air concentration for tritium.

One of the most reliable methods for the detection of possible exposure of humans is urinalysis. The major reason for the use of this method is that a few hours after exposure, the body fluids contain essentially the same concentration of HTO and T_2O. A urine concentration of 28 μCi/liter represents a tritium concentration in body fluids which will deliver the maximum permissible dose of 0.1 rem to the body tissues over a 40-hr week. Many detailed studies[9–16] on the absorption and biological effects of tritium ingestion indicate that extreme caution must be exercised by personnel in areas where T_2O contamination is suspected. From these reports it is evident that a lethal quantity of tritium can be introduced as HTO or T_2O into the human system in a matter of seconds after exposure. Under these extreme conditions, it has been shown[14] that the specific activity of T_2O in the atmosphere would be about 2.6 Ci/cm^3, the same as for pure tritium gas. The emphasis on the critical aspects of tritium exposure is made chiefly from the point of view of neutron generator operation since the potential exists for the release of critical tritium levels.

2. RELEASE MECHANISMS FOR TRITIUM IN NEUTRON GENERATORS

In Chapter 3, the neutron production behavior of tritium targets on both gas-on-metal and gas targets was described. In certain types of neutron generators employing target reloading under beam operation, tritium gas is fed into the ion source and accelerated onto a tritium target. Evaluation of tritium hazards from such targets can be made provided that the basic mechanisms for tritium release are understood. Sealed neutron source tubes can be highly dangerous if accidental rupture of the tube takes place.

It is generally accepted that gaseous tritium is released from targets under deuteron bombardment and is the principal cause for contamination of the accelerator vacuum system. For a typical tritium target containing from 3 to 5 Ci/in.^{2} of deposited tritium, the depletion of the target under bombardment from the average beam current machines is of the order of 10 to 100 mCi of tritium per hour of bombardment. In ion sources employing tritium gas, as much as a curie of tritium per hour is used. Sealed neutron tubes generally contain up to 10 to 20 Ci of tritium, most of it being in the replenisher system. It has been shown[17] that the primary mode of tritium loss from the target is by the displacement process. The loss from the D–T reaction is extremely small because, on the average, for every 10^5 tritium atoms lost, only one neutron is produced. The effects of bombardment of tritium targets has been summarized by Nellis et al.[18] A review of this report issued by the U.S. Department of Health, Education, and Welfare is strongly recommended.

There appears to be some evidence that tritium loss from sputtering can also occur. The distinction between displacement and sputtering is that in the displacement process, the deuterons built up in the target matrix gradually force out the tritium atoms, whereas in the sputtering process, tritium atoms are physically knocked out of the target. At high beam currents (>2.5 mA) tritium losses of over 1 Ci/hr have been observed with a 5-Ci/in.^{2} tritium-in-titanium metal target. The evolution of tritium gas as a direct result of deuteron bombardment is the principal cause of tritium contamination of accelerator components such as the drift tube and the target sections. Other mechanisms for the release of tritium in neutron generator systems, such as titanium tritide formation, are interesting but speculative at the present time.

3. TRITIUM CONTAMINATION

The controlling factor determining the ultimate fate of gaseous tritium released in the bombardment process is the method by which the vacuum within a neutron generator is maintained. A limited amount of information on tritium release indicates that about 5% of the total tritium displaced within the confines of the vacuum system is deposited on the static components, such as the target section, vacuum manifold, isolation valves, drift tube, and accelerator tube. The bulk of the tritium released is either accumulated in the pumping elements of the ion-getter pump or vented to the atmosphere via the roughing pump attached to a diffusion or turbomolecular pump. If an oil diffusion pump is used, a small quantity is found in the diffusion pump oil; a much larger quantity is trapped in the oil of the forepump.

Various procedures have been used to detect tritium contamination on the static components. A measurement of titanium K_α–x-ray emission from the inside wall of drift tubes lends credence to the possibility of the sputtering process. Smear tests on drift-tube sections showing evidence of the titanium K_α–x-ray appear to confirm the idea that recombination of tritium with titanium to form the tritide is a mechanism for removal of tritium from the target. Data on tritium contamination of static components of a Cockcroft–Walton accelerator have been reported.[18] In this study, a specially designed ionization chamber was used to detect the beta rays from tritium decay. Measurements made after several hundred hours of beam operation, during which several tens of curies of tritium had passed through the system, are shown in Table 4.6.

TABLE 4.6. Tritium Contamination on Components of a Cockcroft-Walton Neutron Generator

Component Monitored	Tritium Detected (mCi)
Accelerator tube	8.1
Large manifold for gauge and pump connections	2.1
Small manifold with valve and viewer section	1.0
Drift tube, $1\frac{7}{8}$-in. i.d., 58 in. long	11.0
Bellows section	4.0
Gate valve near target section	1.5
Glass sections	0.4

Although the contamination levels shown in Table 4.6 are specific to the accelerator, mode of operation, and monitoring method used, they do reflect the extent of tritium contamination of static components.

Another survey[19] indicated that, after 30 mA-hr of beam operation with four 5-Ci/in.^{2} tritium targets, the contamination detected was sufficient to warrant careful handling of the components of the accelerator. A summary of the results from this survey is given in Table 4.7. The monitoring locations can be identified in Fig. 4.2. The tritium concentration in the forepump oil should also be noted.

In view of the very limited number of surveys, intercomparison of tritium contamination results is not possible at present. However, evidence exists that over a relatively short time, the total tritium contamination of the static components can amount to several millicuries and therefore presents a health hazard, particularly if the accelerator is opened to the atmosphere for maintenance or repair.

TABLE 4.7. Tritium Contamination of the Neutron Generator after 30 mA-hr of Beam Operation on Four 5-Ci/in.^{2} Tritium Targets

Component Monitored	Monitoring Location (Fig. 4.2)	Tritium Contamination
1. Roughing pump system		
a. Pump oil	1	0.1 μCi/ml
b. Roughing pump hose		
i. External	2	1×10^{-5} μCi/swipe
ii. Internal (pump end)	3	1×10^{-2} μCi/swipe
iii. Internal (generator end)	6	6×10^{-3} μCi/swipe
2. Circulating cooling pump		
a. Inside seal	5	7×10^{-4} μCi/swipe
b. Inside pipe threads	6	5×10^{-4} μCi/swipe
3. Freon cooling tank	7	0.2 μCi/swipe
4. Plastic container—Freon storage	8	3×10^{-2} μCi/swipe
5. Ion source		
a. Inside	9	2×10^{-3} μCi/swipe
b. Electrodes	10	7×10^{-4} μCi/swipe
6. Drift tube		
a. Inside at ion pump connection	11	0.1 μCi/swipe
b. Inside at ion pump flange	12	7×10^{-2} μCi/swipe
7. Ion pump		
a. O-ring (front of pump)	13	4×10^{-4} μCi/swipe
b. O-ring (rear of pump)	14	1×10^{-3} μCi/swipe

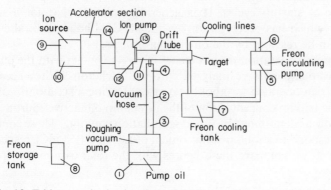

Fig. 4.2. Tritium-monitoring locations of accelerator components (cf. Table 4.6).

It may be recalled from Chapter 3 that a forepump is used to "rough out" the system to a low enough pressure to facilitate starting the main pump. If an ion pump is used, the forepump is valved off after the ion pump is started. The exhaust from diffusion or turbomolecular pumps is discharged directly through the forepump. Over 90% of the total tritium released from the target finds its way either to the ion pump elements or is vented through the diffusion or turbomolecular pump via the forepump and into the atmosphere. Because of the advantages of the ion-getter pump, most neutron generators are equipped with this type. Unfortunately, the unique method of pumping also permits accumulation of tritium. It has been estimated that for conventional 5-Ci/in.^{2} Ti–T metal targets, the tritium content of ion pumps can be several tens of curies over a short period. This accumulation can be extremely hazardous if the pump requires repair or replacement of pump elements.

Neutron generators employing target replenishing collect considerably more tritium than those not possessing this feature. It has been reported[20] that three times as much tritium is required in the ion source than is embedded in the target disk. The two-thirds fraction not usefully deposited is eventually absorbed in the ion pump. Confirmation of this behavior of replenishers is given by Hollister.[21] It must also be remembered that the lower the efficiency of the ion source, the greater is the fraction of tritium taken up by the pump elements. A characteristic of tritium-loaded ion pumps not generally known has been brought into focus by Hoffman.[22] According to Hoffman, considerable outgassing from the pump elements results from the bombardment process. Furthermore, steady operation of the pump above a pressure of 2×10^{-5} torr can raise the pump temperature sufficiently to initiate substantial outgassing of the absorbed tritium. During startup the ion pump can draw currents up to 1 A, and the heat generated also causes the release of tritium. The outgassed products are eventually removed from the system by the forepump. The total amount of tritium release will depend on the existing tritium load in the pump and the frequency of startup cycles. It is generally believed that for a neutron generator which has been run for several hundred hours, several curies of tritium can be released by the outgassing process.

In forepump operation the gas comes in direct contact with the pump oil. Samples of pump oil were analyzed[18] from a Cockcroft-Walton accelerator that had been operated for about two years. During a six-month sampling period, the tritium concentration of the forepump oil increased from 0.75 to 1.08 μCi/ml, as determined by liquid scintillation counting. These concentrations correspond to a total tritium content of 1.06 and 1.53 mCi, respectively, in the oil. If we compare these figures with the total tritium loss from the

III. HAZARDS IN NEUTRON GENERATOR OPERATION

accelerator of several tens of curies, it is obvious that contamination of the forepump oil does not present a serious tritium hazard. The tritium monitoring data[19] given in Table 4.7 are used as an example to calculate the percentage of tritium lost as found in the forepump oil. In this study the initial tritium inventory of 35 Ci can be estimated for the $1\frac{9}{16}$-in. active diameter 5-Ci/in.^{2} T–Ti targets. Since each of the four targets used was depleted by a factor of 5 before disposal, about 28 Ci of tritium was released in $1\frac{1}{2}$ years of operation during which the pump oil was not changed. Assuming that 90% of this amount was absorbed in the ion pump elements, then of the 2.8 Ci remaining, only $0.1 \times 2200 = 220$ μCi of tritium was found in the oil (oil capacity of the pump was 2200 ml). This activity represents approximately 0.01% of the total tritium released. Comparable data reported by Boggs et al.[23] show that after 4 to 6 Ci of tritium had been released from the target, the tritium activity measured in the pump oil was 8 μCi/liter of oil. Assuming a total oil volume of 2 liters, the amount of tritium found in the oil is calculated to be about 0.0004%. It is interesting that Biro et al.[24] have observed the forepump oil contamination to be 10 to 20 times as great as that found in the oil of the diffusion pump. A recent survey conducted by Watson et al.[25] in which 42 samples of pump oil from neutron generator facilities were analyzed, shows that 45% of the samples contained 0.01 to 0.1 μCi T/ml of oil. The histogram shown in Fig. 4.3 summarizes their results.

Of all possible avenues of tritium contamination of the laboratory atmosphere, the exhaust from the forepump is the major one. Substantial quantities of tritium can be vented via this route. Boggs et al.[23] have measured 3.3 μCi T/cm^3 of exhaust gas using a 4-Ci tritium target. By using a tritium scrubber on the forepump exhaust they found that the tritium level was reduced to 1.4×10^{-4} μCi/cm^3 of exhaust gas. Nellis et al.[18] have reported similar experiences. A method of measurement of tritium in the exhaust gases is described in detail in their report.

The hazard associated with the process of taking the generator up to atmospheric pressure by the use of a filling gas such as dry nitrogen is described[20] as serious. Information provided for a 50% depleted 10-Ci target indicates that about 50 μCi of tritium can be displaced from the system to be ultimately removed by the roughing pump. If only the short target section is opened up to atmosphere with the filling gas, about 5% of that amount will be displaced. However, data on this mode of tritium removal are conflicting at present. The number of times the accelerator is brought up to atmospheric pressure is a significant parameter in this instance, as is the location of tritium in the system as a whole.

Although the data given above are sketchy and difficult to compare, it is

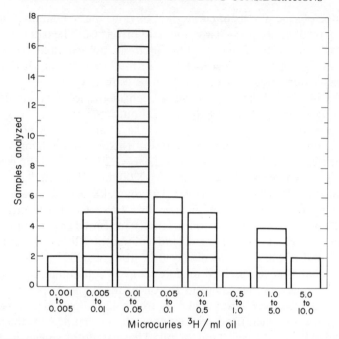

Fig. 4.3. Tritium concentrations in forepump oil from neutron generator facilities.

obvious that the principal hazard associated with the use of tritium targets in neutron generators of the pump type is that associated with the ion-getter pump. Manufacturers generally offer full maintenance and decontamination service for these pumps. However, the removal and transportation of the defective pump loaded with, perhaps, 50 Ci of tritium can create serious problems. It is imperative that users of neutron generators be aware of these difficulties and make arrangements with the manufacturer for possible removal and transportation of contaminated pumps prior to initial installation.

4. HANDLING AND REPLACEMENT OF TRITIUM TARGETS

Many techniques and devices have been developed to eliminate or to reduce drastically the problems associated with the handling of contaminated accelerator components and target changing. Some of these techniques are extremely simple and rapid, whereas others include very elaborate procedures and are quite time-consuming. The problems of tritium target storage and handling procedures can best be appreciated if some critical properties of tritium targets are first understood.

III. HAZARDS IN NEUTRON GENERATOR OPERATION

It is well known that with time, tritium targets lose tritium by gas evolution. Gibson[26] found that, for a freshly prepared 15-Ci target, tritium was lost at the rate of 320 μCi/8 hr. However, gas evolution from older targets is considerably less and varies from about 0.2 to 2μCi/8 hr per Ci of tritium. Assuming the maximum breathing rate for man to be 10^7 cm^3/8 hr, Gibson calculated that the maximum intake of tritium as T_2O would be 50 μCi/8 hr. Therefore, even under extreme conditions, no one person could inhale hazardous quantities of tritium evolved from a single target. If, however, several targets are stored in the same place, tritium intake can approach dangerous levels. To reduce the hazard from inhalation, targets should be stored in well-ventilated hoods and handled with double gloves.

At this point we shall consider the permeability of tritium through protective clothing and gloves. The data provided in Table 4.8 are useful in

TABLE 4.8. Water-Vapor Permeability Test Results at 23°C

Material	Average "P" Value (cm^3/sec)/ (cm^2/cm thick)/(cm Hg pressure)
Plastics	
Polyethylene	0.19×10^{-7}
Polyvinyl chloride	2.2×10^{-7}
Saran	0.07×10^{-7}
Teflon X-100	0.02×10^{-7}
Mylar	0.19×10^{-7}
Natural rubber	
Faultless epiderm surgeons gloves	3×10^{-7}
Rubber overshoes	5×10^{-7}
Neoprene gloves	2×10^{-7}
Butyl gloves	1×10^{-7}

determining the type of clothing that can give the best protection. The method of acquisition of these data is described by Butler and Van Wyck.[27] The permeability constant "P" is a measure of the rate of T_2O transmission per unit thickness of material tested and is therefore independent of the thickness. This constant can be used for the evaluation of materials for tritium handling. Tests reported by Butler and Van Wyck show that the permeation of plastics increases with the temperature. For vinyl plastic at 50°C the rate is six times that at 23°C. The increase in the rate appears to be a function of the nature of the material. For example, at 50°C the permeation through polyethylene is 11 times as great as that at 23°C.

Tritium gas is known to adhere to protective clothing and penetrate it. The degree of penetration increases with the exposure concentration. Since tritium targets are often stored in plastic bags, contamination from tritium permeating through the bags occurs. This diffusion through plastics is a great nuisance and adequate precautions, including multiple bags and frequent bag changes, must be taken to prevent total contamination of the storage space or laboratory. Tritium can also be absorbed through the skin by contact with contaminated surfaces. The gas has the capacity for release from contaminated surfaces. Particulate contamination from tritiated titanium particles is a distinct hazard, since it is extremely difficult to locate and identify this type of particulate contamination. Since titanium is relatively inert, if it is retained in the biological system, the tritium will be evidenced in the biological system for the time period governed by its radioactive half-life of about 12 years.

In a typical facility, the target, cooling water, and sections of the neutron generator become contaminated and target changing must be done with great care. Aberle et al.[28] have suggested an elaborate method for changing tritium targets. Briefly, they recommend the use of a stainless steel glove box or, if the operation is not likely to take more than 30 min, a thick plastic glove box. The use of gloves is mandatory, and double gloves should be worn. These gloves should be changed every 10 min. A procedure for target change described by Battist and Swift[29] virtually eliminates all possibility of external contamination. In spite of the elaborate nature of their method, the procedure is relatively rapid. Essentially, a multilayered container of five or more plastic bags is introduced over the target end of the generator and the assembly is removed to the glove box for target change. After the operation has been completed, the new target is returned to the generator for assembly in the same manner. Another practice that works well is to replace the entire front end or target assembly by another one in which the target is already installed by a commercial supplier. Since the gate valve forms a part of the replaceable front end, the transportation and installation of the front end are performed without exposing the target itself to the atmosphere. Replaceable front ends can be stored under vacuum in plastic bags. In spite of the efficacy of this technique, tritium is known to leak from the assembly via the O-rings and contaminate the storage area. Periodic plastic bag changes, however, eliminate this contamination. No matter which technique is practiced by the operator, it is important to remember that the time involved plays a key role in target change operations. The hazard from tritium contamination appears to be a strong function of the total time of handling regardless of the protective features of a particular technique.

III. HAZARDS IN NEUTRON GENERATOR OPERATION 153

Of the total tritium entering the forepump, the fraction exhausted is perhaps of most consequence. To reduce the tritium levels in the exhaust, various devices and traps have been used. Throwaway molecular seive cartridges installed between the ion pump and the forepump are found to be very effective in reducing the tritium levels in the exhaust. A typical device of this type has been described by Zimmerman et al.[30] Experiments showed that the molecular seive reduced the tritium content in the exhaust by a factor of 500. A very effective method for the reduction of tritium in the system as a whole has already been described in Chapter 3 (Section II.E.1.b), where a bulk titanium sublimator is used during beam operation.

Instruction manuals provided by manufacturers are generally fairly useful in the description of the accelerator itself, but leave much to be desired in drawing the attention of the user to the hazards involved. The manufacturers would do well to revise these manuals to include discussion of some of the serious tritium hazards associated with accelerators which generate neutrons by the use of D–T reactions.

5. SAMPLING AND MEASUREMENT OF TRITIUM

Instrumentation capable of accurate tritium detection and measurement is commercially available. Both liquid scintillation counters and airborne tritium monitors are commonly used in most installations employing tritium in one form or another. However, the lack of survey equipment to assess the surface contamination from tritium directly has complicated its radiological control. The only practical method available at present is the smear technique. Because of its obvious inadequacies, a list of conservative levels for smear tests has been established by a group at Savannah River, Georgia.[27] A summary of their recommendations is furnished as a guide in Table 4.9.

Some of the methods using smear-type tritium collectors have been described by Porter and Slaback.[31] This report illustrates the performance of 12 types of swipe materials used for sample pickup of surface tritium contamination. Counting techniques include the use of a liquid scintillation counter and also proportional and Geiger-Mueller counters. Conclusions drawn show that liquid scintillation counting of swipes can be reproducible, efficient, and accurate for the measurement of surface contamination, and recommendations as to the best swipe material are also given. A suggestion for the use of ethylene glycol on swipe materials to assist pickup of tritium was also made. The basic requirements for the measurement of radioactive surface contamination have been described in detail by Barnes[32] and Dunster.[33] Useful information relevant to this subject can be found in their reports.

TABLE 4.9. Maximum Smearable Tritium Contamination Limits (Dry Smears)

Monitoring Surfaces and Areas	Tritium Activity[a]
Clean areas	$\frac{1}{4}$ background or 200 counts/min, whichever is less
Regulated areas	500 counts/min
Regulated areas, tools, and equipment	500 counts/min
Radiation danger zone, tools	200,000 counts/min
Laundered rubber garments	1000 counts/min
Laundered plastic suits	
Outside	2000 counts/min
Inside	1000 counts/min
Sterilized face masks	
Outside	2000 counts/min
Inside	1000 counts/min

[a] It is presumed here that a gas-flow, windowless proportional counter was used in 50% counting geometry.

Air monitoring of tritium is difficult because of the small range of the beta particles, and normal beta radiation detectors are not adequate for this purpose. The problem is further confused in that the mechanism by which tritium gas mixes with the air is complex and susceptible to control by the air motion prevailing at the time. A study of this mechanism was performed by a Los Alamos group[34] who demonstrated that the effective concentration velocities of tritium gas could be as much as three orders of magnitude greater than those predicted from strict diffusion considerations. It was recommended that, to obtain a representative breathing-zone sample, the detector should be placed as close as possible to a worker's face and positioned to take account of the air flow in the area.

An adequate air-monitoring program is essential for good tritium hazard control operations. One of the most effective devices used for air monitoring of tritium is the Kanne ionization chamber.[35] Tritium concentrations from 10^{-5} to 5 μCi/cm^3 can be measured with chambers used at the Savannah River plant.[36] A Kanne system installed in a central location is ideal for monitoring of individual sites and room air exhaust ducts. A small, portable, self-contained tritium monitor or "sniffer" can also be used. This battery-powered, gamma-compensated monitor weighs about 15 lb and can detect between 1×10^{-5} and 2500×10^{-5} μCi/cm^3 of air.

III. HAZARDS IN NEUTRON GENERATOR OPERATION

As pointed out earlier in this section, tritium is assimilated through the skin as well as through the lungs. This absorbed tritium distributes itself throughout the body fluids and hence constitutes a whole-body irradiator. Personnel involved in neutron generator operation, maintenance, or target changes must be subjected to routine urine tests. Liquid scintillation counting is generally used to determine the tritium concentration in urine samples. A concentration of 20 μCi of tritium per liter of body fluids indicates a body absorption of approximately 1 mCi. The integrated dose from assimilated tritium can be estimated from the following equation[37]

$$D = 0.73 \cdot B \cdot T_b \qquad (4.14)$$

where

D = integrated dose from a single tritium uptake, mrem
B = tritium concentration in urine, μCi/liter
T_b = biological half-life of tritium (12 days).

In addition to the methods described above, various other techniques for tritium counting are available. A special film badge described by Geiger[38] measures tritium exposure. The response of the badge is found to be greater for tritium in water vapor than in the gaseous form. The badge is capable of measuring exposures greater than one body burden of tritium. Another method[39] involves a metallic strip of aluminum immersed in liquid nitrogen and extending well into the atmosphere to be sampled. Water droplets and CO_2 from the atmosphere condense on the upper part of the cold strip. After the liquid nitrogen is exhausted, the cold strip warms up, the ice melts, and the water droplets trickle down into the collector flask. The sample is then counted for tritium. The use of a bubbler-type tritium sampler has been described by Valentine.[40] Briefly, the tritium collector consists of a dry, gas washbottle with a fritted glass strip on the inlet and distilled water as the collecting medium. Tests indicate that the collection characteristics of the bubbler were similar to reported human body uptake rates for tritium gas and tritiated water vapor. Valentine reports a collection efficiency greater than 90% for tritiated water vapor for sampling rates of 8 to 10 liters/min, water temperatures from 25 to 35°C, and water volumes from 30 to 50 cm³. Tritium gas collection efficiencies reported were less than 0.1% under the same conditions but with a water temperature of 18°C. Because of this selectivity, the health physicist is better equipped to assess the concentration of tritiated water vapor in the atmosphere. A very sensitive method of measuring tritium in concentrations as low as 1% of maximum permissible concentration in air has been described by Osborne.[41] This method involves the bubbling of measured volumes of air through water and determining the tritium concentration by liquid scintillation counting. The major disadvantage of this method is that results are obtained some time after the sampling period.

The procedures, techniques, and findings given above are intended to serve only as a guide for tritium monitoring around neutron generator installations. It is strongly recommended that all personnel associated with neutron generator operations investigate these methods and systems and choose those best suited for their individual facility.

B. EXTERNAL HAZARDS FROM NEUTRON GENERATOR OPERATION

In the development of the subject matter related to the shielding of generator facilities, it is necessary to outline some fundamentals pertinent to potential sources of external hazards. In this regard, the discussion will be categorized into three specific areas. First, the expected levels of radiation with and without shielding are given. Where pertinent, the attenuating properties of the radiation in question are included. Second, the shielding considerations applicable to the specific radiation of interest are discussed, and finally, methods for control and protection of personnel are emphasized.

1. NEUTRONS

In contrast with the preponderance of available theoretical and experimental data on shielding of fission neutrons, analogous information on high-energy (e.g., 3 and 14 MeV) neutrons is limited. The problem of shielding a high-intensity, fast neutron source such as a neutron generator is complicated because the radiation field consists of contributions from several sources. Both 3- and 14-MeV neutrons give rise to a complex spectrum of degraded neutron energies in their penetration through shielding materials. This spectrum changes as a function of shield thickness.

To perform sufficiently accurate shielding calculations, detailed information on flux densities and neutron energy spectra is required. A brief treatment on the attenuation of high-energy neutrons and the necessary shielding requirements for hazard control will be given. From a knowledge of these properties, good estimates of anticipated needs for safe design of biological shields around neutron generator facilities can be obtained.

a. *Attenuation of Fast Neutrons.* It may be recalled that the high-energy neutron intensities obtained from neutron generators are of the order of 10^9 and 10^{11} neutrons/sec from the D–D and D–T reactions, respectively. In view of the maximum permissible level of 10 neutrons/(cm^2 sec) (14 MeV),

III. HAZARDS IN NEUTRON GENERATOR OPERATION

the reduction of high-energy neutron flux by a factor of about 10^9 must be achieved by the biological shield and by distance. In terms of dose rates it is interesting to note that a 14-MeV neutron source of the intensity 2.5×10^{11} neutrons/sec will deliver a dose rate of 500 rems/hr at 1 m from the tritium target. On the other hand, a D–D neutron source capable of producing an intensity of 10^9 neutrons/sec will give a dose rate of only 1 rem/hr at 1 m and 2.5 mrem/hr at 20 m. The degree of penetration of neutrons through shielding materials depends on the neutron energy and the composition of shield material. The attenuation takes place by elastic collisions and inelastic scattering, as well as by neutron capture reactions. Since the probability of neutron capture is significant only at thermal energies, a considerable degradation in the fast neutron energy must first take place before the neutron can be removed by capture. Although the process of inelastic scattering results in a large energy loss by the scattered neutron, the process itself is significant only at neutron energies greater than 0.5 MeV. The presence of heavy elements in the shield is most effective in reducing the neutron energy by inelastic scattering. For most of the elements, the spectra of scattered neutrons have a maximum at about 2 MeV, the majority of all neutrons being below 3 MeV in energy. Since hydrogen has no excited state, it cannot contribute to energy degradation of neutrons by the inelastic scattering process. Elastic scattering is therefore necessary to reduce the neutron energy to thermal energy. This billiard-ball collision process involves the transfer of the kinetic energy from the neutron to the struck nucleus, and the fractional energy loss can be deduced from the factor $2A/(A + 1)^2$ where A is the mass number of the struck nucleus. It is seen that only the lightest nuclei contribute significantly to neutron slowing down by the elastic scattering process.

The processes governing the energy transfer of neutrons indicate that the optimum shield must contain both light and heavy nuclei to be effective. Since the hydrogen cross section for elastic scattering increases rapidly with decrease in the neutron energy, the elastic interaction complements the effect of inelastic interactions with heavy elements. Because the highest energy neutrons are most penetrating, the degraded neutron flux reaches equilibrium with that of the source neutrons at some depth in the shield material. The total flux then falls off with a "relaxation" length characteristic of the incident neutron energy. The relaxation length here has the meaning of the distance in the shield necessary for the exponential absorption of neutrons and is characteristic of the neutron energy and shield material. For a point isotropic neutron source it is defined as

$$\phi(r, E) = \frac{S \cdot e^{-x/\lambda(E)}}{4\pi r^2} \qquad (4.15)$$

where

$\phi(r, E)$ = Neutron flux at a distance r from the source, neutrons/(cm² sec)
S = source strength, neutrons/sec
$\lambda(E)$ = relaxation length, cm
x = shield thickness, cm.

Day and Mullender,[42] in their comprehensive study of shielding problems associated with neutron generators, measured the relaxation length λ of 14-MeV neutrons penetrating an interlocking concrete block shield. The accelerator was operated at 200 kV at neutron source intensities of 10^9 to 10^{10} neutrons/sec. A long counter, described by McTaggart,[43] was used for monitoring the neutron flux. Since the relaxation length depends on the energy sensitivity of the detector, the results given by Day and Mullender are pertinent for neutron energies from 0.5 eV to 14 MeV. The results obtained by them are summarized in Table 4.10. The measurements were made directly behind the shield, and the measured relaxation lengths did not appear to change significantly with shield depth. Equation 4.15 was used in the calculation of relaxation lengths.

TABLE 4.10. Attenuation of Neutron Flux by Concrete Shielding

Shield Depth x (cm)	Distance from Source r (cm)	Relaxation Length λ (cm)	Fast Flux [neutrons/(cm² sec)]	Thermal Flux/ Fast Flux
60	810	15.4 ± 0.5	1.9	0.41
71	900	14.7 ± 0.5	0.97	0.51
130	600	15.0 ± 0.5	0.05	1.2

The thermal flux measurements were made with a BF_3 proportional counter. From Table 4.10 it may be seen that the thermal flux buildup as a function of shield thickness is not significant.

One of the common methods used for making shielding calculations is based on the concept of "removal cross section." The macroscopic removal cross section is defined as the inverse of the relaxation length, and is in effect the effective probability for the removal of neutrons from the incident energy range. Numerically the microscopic removal cross section per atom, σ_{rem}, is the sum of the microscopic inelastic and absorption cross sections plus a fraction of the microscopic elastic cross section. The elastic contribution is necessary because in the elastic collision process the path length of the neutron is effectively increased, thus increasing the probability of energy

III. HAZARDS IN NEUTRON GENERATOR OPERATION

loss. As a first approximation the microscopic removal cross section, σ_{rem}, is approximately equal to 0.6 × total microscopic cross section, σ_T, for high-energy neutrons. For 3-MeV neutrons and for elements of proton numbers less than 11, the removal cross section is more closely approximated as 0.75 × total cross section. However, for Z numbers greater than 11 the relationship $\sigma_{rem} = 0.6 \times \sigma_T$ may be used for the estimation of microscopic removal cross sections for 3-MeV neutrons. In both cases the microscopic removal cross section for hydrogen approximates its total microscopic cross section. By using the summation process described in the National Bureau of Standards Handbook 63,[1] the total macroscopic removal cross section, Σ_{rem}, which is numerically equal to the inverse of the relaxation length, can be calculated for any material provided the microscopic removal cross sections for the individual elements are known. The macroscopic removal cross section can be calculated for any element from the relationship

$$\Sigma_{rem} = \frac{0.602 \cdot \sigma_{rem} \cdot \rho}{A} \text{ cm}^{-1}$$

Olive et al.[44] have calculated the macroscopic removal cross section for 14.5-MeV neutrons for a number of shielding materials. These values are shown in Table 4.11. The composition of the shielding materials given in Table 4.12 has been abstracted from reference 45. The atomic removal cross section data used in the calculation of the macroscopic removal cross sections given in Table 4.11 were taken from a table given by Avery et al.[46] An approximate but more comprehensive table of microscopic removal cross sections for most of the elements in the periodic table has been constructed for 3- and 15-MeV neutrons. The relationships between the microscopic removal cross section and the microscopic total cross sections,[47] defined previously, were used in these calculations. This tabulation is shown in Table 4.13. Although only a few elements included in this

TABLE 4.11. Calculated Macroscopic Removal Cross Sections, Σ_{rem}

Shield Material	Σ_{rem} (cm^{-1})	Relaxation Length (cm)
Water	0.0786	12.7
Polyethylene	0.0837	12.0
Ordinary concrete $\rho = 2.3$ g/cm^3	0.0798	12.5
Barytes concrete $\rho = 3.5$ g/cm^3	0.0824	12.1
Iron shot concrete $\rho = 5.9$ g/cm^3	0.105	9.54

TABLE 4.12. Composition of Various Concretes

Concrete	Composition
Ordinary	Portland cement 8.2%; sand 28.7%; gravel 56.4%; water 6.7%
Barytes	Coarse barytes aggregate (1 in.) 45.2%; fine barytes aggregate ($\frac{3}{8}$ in.) 39.2%; Portland cement 9.4%; water 6.2%
Iron shot	Iron shot ($\frac{1}{4}$–1 in. diameter) 50.4%; SAE shot ($\frac{1}{8}$ in. diameter) 22.8%; SAE shot ($\frac{1}{32}$ in. diameter) 15.2%; Portland cement 8.9%; water 2.7%

table are generally employed for shielding, the inclusion of the remaining elements will serve a useful purpose later in the calculation of macroscopic removal cross sections. These cross-section values will find use in the treatment of systematic error analysis to be discussed in Chapter 6. Experimental results obtained by Caswell et al.[48] indicate that the relaxation length of 14-MeV neutrons in a large water tank is 14 cm. The attenuation of 14-MeV neutrons in water in an arrangement where 6 in. of lead was interposed between the target and the water tank was studied by Duggal et al.[49]

Before proceeding with the various methods for shield-design calculations, we shall briefly mention secondary gamma radiation produced in the shield material. Gamma radiation is produced in the biological shield from neutron capture and inelastic scattering reactions. The dose of gamma radiation outside the shield therefore depends on the neutron flux as a function of shield thickness, and the attenuating properties of the shield material for gamma radiation. Multigroup calculations, supported by experimental data, indicate that for a 130-cm-thick concrete shield, the gamma dose rate measured outside the shield is about one-half that due to all neutrons above 0.5 eV. In the case of a water shield, data given by Price et al.[45] for fission neutrons can be used to obtain an approximate measure of the gamma dose outside the shield. This information reveals that for water-shield thicknesses greater than 75 cm, the gamma dose exceeds the fast neutron dose. At 130-cm thickness of water, the gamma dose is almost a factor of 10 greater than the fast neutron dose. Although hydrogenous material is excellent for slowing down fast neutrons, it is also an efficient producer of high-intensity gamma rays of 2.23-MeV energy from the neutron capture reaction ^{1}H(n, γ)^{2}H. Similarly, the prompt gamma emission of about 7.6 MeV from the neutron capture reaction ^{56}Fe(n, γ)^{57}Fe can be a problem if iron is used as shield material. Day and Mullender[42] have

TABLE 4.13. Microscopic Removal Cross-Section of the Elements for 3- and 15-MeV Neutrons

	Microscopic Removal Cross Section (barns)				
Element	3 MeV	15 MeV	Element	3 MeV	15 MeV
H	2.25	0.66	Ru	2.51	2.47
He	2.14	0.59	Pd	2.69	2.47
Li	1.50	0.81	Ag	2.68	2.51
Be	2.33	0.87	Cd	2.69	2.61
B	1.13	0.81	In	2.81	2.70
C	1.35	0.85	Sn	2.76	2.65
N	1.20	0.69	Sb	2.85	2.73
O	0.98	0.96	Te	3.18	3.03
F	1.69	1.05	I	3.18	2.88
Na	1.65	1.02	Cs	3.38	2.97
Mg	1.35	1.08	Ba	3.60	3.09
Al	1.59	1.04	La	3.57	2.91
Si	1.26	1.17	Ce	3.63	2.97
P	1.92	1.17	Pr	3.63	2.97
S	1.76	1.10	Nd	3.66	2.88
Cl	2.04	1.17	Sm	3.78	2.94
K	2.07	1.23	Eu	3.87	3.06
Ca	2.10	1.32	Gd	3.75	3.03
Sc	2.12	1.27	Tb	3.69	3.12
Ti	2.29	1.35	Dy	3.87	3.15
V	2.28	1.34	Ho	3.90	3.15
Cr	2.11	1.37	Er	3.93	3.21
Mn	2.18	1.52	Tm	4.02	3.09
Fe	2.03	1.52	Yb	4.29	3.33
Co	2.05	1.55	Lu	4.17	3.09
Ni	2.01	1.56	Hf	4.11	3.51
Cu	1.98	1.69	Ta	4.01	3.27
Zn	1.99	1.68	W	4.26	3.36
Ga	1.86	1.92	Re	4.29	3.15
Ge	2.04	1.98	Os	4.17	3.12
As	2.22	1.99	Ir	4.32	3.15
Se	2.37	2.04	Pt	4.35	3.27
Br	2.27	1.97	Au	4.41	3.15
Sr	2.94	2.22	Hg	4.56	3.30
Y	2.41	2.27	Tl	4.77	3.48
Zr	2.46	2.26	Pb	4.98	3.33
Nb	2.46	2.33	Bi	4.77	3.39
Mo	2.48	2.34	Th	4.50	3.42
Tc	2.59	2.45	U	4.74	3.54

reported prompt gamma emission up to 10 MeV emanating from capture reactions in the shield material. The mean energy of gamma rays emitted from a low-density concrete shield is, however, about 4 MeV. A comparison of neutron and gamma dose rates as measured by Day and Mullender is shown in Table 4.14.

TABLE 4.14. Comparison of Secondary Gamma and Neutron Dose Rates[a]

Shield Depth (cm)	Distance from Source (ft)	Secondary Gamma Dose Rate (mrem/hr)	Fast Neutron Dose Rate (mrem/hr)	Thermal Neutron Dose Rate (mrem/hr)
60 (concrete)	27	7.0×10^{-2}	48×10^{-2}	0.59×10^{-2}
71 (concrete)	30	1.8×10^{-2}	24×10^{-2}	0.37×10^{-2}
130 (concrete)	20	1.0×10^{-2}	1.25×10^{-2}	0.05×10^{-2}
70 (water)	27	15.0×10^{-2}	380×10^{-2}	3.50×10^{-2}

[a] Background included; neutron output 10^9 neutrons/sec.

In the calculations of the dose rates in the table, Day and Mullender assumed a fast neutron MPD of 30 neutrons/(cm^2 sec) and a thermal neutron MPD of 1000 neutrons/(cm^2 sec). Measurements made by Broerse and Van Werven[50] confirm the fact that the ratio of the gamma to the neutron dose rate increases with shield thickness; however, measurements reported by them and made outside their shield appear to indicate that the gamma dose rate is comparable to the neutron dose. In general, it can be safely assumed that adequate shielding of a 2.5×10^{11}-neutron/sec source will compensate for gamma radiation emanating from the shield material for a concrete shield.

b. *Shielding Calculations.* Several methods have been employed for calculating the shielding requirements for neutron generator facilities. A bibliography referencing shielding reports has been compiled by the USAEC Radiation Shielding Information Center, Oak Ridge National Laboratory, Oak Ridge, Tennessee. This computer-stored compendium can be used for information relevant to shielding problems applicable to neutron generators. Basically, two general methods for shield calculations are employed. The first method employs the concept of removal cross sections and is described by Cloutier[51] and by Nachtigall and Heinzelman.[52] The basic equation used for the attenuation calculation is

$$\phi = \phi_0 \cdot B \cdot \exp(-\sum\nolimits_{\text{rem}} \cdot x) \qquad (4.16)$$

III. HAZARDS IN NEUTRON GENERATOR OPERATION

where

ϕ_0 = neutron flux without shielding, neutrons/(cm² sec)
ϕ = neutron flux with shielding, neutrons/(cm² sec)
B = buildup factor
Σ_{rem} = macroscopic removal cross section for neutrons of 14 MeV, cm^{-1}
x = thickness of shield material, cm.

The neutron flux calculated from Eq. 4.16 can be converted to dose rate by using the appropriate dose rate-neutron flux equivalence factor. If a point isotropic neutron source is considered, then Eq. 4.16 can be modified to include the $1/D^2$ dependence of the neutron flux with distance D. Berger et al.[53] have compiled a list of macroscopic removal cross sections for some common shield materials. Values taken from their report for 14-MeV neutrons are shown in Table 4.15.

TABLE 4.15. Macroscopic Removal Cross Section of Shield Materials for 14-MeV Neutrons

Material	Macroscopic Removal Cross Section (cm^{-1})
Concrete	0.084, 0.075
Water	0.070
Steel	0.15

The disadvantage of using the removal-cross-section method is that both the buildup factor and removal cross section vary with distance in the shield owing to the changing neutron spectrum. A value of 5 for the buildup factor for shield thicknesses greater than 20 cm is recommended.[1]

An alternative method, suggested by Broerse and Van Werven,[50] and Broerse[54] uses experimentally determined constants C and the relaxation length λ in the relationship

$$(DE)_x = (DE)_0 \cdot C \cdot e^{-x/\lambda} \quad \text{for } x \text{ greater than 15 cm} \quad (4.17)$$

where $(DE)_0$ and $(DE)_x$ are the dose equivalent values without and behind the shield, and the constant C accounts for the scattering factor necessary to correct the nonexponential part of the attenuation curve at small shield thicknesses. This constant is not to be confused with the buildup factor; the latter results from the contribution of scattered neutrons, increases with thickness, and is dependent on the neutron energy. Broerse and Van

Werven[50] have experimentally determined the constant C for the attenuation of 3- and 15-MeV neutrons in various materials. Their results are given in Table 4.16.

TABLE 4.16. Experimental Determination of Shielding Constants

Material	Neutron Energy			
	3 MeV		15 MeV	
	λ (cm)	C	λ (cm)	C
Concrete	15.7	1.0	19.7	1.2
Laminated wood	—	—	20.5	1.35
Paraffin	5.5	1.5	17.5	1.3
Borated paraffin	—	—	17.5	1.25
Water	9.3	1.0	19.0	1.0
Concrete + paraffin[a]	—	—	15.0	1.5
Paraffin + concrete[a]	—	—	18.7	1.3
Water + paraffin + concrete[a]	—	—	18.3	1.0

[a] Material mentioned first is located toward the neutron source.

Working with several different materials, Broerse and Van Werven obtained dose equivalent transmission data as shown in Fig. 4.4. The densities of the materials used in their studies are given as: paraffin, 0.9 g/cm^3; concrete, 2.3 g/cm^3; and laminated wood, 1.15 g/cm^3. Figure 4.4 was used to determine the attenuation constants C and λ, as tabulated in Table 4.16. The effect of material location for compound slab structures was also studied. As expected, compound shields with the heaviest material located toward the neutron source proved the most effective. From Fig. 4.4 it can be estimated that for concrete the 15-MeV neutron dose is attenuated by a factor of 10 for each 50 cm of thickness. It is important to note that the measured attenuation characteristics, as observed by Broerse and Van Werven, cannot be arbitrarily related to measurements made by other investigators since radically different shielding geometries may be used. However, their work does show relaxation lengths larger than those measured by Day and Mullender,[42] Caswell et al.,[48] and Duggal et al.[49]

The use of the two methods described thus far depends on the experimental determination of constants such as the relaxation length and the C factor. Clark[55] has performed Monte Carlo calculations and has developed curves for approximate dose reduction factors in concrete for neutrons from 0.7 to 14 MeV.

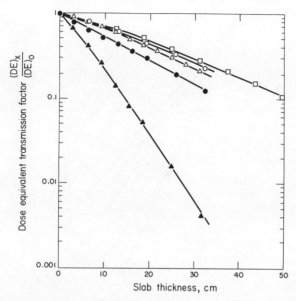

Fig. 4.4. Relative dose equivalent values as a function of slab thickness for 3- and 15-MeV neutrons: □, wood (15 MeV); ○, concrete (15 MeV); △, paraffin (15 MeV); ●, concrete (3 MeV); ▲, paraffin (3 MeV).

Correction factors, such as the C factor and the buildup factor, are necessary for the neutron spectrum change with shield thickness. The neutron distribution from a point isotropic source of 14-MeV neutrons has been obtained from a 10-group calculation performed by Day and Mullender.[42] This distribution in concrete is shown in Fig. 4.5. The theoretical data suggest that neutron attenuation is exponential at distances not too near the shield boundaries. The neutron flux as a function of energy relative to the total neutron flux for a 14-MeV neutron source through a concrete shield about 130 cm thick, as calculated by Berger et al.[53] from data by Day and Mullender,[42] is shown in Table 4.17.

Graphs which may be used for shield design have been given by Prud'homme.[56] These are shown in Figs. 4.6 and 4.7. As a general rule, it can be safely assumed that about 15 in. of concrete and/or water reduces the 14-MeV neutron flux by a factor of 10. About 8 in. of the same material will attenuate a 3-MeV neutron beam by a factor of 10. It must be remembered, however, that the attenuation characteristics shown in Fig. 4.6 can be radically different if complex geometries are used. The calculated values in Fig. 4.6 for concrete are based on a density of 2.3 g/cm³ and a composition of 70% oxygen, 15% silicon, and 15% calcium. Buildup factors

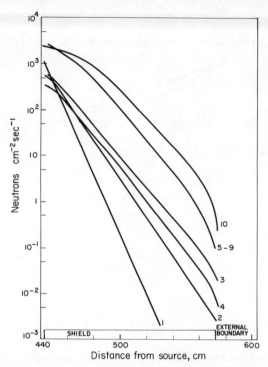

Fig. 4.5. Calculated neutron flux distribution in concrete due to a 14-MeV point source.

Group Number	Energy Range—Lower Limit (MeV)
1	14
2	6.5
3	1.6
4	0.78
5	0.18
6	0.01
7	0.26×10^{-3}
8	15×10^{-6}
9	0.4×10^{-6}
10	Thermal source strength = 2.95×10^9 neutrons/sec

Composition of Concrete
($\rho = 2.4$ g/cm^3)

Element	Wt. %
H	0.63
O	53.6
Al	0.61
Si	36.7
Ca	6.9
Mg	0.45
Fe	1.14

Water content: 5.63% by weight

III. HAZARDS IN NEUTRON GENERATOR OPERATION

TABLE 4.17. **Distribution of Neutron Energies for a 14-MeV Neutron Source through 130-cm Concrete**

Neutron Energy Interval (MeV)	Relative Contribution to Total Flux (%)
> 14	< 1 × 10^{-5}
6.5–14	1.6
1.6–6.5	4.8
0.78–1.6	1.9
0.4 × 10^{-6}–0.78	19.3
Thermal	72.4

have been taken into account in the curves shown in Fig. 4.6. In estimating the dose, it is considered safe to use the factor, 20 fast neutrons/(cm^2 sec)/1 MPD, for hydrogenous shield thicknesses with about 5% water content and varying in thickness from 30 to 60 in.

Earlier in this section a brief reference was made to the secondary gamma emission in the shield material. This gamma radiation buildup can be calculated from the spatial distribution of neutrons in the shield. Day and

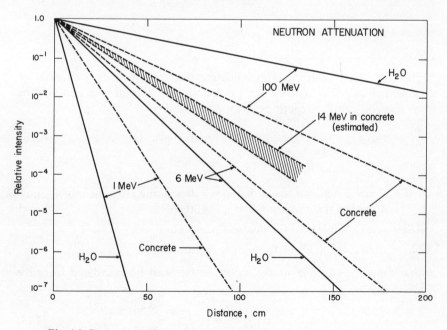

Fig. 4.6. Dose attenuation as a function of thickness for concrete and water.

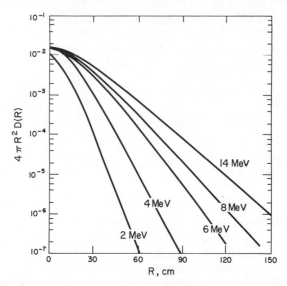

Fig. 4.7. Calculated neutron dose attenuation in water. Neutron source located in the center of a spherical water tank of radius R. The absorbed dose $D(R)$ is in mrep/hr per source neutrons/sec. Absorbed dose due to neutron source S neutrons/sec at distance R'

$$= \frac{S \times \text{value of ordinate for distance } R'}{4\pi(R')^2} \text{ mrep/hr}$$

biological dose $\simeq$ mrep/hr $\simeq$ mrem/hr $\times$ RBE

Mullender[42] have shown that if the neutron distributions are exponential for all shield thicknesses, x, then the gamma source strength per unit volume from neutron-capture reactions is approximated by

$$S_\gamma = \phi \, \Sigma_a \cdot e^{-x/\lambda(E)} \, \gamma (\text{cm}^{-3} \, \text{sec}^{-1}) \quad (4.17)$$

where

ϕ = thermal neutron scaler flux at $x = 0$ assuming exponential attenuation for the condition $0 < x < 130$ cm

$\lambda(E)$ = relaxation length for neutron penetration, cm

Σ_a = macroscopic absorption cross section, cm^{-1}

and the gamma-ray flux at the shield surface can be calculated from the relationship,

$$\phi_\gamma(x) = \phi \frac{\Sigma_a}{2} \times \frac{\left(e^{-\mu \cdot x} - e^{-x/\lambda(E)}\right)}{1/\gamma(E) - \mu} \quad (4.18)$$

III. HAZARDS IN NEUTRON GENERATOR OPERATION

where μ is the linear absorption coefficient for gamma rays produced in the concrete. The dose rate can be calculated by multiplying Eq. 4.18 by the mean gamma energy E_γ and using the conversion factor, 1 MeV(cm^{-2} sec^{-1}) = 2×10^{-3} mrad/hr; therefore

$$D_\gamma(x) = 2\phi_\gamma(x) \cdot E_\gamma \cdot 10^{-3} \text{ mrad/hr} \tag{4.19}$$

This equation only holds for gamma energies greater than 0.1 MeV. Similarly, the gamma-ray dose from inelastic scattering reactions can be calculated by using the appropriate parameters corresponding to 14-MeV neutron attenuation and to lower gamma energies. This effect is of less significance since it contributes about 20% of the total gamma dose. These parameters, as given by Day and Mullender,[42] are listed in Table 4.18.

TABLE 4.18. Nuclear Parameters for Calculation of Gamma Dose Rate for Concrete

Parameter	Thermal Neutron Capture	Inelastic Scattering
$\lambda(E)$	13.8 cm	12.0 cm
Σ_a	0.01 cm^{-1}	0.083 cm^{-1}
E_γ	4 MeV	1 MeV
μ	0.067 cm^{-1}	0.14 cm^{-1}

For a 4.25-ft-thick concrete shield, Day and Mullender[42] obtained the data shown in Table 4.19, in which they have made comparisons between theoretical predictions and experimental results and found good agreement. These results have been normalized to an initial neutron output from a point source of 10^9 neutrons/sec. Also the theoretical relaxation length has been

TABLE 4.19. Comparison of Theoretical and Experimental Attenuation Results for a 4.25-ft-Thick Concrete Shield (Measurements Made at the Shield Surface)

Parameter	Experiment	Theory
Fast neutron relaxation length, $\lambda(E)$	15.0 ± 0.5 cm	13.8 cm ($x \geq 20$ cm)
Fast neutron flux, $\phi_1(r, E)$	0.05 neutrons/(cm^2 sec)	0.038 neutrons/(cm^2 sec)
Thermal neutron flux, $\phi_2(\gamma)$	0.06 neutrons/(cm^2 sec)	0.075 neutrons/(cm^2 sec)
Secondary gamma dose rate, $D_\gamma(x)$	5.0×10^{-3} mrad/hr	6.8×10^{-3} mrad/hr

defined as the distance in which the neutron flux in a particular energy group is reduced by a factor e. The calculated values refer only to the exponential part of the neutron distribution and are constant for all shield depths (x) greater than 20 cm.

c. "*Sky Shine*." From the discussions thus far it can be readily understood that rather large amounts of shielding are necessary to reduce the dose to acceptable levels. For best results, a 4π shield configuration is optimum. However, laboratory conditions, more often than not, only permit the construction of "direct-line" barrier shields, the thicknesses of which vary in all directions according to the experimental requirements and space available. In these cases air scattering can be a problem and can contribute significantly to the total dose. Approximate calculations[57] for "sky shine" show that for distances less than 100 ft, the ratio of the air-scattered flux to the direct flux at a distance r feet from the source is approximately $(\Omega/4\pi) \times (r/200)$, where Ω is the solid angle into which the source radiates. This relationship holds true for $\Omega \simeq 2\pi$. Therefore, for $\Omega = \pi$, the sky shine is about 5% of the direct flux at 40 ft from the source. It is reasonable, therefore, that shielding of the direct flux by several orders of magnitude is pointless unless a suitable roof shield is provided to attenuate the sky shine. Additional methods for sky-shine calculations are also available.[58] The presence of sky shine can generally be detected by taking measurements at distances away from the outer shield wall. An increase in the dose with increasing distance is a good indicator of sky shine.

It must be recognized that the material presented thus far on the subject of neutron shielding is perhaps an oversimplification of the problem. However, the methods described do give meaningful data, as can be seen from comparisons of experimental and theoretical data. For more detailed treatments of this subject, the reader is referred to some classical texts and compendiums on shielding.[45, 58-61] In summary it can be said that if concrete is utilized for neutron generator shielding in which D–T neutrons are produced isotropically, then with the power of machines available today, about 70 in. of 150-lb/ft^3-density concrete is necessary to reduce the attenuated dose down to one MPD. This thickness of concrete and/or water or paraffin will also attenuate the secondary gamma emission to acceptable levels. Most shielded facilities do not have this degree of extensive shielding in all directions. Some of these facilities will be described, as well as the radiation-monitoring results obtained. From these data, it is possible to evaluate a particular shielding configuration being considered for construction. In a detailed consideration of neutron production, the anisotropy of the 14-MeV flux may be noted. With higher voltage machines, the degree of anisotropy can be significant since flux peaking in the forward direction takes place. In such

III. HAZARDS IN NEUTRON GENERATOR OPERATION

cases, the shielding in the forward direction will be greater than in the backward direction. The effect of stacked-block shield versus poured concrete shielding should be taken into account. The presence of ducts, voids, passageways, safety doors, and so on, all require careful consideration in the design of the final facility.

d. *Neutron Detectors.* In the area of nuclear radiation detection and measurement, neutron counting has been the subject of considerable research and development. For purposes of monitoring biological shields surrounding neutron generators, only a brief treatment of neutron detection instrumentation will be offered. Literature on nuclear radiation detection is extremely rich, and detailed descriptions of detector systems can be readily obtained.

It is important to emphasize that the presence of neutrons can be revealed by two basic processes: first, by some instantaneous interaction, and second, by some delayed effect from a primary interaction. In most cases, the process of ionization is utilized for the measurement of neutrons; examples of instantaneous detection processes include the detection of ionization caused by fission recoils in a counter or ion chamber, measurement of ionization by recoil particles other than fission fragments from direct neutron interactions, and detection of prompt gamma emission from neutron capture processes. Detection of neutrons by observation of delayed events from neutron interactions includes measurement of radioactive disintegrations from fission fragment decay or from other neutron capture nuclides.

Neutron counters employ the same type of basic instrumentation as those used for counting of other nuclear radiation, for example, ionization chambers, photographic emulsions, and recoil counters. Neutron detectors are specifically designed to discriminate against high fields of gamma radiation which invariably accompany neutrons in the passage of the latter particles through matter. This discrimination is relatively easy to accomplish electronically since the electronic pulse from an ionization burst from recoiling nuclei is much higher in amplitude than any burst that is likely to be recorded as a result of gamma interaction in the detection system. Those counting devices employing the delayed-event principle are reliable in providing this type of discrimination and can be specifically designed to provide additional information regarding the neutron energy spectrum impinging upon the sensitive detector volume.

In the simple case of monoenergetic neutron counting, the neutron flux may be calculated from a knowledge of the activation parameters related to the specific target used as the detector. If it is necessary to measure the neutron flux of an unknown energy spectrum, meaningful data can only be obtained if the efficiency of the detection device is independent of neutron energy. This important requirement is met by a few detection systems. A qualitative description of some basic neutron counters is given below.

The *gas recoil counter* is essentially a chamber which is filled with a hydrogenous gas to a pressure dependent on the incident energy of the impinging neutrons. Above 3 MeV, high-pressure or large-diameter chambers are used, so that the range of recoil particles is much less than the counter diameter. Since, for a given energy, the most energetic recoils are those of hydrogen nuclei, the counter gas is usually a hydrocarbon such as methane, propane, or hydrogen. This type of counter is generally operated in the proportional region and is useful for the detection of neutrons in the 20-keV to 3-MeV energy range. During counting, the axis of the counting chamber is aligned in the general direction of the neutron-emitting source so that end or wall effects resulting from the predominant forward scattering of protons are minimized. The main advantage in the use of this counter is its relative insensitivity to thermal neutron and gamma-ray responses.

Scintillation detectors containing hydrogen make efficient neutron detectors. Organic scintillators like anthracene, stilbene, plastics, and liquids are good sources for the production of copious quantities of recoil protons. The size of the scintillator is important since unduly large scintillators, unfortunately, also make good gamma-ray and charged-particle detectors.

Coincidence and anticoincidence counting techniques are also used to obtain good neutron discrimination from the background. The *recoil telescope* detector falls in this category. A typical proton recoil telescope consists of a hydrogenous radiator with two or more counters in tandem, operating in the coincidence or anticoincidence mode. By this method the protons which recoil into a small solid angle, and at some angle to a collimated neutron beam, can be detected. If the (n, p) scattering cross section is known, the neutron detection efficiency can be calculated from a knowledge of the radiator's composition and the solid angle of the detector. Many varieties of such telescopes have been developed for neutron spectroscopy studies.

The technique of *photographic plate* detection uses proton recoils in nuclear emulsions for the measurement of neutron energy and intensity. A large variety of nuclear emulsions varying in grain size and sensitivity are available. The basic properties of the emulsions set the limits on the type and energy of particles that can be detected. Since the track left by a recoil particle in a photographic nuclear emulsion is typically $1\ \mu$ wide, special optical methods are needed for track counting, range, and energy measurements.

Perhaps the most important types of counters for neutron detection are those which provide a relatively flat response between the energy range of 10 keV and 10 MeV. An added requirement is that they be relatively insensitive to high gamma radiation fields. These desirable conditions are closely approximated by the so-called *long counter* developed by Hanson and McKibben.[62] This counter is extremely reliable for neutron flux measurements provided allowances are made for the deviations in the nonlinear regions. The sensitivity versus neutron energy of the long counter of many sizes has been studied extensively. The principle of operation consists essentially of thermalization of the fast neutrons in a paraffin moderator surrounding the chamber. The thermal neutrons then react with ^{10}B of the BF_3 gas filling by the (n, α) reaction. The α-particles cause ionization, which is recorded.

The *fission pulse* chamber is another useful flat response counter and makes use of the ionization resulting from fission fragment recoils. Although the inherent efficiency of this counter is as high as 90%, compared to 10 to 20% of the long counter, it operates at about 1% efficiency because of the large background from α-particles from the decay of the active U-235 used in the counter. The bias required for complete discrimination against α-particles is set very high to achieve accuracy for neutron counting.

Scintillation detectors utilizing *solid inorganic scintillators* make efficient fast neutron detectors over a wide energy range. Phosphors like ^{6}LiI doped with about 0.06% Eu make use of the ^{6}Li(n, α)^{3}H reaction with thermal neutrons produced from the thermalization of fast neutrons by a large polyethylene or paraffin sphere surrounding the scintillator. A Lucite light pipe is used to transmit the pulses to a magnetically shielded photomultiplier which uses integral discrimination circuitry. The neutron energy versus sensitivity response of this counter is essentially flat up to about 0.1 MeV, then rises to a maximum at about 1 MeV, and remains essentially so up to 8 MeV. It then falls somewhat with increase in neutron energy. Because of the high Q value of the reaction (4.64 MeV), the counter discriminates very well against gamma radiation.

The reader is advised to consult literature specializing in neutron counting for suitable selection of counting equipment for neutron measurement. An excellent treatment of the many counters, their individual characteristics, and responses is available.[59]

2. BREMSSTRAHLUNG FROM TARGET AND TARGET BOMBARDMENT

Bremsstrahlung production from tritium targets and from the bombardment of any target results in a source of radiation which can be significant in terms of the hazard involved. Although the low-energy beta rays from the decay of tritium present no external hazard because of their extremely limited penetrating powers, their interaction with the titanium matrix resulting in bremsstrahlung production presents a source of radiation exposure. Significant bremsstrahlung production can also take place during the acceleration process. Bombardment of machine components and targets releases electrons which are accelerated in the reverse direction. These backstreaming electrons, upon striking the positive high-voltage terminal, produce a continuous spectrum of gamma rays with an energy maximum equivalent to the energy of the incident bombarding deuteron or hydrogen beam. The characteristics of these x-rays depend on the acceleration potential, beam current, component and target configurations, and of course the materials being bombarded by the electrons.

a. *Radiation Levels.* Radiation measurements made[26, 28] at 10 cm from the surface of a tritium target have shown exposure rates of 1.7 mR/(hr Ci) of tritium. The exposure rate from a 20-Ci tritium target, $\frac{1}{4}$ in. from its surface, was measured to be as high as 7 R/hr.

With bremsstrahlung produced from backstreaming electrons, typical dose rates of 750 mR/hr have been measured at the high-voltage terminal.[23] Other workers[42, 51, 56, 63] have reported bremsstrahlung levels of about 200 mR/hr. Using a 200-kV accelerator and bombarding a blank copper target with hydrogen, Suddueth and Nargolwalla[19] studied bremsstrahlung production at specific conditions of beam operation. Dose rates as high as 2.5 R/hr were measured* in the vicinity of the oil-filled, insulated, high-voltage terminal. Their results are given in Table 4.20. The location of the monitoring points around the accelerator is depicted in Fig. 4.8.

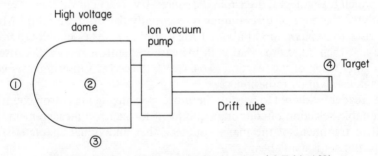

Fig. 4.8. Bremsstrahlung monitoring locations (cf. Table 4.20).

b. *Shielding and Control.* Tritium targets must be handled with special care. Direct physical contact with the tritiated surface, in particular, must be avoided. Special tools can be used to facilitate their removal and handling. The normal shielding provided for protection against high-energy neutrons is more than adequate for shielding of the bremsstrahlung dose. The backstreaming electron density can be substantially reduced by the inclusion of a

---

* X-Ray dose rates depend on vacuum conditions as well as high voltage and ion current.

TABLE 4.20. Bremsstrahlung Dose Measurements as a Function of Beam Current

Location of Dose Measurement (cf. Fig. 4.8)	Beam Current (mA)	Dose Rate (mrem/hr at Contact)
1. High-voltage terminal		
(a) Back (1)	0.5	15
(b) Side (2)	0.5	100
(c) Bottom (3)	0.5	20
(a) Back	1.0	40
(b) Side	1.0	310
(c) Bottom	1.0	50
(a) Back	1.5	80
(b) Side	1.5	1750
(c) Bottom	1.5	100
(a) Back	1.75	175
(b) Side	1.75	2500
(c) Bottom	1.75	200
2. Target end between suppressor and cooling cap (4)		
	0.5	0.6
	1.0	1.4
	1.5	2.8
	1.7	3.0
	2.0	5.8
	2.35	9.5

suppressor ring with which most accelerators today are usually fitted as a standard item. The negative voltage impressed on the suppressor ring should be checked periodically, particularly if the potential is supplied via dry batteries. Loss of this voltage can substantially increase the X-ray production. If tuning-up operations necessitate the presence of the operator inside the biological shield, a $\frac{1}{4}$-in.-thick lead sheet can give efficient protection from the bremsstrahlung dose.

3. INDUCED RADIOACTIVITY IN MATERIALS OF CONSTRUCTION

Owing to the large amount of material in direct contact with and close to the tritium target, considerable radioactivation of these bulk materials occurs. Generator components, target assembly, sample holders, and coolants used will undoubtedly be activated during the irradiation process. Personnel approaching or handling these systems are subject to radiation exposure from the decay of the activated nuclides in the bulk materials.

Delayed radiation from the tritium target holder is the major source of hazard.

If the target holder is constructed from aluminum, the major nuclear reactions resulting from 14-MeV neutron irradiation are ^{27}Al(n, p)^{27}Mg, ^{27}Al(n, α)^{24}Na, and ^{27}Al(n, γ)^{28}Al. The half-lives of ^{27}Mg, ^{24}Na, and ^{28}Al are 9.5 min, 15 hr, and 2.3 min, respectively. If copper is used as the backing surface for the tritium target, the predominant reactions to be considered are ^{63}Cu(n, 2n)^{62}Cu, ^{65}Cu(n, 2n)^{64}Cu, ^{63}Cu(n, γ)^{64}Cu, and ^{65}Cu(n, γ)^{66}Cu. The half-lives of the copper isotopes are 5.1 min (^{66}Cu) and 12.8 hr. (^{64}Cu). Activation of other components in the vicinity of the target is less significant. The 7.2-sec ^{16}N activity generated in the cooling water will have decayed before personnel can approach the target assembly. If, however, Freon is used as the coolant and long-term irradiations are performed, then the ^{18}F activity buildup in the coolant from the ^{19}F(n, 2n)^{18}F reaction can be significant. The 110-min ^{18}F activity can build up to major proportions and a cooling period of several hours is necessary before maintenance work can be performed on the system.

Prud'homme[56] has reported expected levels of radiation from radioactivation of target components for 1 hr of beam operation at 4×10^{10}-neutrons/sec output. Exposure rates up to 200 mR/hr from ^{27}Mg decay at 10 cm from the target are quoted. A total radiation dose rate from an aluminum target holder of about 300 mR/hr immediately after shutdown has been estimated by Prud'homme. If, however, stainless steel is employed for target holders, the radiation dose under similar conditions is lower by a factor of almost 10.

It is obvious that the normal shielding requirements of a facility are adequate for protection from induced activated products. Work on the target assembly should, however, be delayed by a few hours following an irradiation to permit the decay of products induced in aluminum. Maintenance and recharging of Freon in systems employing this coolant should be delayed by several hours if irradiations greater than 10 min have been undertaken. An overnight delay is necessary if the irradiations are over 1 hr.

4. OTHER RADIATION HAZARDS FROM BEAM OPERATION

a. Prompt Gamma Radiation. In addition to the primary hazard from neutron production, the operation of the generator causes the production of other radiations which contribute significantly to the overall biological hazard. One such source of radiation is due to the production of prompt gamma emission from neutron capture and inelastic scattering processes in the shield material. The discussion of this radiation has been described in detail in Section III.B.1.b.

b. *Radiation Hazard from "Blank" Target Experiments.* The second major source of radiation is from the bombardment of blank targets with a beam of hydrogen or deuterium ions during beam alignment or tune-up operations. These "blank" target experiments are a necessary part of operator training, evaluation of beam characteristics of new installations, and general tune-up operations that have to be conducted as a part of preventive maintenance measures for most installations. D–D neutron buildup with integrated beam on tritium targets and blank targets has been semiquantitatively evaluated in Chapter 3 (Section II.F.2.e), from which it can be deduced that serious exposure from energetic neutrons can occur in the vicinity of blank targets bombarded by deuterons. For this reason, tune-up experiments must always be conducted using hydrogen gas. Although the beam characteristics with hydrogen vary slightly when compared with those when deuterium is used, the small inconvenience experienced is more than outweighed by the hazard associated with D–D neutrons from deuterium bombardment. However, since normal hydrogen supplied contains 0.015% of deuterium as part of its normal isotopic content, it is possible to estimate the approximate D–D neutron production rate from the bombardment of hydrogen on a blank target upon which trace quantities of deuterium are building up with time. From data provided in Chapter 3 (Section II.F.2.e) it is possible to make some theoretical estimates.

As may be recalled, the D–D flux at about 4 cm from a target after about 2 mA-hr of beam operation reached a saturation value of about 5×10^5 neutrons/(cm^2 sec) operating at a constant beam current of 1 mA. For hydrogen as the bombarding gas, the D–D saturation value will be proportional approximately to the trace deuterium content of the impinging beam and the trace deuterium content embedded in the target material. It is expected that the D–D flux at saturation would approximate $5 \times 10^5 \times 0.015 \times 0.015 \times 10^{-4}$, or 0.01 neutrons/(cm^2 sec). If, however, a previously saturated deuterium target is used and a beam of hydrogen is permitted to bombard it, the neutron flux will be approximately 75 neutrons/(cm^2 sec). This flux is about five times the maximum permissible level. Although this flux may appear insignificant when compared to that generated by the use of a deuterium beam, it is not generally recognized or mentioned in instruction manuals provided with neutron generators. It is not entirely improbable that tune-up experiments could take several hours, and therefore care should be exercised to ensure that blank targets used have not been previously bombarded with deuterium gas. In general the use of blank targets for experiments in which hydrogen is accelerated does not eliminate neutron production. Generators that have been exposed to deuterium and tritium retain sufficient residues of these gases that neutron production is observed in so-called *blank* experiments. Therefore, extreme caution must be used in beam-viewing experiments when these become necessary. Remote beam-viewing or monitoring systems will of course eliminate exposure from this source of neutrons; however, they are generally expensive to install and most facilities rely on beam studies using the human eye.

5. NONNUCLEAR HAZARDS

Nonnuclear hazards contribute to those already described. All installations of this type are potential sources of danger other than that from beam operation. The mere presence of high electrical potentials of several hundred thousand volts requires that extreme precautions be taken by all personnel engaged in operating neutron generators. Since hydrogen and deuterium gas are routinely used, the obvious danger of explosion and fire is always present. Appropriate precautionary measures must be brought to the attention of necessary personnel and safety programs must be rigidly enforced. Warning signs and light beacons should be installed to warn of high-voltage hazards. A fail-safe interlock system is a mandatory requirement, so that inadvertent exposure to radiation is prevented. These interlocks must never be bypassed. The generation of neutrons should be indicated by a warning light or rotating beacon in full view of anyone approaching the area. No accelerator facility should be operated without adequate and exhaustive safety procedures. An excellent summary of safety measures suggested for particle accelerator facilities has been issued by the U.S. Department of Health, Education and Welfare.[64] This manual is recommended to all users of neutron generator facilities for setting up a safety program best suited for the individual laboratory. It is most important that personnel being trained for neutron generator operation and maintenance be thoroughly conversant with all aspects of nuclear and nonnuclear hazards associated with particle accelerators.

There is sufficient evidence to indicate that accidents involving neutron generator systems have in the main been caused by nonobservance of safety procedures. Bypassing of the interlock circuit has been the cause of a significant proportion of these accidents. In view of the inherent nature of the particle accelerator, an accident is generally serious and possibly fatal.

IV. DESIGN AND CONSTRUCTION OF BIOLOGICAL SHIELDS

In the earlier sections of this chapter, the basic physics of shielding has been emphasized and very little has been said regarding the practical difficulties in design and construction of biological shields. We must now consider some of the important engineering aspects of shield design and construction since it is often necessary to reach a compromise between the

IV. DESIGN AND CONSTRUCTION OF BIOLOGICAL SHIELDS   179

optimum shield design and practical engineering principles. Such compromises are necessary because invariably the space required for optimum design and maximum protection is not available. Compromises are possible because the duty cycle or work load of a typical facility seldom approaches 40 hr of neutron production per week. The shield designer must therefore rely on his knowledge of the physics of radiation protection and the anticipated use of the neutron source to decide the best design that will meet the safety requirements.

A. DESIGN CONSIDERATIONS

The approach to shield design does not follow any hard and fast rules. However, some general ideas are useful and are considered here. First, a decision has to be made as to the level of radiation that will be permissible outside the biological shield. If the generator is to be located in an area accessible to personnel not involved directly with radiation, then the design must satisfy the minimum requirement of one MPD at full power operation at these locations. However, it is conceivable that, at points rarely visited by the operating personnel and inaccessible to all others, the maximum radiation level could be as high as 1000 MPD. Under no circumstances should the design permit radiation doses greater than 1000 MPD in any area accessible to personnel during generator operation. Second, the location and number of all penetrations through the biological shield which will be required by the experimenters must be decided. These penetrations are necessary for the installation of pneumatic tubes, insertion of power cables, remote beam-viewing hardware, and the like. Third, the available shielding materials must be categorized in terms of their characteristics and cost. Having arrived at a series of possible materials, one must direct attention to the layout of the shield structure within the configuration of the available space. Finally, when the shield is constructed, a detailed radiation survey must be carried out immediately for the detection of possible weaknesses in the shielded facility. The results from the survey should indicate locations where additional shielding may be required.

B. CONSTRUCTION OF THE BIOLOGICAL SHIELD

1. GENERAL CONSIDERATIONS

With the possible exception of the neutron generator itself, the biological shield is the most expensive item of the activation analysis facility, and if a special room must be constructed, the cost can be much greater than that of the accelerator itself. Therefore, a judicious use of the available space, careful construction techniques, and the utilization of any available natural shielding are of paramount importance for good economics. If a concrete-block structure is to be adopted, extra care must be exercised in the stacking process so that, as far as possible, both a vertical and horizontal stagger configuration is maintained. A shield so constructed is almost 90% as efficient as a poured concrete structure of equivalent thickness. The stability of the structure can be considerably enhanced if, during construction, a little care is taken to "tie" in the corners of the shield walls. The use of well-designed access labyrinths can significantly reduce the scattered neutron dose. Once completed, the biological shield should be painted to minimize dust problems. During construction, allowances should be made for the future installation of pneumatic tubes and cables and the ducting requirements for proper air circulation or humidity control. Excessive moisture in the vicinity of the neutron generator high-voltage terminal can cause arc-downs, particularly if the terminal is air-insulated. Adequate space must be available for anticipated generator movement, for all possible maintenance work, and experimental requirements.

2. TYPICAL BIOLOGICAL SHIELD STRUCTURES

It is not the intention of this text to describe all the engineering requirements such as floor loading and beam structural strength calculations, or to deal with the labor ramifications of actual construction. Instead, a selection of approved existing structures has been made which can serve as examples for a wide variety of biological shields currently in use. Information given in each example can be used for the construction of exact copies or adaptations. The selection has been governed by the divergence of designs rather than by random selection. Perhaps many excellent shielding configurations have been omitted; however, it is hoped that the examples given will satisfy most normal laboratory conditions.

The neutron output data given are those used for the normalization of the radiation dose measurements and do not imply that they are the highest available with the particular neutron source used. Each example is prefixed with a short description of the location or includes some important points to be considered in the interpretation of the efficacy of the structure quoted.

Example No. 1. This shield structure was constructed in the basement of a three-level engineering building at the University of Toronto. The plan and side views of the biological shield are shown in Figs. 4.9a and b. The main structure is composed primarily of stacked concrete blocks staggered both vertically and horizontally and placed around a nominal-thickness poured concrete wall framework. The density of the stacked blocks is about 130 lb/ft^3. The roof is constructed from stacked blocks placed on hollow

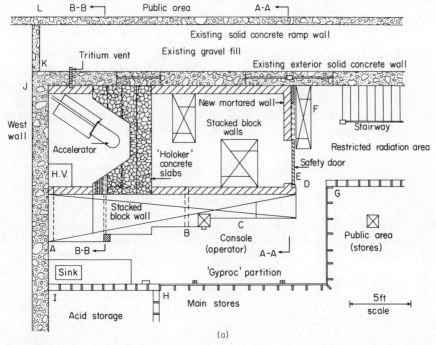

Fig. 4.9a. Neutron generator facility at the University of Toronto (adapted from reference 65), plan view.

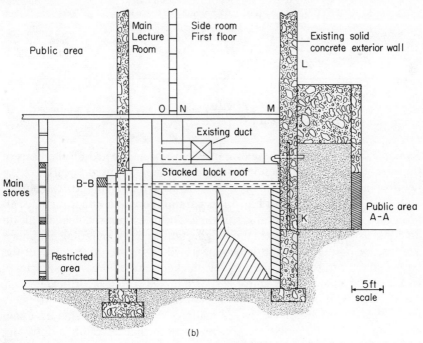

Fig. 4.9b. Neutron generator facility at the University of Toronto (adapted from reference 65), side view.

concrete beams filled with cement. As can be observed from the radiation monitoring data given in Table 4.21, the biological shield is effective for about 10 hr/wk of beam operation at the maximum flux capability of the generator. However, significant radiation can be measured in the student classroom located directly above the shield roof, and therefore irradiations performed during the regular school hours are limited.

TABLE 4.21.[a] Radiation Survey Results (Number of mpd)[b]

| Location | Gamma | Neutrons | | | Total MPD |
		Thermal	Epithermal	Fast	
A	1.6	0.4	1.9	5.5	9.4
B	1.2	0.2	2.1	3.1	6.6
C	0.8	0.2	0.8	1.8	3.6
D	3.0	1.0	3.9	6.1	14.0
E	8.0	2.6	12.8	15.2	38.6
F	0.4	0.2	0.5	1.0	2.1
G	0.4	0.3	0.5	1.5	2.7
H	0.3	0.1	0.4	1.0	1.8
I	0.4	0.1	0.4	1.0	1.9
J		Below detectable limits			
K	130	100	650	6120	7000
L		Below detectable limits			
M	6.0	3.0	9.8	14.5	33.3
N	1.7	0.7	2.0	4.0	8.4
O	1.0	0.4	1.8	3.2	6.4

[a] Cf. Figs. 4.9a and 4.9b.
[b] Acceleration voltage, 150 kV; 14.7-MeV neutron output, 2.5×10^{11} neutrons/sec. (Table 4.5 is used for dose-flux equivalence factors.)

Example 2 (a, b, c, d). These examples have been selected on the basis of the wide variety of constructional designs and materials used. In all cases the neutron radiation levels have been normalized to a constant neutron output of 4×10^{10} neutrons/sec. In Fig. 4.10a, paraffin, concrete, and water have all been used to provide acceptable shielding. The measurements made indicate a very satisfactory experimental configuration in terms of biological safety. The use of borated paraffin is demonstrated in Fig. 4.10b. In view of the large-diameter water tank, the concrete thicknesses used are relatively small. In Fig. 4.10c a neutron generator is shown linked with a subcritical nuclear reactor facility. It is obvious from the plan view that this facility is extremely well shielded. In Fig. 4.10d, use is made of dirt fill to provide adequate shielding. As can be noted, the concrete wall in that direction is only 8 in.

IV. DESIGN AND CONSTRUCTION OF BIOLOGICAL SHIELDS   183

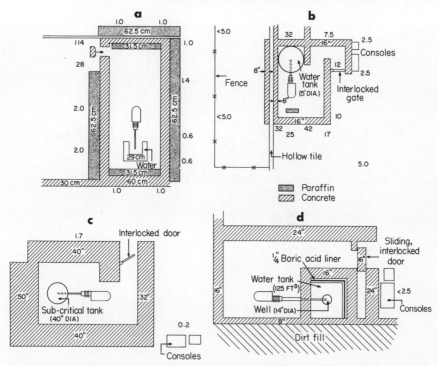

Fig. 4.10. Examples of neutron generator installations.[53] Neutron radiation levels are given in mrem/hr normalized to a 14-MeV neutron yield of 4×10^{10} neutrons/sec. (*a*) Adapted from Broerse and Van Werven.[50] (*b*) Adapted from Cloutier.[51] (*c*, *d*) Adapted from Prud'homme.[56]

thick. The system has been designed for both thermal and fast neutron irradiations.

Example 3. This example was taken from the laboratory facility at the Institute for Nuclear Techniques, Krakow, Poland. The design is very compact and self-contained. The shielding for the detector systems in Fig. 4.11 should be noted. The use of a labyrinth effectively reduces the neutron flux to acceptable levels, as indicated in the radiation survey shown in Table 4.22.

Example 4. The schematic shown in Fig. 4.12 illustrates the facility at the Department of Energy Mines and Resources, Ottawa, Canada. This is an example of perhaps the most popular type of construction. Again the presence of the water tank and paraffin wax shroud placed around the beam tube has permitted the use of minimal concrete shielding for a high-source-strength machine.

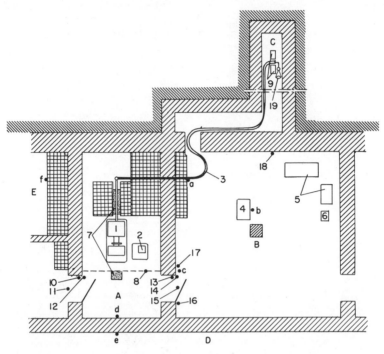

Fig. 4.11. Neutron generator facility at the Institute of Nuclear Techniques, Academy of Mining and Metallurgy, Krakow, Poland (adapted from reference 66). Acceleration voltage = 150 kV; neutron output = 10^{10} neutrons/sec.

Example 5. A design of the shielding structure at the Eastman Kodak Research Laboratories, Rochester, New York, is shown in Fig. 4.13. The cavity which houses the Cockcroft-Walton generator is located underground outside of the building proper and has a road and sidewalk passing over it. The walls of the cavity are constructed of 2-ft-thick high-density poured concrete. The floor of the cavity is 1-ft-thick poured concrete resting on bedrock. The ceiling is poured concrete, 2 ft thick with $9\frac{1}{3}$ ft of earth fill above it. There exists $7\frac{1}{2}$ ft of earth fill between the cavity wall and the control wall in the horizontal plane. The shortest diagonal between the cavity (ceiling) and the control room (floor) runs a distance of $8\frac{1}{2}$ ft through the earth fill. The dimensions of the cavity are $15 \times 24 \times 9$ ft high. The specific locations marked in Fig. 4.13 are points at which dosimetric measurements were made. The measured values at these locations are tabulated in Tables 4.23 and 4.24 for the production of 2.8- and 14-MeV neutrons, respectively.

IV. DESIGN AND CONSTRUCTION OF BIOLOGICAL SHIELDS

TABLE 4.22.[a] **Radiation Survey Results**

Point of Measurement	Neutron Flux [neutrons/(cm² sec)]		
	Calculated	Measured	Permissible[b]
a	0.2	3	100
b	—	6	100
c	—	12	100
d[c]	3000[d]	180	—
e	30	22	100
f[c]	35	—	—
g[c,e]	136	90	—

[a] Cf. Fig. 4.11:
A, neutron generator room; B, measuring room; C, shielded cave (tunnel); D, hole; E, stores. 1, neutron generator; 2, 150-kV power supply; 3, pneumatic tube; 4, control console; 5, spectrometric, monitoring, and time control measuring units; 6, 400-channel pulse height analyzer; 7, paraffin wax stopper; 8, safety net; 9, scintillation counters; 10, 13, electromechanical locks; 11, 15, warning lamps (HV), 12, 14, door-closing switches; 16, electric bell; 17, emergency switch, 18, warning lamps (neutrons); 19, blower.

[b] Based on duty cycle of 4 hr/wk and radiation dose of 10 neutrons/(cm² sec) per 40-hr week.

[c] Measuring points are in rooms uninhabited by personnel during generator operation.

[d] Calculations without considering the effect of the paraffin stopper (position 7 in Fig. 4.11).

[e] Measuring point is directly above the generator in a room located on a higher floor.

Locations G and F on the schematic drawing of the cavity are points to which personnel have access when the neutron generator is producing neutrons. In addition to these locations within the laboratory, sidewalk and lawn areas above the cavity outside the building are also points of uncontrolled closest approach to the cavity. Dosimetry measurements at all of these locations fail to give any detectable indication of the presence of neutrons when the generator is operating at its highest output, producing either 14- or 2.8-MeV neutrons. Beyond the gate leading into the labyrinth in the cavity, the neutron field increases the closer one approaches the cavity. Measurements were made to indicate the level of activity in these areas so that any changes in protection factors in the future might be detected. It is clear that the concrete and earth shielding around the cavity is effective in reducing the neutron field to nondetectable levels. The primary source of

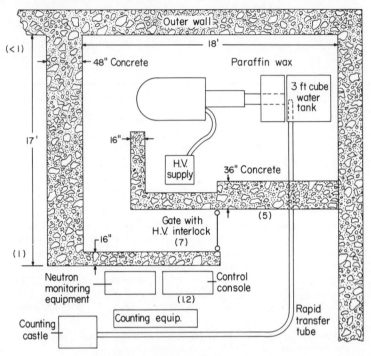

Fig. 4.12. Neutron generator laboratory at the Mines Branch, Department of Energy Mines and Resources, Ottawa (adapted from reference 67). Roof shielding = 36 in. wood, 8 in. concrete, 10 in. water. Acceleration voltage = 150 kV; 14-MeV neutron output = 10^{11} neutrons/sec. Numbers in parentheses are fast neutron dose rates in mrem/hr.

neutrons in the labyrinth leading to the cavity appears to be the scattering of neutrons from the walls of the cavity and labyrinth. Should a higher output neutron generator be used in the future, it may become necessary to place several baffles in the corridor in order to minimize the scattered neutron dose.

The shielding of the cavity for the neutron generator provides complete protection from x-rays produced by the neutron generator during its operation. Another source of gamma radiation, however, is from the induced radioactivity in the cavity walls and the equipment in the cavity. A series of measurements were made at different times after neutron production. When 14-MeV neutrons were produced for 5 min at a beam current of 2.8 mA [approximately 3×10^9 neutrons/(cm² sec)], the wall immediately behind the target assembly gave a reading of 12 mR/hr within several minutes after irradiation. The sample rotating assembly gave a reading of 500 mR/hr. It was concluded that a substantial portion of this radiation was due to ^{28}Al

IV. DESIGN AND CONSTRUCTION OF BIOLOGICAL SHIELDS 187

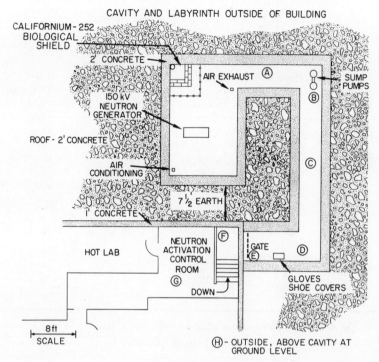

Fig. 4.13. Neutron generator facility at the Eastman Kodak Company Research Laboratories, Rochester, New York.

activity produced from the aluminum target holder and silicates in the brick. After 1 hr the activity in the wall had decayed to a nondetectable level, whereas the target assembly still gave 70 mR/hr. Further investigation demonstrated that this latter activity was caused by the decay of ^{24}Na, presumably produced from the magnesium and aluminum in the alloy of construction for the sample rotating assembly. This appears to be a very poor design feature since it prevents access for maintenance work when such is required on the sample assembly itself. This equipment will have to "cool" for two to three days after continued use in irradiations to permit the activity to decay to an acceptable level before repairs involving substantial amounts of contact time with the sample assembly are carried out.

Similar studies carried out with 2.8-MeV neutrons produced very little detectable activity. A 1-min irradiation at 3 mA produced no activity in the surrounding walls after 1 min and only 1 mR/hr in the target assembly itself.

Example 6. The plan and side view of the neutron generator laboratory at the National Bureau of Standards, Washington, D.C. (Fig. 4.14), serves as an

TABLE 4.23. 2.8-MeV Neutron Dosimetry

Monitor Location[a]	Beam Current (mA)	Target Neutron Yield (neutrons/sec)	Monitors	
			BF$_3$ Survey Meter (counts/min)	P–N Sphere Dosimeter (mrem/hr)[b]
A	0.5	2.60×10^6	350	—
	3.0	3.40×10^7	1000	1.02
B	0.5	2.60×10^6	75	—
	1.0	9.86×10^6	100	—
	3.0	3.38×10^7	250	0.25
		5.08×10^7	250	—
C	0.5	2.60×10^6	25	—
	3.0	5.08×10^7	0	0
D	3.0	5.08×10^7	0	—
E,F,G	0.5	2.60×10^6	0	—
	3.0	5.08×10^7	0	—
H	3.0	5.08×10^7	0	—

[a] See Fig. 4.13 for monitor location relative to neutron generator.
[b] Average weekly dose (40 hr) shall not exceed 100 mrem.

TABLE 4.24. 14 MeV Neutron Dosimetry

Monitor Location[a]	Beam Current (mA)	Target Neutron Yield (neutrons/sec)	BF$_3$ Survey Meter (counts/min)	P–N Sphere Dosimeter (mrem/hr)[b]
C	0.5	2.38×10^{10}	250	0.24
	1.0	4.74×10^{10}	550	0.48
D	1.0	4.7×10^{10}	—	0.05
E	3.0	1.64×10^{11}	200	0.10
F	3.0	1.64×10^{11}	0	0
G	3.2	3.58×10^{11}	0	0
H	3.0	1.82×10^{11}	0	—

[a] See Fig. 4.13 for monitor location relative to neutron generator.
[b] Average weekly dose (40 hr) shall not exceed 100 mrem.

IV. DESIGN AND CONSTRUCTION OF BIOLOGICAL SHIELDS

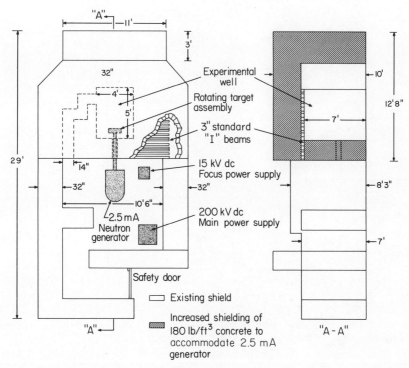

Fig. 4.14. Neutron generator activation analysis laboratory at the National Bureau of Standards, Washington, D.C. (adapted from reference 69). Acceleration voltage = 200 kV; 14-MeV neutron output = 2.5×10^{11} neutrons/sec.

example of optimum utilization of limited space. The radiation-monitoring-data locations are indicated in Fig. 4.15 and the results are shown in the squares. It may be noted that the high-powered neutron generator was contained in a shield which had only half a roof on top. The high-density shielding added contains almost 50% iron. The facility has been approved for up to 10 hr of operation per week. This is possible since the tritium target half-life for neutron production under normal conditions of operation is of the order of 1 mA-hr. During the four years this facility has been in operation, no worker associated with its operation has received a dose in excess of one MPD. The entire area shown in Fig. 4.15, however, is designated as restricted, and appropriate measures to be followed in such areas are applied. The increased shielding noted in Fig. 4.14 was necessary for the installation of the 2.5-mA generator, which replaced the earlier 500-μA machine initially installed.

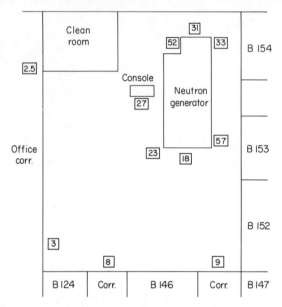

Fig. 4.15. Location of shield structure in high bay area (cf. Fig. 4.14). Numbers in parentheses are the total biological dose rate in mrem/hr from gamma rays and neutrons normalized to a neutron output at the target of 2.5×10^{11} neutrons/sec.

Example 7. A radiation facility surveyed in detail by Day and Mullender[42] is shown in Figs. 4.16 and 4.17. This survey is of special interest since the radiation measurements have been broken down into fast neutron, thermal neutron, gamma-, and x-ray doses. This construction shows a rather elaborate shield. The efficacy of the shielded structure is evident from the low dose rates measured at locations that personnel would normally occupy for long periods.

Example 8. An example of one of the early shielded enclosures is depicted in Fig. 4.18. This facility was specially designed for activation analysis, and allowed great flexibility of operation. A shield efficacy survey carried out after construction yielded the data given in Table 4.25, and the three locations monitored indicate that full utilization of this generator facility is possible without any danger of overexposure.

Example 9. A facility featuring an adequate monitoring system is shown in Fig. 4.19. A close-up of the activation tank is illustrated in Fig. 4.20. This is an example of a conventional type of shielding arrangement for which dose

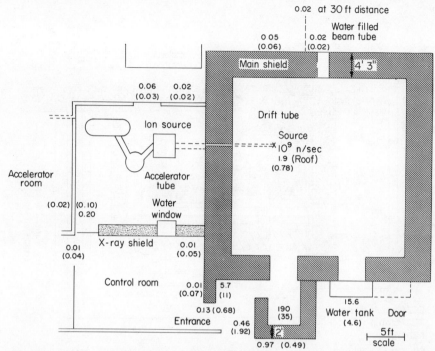

Fig. 4.16. Neutron generator facility at U.K. Atomic Energy Authority, Aldermaston, England (adapted from reference 42). Numbers refer to fluxes in neutrons/(cm² sec) for a source strength of 10^9 neutrons/sec at the target and parentheses indicate thermal neutrons.

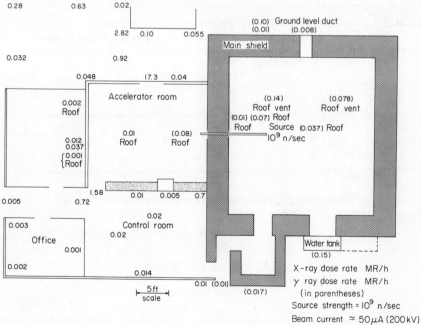

Fig. 4.17. Neutron generator facility at U.K. Atomic Energy Authority, Aldermaston, England. Gamma- and X-ray levels near the accelerator (cf. Fig. 4.16).

191

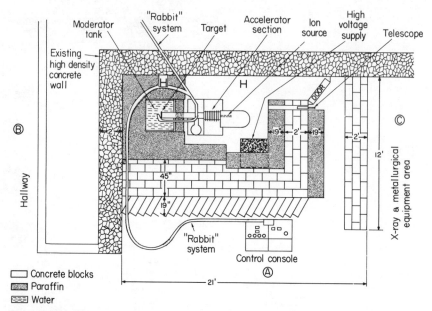

Fig. 4.18. Shielding arrangement for neutron generator operating at 150 kV at the University of Michigan, Ann Arbor, Michigan (adapted from reference 70).

TABLE 4.25.[a] **Neutron Monitoring [neutrons/(cm² sec)] Data at a Fast Neutron Flux of Approximately 1×10^9 neutron/(cm² sec) at the Target Holder**[b]

	With Water in Tank			Without Water in Tank		
Position	Fast (> 0.1 MeV)	Epithermal (> 0.14 eV)	Thermal	Fast (> 0.1 MeV)	Epithermal (> 0.14 eV)	Thermal
A (console)	0	25	30	9	100	90
B (hallway)	0	5	5	3	10	10
C (x-ray room)	0	10	10	5	100	100

[a] Cf. Fig. 4.18.

[b] Acceleration voltage = 150 kV, beam current = 500 μA, target age = 30 mA-min.

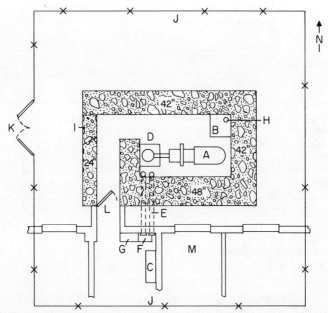

Fig. 4.19. Neutron activation facility at the National Lead Company of Ohio, Cincinnati, Ohio (adapted from reference 71). Acceleration voltage = 150 kV; neutron output = 5×10^{10} neutrons/sec. A, accelerator; B, HV supply; C, console; D, tank; E, underground U-tubes; F, counting equipment; G, sequential programming timer; H, helium-3 monitor; I, concrete-filled door; J, personnel fences; K, gate; L, access door; M, radiochemical laboratory.

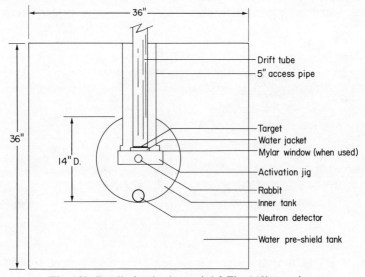

Fig. 4.20. Detail of activation tank (cf. Fig. 4.19), top view.

TABLE 4.26. Radiation Dose Levels[a] (mrem hr)

Position of Measurement	Area	Inner Tank Filled			Inner Tank Empty		
		Neutron		Gamma	Neutron		Gamma
		Fast	Thermal		Fast	Thermal	
Personnel door	L	46.9	1.0	0.7	41.7	1.0	0.6
Operating position	F	1.6	0.08	0.2	3.3	0.1	0.2
Radiochemical laboratory	M	0.3	0.02	< 0.2	1.0	0.03	< 0.2
Fence gate	K	1.0	0.01	< 0.2	1.8	0.02	< 0.2
Lower fence	J	1.3	0.01	< 0.2	2.1	0.02	< 0.2
Upper fence	J	2.8	0.04	< 0.2	5.8	0.06	< 0.2

[a] Acceleration voltage = 150 kV; neutron output = 5×10^{10} neutrons/sec.

rate data are shown in Table 4.26. Dose measurements were taken with the accelerator operating at 500-μA beam current on a new tritium target containing about 3 to 5 Ci/in.^{2} of tritium. This would correspond to a fast neutron output of about 5×10^{10} neutrons/sec at the target. The reason for the small reduction in the fast neutron dose at the location marked "Personnel Door" when the activation tank was filled with water has not been elaborated upon by the author.[71]

Examples 10 and 11. Two examples of acceptable shielding for sealed-tube neutron sources are shown in Figs. 4.21, 4.22, and 4.23. No shield survey data are available for these structures. However, from a critical inspection, it

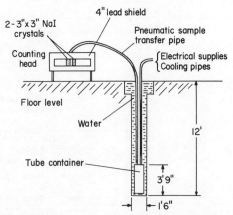

Fig. 4.21. Approved shielding configuration for sealed-tube neutron source (adapted from reference 72). Accelerating voltage = 120 kV; neutron output = 10^{10} neutrons/sec.

Fig. 4.22. The A-711 sealed-tube neutron source on down-hole mount. Source strength, 10^{11} neutrons/sec; maximum voltage, 190 kV. (Courtesy of Kaman Nuclear, Colorado Springs, Colorado.)

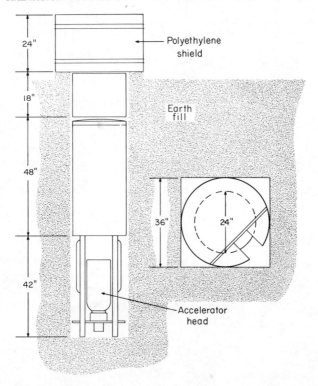

Fig. 4.23. Shielding requirements for A-711 sealed-tube neutron source (cf. Fig. 4.22).

is not too difficult to conclude that they represent safe construction in terms of biological dose. The hole-in-the-ground approach for shielding of these compact sources is particularly useful in terms of cost reduction.

V. SHIELDING OF DETECTOR SYSTEMS

Detector systems employing sodium iodide scintillators are mostly located near the generator facility. As such, they can be sensitive to high-energy prompt gamma emission during beam operation and, if poorly shielded, can also be activated by thermal neutrons. Since during counting the generator is usually turned off, the prompt gamma background causes no interference. This background during beam operation is fairly significant[65] and difficult to attenuate completely. On the other hand, activation of the detector itself is a much more serious problem. The principal activity produced from thermal or epithermal neutrons is in the iodine content of the sodium iodide crystal. The 25-min ^{128}I activity is a considerable nuisance in the gamma energy range up to about 1 MeV, and must be accounted for after allowance has been made for the extrapolation of this background as a function of neutron flux and time of sample measurement. When it is necessary to have the detector system close to the neutron source, a properly graded shield can reduce this source of background to insignificant proportions. Dibbs[67] has described a counting shield built especially for the reduction of the neutron-induced background. A schematic of this shield is shown in Fig. 4.24.

In summary, we must emphasize that after the biological shield has been designed and constructed, the entire area around the structure and in locations likely to be inhabited must be carefully surveyed. In many cases, errors in activation analysis with neutron generators have been attributed to higher neutron or gamma background than expected from shield calculations. When considering the fact that a great deal of thought and effort goes into the designing and developing of an analytical technique, prior knowledge of possible sources of errors resulting from weaknesses in the biological shield can save much time and effort. A review of literature on the many applications of the neutron generator in the field of activation analysis indicates that, in some analyses, the determination of generator-induced background was the key to the success of the analysis. Any major design change after initial construction should be treated as a completely new structure and duly evaluated for safety.

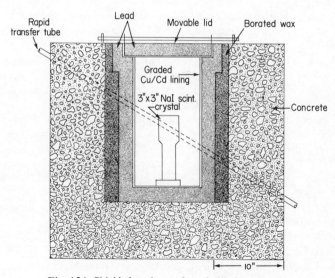

Fig. 4.24. Shielded enclosure for scintillation counter.

REFERENCES

1. "Protection against Neutron Radiation up to 30 Million Electron Volts," National Bureau of Standards Handbook 63, April 1967.
2. N. G. Goussev, "Relationship between Dose Equivalent (Absorbed Dose) and Fluence (Flux Density)" in *Engineering Compendium on Radiation Shielding*, Vol. I, R. G. Jaeger (Editor-in-Chief), Springer-Verlag, New York, 1968, p. 12.
3. "Standard for Protection against Radiation," *Federal Register*, Title 10, Atomic Energy, Part 20, November 17, 1960.
4. A. T. Nelms, "Energy Loss and Range of Electrons and Positrons," National Bureau of Standards Circular 577, 1956, and Supplement, 1958.
5. "Reactor Physics Constants," Argonne National Laboratory, USAEC Report ANL-5800, 2nd ed., 1963, p. 695.
6. D. G. Jacobs, "Sources of Tritium and Its Behavior upon Release to the Environment," Critical Review Series, USAEC Report TID-24635, 1968.
7. "Report of ICRP Committee II on Permissible Dose for Internal Radiation, 1959," *Health Phys.*, **3**, 1 (1960).
8. *Federal Register*, Title 10, USAEC, Part 20, 1962.
9. E. A. Pinson and W. H. Langham, *J. Appl. Physiol.*, **10**, 108 (1957).
10. E. A. Pinson and E. C. Anderson, "The Absorption, Distribution and Excretion in Tritium in Man and Animals," USAEC Report AECU-937, 1950.
11. E. A. Pinson, Los Alamos Scientific Laboratory, Report LA-1464, 1952.
12. J. N. P. Lawrence, Los Alamos Scientific Laboratory, Report LA-2163, 1958.
13. E. A. Pinson, Los Alamos Scientific Laboratory, Report LA-1469, 1952.
14. E. C. Anderson and W. H. Langham, Los Alamos Scientific Laboratory, Report LA-1646, 1954.
15. E. A. Pinson, Los Alamos Scientific Laboratory, Report LA-1218, 1951.
16. T. T. Trujillo, E. C. Anderson, and W. H. Langham, Los Alamos Scientific Laboratory, Report LA-1986, 1955.
17. L. K. Akers and H. E. Banta, "Solid Tritium Target Quality Studies," ORINS Internal Report, Oak Ridge National Laboratory, Oak Ridge, Tennessee, 1963.
18. D. O. Nellis, E. L. Hudspeth, I. L. Morgan, P. S. Buchanan, and R. F. Boggs, "Tritium Contamination in Particle Accelerator Operation," U.S. Department of Health, Education and Welfare, Public Health Service Publication No. 99-RH-29, 1967.
19. J. E. Suddueth and S. S. Nargolwalla, "Activation Analysis with the Cockcroft-Walton Neutron Generator," in *Activation Analysis Section: Summary of Activities July 1968 to June 1969*, P. D. LaFleur (Ed.), NBS Technical Note 508, issued July 1970.
20. "Operation Manual for Model A-1001 Neutron Generator," Kaman Nuclear Corporation, Colorado Springs, Colorado, 1965.
21. H. Hollister, *Nucleonics*, **22**, 68 (1964).
22. V. E. Hoffman, *Ind. Res.*, **6**, 46 (1964).
23. R. F. Boggs, R. Jacobs, and R. L. Leupold, "Radiological Health Problems Associated with the Operation of Neutron Generators," 11th Annual Meeting, Health Physics Society, Houston, Texas, June 1966, unpublished.

REFERENCES

24. J. Biŕo, I. Fehér, and T. Szarvas, "The Tritium Incorporation Hazard Involved in the Operation of Neutron Generators," Proceedings of the Second Symposium on Health Physics, Pecs, Hungary, September 1966, Vol. 1, Budapest, Central Research Institute for Physics, 1966, pp. 87–92.
25. E. E. Watson, R. G. Cloutier, and J. D. Berger, Health Phys., **17**, 739 (1969).
26. J. A. B. Gibson, AERE (G.B.), Report AERE-M 1169, 1963.
27. H. L. Butler and R. W. Van Wyck, "A Synopsis of Studies Related to Tritium Monitoring and Personnel Protective Techniques," E. I. du Pont de Nemours, Report DP-329, 1959.
28. J. A. Aberle, A. R. Macdonald, and S. W. Porter, Jr., "A Procedure for Minimizing Personnel Hazard while Changing Tritium Targets in Accelerators," Armed Forces Radiobiology Research Institute, Bethesda, Maryland, Report AD-636, 185, 1966.
29. L. Battist and J. Swift, "Procedure for Tritium Target Replacement in a Cockcroft-Walton Positive Ion Accelerator," Proceedings, Conference on the Use of Small Accelerators for Teaching and Research, Oak Ridge Associated Universities, Oak Ridge, Tennessee, April 1968, USAEC Report CONF-680411, 1968, p. 144.
30. R. L. Zimmerman, R. M. Boyd, J. R. Wright, and M. W. Mitchell, Proceedings, Conference on the Use of Small Accelerators for Teaching and Research, Oak Ridge Associated Universities, Oak Ridge, Tennessee, April 1968, USAEC Report CONF-680411, 1968, p. 138.
31. S. W. Porter, Jr., and L. A. Slaback, Jr., "Measurement of Tritium Surface Contamination," Armed Forces Radiobiology Research Institute, Bethesda, Maryland, Report AFRI SP 66-13, 1966.
32. D. E. Barnes, "Basic Criteria in the Control of Air and Surface Contamination," Symposium on Health Physics in Nuclear Installations, Riso, Sweden, 1959, pp. 47–52.
33. H. J. Dunster, Health Phys., **8**, 353 (1962).
34. M. J. Engelke and E. A. Bemis, Jr., Los Alamos Scientific Laboratory, Report LA-2671, 1962.
35. J. E. Hoy, Health Phys., **6**, 203 (1961).
36. R. C. Cooley, Sr., "Controlling Tritium Hazards around Heavy Water Moderated Reactors," E. I. de Pont de Nemours, Report DP SPU 66-30-13, 1968.
37. W. C. Reinig and E. L. Albenesius, Am. Ind. Hyg. Assoc. J., **2** (3), 276 (1963).
38. E. L. Geiger, Health Phys., **14**, 51 (1968).
39. T. S. Iyengar, S. H. Sadarangani, S. Somasundaram, and P. K. Vaze, Health Phys., **11**, 313 (1965).
40. A. M. Valentine, Los Alamos Scientific Laboratory, Report LA-3916, Los Alamos, New Mexico, 1968.
41. R. V. Osborne, Atomic Energy of Canada Limited, Report AECL-2699, Ottawa, Canada, 1967.
42. L. R. Day and M. L. Mullender, AERE (G.B.), Report AWRE NR-1/63, 1963.
43. M. H. McTaggart, AERE (G.B.), Report AWRE NR/A-1/59, 1959.
44. G. Olive, J. F. Cameron, and C. G. Clayton, AERE (G.B.), Report AERE-R3920, 1962.

45. B. T. Price, C. C. Horton, and K. T. Spinney, *Radiation Shielding*, Pergamon Press, London, 1957.
46. A. F. Avery, D. E. Bendall, J. Butler, and K. T. Spinney, AERE (G.B.), Report AERE-R3216, 1960.
47. Brookhaven National Laboratory, "Neutron Cross Sections," Report BNL-325, 2nd ed., Supplement 1 (1960) and Supplement 2, five volumes, 1964–1966.
48. R. S. Caswell, R. F. Gabbard, D. W. Padgett, and W. P. Doering, *Nucl. Sci. Eng.*, **2**, 143 (1957).
49. V. P. Duggal, S. M. Puri, and K. S. Ram, *Nucl. Sci. Eng.*, **5**, 199 (1959).
50. J. J. Broerse and F. J. Van Werven, *Health Phys.*, **12**, 83 (1966).
51. R. G. Cloutier, *Am. Ind. Hyg. Assoc. J.*, **24**, 497 (1963).
52. D. Nachtigall and M. Heinzelman, University of California Radiation Laboratory, Report UCRL-Trans-10128, 1966.
53. J. D. Berger, E. E. Watson, and R. J. Cloutier, "Health Physics Aspects of Low-Energy Accelerators," Proceedings, Conference on the Use of Small Accelerators for Teaching and Research, Oak Ridge Associated Universities, Oak Ridge, Tennessee, April 1968, USAEC Report CONF-680411, 1968, p. 112.
54. J. J. Broerse, *Kerntechnik*, **9**, 446 (1967).
55. F. H. Clark, Oak Ridge National Laboratory, Report ORNL-TM-1655, 1966.
56. J. T. Prud'homme, "Texas Nuclear Corporation Neutron Generators," Texas Nuclear Corporation, Austin, Texas, 1962.
57. S. T. Lindenbaum, Proceedings, Conference on Shielding of High-Energy Accelerators, New York, USAEC Report TID-7545, 1957, p. 101.
58. R. H. Thomas, "Proton Synchrotron Accelerators," in *Engineering Compendium on Radiation Shielding*, Vol. I, R. G. Jaeger (Editor-in-Chief), Springer-Verlag, New York, 1968, p. 65.
59. C. F. Cook and T. R. Strayhorn, "Laboratory Shielding," in *Fast Neutron Physics*, Part II, J. B. Marion and J. L. Fowlers (Eds.), Wiley, New York, 1953, p. 807.
60. T. Rockwell, III, Ed., *Reactor Shielding Design Manual*, Van Nostrand, Princeton, New Jersey, 1956.
61. H. Goldstein, *Fundamental Aspects of Shielding*, Addison-Wesley, Reading, Massachusetts, 1959.
62. A. O. Hanson and J. L. McKibben, *Phys. Rev.*, **72**, 673 (1947).
63. J. Hacke, *Int. J. Appl. Radiat. Isot.*, **18**, 33 (1967).
64. "Particle Accelerator Safety Manual," prepared by W. M. Brobeck and Associates, California, National Center for Radiological Health, Department of Health, Education and Welfare, Report MORP 68-12, Rockville, Maryland, 1968.
65. S. S. Nargolwalla, Ph.D. Thesis, University of Toronto, Toronto, Canada, 1965.
66. A. Barwiński, L. Górski, J. Janczyszyn, K. Korbel, S. Kwieciński, L. Loska, and J. Ostachowicz, *Sonderdruck aus "Isotopenpraxis"*, **8**, 322 (1966).
67. H. P. Dibbs, "Activation Analysis with a Neutron Generator," Department of Energy Mines and Resources, Mines Branch, Ottawa, Canada, Research Report R155, February 1965.
68. E. P. Przybylowicz, private communications, July 31, 1969 and December 11, 1970.

69. J. E. Suddueth and S. S. Nargolwalla, "Activation Analysis with 14 MeV Neutrons," in *Activation Analysis Section: Summary of Activities July 1967 to June 1968*, P. D. LaFleur (Ed.), National Bureau of Standards Technical Note 458, issued March 1969.
70. W. W. Meinke and R. W. Shideler, Michigan Memorial Phoenix Project, Report MMPP-191-1, Ann Arbor, Michigan, 1961.
71. B. L. Twitty, "Neutron Activation Facility at the National Lead Company of Ohio, Cincinnati, Ohio," Report NLCO-955, Summary Technical Report, April 1, 1965 to June 30, 1965, issued August 11, 1965.
72. J. D. L. H. Wood, D. W. Downton, and J. M. Bakes, "A Fast-Neutron Activation System with Industrial Application," Proceedings, 1965 International Conference on Modern Trends in Activation Analysis, College Station, Texas, 1965, pp. 172–174.

CHAPTER

5

PREPARATION AND TRANSPORTATION OF SAMPLES

I. INTRODUCTION

The accuracy of any method of chemical analysis depends to a large measure on the care taken by the experimenter in selecting and preparing the sample. In most nondestructive nuclear activation techniques the added requirement of sample encapsulation introduces a possible source of error. Meaningful data can be obtained only if the analyst is fully aware of the sample history and reasonably sure of its integrity in terms of approved sampling procedures, eliminating contamination of the sample after sampling and, if relevant, satisfactory sample encapsulation techniques. The unique advantages offered by these methods of ultrasensitivity and selectivity for most of the elements in the periodic table can be drastically affected by inadequate sampling and encapsulation techniques. Although precautions applicable to sample contamination and encapsulation relate to nuclear activation methods in general, the specifics of sample preparation methods and sample containment, and sometimes preirradiation cleaning procedures, vary widely, depending on the analysis being carried out.

Since the activation method has often been referred to as one in which "blank" problems are nonexistent, it is necessary only to exercise care in the preirradiation treatment of the sample. After irradiation, the sample may be handled as deemed appropriate for the particular analytical method.

The application of neutron generators to activation analysis has led to the development of numerous sample preparation techniques, some of which are unique to this method of analysis. In this chapter the essential criteria for the selection of suitable sample containers are discussed. Methods for encapsulation of solids, powders, metallic rods and disks, liquids, and reactive metals are described. Particular emphasis is given to special procedures for sample preparation and to apparatus required for performing these functions. It is most important to emphasize that, although the neutron generator has been demonstrated as a highly accurate and specific tool for compositional analysis, developments in reproducible sample preparation techniques, encapsulation procedures, storage, and irradiation under controlled environmental conditions have helped much to minimize systematic errors in analyses.

Activation analysis with neutron generators has been almost completely devoted to the use of nondestructive analytical methods. In view of the limited life of the tritium target for neutron production, radioactive nuclides of relatively short half-lives are invariably used. Since a biological shield is necessary, and complicated interlock safety considerations often make manual sample retrieval a time-consuming task, after irradiation it is necessary to remove the sample rapidly by remote means for the measurement of its induced activity. In general, therefore, neutron generator facilities are equipped with some mechanism for rapid insertion and removal of the sample to and from the irradiation site, respectively. In later sections of this chapter, the principal methods of sample transfer are described. Important considerations in the construction of the transfer apparatus are emphasized. Finally, methods for automation of analytical procedures are discussed, particular attention being paid to the utilization of sequence programmers and associated pneumatic tube system hardware.

II. SAMPLE PREPARATION

A. SAMPLE CONTAINER

Selection of suitable materials for container fabrication, and the geometrical configuration of the sample container itself often establish the design limitations of a number of key components of a neutron generator facility, for example, pneumatic tube size, sample holders for irradiation and counting, and the like. Further, it is important to use sample containers whose integrity is guaranteed during the course of several analytical cycles necessary for repetitive nondestructive analysis of a single sample. In many pneumatic tube systems the sample travels at velocities in the region of 100 ft/sec. The momentum transfer, particularly from heavy samples, at the terminal ends of the tube system can be severe enough to damage sensitive devices at these locations seriously. The selection of a container material, therefore, is based not only on the desired low activation response from the container but also on its structural properties. In addition to these features, acceptable containers must satisfy the criteria of versatility and adaptability, such that various shapes and sizes of samples can be accommodated using a minimum number of container types.

1. SELECTION OF CONTAINER MATERIAL

Favorable structural properties for use in high-speed pneumatic transfer systems and negligible radioactivation by high-energy neutrons have been the deciding factors for the almost universal adoption of polyethylene sample containers. Since 14-MeV neutron activation analysis lends itself readily to nondestructive analysis for trace oxygen, special efforts have been made to manufacture polyethylene containers with negligible amounts of oxygen as a contaminant. Early attempts by Anders and Briden[1] to mold "low-oxygen" polyethylene containers were prompted by the need to analyze oxygen at the 10-ppm level in cesium. By carefully controlling conditions during the molding and extrusion processes, and avoiding the use of release agents, they succeeded in manufacturing about 8-cm^3-volume polyethylene vials weighing about 3.5 g and containing only about 0.15 mg of oxygen. The oxygen content of these vials was about one-twentieth of that found in similar vials commercially available at that time. Since then, polyethylene containers with oxygen contents comparable to those fabricated by Anders and Briden have become available at about 15 cents each in lots of several thousands. If the vials are sealed in multiple polyethylene bags under nitrogen atmosphere, the limited information available appears to indicate that long-term stability of these vials during storage is satisfactory. For all other analyses except oxygen, polyethylene containers similar to those used by Anders and Briden are ideally suited as sample carriers. These vials contain as much as 3.5 mg of oxygen; however, their price is about 80% less than the low-oxygen vials and therefore they are economically suitable for use in pneumatic transfer systems. Because of their resiliency, polyethylene sample containers are also suitable for rapid transfer of samples without danger of rupture. In addition, because of the low melting point of polyethylene, the vials can be heat-sealed easily. These containers can handle liquid samples and in some cases even reactive metals. The standard vial most commonly used has a volume of about 8 cm^3 and can hold steel samples for oxygen determination weighing as much as about 60 g.

Except for the special case of oxygen analysis at the 50-μg level or less, and in certain cases of oxygen analysis in reactive metals, the vials described above are extremely satisfactory. For oxygen levels below 50 μg, particularly in reactive metals, aluminum, molybdenum, or niobium containers must be used. Steele and Lukens[2] used aluminum capsules (2S Type 1100) containing 50 μg of oxygen, with a special scaling technique to analyze for trace oxygen in cesium.

Copper containers[3] have also been employed for the encapsulation of potassium for the determination of its trace oxygen content. However, Harris[4] has pointed out that in the case of aluminum containers, the special sealing process could add to the initial oxygen content of the container. Methods for using these containers are described in somewhat more detail later.

2. GEOMETRICAL CONFIGURATION OF CONTAINER

For convenience in transportation in pneumatic tubes, the cylindrical sample container is best, although rectangular samples without containers

II. SAMPLE PREPARATION

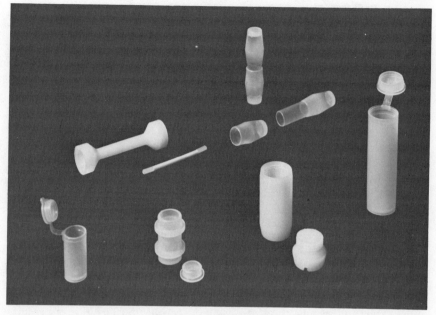

Fig. 5.1. Sample containers.

have been propelled in rectangular-shaped pneumatic tube systems[5] satisfactorily. Various diameters and lengths of containers have been employed, depending on the diameter of the tube system and tube curvatures used. Cylindrical snap-cap polyethylene vials are used widely at the majority of facilities throughout the world. Some analysts prefer a screw-cap vial to the snap-cap type. Examples of some typical shapes and sizes of polyethylene containers generally used are illustrated in Fig. 5.1. It need only be emphasized that efficient sample transport is effected by minimizing the difference in diameters of the container and the inside wall of the tube system. If the inner diameter of the pneumatic tube is much larger than the outer container diameter, the sample container has a tendency to spin in the tube, resulting in negligible or nonreproducible linear motion.

It is often necessary to modify sample containers to facilitate or to improve a certain analysis, or to accommodate an odd-shaped sample. For example, by perforating a standard 8-cm^3 polyethylene vial, it is possible to eliminate the effect of environmental oxygen, which is a source of systematic error, in the trace determination for oxygen at levels below 100 ppm. This "flow-through" container, developed at the National Bureau of Standards,[6] was used in the certification of a standard reference steel for oxygen. The advantages of this design were twofold. First, the weight of a standard vial

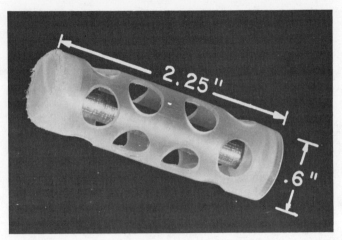

Fig. 5.2. Flow-through sample container.

was reduced substantially by the perforations, thus improving the signal-to-noise ratio; and second, the air entering the capsule during either encapsulation or transport was completely displaced by the nitrogen gas used as the propellant during the first pass of the sample container through the tube system. Such vials are only useful for the encapsulation of rods and disk-like shapes. However, powders and liquids can also be contained in an inner liner which is sealed at both ends and then centrally located in a "flow-through" capsule. Since the inner liner is completely filled, the problem of environmental oxygen leaks does not arise. A typical "flow-through" irradiation container in which a $\frac{1}{4}$-in.-diameter steel rod is located is shown in Fig. 5.2. The weight of this container, which has a nominal volume of 8 cm^3, is approximately one-half that of a similar unperforated capsule. The $\frac{1}{4}$-in.-diameter steel rod weighs about 12 g. Using "low-oxygen" vials with a nominal oxygen content of 50 ppm, a perforated vial weighing 1.5 g contains about 75 μg of oxygen. Under these conditions, steel rods weighing 12 g and containing oxygen at the 5-ppm level can be analyzed at the normal sensitivity capability of a typical generator facility of approximately 2 counts/μg of oxygen.

The obvious answer to the container blank problem in oxygen analysis is, of course, to eliminate the use of the container altogether, and some work[7] has been reported along these lines. More sophisticated techniques have also been developed to strip the capsule from the sample after irradiation and before counting.[8] Although the improvement in the signal-to-noise ratio is obvious, certain features of these methods impose limitations on the precision and accuracy that can be expected from following the "no container" approach. For example, handling of the sample, after a perhaps difficult initial sample-cleaning procedure, between replicate analyses is unavoidable. This no-container procedure, therefore, increases considerably the probability of sample contamination. In addition, the possibility that the sample may pick up ^{16}N recoils from the ^{16}O(n, p)^{16}N reaction with oxygen in the container is significant. This recoil effect is considerably augmented if air is used as the propelling gas.

B. SAMPLE ENCAPSULATION

A variety of techniques have been used for sample encapsulation. The mechanics of each technique depend on the nature and the physical characteristics of the sample. As mentioned earlier, great importance is given to this aspect of any analytical procedure since the ultimate precision of the analytical result can be markedly affected if suitable precautions are not taken at this stage. Some of the common sample preparation and encapsulation methods employed in activation analysis with neutron generators are described below.

1. POWDER SAMPLES

In the majority of cases, the sample to be analyzed is usually in the form of a finely ground powder. Care is therefore necessary to ensure homogeneity of density in the packaging of sample powders in polyethylene vials. Much patience is required to compact only small amounts of the total sample at a time. During each compaction step, the pressure applied to the sample must be uniformly distributed over the entire sample area, since the irradiation and counting processes are strongly geometrically dependent, and the amount of sample material per unit volume should be constant. In addition, sufficient pressure must be applied to minimize or eliminate segregation during sample transfer. In their efforts to obtain very high precisions, Dyer et al.[9] prepared pressed pellets or cylinders of calcium fluoride. A compaction technique was effectively used for uniform sample preparation for most of the analytical studies carried out in the neutron generator laboratory at the National Bureau of Standards.[10] The need for proper sample compaction was recognized by Anders and Briden[11] when examination of samples between irradiations indicated segregation from pneumatic tube operations. Errors in precision of up to 20% were noted when loosely packed samples were used. Their ultimate approach was to compact all samples with a levered pestle-type device. In the discussion above, it has been assumed that the element to be determined is homogeneously distributed within the sample. If this is not the case, loss of precision can occur from sample orientation during irradiation or counting. The effect of segregation, as described by Twitty and Fritz,[12] appears to indicate that errors due to segregation can be observed readily. In the analysis for oxygen in magnesium chips, a significant analytical bias was observed due to the difference in the physical characteristics of the magnesium chips. Twitty and Fritz have concluded that the total precision in the analysis of highly segregated chip-type samples can be improved simply by using a better sampling technique.

2. LIQUID SAMPLES

Unlike powders, liquid samples can be encapsulated rather easily. However, certain difficulties are experienced in sealing which could affect the integrity of the sample during multiple passes through a pneumatic transfer system. In particular, air bubbles must be eliminated during encapsulation. Several methods for sealing liquid samples in polyethylene vials have been developed. However, to avoid contamination of the pneumatic system, it is strongly recommended that samples be examined for possible rupture after each irradiation. The methods described here are by no means the only ones available. However, they are adequate for the efficient encapsulation of aqueous solutions, reactive-element castings, photographic emulsions, hydrocarbon oils, and slurries.

Two basic methods are generally employed. In the first technique the sample container is capped while it is immersed in the liquid sample. The filled sample container is then removed from the liquid reservoir and sealed with a hot iron. In most instances a small air bubble is introduced during the sealing. For most liquids this simple approach yields samples which are adequate for routine analysis where the precision required is only of the order of 5 to 10%. In some instances, however, where corrosive or reactive liquids are used, this technique is difficult to apply and inert atmosphere hoods or glove boxes are necessary for encapsulation.

The second technique is a longer procedure; however, the results obtained more than compensate for the extra time. The polyethylene vial is first sealed effectively. The liquid sample is then drawn from the sample supply by a fine hypodermic needle and injected into the vial. A small hole must be drilled in the vial cap to allow passage of air from the vial to the atmosphere. After the vial has been completely filled, the two small holes can be sealed carefully with a hot iron. Air bubbles, if any, are extremely small. A variation of this method is to place the sample and capsule in a vacuum desiccator and draw in the liquid from a container by producing a vacuum in the desiccator through a small hole in the container cap. The sample is then removed and the small hole sealed with a hot iron. In the analysis of photographic emulsions, Przybylowicz et al.[13] prepared homogeneous photographic emulsions in sample vials by melting the gelatin-containing silver halide emulsions at about 50°C and immersing the vials in an ultrasonically agitated water bath to ensure homogeneity. The samples were then introduced into the polyethylene capsules with a hypodermic syringe.

Another method of sealing polyethylene snap-cap containers is described by Anders and Briden[11] and Nargolwalla.[14] This method lends itself very well to powdered and gelatinous-type materials. It is especially useful if the sample capsule is being spun by an air jet as it rests on top of an arrestor pin during irradiation and counting. To permit proper spinning, the cap end of the cylindrical container is molded into a cone. Therefore the sealing operation has to be such as to produce a good seal and mold the capsule to this shape. A brass cup heated electrically or on a Bunsen flame is used. The filled capsule is inverted and pressed into the cup of the brass sample sealer. The seal is effected as the end of the sample container softens and moves into the cup assuming a conical end suitable for sample spinning. With a little practice the experimenter can perform many such seals very rapidly without destroying the capsule. Polyethylene capsules can also be sealed efficiently by placing an electrically heated metal ring around the snap cap. This technique is also satisfactory with liquid samples.

3. METALLIC RODS AND DISKS

The principal application of activation analysis using neutron generators has been in metallurgy. For the analysis of oxygen in steels, metal rods and disk-shaped samples are generally used. However, the diameter of the rods seldom allows a snug fit inside the normal $\frac{5}{8}$-in.-diameter polyethylene vial and polyethylene inserts therefore have to be used to center the particular rod. Small polyethylene pads or disks are often necessary to reinforce the ends of the capsule, since during pneumatic transfer a rod whose diameter is less than the inside diameter of the capsule has a tendency to puncture the vials. Metal, disk-shaped samples can be stacked inside the vial and have been used extensively in the study of inhomogeneity effects on the analysis for oxygen in steels.

4. REACTIVE METALS

The need for oxygen analysis in reactive metals like sodium, potassium, cesium, and barium has placed special requirements on encapsulation procedures. Steele and Lukens[2] used aluminum-vial encapsulation in a controlled atmosphere. Experiments conducted by them on cesium samples encapsulated in aluminum containers sealed by a patented process showed that runs performed over a two-month period showed no air leaks. In such cases, polyethylene vials are of limited use. Steele et al.[15] used polyethylene, stainless steel, or hydrogen-fired OFHC copper capsules to analyze for oxygen in potassium. All sample handling was performed in a specially constructed inert atmosphere chamber filled with argon gas. Polyethylene capsules have also been used by Anders and Briden[1] for oxygen analysis at the 10-ppm level in cesium. Nargolwalla and Suddueth[16] used low-oxygen polyethylene vials for the analysis of barium; the samples were prepared under a bath of mineral oil. For oxygen analysis in reactive metals, low-oxygen polyethylene vials are satisfactory, provided extreme care is taken to ensure against leaks or rupture. For long-term storage, contamination of the sample may occur through diffusion. Polypropylene vials, however, are better suited for long-term storage of reactive-metal standards. The analysis for aluminum and chlorine in organic solutions was also carried out by Hull[17] and, because the organic solvents tended to penetrate the polyethylene vials, analyses had to be performed within a few hours after encapsulation in an inert atmosphere.

The various types of sample containers and encapsulation techniques described above serve only as examples, and are by no means the only ones available. In fact, it is strongly recommended that the experimenter develop methods best suited to his needs. The many reported methods and materials used have resulted from innovative experiments on the part of individual researchers.

C. SAMPLE HANDLING AND CLEANING

The phase of sample handling and cleaning plays a significant part in the preparation of samples. For example, in the determination for trace oxygen in metals, special care has to be taken to guard against contamination of the sample both during and after the cleaning procedure. It is important that, during sample cutting, heat generation be minimal. A clean hand-hacksaw serves the purpose well. Any filing operation must be carried out with a new file, which must be cleaned carefully before and after use. Steel samples can be cleaned efficiently in ultrasonic baths charged with detergent solution or trichloroethylene. In the certification of the oxygen contents of standard steels, samples were prepared using the procedures mentioned above with excellent results.[18] The cleaned samples were subsequently encapsulated in special containers under a nitrogen atmosphere and analyzed. The efficacy of the sample preparation techniques described above was evident from the examination of comparative and of analytical data for oxygen in valve steel standards,[6] using various approaches.

It is often not easily realized that sample storage is an important aspect of the analytical procedure. Since fast neutron analyses with neutron generators employ short-lived nuclides, the carefully prepared standards can be used repeatedly. It is therefore important that steel standards be stored in an inert gas atmosphere if they are to be used again.

III. SAMPLE TRANSPORT

In tracing the many developments in the field of neutron activation analysis, two obvious conclusions can be drawn. First, the method itself is inherently precise; and second, the technique has been inhibited owing to the lack of adequate sample-handling instrumentation. In examining the wide variety of equipment, both mechanical and electronic, required by the activation analyst, the sample-handling hardware has proved slow in developing. In view of this deficiency, considerable efforts have been made by manufacturers of neutron generators and analysts to bring this necessary and important aspect of instrumentation to a high state of reliability. Today, computer-coupled facilities are available which embody a variety of sophisticated units programmed by the analyst.

Much of the recent impetus in the development of sample transport equipment has been a result of thorough studies of major sources of analytical error. When optimal systems are used, it can be generally concluded that the precision of the analysis with neutron generators is almost entirely dictated by Poisson counting statistics and, under certain conditions, with analytical biases which are small in comparison to the experimental standard deviation.

It is of interest to emphasize some of the key sources of analytical error which should be considered when designing sample handling and transport systems. These are as follows:

1. Imprecision in irradiation, decay, and counting times.
2. Nonreproducibility in sample placement at the irradiation and counting stations.
3. Variation in the sample matrix.
4. Inhomogeneity of the sample.
5. Instability or drift in electronic circuitry.

Because a comprehensive treatment of the major sources and elimination of (or correction for) systematic errors is presented in Chapter 6, only the mechanical aspects of sample transport systems are described here.

A. SAMPLE TRANSPORT SYSTEMS

Several approaches have been adopted for rapid and remote transfer of sample containers to and from the irradiation site. Since activation analysis with neutron generators has been predominantly limited to the generation and measurement of short-lived radioisotopes, the emphasis in the development of suitable sample transfer systems has been on speed. The majority of such systems have been specifically tailored to the important application of oxygen determination using the reaction

$$^{16}O(n, p)^{16}N \ (T_{1/2} = 7.2 \text{ sec})$$

An automatic system for the determination of oxygen in beryllium metal components, as described by Byrne et al.,[19] includes a sample transfer mechanism operating on the pendulum principle. Essentially it consists of a square platform constructed from thick polyethylene, and supported by special arms. The platform, acting as a carriage, is moved by an electric motor to the activation position directly below a vertically mounted neutron generator. A clutch on the drive motor is disengaged before irradiation, thus allowing the carriage to swing through an angle of approximately 170° from the irradiation to the counting location at the end of irradiation.

A rather elaborate system of sensing switches, latches, and cams is included to make the transfer system fully automatic. The pendulum system was specially designed to handle 500 to 2000 g of beryllium components for oxygen determinations at the 1% level. Analysis of a selected beryllium component over a period of time gave a precision of about 3%. The number of determinations made during this time study varied from 20 to 60.

One of the earlier and highly efficient sample transport systems used with a neutron generator has been described by Meinke and Shideler.[20] This system permitted the transfer of a sample container ("bunny" or "rabbit") from the irradiation to the counting location in about 3 sec, a distance of approximately 100 ft. The pneumatic tube was $\frac{1}{2}$-in.-i.d. aluminum. The rabbit was propelled by a vacuum cleaner suction motor drawing about 6 in. of water pressure. The sample was loaded through an entry port adjacent to the neutron generator console and was stopped in front of the detection system by a pair of spring-loaded fingers operated by a solenoid. The arrival of the sample was sensed by the interruption of a light beam, which triggered a photoelectric relay shutting off the vacuum motor and initiating counting of the sample.

A simple pneumatic tube sample transfer system, described by Cerrai and Gadda,[21] is shown schematically in Fig. 5.3. This system was utilized for oxygen determinations in zirconium metal by 14-MeV neutron activation. In the diagram, the transfer system is powered by a blower and the total

III. SAMPLE TRANSPORT

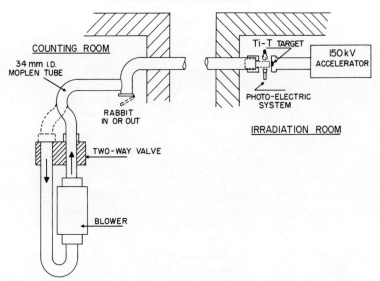

Fig. 5.3. Pneumatic transport system (cf. Cerrai and Gadda[21]).

distance of sample travel is about 100 ft. A two-way valve switches the sample path from the compression to the vacuum side of the blower to facilitate sending and receiving of the sample container to and from the irradiation location, respectively. During irradiation, the rabbit is firmly lodged against a metal seat by continuous air pressure. A photocell located at the irradiation site is used to start a preset irradiation timer. A similar system has also been described by Veal and Cook.[22]

In a classic study of self-shielding effects inherent in thick-sample analysis with neutron generators, and specifically related to oxygen analysis, Anders and Briden[11] have described an efficient sample transfer system. This system, (Fig. 5.4) employs special devices to reduce errors resulting from the nonhomogeneous and nonuniform nature of the fast neutron irradiation flux. After the sample has been packaged in a 2-dram (~ 8 cm^3) polyethylene snap-cap vial, the container is introduced at the sending station, S. Air pressure is applied from a pressure-regulated ballast tank, B, at about 18 psi, by a solenoid valve to transport the sample to the irradiation site, I. The sample is stopped by a solenoid-operated retractable pin, P_1, inside a cadmium-covered aluminum tube, directly in front of the neutron generator target. At the irradiation site, the sample is spun by an air cyclone set up by a laterally applied air jet. After irradiation, the sample is released by retracting the arrestor pin and transferred by a second air blast to the counting station,

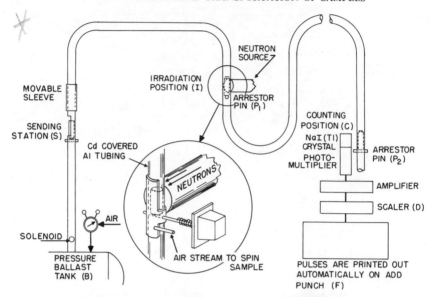

Fig. 5.4. Pneumatic transport system (cf. Anders and Briden[11]).

C. While the first sample is in transit, the arrestor pin, P_1, is returned to accept a second sample. The first sample arrives at the counting station C in about 1 sec, and is stopped there by a second arrestor pin, P_2, and a scaler is turned on 2 sec after sample arrival for 20 sec. After counting is complete, the total number of counts is printed out on an add-punch, F, and the scaler is reset for the next sample. After another 30 sec, pins P_1 and P_2 are withdrawn. The first sample is ejected and the second sample is transferred to the counter. Anders and Briden were able to process about 30 samples every hour using this system. Automatic sequencing of the various steps was performed by a multicam timer. To improve reproducibility, a sample spinner was installed at the counting position also. Although, initially, some difficulty was experienced in spinning, proper location of the air jet apparently solved this problem. At a spinning air pressure of about 15 psi, samples of 1.9 cm i.d. and weighing up to 50 g could be spun.

A simple single-tube pneumatic transfer system (Fig. 5.5), as described by Dibbs,[23] utilizes compressed air for sending and receiving samples. Solenoid valves are used to blow air at 60 psi. The air, in each case, is admitted for a time controlled by a preset timer to ensure arrival of the sample. The transfer tubes are of $\frac{3}{4}$-in.-i.d. rigid polyvinyl chloride. Transfer times of about 1 sec over a distance of about 35 ft were maintained for different sample weights by varying the air pressure. To ensure reproducible positioning of the

III. SAMPLE TRANSPORT

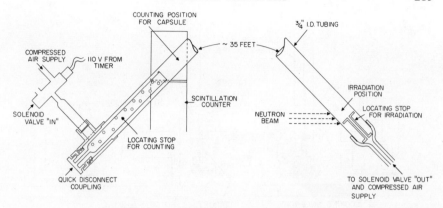

Fig. 5.5. Pneumatic transport system (cf. Dibbs[23]).

sample containers in the neutron beam and at the counting station, the transfer tubing in these locations was inclined at an angle of about 35°. A multicam timer system was used to program the various steps necessary for a typical irradiation-count cycle.

Efforts to develop acceptable routine analytical methods for industrial application led to several extensive studies of factors affecting precision. As mentioned previously, Anders and Briden[11] clearly illustrated the improvement in precision by rotation of a single sample on an axis perpendicular to the direction of the deuteron beam from a neutron generator. Mott and Orange[24] developed an irradiation system consisting of two pneumatic tubes, one for transferring a standard between the irradiation position and the detector, and the other for transferring the unknown sample. Both samples were rotated simultaneously at 300 rev/min around an axis parallel to the beam axis during irradiation. Within a second after the end of irradiation, the rotating assembly could be stopped automatically by a magnetic brake and locked into an ejection position by a solenoid. After being secured, the samples could be returned to their respective counters. A schematic diagram of their design is shown in Fig. 5.6.

The first dual sample-biaxial rotating assembly coupled to a completely automated pneumatic transfer system was reported by Wood et al.[25] Since a patent on the assembly is pending the design details of the sample rotator were not given. Priest et al.[26] have described a unique sample irradiation assembly in which several samples could be rotated simultaneously about a common axis, their own axes, and moved laterally through the neutron flux. This motion was designed to compensate for any irradiation asymmetry.

Air-driven biaxial rotation of several samples has been developed by Dyer

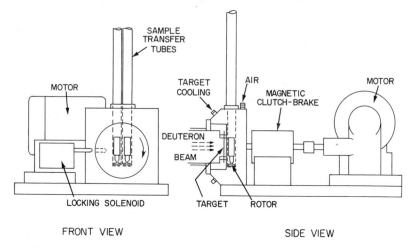

Fig. 5.6. Dual-sample rotator with pneumatic tube (cf. Mott and Orange[24]).

et al.[9] and their design is shown in Fig. 5.7. The samples are loaded and retrieved manually. The sample rotator consists of two concentric plastic cylinders, the inner cylinder holding the samples. Both cylinders are rotated, the outer cylinder end-over-end by a motor-driven hollow shaft attached at right angles to the outer well. The inner cylinder is rotated on its own axis by an air jet and its speed of rotation can be adjusted to several hundred revolutions per minute by a needle valve in the air-handling system. The outer cylinder rotational speed is 60 rev/min.

To improve rotating sample assemblies further, Lundgren and Nargolwalla[10] developed a variable-speed rotating assembly with a high sample capacity. This work is reported in considerable detail[10] and includes a pneumatic tube system adapted for simultaneous irradiation and sequential counting of samples by the same detector system. Pneumatically propelled samples are reproducibly located by special springs inside each cylindrical sample holder in the main rotor housing. A cross section of the rotor assembly is shown in Fig. 5.8. Figure 5.9 is a photograph of the entire assembly.

Recently an elaborate system capable of irradiating, transferring, and counting samples for neutron activation analysis of short-lived nuclides in inhomogeneous samples was reported by Priest et al.[27] In addition to dual sample-biaxial rotation, the entire assembly also traverses the useful neutron beam. The samples are manually loaded and retrieved by a pneumatic system. Before being counted, the samples are removed automatically from their respective containers. It is to the credit of the designers of this system

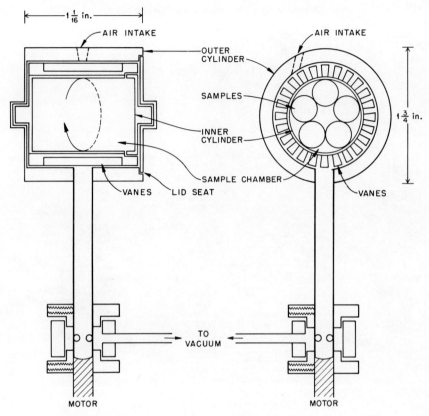

Fig. 5.7. Sample rotator for 14-MeV neutron irradiation (cf. Dyer et al.[9]).

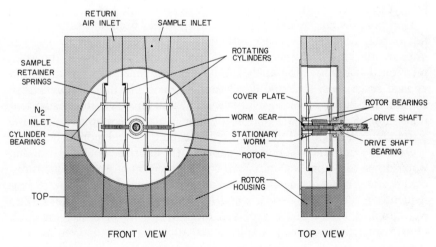

Fig. 5.8. Sample rotating mechanism, cross section (cf. Lundgren and Nargolwalla[10]).

Fig. 5.9. Dual sample-biaxial rotating assembly (cf. Lundgren and Nargolwalla[10]).

that they have spared no effort in cost or ingenuity to search for improved precision, especially in the case of inhomogeneous samples.

Up until now, discussion of sample transport has been limited to solid or liquid samples sealed in small containers and propelled by applying a gas pressure differential between the two ends of the transport capsule. It may be mentioned that activation analysis with neutron generators has also been applied successfully to liquid flow systems. One of the first comprehensive studies was carried out by Nargolwalla[14] and involved the design and utilization of a twin-flow liquid irradiation assembly as illustrated in Figs. 5.10 to 5.12. With carefully controlled flow conditions and monitoring of the ^{19}O activity from the reaction ^{19}F(n, p)^{19}O, the fluoride content of drinking water down to 1 ppm was measured with a precision of $\pm 15\%$. Extensions of this study using a single-cell irradiation assembly (Fig. 5.13) were subsequently made by Longworth[28] and Al-Shahristani.[29] These latter studies were applied to the analysis of vanadium and ^{238}U in process streams. On-stream activation using sample recirculation has been reported by Ashe

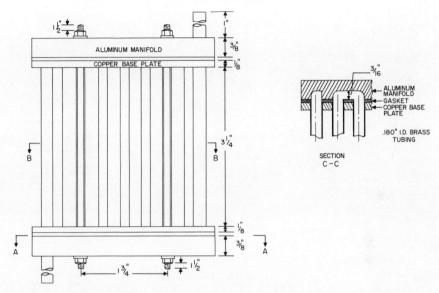

Fig. 5.10. Twin-stream liquid irradiation assembly.

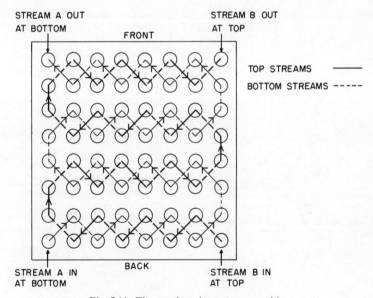

Fig. 5.11. Flow-path, twin-stream assembly.

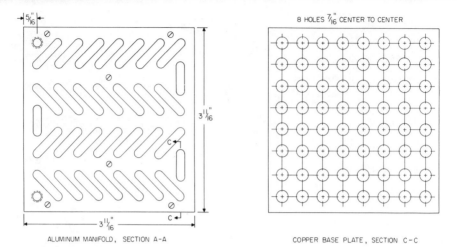

Fig. 5.12. Plan view, twin-stream assembly.

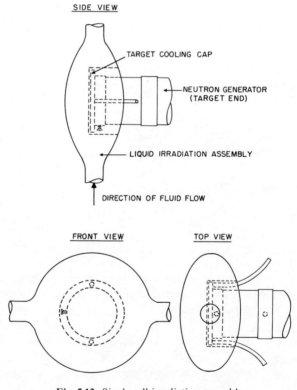

Fig. 5.13. Single-cell irradiation assembly.

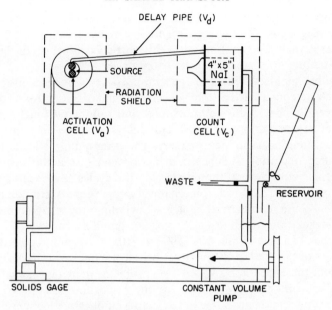

Fig. 5.14. Experimental activation loop.

et al.[30] A schematic diagram of their flow system is shown in Fig. 5.14. A pilot-plant study to develop on-line analytical capability for industrial installations, as reported by Wood,[31] suggests several promising applications. In this comprehensive report, details are given describing the irradiation cell and the general flow system requirements. A description of activation analysis techniques in process control has also been summarized by Morgan.[32]

B. AUXILIARY SAMPLE TRANSPORT EQUIPMENT

The outcome from the great impetus given to the development of efficient sample transport systems is apparent in the sophisticated electrical programmers built to control the various analytical functions in a reproducible manner. Earlier programmers, similar to those used by Dibbs[23] and by Anders and Briden,[11] consisted of a multicam timer with a gear train. Several adjustable cams are mounted on a constant-speed rotating shaft which opens and closes a bank of microswitches. An electric motor is started by a momentary current pulse, energizing a starting magnet. Shaft rotation is stopped after one complete revolution. If the sample has to be recycled, the starting magnet is left switched on for as many cycles as desired. Each cam-microswitch combination initiates a particular function in the general pneumatic tube irradiation-transfer-count cycle.

A more sophisticated and general-purpose sequence programmer, designed and constructed at the National Bureau of Standards, is described in a recent report.[33] Both the manual and automatic functions of this programmer are performed by several relays actuated in sequence by a multipole, multiposition stepper relay. The programmer was designed for trouble-free operation and has been performing satisfactorily since 1966. During this period several thousand analytical cycles have been carried out.

Both solid-state and electromechanical programmers are commercially available. Good commercial systems, including the programmer, dual-sample irradiation capability, necessary air-handling hardware, and pneumatic tubes, can cost as much as $8000 to $10,000. In many cases, it is to the advantage of the analyst to construct a system most suited to his demands.

A recent presentation by Wood[34] on problems in precision analysis with neutron generators best summarizes the important role of an efficient sample transport system. Some of the factors emphasized include effects due to poor timing, irreproducible sample placement, and sample preparation. It is recommended that in the design of transport systems, simplicity, reliability, and flexibility be the chief factors in the selection of both materials and devices from the many available in the electronic, pneumatic, and electromechanical fields of engineering.

REFERENCES

1. O. U. Anders and D. W. Briden, *Anal. Chem.*, **37**, 530 (1965).
2. E. L. Steele and H. R. Lukens, Jr., " Development of Neutron Activation Analysis Procedures for the Determination of Oxygen in Potassium," General Atomics Division, General Dynamics Corporation, Report GA-4855, 1964.
3. R. F. Gahn and L. Rosenblum, *Anal. Chem.*, **38**, 1014 (1966).
4. W. F. Harris, *Talanta*, **11**, 1376 (1964).
5. J. Hoste, D. DeSoete, and A. Speecke, " The Determination of Oxygen in Metals by 14 MeV Neutron Activation Analysis," Euratom Report EUR-3565e, Brussels, September 1967.
6. S. S. Nargolwalla, E. P. Przybylowicz, J. E. Suddueth, and S. L. Birkhead, *Anal. Chem.*, **41**, 168 (1969).
7. K. G. Broadhead and H. H. Heady, *Anal. Chem.*, **37**, 759 (1965).
8. G. H. Anderson, "The Determination of Oxygen in Titanium and Refractory Metals by Activation Analysis," in *Nucleonics in Aerospace*, P. Polishuk (Ed.), Plenum Press, New York, 1968, pp. 317–322.

9. F. F. Dyer, L. C. Bate, and J. E. Strain, *Anal. Chem.*, **39**, 1907 (1967).
10. F. A. Lundgren and S. S. Nargolwalla, *Anal. Chem.*, **40**, 672 (1968).
11. O. U. Anders and D. W. Briden, *Anal. Chem.*, **36**, 287 (1964).
12. B. L. Twitty and K. M. Fritz, "A Rapid Determination of Oxygen in High-Purity Magnesium Chips by Neutron Activation Analysis," National Lead Company of Ohio, Cincinnati, Ohio, Report NLCO-973, May 1966.
13. E. P. Przybylowicz, Gilbert W. Smith, J. E. Suddueth, and S. S. Nargolwalla, *Anal. Chem.*, **41**, 819 (1969).
14. S. S. Nargolwalla, Ph.D. Thesis, University of Toronto, Toronto, Canada, 1965.
15. E. L. Steele, H. R. Lukens, Jr., and V. P. Guinn, "Neutron Activation Analysis Procedures for the Determination of Oxygen in Potassium," General Atomics Division, General Dynamics Corporation, Report GA-5982, 1964.
16. S. S. Nargolwalla and J. E. Suddueth, "Activation Analysis with the Cockcroft-Walton Neutron Generator," in *Activation Analysis Section: Summary of Activities, July 1969 to June 1970*, P. D. LaFleur and D. A. Becker (Eds.), National Bureau of Standards Technical Note 548, issued December 1970.
17. R. L. Hull, "Fast-Neutron Activation Analysis Using Small Accelerators," *Proc. Am. Pet. Inst.*, **44** (III), 264 (1964).
18. S. S. Nargolwalla, J. E. Suddueth, M. R. Crambes, E. P. Przybylowicz, and Gilbert W. Smith, "Activation Analysis with 14-MeV Neutrons" in *Activation Analysis Section: Summary of Activities, July 1967 to June 1968*, P. D. LaFleur (Ed.), National Bureau of Standards Technical Note 458, issued March 1969.
19. J. T. Byrne, C. T. Illsley, and H. J. Price, Proceedings, 1965 International Conference on Modern Trends in Activation Analysis, College Station, Texas, 1965, pp. 304–309.
20. W. W. Meinke and R. W. Shideler, Michigan Memorial Phoenix Project, University of Michigan, Ann Arbor, Michigan, Report MMPP-191-1, December 1961.
21. E. Cerrai and F. Gadda, *Energ. Nucl.*, **9**, 317 (1962).
22. D. J. Veal and C. F. Cook, *Anal. Chem.*, **34**, 178 (1962).
23. H. P. Dibbs, "The Determination of Oxygen by Fast Neutron Activation Analysis," Department of Energy, Mines and Resources, Ottawa, Canada, Mines Branch Technical Bulletin TB 55, March 1964.
24. W. E. Mott and J. M. Orange, *Anal. Chem.*, **7**, 1338 (1965).
25. D. E. Wood, P. L. Jessen, and R. E. Jones, Pittsburgh Conference on Analytical Chemistry and Applied Spectroscopy, Pittsburgh, Pennsylvania, February 1966.
26. G. L. Priest, F. C. Burns, and H. F. Priest, *Anal. Chem.*, **39**, 110 (1967).
27. H. F. Priest, F. C. Burns, and G. L. Priest, *Anal. Chem.*, **42**, 499 (1970).
28. J. M. Longworth, B. A. Sc. Thesis, University of Toronto, Toronto, Canada, 1967.
29. H. Al-Shahristani, Ph.D. Thesis, University of Toronto, Toronto, Canada, 1969.
30. J. B. Ashe, P. F. Berry, and J. R. Rhodes, Third International Conference on Modern Trends in Activation Analysis, National Bureau of Standards Special Publication 312, Vol. II, 1969, pp. 913–917.
31. D. E. Wood, "Development of Fast Neutron Activation Analysis for Liquid Loop Systems," Kaman Nuclear Corporation, Colorado Springs, Colorado, Report KN-67-458(R), July 1967.

32. I. L. Morgan, *Trans. Am. Nucl. Soc.*, **10**, 41 (1967).
33. J. E. Suddueth and S. S. Nargolwalla, in *Activation Analysis Section: Summary of Activities, July 1967 to June 1968*, P. D. LaFleur (Ed.), National Bureau of Standards Technical Note 458, issued March 1969.
34. D. E. Wood, Proceedings, Conference on the Use of Small Accelerators for Teaching and Research, Oak Ridge Associated Universities, Oak Ridge, Tennessee, USAEC Report CONF-680411, 1968.

CHAPTER

6

SOURCES AND REDUCTION OF SYSTEMATIC ERROR

I. INTRODUCTION

In activation analysis, the precision index expressing the expected variation of the final analytical result is invariably computed from the counting rates using Poisson statistics. It is frequently assumed that such a computed index can be improved *ad infinitum* by increasing the number of counts accumulated. This assumption is correct only to a point. Additional sources of non-Poisson random error contributing to the total random error do exist, and ultimately become controlling in the precision and accuracy of a measurement. These additional random errors result from such factors as inherent fluctuations in the line voltage which cause measurement changes, changes in the background of a counter while a sample count is being measured, and the like. Therefore it is important to consider non-Poisson random effects in estimating individual parameters and their standard errors. If random error (Poisson) were the only cause for experimental variance, then elimination or rigid control of excess or non-Poisson variance would determine the limit of ultimate precision, that is, that controlled by the statistics of the radioactivity process itself. A succinct treatment describing the significance of this excess (non-Poisson) error in activation analysis, was reported by Currie.[1]

In physical measurement, two types of errors ultimately determine the reliability of the final result, random and systematic. In this chapter we are concerned principally with the second type, that is, systematic errors, their sources, and experimental solutions. Systematic errors (other than obvious experimental mistakes) arise from a myriad of sources. For example, the incorrect calibration of a counter or utilization of doubtful nuclear data can lead to a significant difference between the measured and "true" values. Many sources of systematic error lie in improper or unsound assumptions. In this chapter, the efforts of researchers and analysts to reduce experimental errors in activation analysis with neutron generators are described. Where necessary, critiques and controversies are given, so that the reader is presented with a broad perspective of this problem and the interdependencies of the individual sources of error. Continuing efforts in error analysis have yielded promising results. Today, many analytical applications using neutron generators provide analyses in which the precision index is almost entirely dictated by random counting statistics, and in specific cases, the neutron generator is considered superior to all other analytical techniques.

II. EXISTENCE AND EVALUATION OF SYSTEMATIC ERROR

A review of the literature pertaining to the application of neutron generators allows the following conclusions to be drawn. An evaluation of numerous methods shows that more than 50 elements have been determined successfully in the concentration range 10 ppm to 50%. However, the area of analytical interest is clearly depicted in Fig. 6.1, abstracted from a recent

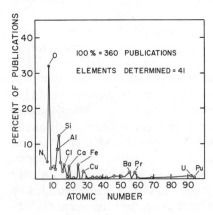

Fig. 6.1. Frequency plot of element determination in the literature (cf. reference 2).

review.[2] This plot indicates that about 45% of the analytical activity has been devoted to the determination of oxygen and silicon, oxygen being far in the lead. The analytical activity in the reduction of experimental error in the case of oxygen and silicon can best be summarized from Fig. 6.2, also abstracted from reference 2. This figure was constructed from a vast number of reported analytical data and, as far as possible, it attempts to illustrate the improvement in analytical precision over a period of time. To a large measure the improvement indicated is a direct result of, first, recognition of error sources, and second, in-depth studies to evaluate them. In this chapter we attempt to classify each principal source of systematic error and then describe a method or methods for its determination and correction. Where necessary, discussion relating to or describing actual causes or parameters responsible for errors has also been given for the sake of continuity and

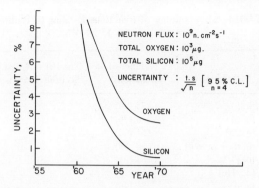

Fig. 6.2. Improvement with time in calculated uncertainties of reported analyses for oxygen and silicon (cf. reference 2).

completeness. The information presented here is intended to generate interest in the search for systematic errors rather than to indicate any approved method or practice.

A. NUCLEAR CONSTANTS

As mentioned in Chapter 1 (Section II) activation parameters that are not specifically or readily determined include detection efficiency, activation cross section, and impinging particle flux. Therefore, a comparative approach is generally used to determine elemental concentration in unknown samples. At times, however, it is not possible to follow such a procedure and the analyst has to resort to a complete dependency on published nuclear data for the measurement of some of these factors. In such instances, the accuracy of the analytical results will depend on the degree of confidence placed on the nuclear data used. Csikai et al.[3] have systematically examined nuclear data for neutron activation analysis and have arrived at definite conclusions, which we shall summarize. An evaluation of reaction cross sections for incident neutron energies of approximately 15 MeV gave the results listed in Table 6.1. This survey indicates the general degree of reliability and/or confidence with which the analyst has to contend. The number of cross sections not known or yet evaluated represents a significant proportion of the total number of reactions in each case. Even for some relatively high yield (n, p), (n, α), and (n, 2n) reactions, there appears to be a lack of reliable cross-section data.

Any attempt to summarize similar data for D–D neutrons would not be meaningful owing to the sparsity of available information. The lack of cross-section data, as well as the great spread in experimentally determined

TABLE 6.1. Number of Reactions Leading to Radio-isotopes at $E_n \simeq 15$ MeV[a]

Reaction	Number of Reactions[b]	Number of Cross Sections Known
(n, α)	175 (74)	95
(n, p)	226 (78)	135
(n, 2n)	167 (94)	96
(n, d) + (n, n'p)	158 (53)	24
(n, γ)	184 (91)	31
(n, ³He)	122 (32)	4
(n, n'α)	85 (46)	6
(n, ³H)	193 (97)	4
(n, n'γ)	64	4

[a] Cf. reference 3.
[b] The numbers in parentheses denote isomers.

values revealed by Csikai et al.,[3] tends to suggest not only the necessity for further comprehensive experimental studies but also the potential inaccuracies of those elemental determinations based on the utilization of nuclear constants such as activation cross sections.

Other nuclear constants important in noncomparative activation analysis are those of half-life and decay scheme characteristics of the product nucleus being measured. In most cases the half-lives of radioactive species produced from bombardment of elements with neutrons of energy $\simeq 15$ MeV fall between 1 sec and one year. Csikai et al.[3] have examined the reported accuracy of half-life measurements for about 600 radioisotopes and have grouped them by error classification. According to these authors, the error in the half-life is 10% or less for 66% of all radioisotopes; 10% to 50% for 12%; 50% to 100% for 17%; and more than 100% for 5% of all radioisotopes. The 24 radioisotopes in the last group are listed in Table 6.2.

Assembling the decay schemes in a somewhat arbitrary fashion, Csikai et al.[3] conclude that 63% of the radioisotopes evaluated have known decay schemes, for 25% the decay schemes are known to some extent, and for the remaining 12% the decay characteristics are not known. The radioisotopes of this last group are listed in Table 6.3. The availability of high-resolution semiconductor detectors has given considerable impetus to a reevaluation of decay schemes. No doubt many of the discrepancies that exist in decay scheme listings will be removed when a comprehensive compilation of such results becomes available to the activation analyst.

II. EXISTENCE AND EVALUATION OF SYSTEMATIC ERROR

TABLE 6.2. Radioisotopes with an Unknown Half-life That May Be Produced in Neutron Reactions[a]

Isotope	Half-life	Isotope	Half-life
^{32}Si	100–710 years	^{178}Lu	"Orders of minutes"
^{52}Ti	< 1 day or > 100 days	^{179}Ta	1.6 years, ~600 years
^{53}Mn	140 years, > 1.3×10^5 years, 2×10^6 years	^{194}Os	5.8 years, 1.9 years
^{93g}Mo	> 100 years, 10^4 years	^{196}Ir	< 5 years
^{101g}Rh	5 years, > 3 years, (10 ± 3) years	^{193g}Pt	< 500 years
^{108}Ag	≥ 5 years	^{197m}Pt	2.8 hr?
^{113m}Cd	5.1 years, 14 years	^{206}Hg	7.5 min, < 10 hr or > 2 years, < 15 hr or > 100 years, "estimated value of 140 years"
^{121m}Sn	> 5 years, 25 years	^{229}Ra	< 5 min, "short"
^{162}Gd	> 1 year	^{230}Ac	< 1 min
^{157}Tb	> 30 years, < 10 min or 725 years, 4.7 days	^{235}Th	< 5 min
^{173}Er	2.0 min?	^{237}Pa	10 min, 39 min
^{174g}Lu	~165 days, (1300 ± 150) days, > 800 days, > 1000 days	^{76}Zn	< 2 min

[a] Cf. reference 3.

TABLE 6.3. Nuclides with Unknown Decay Schemes That May Be Produced in Neutron Reactions[a]

^{46}K	^{78}As	^{84}Br	^{86}Br	^{98}Zr	^{100}Nb
^{102}Mo	^{102}Tc	^{103}Tc	^{104}Tc	^{109}Rh	^{113}Pd
^{114}Pd	^{104g}Ag	^{105}Cd	^{118}Cd	^{119}Cd	^{121}Cd
^{110m}In	^{112m1,m2}In	^{120}In	^{124}In	^{126}Sn	^{127}Sn
^{128}Sn	^{128}Sb	^{129}Sb	^{130}Sb	^{133}Te	^{134}Te
^{137}Xe	^{129}Ba	^{135}Ce	^{147}Ce	^{148}Ce	^{148}Pr
^{149}Pr	^{152}Pm	^{153}Pm	^{154}Pm	^{156}Sm	^{157}Sm
^{157}Eu	^{158}Eu	^{159}Eu	^{160}Eu	^{162}Gd	^{156m}Tb
^{162}Tb	^{163}Tb	^{164}Tb	^{167}DY	^{159}Ho	^{173}Er
^{176}Tm	^{178}Lu	^{182}Hf	^{183}Hf	^{191}Re	^{194}Os
^{197}Pt	^{200}Pt	^{201}Pt	^{202}Au	^{203}Au	^{206}Hg
^{229}Ra	^{230}Ra	^{230}Ac	^{231}Ac	^{235}Th	^{237}Pa

[a] Cf. reference 3.

B. INTERFERING NUCLEAR REACTIONS

Nuclear reaction interference experienced in activation analysis depends on the composition of the sample, the nature and energy of incident radiation, experimental irradiation and decay times selected, and the type of detector system used. For a reliable estimate of possible nuclear reaction interferences, it is best to assume that the sample contains all possible elements leading to interfering reactions.

Reaction interferences can be classified into three main categories. First, primary reactions can lead to the same radioactive nuclide as that from the element being analyzed. Such reactions are produced from other elements in the sample by different reactions than that occurring in the element of interest. In 14-MeV neutron activation, reactions such as (n, p), (n, 2n), (n, α), or (n, n'γ) can all be produced. Therefore a given radioactive species can be produced by activation of the element of interest or by activation of a nuclide of an element with an atomic number one or two units higher. For example, in the analysis for oxygen by the ^{16}O(n, p)^{16}N reaction, if the element fluorine is also present, then the reaction ^{19}F(n, α)^{16}N also results in the same product, ^{16}N.

Second, secondary reactions can be produced in the sample matrix by particles different from those comprising the primary incident beam. In neutron generator activation analysis for nitrogen by the ^{14}N(n, 2n)^{13}N reaction, ^{13}N is produced in matrices containing principally hydrogen-oxygen or hydrogen-carbon. This type of interference can affect the measured nitrogen concentration seriously at the 0.1% level. This interference has already been discussed in Chapter 2 (Section IV.B). Finally, second-order reactions induced in the radioactive products from the principal reaction can result in errors due to depletion of the principal product during the irradiation time. Except for a few nuclides, this effect as discussed by Ricci and Dyer[4] is only important in high-cross-section thermal neutron reactions when long irradiation times are used. In addition to these three types of interferences, the inability of the measuring device to separate gamma energies effectively can lead to serious error. This type of instrumental error can, however, be controlled or minimized by judicious selection of the detector system and the gamma energy used for analysis and control of irradiation and decay time.

To summarize data on nuclear reaction interferences, we have selected two sets of data on the basis of simplicity and completeness. The table from Csikai et al.,[3] which lists comprehensively the expected interferences from competing nuclear reactions induced in the sample matrix, is included in Appendix II. We have also selected an excellent compilation by Mathur and Oldham[5] for inclusion in this discussion. This compilation is given in Table 6.4 where the type 1 and type 2 interferences are both defined in the first category listed above, and where type 2 interference is that resulting from instrumental effects. The extent of instrumental effects is based on the assumption that all nuclides listed under "type 2 interference" have half-lives in the range of half to twice the half-life of the nuclide of interest and in which the gamma energies are unresolvable by the NaI detector system. Mathur and Oldham have also assumed that adequate resolution of two gamma-ray peaks cannot be achieved if their energies lie within the following limits:

II. EXISTENCE AND EVALUATION OF SYSTEMATIC ERROR

Energy Range	Limits
$E = 0$–0.5 MeV	$E = \pm 20\%$
$E = 0.5$–1.0 MeV	$E = \pm 15\%$
$E = \geq 1.0$ MeV	$E = \pm 10\%$

In addition to these assumptions Table 6.4 neglects Compton and backscatter interferences. In the last column of this table, additional useful observations can be noted and several additional references[6–10] are given. Although Table 6.4 does not represent all the reactions possible with 14-MeV neutrons, the principal ones listed do give a ready reference for a rapid feasibility evaluation of a particular analysis being contemplated without radiochemical separation. Selected information from this reference has been included in the working curves in Chapter 7.

The reader is reminded that interfering nuclear reactions resulting from D–D neutrons produced from interaction of the primary deuteron beam with deuterons embedded in the tritium target from previous irradiations must also be considered. A detailed discussion of this effect is given in Chapter 3 (Section II.F.2.e).

C. NUCLEAR REACTION RECOIL EFFECT

A possible source of systematic error in noncomparative activation analysis with neutron generators results from the recoiling of nuclei within either the sample, the sample container, or the atmosphere surrounding the sample. Csikai et al.[11] have measured such effects in the case of thin foils and found them significant. Their results, as abstracted from their report,[3] are shown in Table 6.5. In Fig. 6.3, also abstracted from this report,[3] the ratio of the radioactivity deposited on the catcher foil to the total (target plus catcher) is plotted as a function of the aluminum target foil thickness. The plot shows that for foil thicknesses greater than the range of recoil nuclei, the relationship $A = A_f X/(X - C)$ holds between the total activity A and the sample activity A_f, where X is the sample thickness in mg/cm^2 and C is dependent on the neutron energy and the type of reaction under consideration.

The alternative case in which recoil nuclei are deposited from sample containers or from the activation of the environment around the sample itself, was demonstrated by Anders and Briden[12] for oxygen analysis of samples encapsulated in polyethylene containers. These workers developed a method of measuring oxygen at levels between 5 and 500 ppm in highly reactive metals such as cesium. In their evaluation of the low-oxygen-background containers used to encapsulate the sample, the effect of recoiling ^{16}N nuclei from the atmosphere surrounding the sample vial was studied in detail. They showed that the effect of such recoiling nuclei was proportional

TABLE 6.4. Interferences Encountered in 14-MeV Neutron Activation Analysis

Isotope Determined and Percentage Abundance	Reaction, Its Cross Section (mb), and Product Half-life	Principal γ-Energy (MeV) Used for Detection and Intensity	Type 1 Interference	Type 2 Interference	Additional Observations
^{14}N (99.63)	^{14}N(n, 2n)^{13}N 8; 10 min	0.51 annihilation radiation	None	Positron emitters, namely, ^{78}Br, ^{38}K, ^{143}Sm, ^{53}Fe, ^{136}La, ^{62}Cu, ^{18}F, ^{162m}Ho, ^{112}In, ^{91}Mo, ^{120}Sb, and ^{80}Br	In organic matrix, knock-on protons are produced which react with carbon and oxygen of the matrix by the reactions ^{13}C (p, n) ^{13}N and ^{16}O (p, α) ^{13}N. These interferences constitute an activity of the order of a few hundred ppm[a]
^{16}O (99.76)	^{16}O (n, p) ^{16}N 33; 7.14 sec	> 4.0	^{19}F (n, α) ^{16}N	None	High-energy emitters such as ^{11}Be, ^{34}P, ^{20}F, ^{23}Ne, and fission products of U and Th cause minor interference.[b] ^{19}F can be estimated by the reaction ^{19}F (n, p) ^{19}O and its contribution subtracted[c]
^{19}F (100)	(i) ^{19}F (n, p) ^{19}O 14; 29 sec	0.2 (96%) 1.366 (54%)	^{22}Ne (n, α) ^{19}O, in view of the rarity of neon in nature, may be neglected	None	
	(ii) ^{19}F (n, α) ^{16}N 23; 7.14 sec	> 4	^{16}O (n, p) ^{16}N	None	See oxygen for discussion
	(iii) ^{19}F (n, 2n) ^{18}F 60; 110 min	0.51 annihilation radiation	None	Positron emitters, namely, ^{105}Cd, ^{68}Ge, ^{77}Kr, ^{95}Ru, ^{123}Xe, ^{141}Nd, ^{129}Ba, ^{161}Er, ^{45}Ti, and ^{190}Ir	
^{23}Na (100)	^{23}Na (n, p) ^{23}Ne 37; 38 sec	0.436 (33%)	^{26}Mg (n, α) ^{23}Ne	None	
	^{23}Na (n, α) ^{20}F 170; 11 sec	1.63 (100%)	^{20}Ne (n, p) ^{20}F, but in view of rarity of neon in nature this may be neglected	None	

Target (abundance %)	Reaction; half-life	Energy MeV (%)	Interfering reaction	Remarks
^{24}Mg (78.70)	^{24}Mg (n, p) ^{24}Na 180; 15 hr	2.75 (100%) 1.38 (100%)	^{27}Al (n, α) ^{24}Na	
^{25}Mg (10.13)	^{25}Mg (n, p) ^{25}Na 60; 60 sec	0.40 (15%), 0.58 (15%) and 0.98 (15%)	None	
^{26}Mg (11.17)	^{26}Mg (n, α) ^{23}Ne 3; 38 sec	0.436 (33%)	^{23}Na (n, p) ^{23}Ne	
^{27}Al (100)	^{27}Al (n, p) ^{27}Mg 70; 9.5 min	0.834 (68%) 1.01 (29%)	^{30}Si (n, α) ^{27}Mg	Low sensitivity, only about 5% of the previous reaction
	^{27}Al (n, α) ^{24}Na 110; 15 hr	2.75 (100%) 1.38 (100%)	^{24}Mg (n, p) ^{24}Na	
^{28}Si (92.21)	^{28}Si (n, p) ^{28}Al 200; 2.3 min	1.78 (100%)	^{31}P (n, α) ^{28}Al	Serious interference, if phosphorus present
^{29}Si (4.70)	^{29}Si (n, α) ^{29}Al 100; 6.56 min	1.28 (85%) 2.43 (15%)	None	Low sensitivity but interference free
^{30}Si (3.09)	^{30}Si (n, α) ^{27}Mg 45; 9.5 min	0.834 (68%) 1.01 (29%)	^{27}Al (n, p) ^{27}Mg	Low sensitivity cannot be used if Al present
^{31}P (100)	^{31}P (n, 2n) ^{30}P 10; 2.56 min	0.51 annihilation radiation (100%)	None	Positron emitters, namely, ^{108}Ag, ^{140}Pr, and ^{89m}Zr
	^{31}P (n, α) ^{28}Al 140; 2.3 min	1.78 (100%)	^{28}Si (n, p) ^{28}Al	Low sensitivity, but should be considered if silicon is present in the sample Serious interference from silicon
^{34}S (4.22)	^{34}S (n, p) ^{34}P 85; 12.4 sec	2.1 (25%)	^{37}Cl (n, α) ^{34}P	Beta counting has to be resorted to if sulfur has to be determined by fast neutron activation analysis
	^{34}S (n, α) ^{31}Si 126; 2.62 hr	1.26 (0.07%)	^{31}P (n, p) ^{31}Si	
^{35}Cl (75.53)	^{35}Cl (n, 2n) ^{34m}Cl 8; 32.4 min	2.1 (43%)	None	^{38}Cl, ^{84}Br, ^{116}In, and ^{138}Cs produced from K, Rb, Sn, and Ba, respectively
		0.51 annihilation radiation (100%)	None	Positron emitters, namely, ^{80}Br, ^{106}Ag, ^{130}Cs, ^{111}Sn, ^{63}Zn, ^{44}Cr, ^{73m}Se, and ^{105}Cd These interferences may in most cases be of little practical importance More sensitivity is obtained than by counting in the 2.1 MeV channel

Continued

[a] Cf. references 6, 7, and 8.
[b] Cf. reference 9.
[c] Cf. references 7 and 10.

Table 6.4—continued

Isotope Determined and Percentage Abundance	Reaction, Its Cross Section (mb), and Product Half-life	Principal γ-Energy (MeV) Used for Detection and Intensity	Type 1 Interference	Type 2 Interference	Additional Observations
^{37}Cl (24.47)	^{37}Cl (n, α) ^{37}S 25; 5.04 min	3.13 (100%)	^{40}A (n, α) ^{37}S	None	In view of the rarity of argon may be neglected
	^{37}Cl (n, α) ^{34}P 50; 12.5 sec	2.1 (25%)	^{34}S (n, p) ^{34}P	None	Cl in the presence of S cannot be determined by this reaction
^{39}K (93.10)	^{39}K (n, 2n) ^{38m}K 3; 7.7 min	0.51 annihilation radiation	None	Positron emitters ^{89m}Zr, ^{78}Br, ^{143}Sm, ^{53}Fe, ^{136}La, ^{62}Cu, ^{13}N, ^{162m}Ho, ^{112}In, ^{91}Mo, and ^{120}Sb	
^{41}K (6.88)	^{41}K (n, p) ^{41}A 70; 110 min	2.16 (100%) 1.29 (99%)	^{44}Ca (n, α) ^{41}A	None	
	^{41}K (n, α) ^{38}Cl 30; 37.7 min	2.15 (47%) 1.60 (31%)	^{38}A (n, p) ^{38}Cl	^{81}Br (n, α) ^{78}As and ^{112}Cd(n, p) ^{112}Ag	The interference may be neglected due to rarity of ^{38}A
^{44}Ca (2.06)	^{44}Ca (n, p) ^{44}K 25; 21 min	1.13 (100%)	None	None	
	^{44}Ca (n, α) ^{41}A 35; 110 min	1.27 (99%)	^{41}K (n, p) ^{41}A	None	
^{45}Sc (100)	^{45}Sc (n, 2n) ^{44}Sc 130; 3.92 hr	0.51 annihilation radiation (95%)	None	Positron emitters, namely, ^{123}Xe, ^{141}Nd, ^{129}Ba, ^{161}Er, ^{45}Ti, ^{190}Ir, and ^{73}Se, ^{106}Rh, ^{65}Ni, and ^{117m}Cd	
	^{45}Sc (n, α) ^{42}K 60; 12.5 hr	1.16 (99%) 1.52 (18%)	^{42}Ca (n, p) ^{42}K	None	Low isotopic abundance of ^{42}Ca reduces this interference to negligible proportions
^{46}Ti (7.93)	^{46}Ti (n, 2n) ^{45}Ti 25; 3.08 hr	0.51 annihilation radiation (85%)	None	Positron emitters, namely, ^{18}F, ^{44}Sc, ^{95}Ru, ^{141}Nd, and ^{190}Ir	

Isotope (abundance)	Reaction; half-life	Energy (%)	Interference	Notes
^{48}Ti (73.94)	^{48}Ti (n, p) ^{48}Sc; 44 hr	1.31, 1.04, 0.98 (all 100%)	^{51}V (n, α) ^{48}Sc	Low sensitivity
^{50}Ti (5.34)	^{50}Ti (n, p) ^{50}Sc; 1.8 min	1.56, 1.11, and 0.51 (all 100%)	None	^{76}As and ^{69}Ge (in the 1.31 channel) and ^{44}K, ^{69}Ge, and ^{82}Br in the 1.04 and 0.98 channels
				^{52}V in the 1.56 channel ^{53}V, ^{71}Zn, and ^{91m}Mo in the 1.11 channel ^{25}Na, ^{30}P, ^{71}Zn, ^{86m}Rb, ^{91m}Mo, and ^{140}Pr
^{51}V (99.76)	^{51}V (n, p) ^{51}Ti 50; 5.8 min	0.32 (95%)	Minor interference from ^{54}Cr (n, α) ^{51}Ti	^{53}Fe and ^{161}Gd
	^{51}V (n, α) ^{48}Sc 20; 44 hr	1.31, 1.04, and 0.98 (100%)	^{48}Ti (n, p) ^{48}Sc	^{76}As and ^{69}Ge (in the 1.31 channel) and ^{44}K, ^{69}Ge, and ^{82}Br in the 1.04 and 0.98 MeV channels
				Low sensitivity
^{52}Cr (83.76)	^{52}Cr (n, p) ^{52}V 100; 3.76 min	1.43 (100%)	^{55}Mn (n, α) ^{52}V	^{84}Br and ^{89m}Zr. High sensitivity, in the absence of Mn
^{53}Cr (9.55)	^{53}Cr (n, p) ^{53}V 37; 2 min	1.00	None	^{25}Na and ^{71}Zn. Lower sensitivity but no direct interference
^{55}Mn (100)	^{55}Mn (n, α) ^{52}V 30; 3.76 min	1.43 (100%)	^{52}Cr (n, p) ^{52}V	^{84}Br and ^{89m}Zr. In the presence of Cr, cannot be determined
^{54}Fe (5.82)	^{54}Fe (n, 2n) ^{53}Fe 10; 8.9 min	0.38 (ca. 30%) 0.51 annihilation radiation	None	^{51}Ti from ^{51}V and ^{54}Cr. Positron emitters, namely, ^{78}Br, ^{38}K, ^{143}Sm, ^{136}La, ^{62}Cu, ^{13}N, ^{162m}Ho, ^{112}In, ^{91}Mo, ^{120}Sb, and ^{80}Br
^{56}Fe (91.66)	^{56}Fe (n, p) ^{56}Mn 100; 2.58 hr	0.845 (99%)	^{59}Co (n, α) ^{56}Mn	^{142}La, ^{92}Y, and ^{150}Pm. More sensitive but Cobalt produces rather serious interference
^{59}Co (100)	^{59}Co (n, α) ^{56}Mn 30; 2.58 hr	0.845 (99%)	^{56}Fe (n, p) ^{56}Mn	^{142}La, ^{92}Y, and ^{150}Pm. Serious interference from iron
^{58}Ni (67.88)	^{58}Ni (n, 2n) ^{57}Ni 30; 36 hr	1.368 (86%) 0.51 annihilation radiation	None	None. Uneconomical to employ fast neutron activation
			None	Positron emitters: ^{83}Sr, ^{69}Ge

Continued

Table 6.4—continued

Isotope Determined and Percentage Abundance	Reaction, Its Cross Section (mb), and Product Half-life	Principal γ-Energy (MeV) Used for Detection and Intensity	Type 1 Interference	Type 2 Interference	Additional Observations
^{61}Ni (1.19)	^{61}Ni (n, p) ^{61}Co 22; 1.65 hr	0.072 (ca. 100%)	None	None	
^{62}Ni (3.66)	^{62}Ni (n, p) ^{62}Co 22; 13.9 min	1.17 (82%)	^{65}Cu (n, α) ^{62}Co	^{29}Al, ^{62}Cu, ^{70}Ga, ^{120}Sb, and ^{151}Nd	
^{63}Cu (69.09)	^{63}Cu (n, 2n) ^{62}Cu 500; 9.73 min	0.51 annihilation radiation (98.2%)	None	Positron emitters, namely, ^{78}Br, ^{38}K, ^{143}Sm, ^{53}Fe, ^{136}La, ^{13}N, ^{162m}Ho, ^{112}In, ^{91}Mo, ^{120}Sb, and ^{80}Br	
^{65}Cu (30.91)	^{65}Cu (n, 2n) ^{64}Cu 1000; 12.8 hr	0.51 annihilation radiation (19%)	^{64}Zn (n, p) ^{64}Cu	Positron emitters, namely, ^{73}Se, ^{101}Pd, and ^{119}Te	
	^{65}Cu (n, p) ^{65}Ni 25; 2.56 hr	1.48 (24%)	^{68}Zn (n, α) ^{65}Ni	^{92}Y	
^{64}Zn (48.89)	^{64}Zn (n, 2n) ^{63}Zn 100; 38.1 min	0.51 annihilation radiation (90%)	None	Positron emitters: ^{106}Ag, ^{130}Cs, ^{34m}Cl, ^{111}Sn, ^{49}Cr, ^{73m}Se, ^{105}Cd, and ^{68}Ga	
	^{64}Zn (n, p) ^{64}Cu 250; 12.8 hr	0.51 annihilation radiation (19%)	^{65}Cu (n, 2n) ^{64}Cu	Positron emitters: ^{73}Se, ^{101}Pd, and ^{119}Te	
^{66}Zn (27.81)	^{66}Zn (n, p) ^{66}Cu 80; 5.1 min	1.04 (9%)	^{69}Ga (n, α) ^{66}Cu	^{27}Mg, ^{51}Ti, and ^{62}Cu	Low sensitivity
^{68}Zn (18.57)	^{68}Zn (n, p) ^{68}Cu 25; 30 sec	1.08 (95%)	None	^{25}Na	
	^{68}Zn (n, α) ^{65}Ni 50; 2.61 hr	1.48 (24%)	^{65}Cu (n, p) ^{65}Ni	^{92}Y	
^{69}Ga (60.4)	^{69}Ga (n, 2n) ^{68}Ga 550; 68 min	0.51 annihilation radiation (85%)	None	Positron emitters: ^{111}Sn, ^{63}Zn, ^{49}Cr, ^{105}Cd, ^{95}Ru, and ^{18}F	
	^{69}Ga (n, p) ^{66}Cu 100; 5.1 min	1.08 (100) 1.04 (9%)	^{66}Zn (n, p) ^{66}Cu	^{142}La and ^{106m}Rh ^{27}Mg, ^{51}Ti, and ^{62}Cu	Low sensitivity, and interference from zinc

Isotope (abundance)	Reaction	Energy (MeV, %)	Competing reaction	Interferences	Remarks
^{71}Ga (39.6)	^{71}Ga (n, 2n) ^{70}Ga 700; 21 min	0.17 (0.2%) 1.04 (0.5%)	^{70}Ge (n, p) ^{70}Ga	^{34m}Cl, ^{186}Ta, ^{199}Pt, and ^{199}Hg (in the 0.17 channel) ^{34m}Cl, ^{62}Co, ^{63}Zn, ^{84}Br, ^{88}Rb, and ^{120}Sb (in 1.04 channel)	Sensitivity low, unless beta counting is resorted to
^{70}Ge (20.52)	^{70}Ge (n, 2n) ^{69}Ge 700; 40 hr	0.51 annihilation radiation	None	^{83}Sr and ^{57}Ni	Unsuitable for generator-produced neutrons
	^{70}Ge (n, p) ^{70}Ga 120; 21 min	0.17 (0.2%) 1.04 (0.5%)	^{71}Ga (n, 2n) ^{70}Ga	^{34m}Cl, ^{186}Ta, ^{199}Pt, and ^{199}Hg (in 0.17 channel) ^{34m}Cl, ^{62}Co, ^{63}Zn, ^{84}Br, ^{88}Rb, and ^{120}Sb (in 1.04 channel)	Sensitivity low, unless beta counting is resorted to
^{72}Ge (27.43)	^{72}Ge (n, p) ^{72}Ga 60; 14.1 hr	0.84 (88%)	^{75}As (n, α) ^{72}Ga	^{77}Ge, ^{91}Sr, and ^{130}I	
^{74}Ge (36.54)	^{74}Ge (n, α) ^{71}Zn 40; 2.2 min	0.51 (100%)	None	Positron emitters: ^{108}Ag, ^{30}P, ^{140}Pr, and ^{89m}Zr	Highest sensitivity
^{76}Ge (7.76)	^{76}Ge (n, 2n) ^{75}Ge 1800; 82 min	0.27 (11%)	^{75}As (n, p) ^{75}Ge	^{149}Nd	
^{75}As (100)	^{75}As (n, p) ^{75}Ge 30; 82 min	0.27 (11%)	^{76}Ge (n, 2n) ^{75}Ge	^{149}Nd	
	^{75}As (n, α) ^{72}Ga 20; 14.1 hr	0.84 (88%)	^{72}Ge (n, p) ^{72}Ga	^{77}Ge, ^{91}Sr, and ^{130}I	Sensitivity low and interference from Ge
^{80}Se (49.82)	^{80}Se (n, α) ^{77}Ge 38; 11.3 hr	0.27 (57%)	None	^{184}Ta and ^{197}Pt	
^{82}Se (9.19)	^{82}Se (n, 2n) ^{81m}Se 1800; 57 min	0.100 (8%)	None	None	More sensitive than the previous reaction
^{79}Br (50.54)	^{79}Br (n, 2n) ^{78}Br 800; 6.4 min	0.51 annihilation radiation	None	Positron emitters: ^{140}Pr, ^{89m}Zr, ^{38}K, ^{143}Sm, ^{53}Fe, ^{136}La, ^{62}Cu, ^{13}N, and ^{162m}Ho	
^{79}Br (n, α) ^{76}As 10; 26.5 hr	0.56 (45%)	None	^{69}Ge, ^{72}Ga, and ^{82}Br	Sensitivity very poor
^{81}Br (49.46)	^{81}Br (n, 2n) ^{80}Br 700; 17.6 min	0.62 (14%) 0.51 annihilation radiation (8%)	^{80}Kr (n, p) ^{80}Br	^{62}Cu, ^{101}Tc, ^{135m}Xe, and ^{138}Cs	Interference from Kr can be neglected due to its rarity in nature
	^{81}Br (n, α) ^{78}As 100; 91 min	0.62 (100%)	^{78}Se (n, p) ^{78}As	^{71m}Zn, ^{75}Ge, ^{78}As, ^{97}Nb, ^{142}La, ^{142}Nd, and ^{195}Ir	

Continued

Table 6.4—concluded

Isotope Determined and Percentage Abundance	Reaction, Its Cross Section (mb), and Product Half-life	Principal γ-Energy (MeV) Used for Detection and Intensity	Type 1 Interference	Type 2 Interference	Additional Observations
^{85}Rb (72.15)	^{85}Rb (n, 2n) ^{84m}Rb 800; 23 min	0.217 and 0.25	^{84}Sr (n, p) ^{84m}Rb	None	
^{87}Rb (27.85)	^{87}Rb (n, 2n) ^{86m}Rb 1000; 61 sec	0.56	^{86}Sr (n, p) ^{86m}Rb ^{89}Y (n, α) ^{86m}Rb	^{91m}Mo, ^{104}Rh, and ^{106}Rh	
	^{87}Rb (n, α) ^{84}Br 50; 32 min	0.88	None	^{63}Zn, ^{68}Ga, and ^{88}Rb	^{82}Br and ^{84}Br are also produced by the (n, p) reactions of krypton but this interference may be neglected due to the rarity of Kr
	^{85}Rb (n, α) ^{82}Br 140; 36 hr	0.55	None	^{69}Ge, ^{71m}Zn, ^{76}As, ^{83}Sr, and ^{57}Ni	
^{88}Sr (82.56)	^{88}Sr (n, p) ^{88}Rb 18; 18 min	0.90	None	^{27}Mg, ^{62}Cu, ^{70}Ga, ^{84}Br, and ^{84m}Rb	
	^{88}Sr (n, α) ^{85m}Kr 87; 4.4 hr	1.84 0.15	None	None None	
^{90}Zr (51.46)	^{90}Zr (n, 2n) ^{89m}Zr 80; 4.18 min	0.59 (94%)	None	^{108}Ag, ^{140}Pr, ^{78}Br, and ^{38}K	
^{94}Zr (17.40)	^{94}Zr (n, p) ^{94}Y 7; 17.0 min	0.92 (43%)	None	^{27}Mg, ^{84}Br, and ^{88}Rb	Low sensitivity
^{93}Nb (100)	^{93}Nb (n, α) ^{90m}Y 8; 3.14 hr	0.203	^{90}Zr (n, p) ^{90m}Y	^{92}Y and ^{149}Nd	
^{92}Mo (15.84)	^{92}Mo (n, 2n) ^{91}Mo 100; 15.6 min	0.51 annihilation radiation (94%)	None	Positron emitters: ^{143}Sm, ^{53}Fe, ^{136}La, ^{62}Cu, ^{13}N, ^{162m}Ho, ^{112}In, ^{120}Sb, and ^{80}Br	
^{97}Mo (9.46)	^{97}Mo (n, p) ^{97}Nb 110; 72 min	0.66	None	^{78}As	

Target (%)	Reaction; half-life	Gamma energies (MeV)	Competing reactions	Interferences
^{96}Ru (5.51)	^{96}Ru (n, 2n) ^{95}Ru 480; 1.65 hr	0.51 annihilation radiation 0.34	None	Positron emitters: ^{105}Cd, ^{68}Ga, ^{77}Kr, ^{18}F, ^{123}Xe, ^{141}Nd, ^{129}Ba, ^{161}Er, ^{45}Ti, and ^{190}Ir
^{107}Ag (51.82)	^{107}Ag (n, 2n) ^{106m}Ag 880; 24 min	0.51 annihilation radiation (61%)	None	Positron emitters: ^{112}In, ^{91}Mo, ^{120}Sb, ^{80}Br, ^{130}Cs, ^{34m}Cl, ^{111}Sn, ^{63}Zn, ^{49}Cr, and ^{73m}Se
^{121}Sb (57.25)	^{121}Sb (n, 2n) ^{120}Sb 1000; 15.2 min	0.51 annihilation radiation (44%)	None	Positron emitters: ^{38}K, ^{143}Sm, ^{53}Fe, ^{136}La, ^{62}Cu, ^{13}N, ^{162m}Ho, ^{112}In, ^{91}Mo, ^{80}Br, ^{106}Ag, and ^{130}Cs
^{138}Ba (71.66)	^{138}Ba (n, 2n) ^{137m}Ba 1250; 2.6 min	0.662 (100%)	^{140}Ce (n, α) ^{137m}Ba (minor due to small cross section)	^{89}Zr and ^{108}Ag
^{140}Ce (88.48)	^{140}Ce (n, 2n) ^{139m}Ce 1200; 55 sec	0.74	^{142}Nd (n, α) ^{139m}Ce	^{97m}Nb
^{141}Pr (100)	^{141}Pr (n, 2n) ^{140}Pr 1400; 3.4 min	0.51 annihilation radiation (54%)	None	Positron emitters: ^{108}Ag, ^{30}P, ^{89}Zr, and ^{78}Br
^{142}Nd (27.11)	^{142}Nd (n, 2n) ^{141}Nd 2000; 64 sec	0.76	None	^{91}Mo, ^{97m}Nb, and ^{139}Ce
^{145}Sm (3.09)	^{145}Sm (n, 2n) ^{143}Sm 1200; 9.0 min	0.51 annihilation radiation	None	Positron emitters: ^{78}Br, ^{38}K, ^{53}Fe, ^{136}La, ^{62}Cu, ^{13}N, ^{162m}Ho, ^{112}In, ^{91}Mo, ^{120}Sb, and ^{80}Br
^{197}Au (100)	^{197}Au (n, n') ^{197m}Au —; 7.3 sec	0.13 (99%) 0.28 (99%)	None	None
^{199}Hg (16.84)	^{199}Hg (n, n') ^{199m}Hg —; 44 min	0.16 0.37	None	^{34m}Cl ^{75}Ge
^{204}Pb (1.48)	^{204}Pb (n, n') ^{204m}Pb —; 67 min	0.38 0.90	None	^{199m}Hg ^{84}Br and ^{142}La

TABLE 6.5. The C Values for Calculation of the Produced Activity

Reaction	Maximum Recoil Energy (MeV)	C
^{27}Al(n, α)^{24}Na	3.814	0.245
^{27}Al(n, p)^{27}Mg	1.951	0.197
^{63}Cu(n, 2n)^{62}Cu	0.687	0.040
^{65}Cu(n, 2n)^{64}Cu	0.728	0.040
^{65}Cu(n, p)^{65}Ni	0.853	0.050
^{107}Ag(n, 2n)^{106}Ag	0.471	0.014
^{109}Ag(n, 2n)^{108}Ag	0.476	0.012

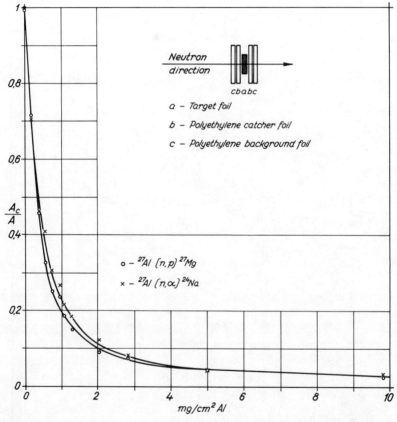

Fig. 6.3. Ratio of the escape, A_c, and produced total $A = A_c + A_f$ activity versus sample thickness (cf. reference 3).

to the surface area of the vial in contact with air during the irradiation process. These recoil nuclei were embedded in the surface of the container. Several important conclusions can be drawn from these experiments. The presence of air inside the sample container greatly influenced the apparent oxygen content of the vial. The excess oxygen value resulting from recoiling of ^{16}N nuclei generated in the air was correlated with the surface area in contact with air during activation. Their data indicated that oxygen excess of samples activated in contact with air was almost an order of magnitude larger than the oxygen found as dissolved gas in the polyethylene vial. This entire recoiling effect can be nearly eliminated if the gas used for propulsion of the capsule is nitrogen. In this case, the recoiling ^{13}N nuclei from the ^{14}N(n, 2n)^{13}N reaction emits radiation of much lower energy than that (> 4 MeV) generally used as a measure of oxygen activity. Nargolwalla et al.,[13] in their study of blank problems in the 14-MeV neutron activation analysis for trace oxygen, used a pneumatic transfer system which employed, in addition to dry nitrogen as the propelling gas, a continuous nitrogen bleed into the biaxial sample rotator during irradiation to preclude the entry of air and eliminate the effect of ^{16}N recoils. Their efforts in the minimization of the blank error are discussed in the next section of this chapter.

D. ERROR FROM SAMPLE BLANK

Activation analysis with neutron generators is invariably carried out on samples encapsulated in polyethylene vials propelled in a pneumatic tube system. Therefore, the effect of the blank or container is significant only in oxygen determinations in the microgram range. The accuracy of trace oxygen determinations can be radically affected if an inaccurate evaluation of the container blank is made. A comprehensive study[13] carried out at the National Bureau of Standards was prompted by the need for accurate steel standards for trace levels of oxygen in the range 20 to 500 ppm. Prior to this study several workers attempted to provide a solution to this problem.[14-17] However, the solutions provided were specific to individual systems and not generally applicable. The National Bureau of Standards' study tried to provide a general solution to this problem in which errors as great as 100% could be experienced if the count from the capsule blank was merely subtracted from the gross (sample plus container) count.

Some difficulty was always experienced in the encapsulation technique itself. There was always some doubt whether the encapsulation of steel samples in a dry nitrogen environment could effectively ensure the containment of the nitrogen environment around the sample itself. To eliminate this uncertainty a special "flow-through" container, described in Chapter 5 (Fig. 5.2), was used. Basically, this study involved the determination of and correction for the attenuation of the activity from the container by sample rods of different diameters. This attenuation results from a combination of attenuated incident neutron flux and the counted gamma radiation. A simple geometrical model was developed and the theoretical prediction of the correction factor was confirmed by experiment. The effectiveness of the "flow-through" container was indicated by the reproducibility of blank determinations, and the improvement in methodology was shown by comparative results using the general practice of blank subtraction from the gross count. The results shown in Table 6.6 were obtained for seven rods of valve steel which had previously been checked against known standards for acceptable homogeneity.

TABLE 6.6. Comparison of Methods Using the Standard Capsule and the "Flow-Through" Container

Rod Number	Oxygen Concentration (ppm)[a]		
	Uncorrected for Attenuation of Blank Container[b]	Corrected for Attenuation of Blank Container[c]	Corrected for Attenuation of Blank Container Using "Flow-Through" Container[d]
1	50 ± 4	94 ± 5	58 ± 5
2	30 ± 4	158 ± 9	61 ± 5
3	34 ± 4	83 ± 6	51 ± 5
4	44 ± 4	92 ± 8	55 ± 5
5	22 ± 5	77 ± 6	63 ± 5
6	33 ± 4	84 ± 6	66 ± 5
7	28 ± 5	82 ± 7	66 ± 5

[a] Concentrations are weighted means of eight determinations on each rod, with the weighting factor equal to the reciprocal of estimated variances; estimated variances are based on Poisson counting statistics and propagation of error formulas. Errors are standard errors of weighted means.

[b] Nitrogen encapsulation method using 8-cm^3-volume standard polyethylene snap-cap vials.

[c] Nitrogen encapsulation method using standard vial and corrected for sample attenuation caused by a $\frac{1}{4}$-in.-diameter steel rod.

[d] "Flow-through" encapsulation method with correction for sample attenuation caused by a $\frac{1}{4}$-in.-diameter steel rod.

Footnotes c and d appended to Table 6.6 are explained later in this chapter in sections dealing with sample attenuation errors.

The procedures developed in this study can be generally applied to any situation where the container blank is significant compared with the element

E. ERRORS FROM NEUTRON FLUX NORMALIZATION METHODS

In 1965, at the second conference on "Modern Trends in Activation Analysis" held in Texas, an informal discussion was initiated to consider industrial application of neutron generators. However, the principal topic discussed was related to the measurement of the neutron flux from neutron generators. It appeared necessary at that time to standardize upon an accepted procedure for flux determinations so that a meaningful comparison of both activation data and generator performance from different facilities could be made. Faced with a choice between the thin-foil activation technique and the dispersed activation monitor comparable to the geometrical shape of the sample, the conferees selected the former method. Since the fast neutron reaction ^{63}Cu(n, 2n)^{62}Cu has a threshold at about 12 MeV, and is therefore insensitive to degraded neutron energies, this reaction was considered to be the most suitable in estimating 14-MeV outputs from neutron generators. A secondary reason for its selection was that the excitation function of the (n, 2n) reaction with ^{63}Cu closely approximated that for ^{16}O(n, p)^{16}N—a reaction with which almost all generator users were closely involved at that time. From proposals made, a procedure for flux monitoring of neutron generator outputs was set up and called the "Texas Convention on the Measurement of 14-MeV Neutron Fluxes from Accelerators." The procedure was endorsed by many of the leading workers in the field present at the conference. Briefly the procedure consists in exposing high-purity (99.9%) copper disks (0.25-mm thickness and 1-cm and/or 2.5-cm diameter) for 1 min to the neutron flux to be measured. After a cooling period of 1 min, so that the sample could be transferred to a counter and to allow for the decay of any ^{16}N activity from any oxygen in the sample, the 0.51-MeV annihilation radiation is counted. From this count, the absolute disintegration rate of ^{62}Cu activity is determined and extrapolated to that produced at the end of the irradiation time. The 14-MeV neutron flux is expressed in units of disintegrations per minute per gram of copper, and the size of the copper disk used is noted. As a recommendation, it was suggested that the copper disk be placed at a distance of 3 cm from the center of the top surface of a right-circular cylindrical NaI(Tl) scintillation detector with a 1-g/cm^2 plastic beta absorber between the sample and the detector face. By correcting for the solid angle subtended by the source and by using the appropriate detector efficiency and photofractions as described by Heath,[18] the net area under the 0.51-MeV photopeak is calculated in counts per minute. The error in the calculations related to the photopeak area determination is estimated to be ±5%.

An obvious and perhaps the most critical assumption made in the use of this procedure was the acceptance of a cross-section value for the ^{63}Cu(n, 2n)^{62}Cu reaction for identical-output neutron generator tritium targets. The errors resulting from this assumption on activation and neutron flux monitoring data were reported by Ricci[19] and are discussed in Chapter 2 (Section III.D). As a matter of fact, a short summary on this subject was reported by Ricci[20] almost a year before the Texas convention. Ricci used simple concepts of nuclear particle ranges in materials and fast neutron scattering data to propose an approximate and rapid method for calculating 14-MeV neutron flux distributions as obtained from neutron generators. This method eliminates the basic assumption of a monoenergetic neutron source at the sample which is generally made in the determination of 14-MeV neutron outputs. Neutron flux measurements are used for normalization of activation data obtained from the irradiation and counting of samples by comparison with counting data obtained from the measurement of a copper foil or other elemental foil activation data. Measured flux values are also used to derive experimental sensitivities of elements. Errors in such measurements can be evaluated and corrected for by using the Ricci approach described below.

In his theoretical treatment, Ricci used simple geometrical considerations to describe the neutron source-sample irradiation geometry. For neutron scattering media, an aluminum water-cooling jacket and water as coolant were considered most common and therefore most appropriate for scattering calculations, the scattering media being situated between the tritium target and the sample surface. It must be noted that Ricci's neutron scattering derivations and error evaluations apply only to scattering processes occurring outside the sample, that is, in the water-cooling jacket and the coolant. Although precise neutron flux-energy distribution data can also be obtained by using neutron spectrometry, such methods necessarily involve very expensive equipment, complex data reduction techniques, and instrumental problems that an analytical chemist would probably wish to avoid. The importance of Ricci's approach to accurate activation analysis can best be revealed by a detailed description of the method and analytic examples. The reader is reminded to consider all aspects of neutron scattering in the evaluation of neutron flux monitoring methods to be described later.

The simple 14-MeV neutron generator target-sample configuration assumed by Ricci for neutron scattering calculations is shown in Fig. 6.4. The irradiation geometry is also shown. In this figure the sample is a cylindrical pellet in contact with the aluminum cooling cap. The 150-keV deuteron beam is 1.25 cm in diameter and bombards the target at its center, and normal to the target face. Ricci's calculations are based on the angular

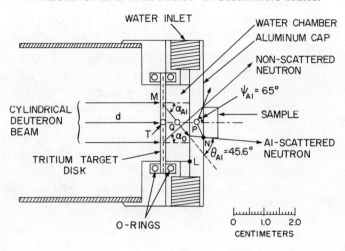

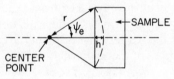

Fig. 6.4. Tritium target of a 14-MeV neutron generator. Irradiation geometry assumed in the calculation is shown (cf. reference 19).

distribution of neutrons from the ^{3}H(d, n)^{4}He reaction as given by Fowler and Brolley[21] (see Table 2.4, Chapter 2, Section III.B.2). The neutrons striking the sample are divided into four principal energy groups.

a. Unscattered neutrons (i.e., unperturbed in energy from birth to interaction with the sample), group "D."

b. Singly scattered neutrons after collision with the aluminum parts of the water-cooling jacket, group "Al."

c. Singly scattered neutrons after collision with the oxygen in the cooling water, group "O."

d. Singly scattered neutrons after collision with the hydrogen in the cooling water, group "H."

Ricci's method involves calculating successively the following quantities:

A. Energy range for:
  1. unscattered neutrons, group "D."

2. singly scattered neutrons, groups "Al," "O," and "H."
B. Relative abundance of neutrons in each of the four groups.

For the geometrical considerations in Fig. 6.4, the energy range of unscattered neutrons reaching the sample can be computed. If α is the angle formed between the directions of the deuteron beam striking the tritium target and the emitted neutron, then for unscattered neutrons striking the sample, α_D is limited to the range $0 \leq \alpha_D \leq 41.5°$. From angular distribution tables,[21] the corresponding range of the kinetic energy of the neutrons for the ^{3}H(d, n)^{4}He reaction is estimated as 14.9 MeV $\gg E_D \gg$ 14.7 MeV for an incident deuteron energy of 0.15 MeV.

The kinetic energy, E'_e, of neutrons scattered by a nucleus of element e can be computed from the nonrelativistic equation[22]:

$$E'_e = E_e \frac{A_e^2 + 2A_e \cos \theta_e + 1}{(A_e + 1)^2} \tag{6.1}$$

where

e = Al, O, and H
E_e = energy of neutrons produced from the tritium target
θ_e = scattering angle formed by the directions of the neutrons before and after the scattering process in the center-of-mass (C) system.
A_e = mass number of the scattering nucleus.

For mass numbers $A_e > 10$, θ_e in the (C) system can be closely approximated by the scattering angle in the laboratory system (Fig. 6.4). Considering first-order (single) scattered neutrons only, their energies after collision with aluminum, oxygen, and hydrogen are calculated successively by Ricci as follows:

From Eq. 6.1, the neutron energy ranges for

Group "Al": 14.3–14.9 MeV
Group "O": 13.5–14.9 MeV
Group "H": 0–14.9 MeV

Using probability calculations, Ricci has computed the fractions or percentages of unscattered and scattered neutrons reaching the sample:

fractions, $F_{Al} = 0.272$, $F_O = 0.068$, $F_H = 0.076$, and $F_D = 0.584$

These values indicate that 58.4% of all neutrons incident on the sample suffered no collisions. Percentages of singly scattered neutrons from aluminum, oxygen, and hydrogen are 27.2, 6.8, and 7.6, respectively. Individual parametric values, derivations, and methods of calculation of the results above are described in detail in the original reference.[19] The angular distributions of neutrons scattered by hydrogen, oxygen, and aluminum in the center-of-mass system were those from the work by Goldberg et al.,[23] and total neutron cross-section values were taken from a compilation by Hughes and Schwartz.[24] The combined results as reported by Ricci[19] are given in Table 6.7.

TABLE 6.7. Groups of Neutrons That Reach the Sample[a]

Group	Scattering Element	Relative Neutron Abundance	Energy Range (MeV)	Energy Interval (MeV)	Relative Abundance per MeV at Midinterval
Al	Aluminum	0.272	14.3–14.9	0.6	0.453
O	Oxygen	0.068	13.5–14.9	1.4	0.049
H	Hydrogen	0.076	0 –14.9	14.9	0.005
D	(No scattering)	0.054	14.7–14.9	0.2	2.92

[a] Irradiation geometry as in Fig. 6.4.

A histogram abstracted from Ricci's work[19] in which the area of a triangle or rectangle is proportional to the third column in Table 6.7 is shown in Fig. 6.5. For clarification, triangles with hypotenuses of positive slope represent groups "Al" and "O" because the scattering probability of these elements increases rapidly for small angles θ_e,[23] thus leading to high energies E'_e (Eq. 6.1). The bases of rectangles and triangles are the energy intervals given in columns 4 and 5 in Table 6.7. The heights of the rectangles for groups "H" and "D" and the half-heights of the triangles for groups "Al" and "O" are given in column 6 of Table 6.7. These are obtained by dividing the areas (column 3, Table 6.7) by the corresponding bases (column 5, Table 6.7). Therefore the graphical sum (solid lines, Fig. 6.5) of the areas of the rectangles and triangles represents the relative neutron distribution or spectrum in the units of neutrons/MeV, as incident on the sample in Fig. 6.4. The solid line is therefore directly proportional to the neutron energy distribution curve for the neutron flux. For convenience Ricci has expressed the ordinate in Fig. 6.5 to represent the neutron flux, $\phi(E)$, incident on the sample. Since the integrated area under the solid line is unity, $\phi(E)$ is the normalized neutron flux, or

$$\phi_I = \int_0^{14.9 \text{ MeV}} \phi(E) \, dE = 1 \text{ neutron}/(\text{cm}^2 \text{ sec}) \qquad (6.2)$$

where ϕ_I is the integrated flux.

In the discussion on the application of his method for estimating cross sections in activation analysis, Ricci shows that for materials such as Be, C,

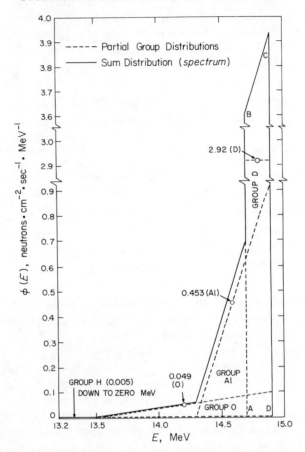

Fig. 6.5. Histogram showing the neutron distribution calculated for the geometry in Fig. 6.4 (cf. reference 19).

N, Mg, Fe, Cu, Zn, Ag, Cd, W, or Pb, which show marked peaks for acute scattering angles (similar to those for aluminum and oxygen), the approximations made are also applicable. Ricci has also shown that second-order scattering effects for irradiation geometries normally used in neutron generator irradiations are seldom significant. Calculations given by him show that errors in flux measurement based on the assumption that all neutrons striking the sample are monoenergetic can be significant. Examples given indicate that the error for the reactions ^{27}Al(n, α)^{24}Na, ^{63}Cu(n, 2n)^{62}Cu, and ^{28}Si(n, p)^{28}Al can be as high as −14, −8, and −14%, respectively.

Insofar as errors in activation analysis and cross-section measurements are concerned, Ricci has shown that when a monitor is used in comparative-type analysis, errors can also result. For example, in the analysis for aluminum using a copper monitor, an error of -7% can be expected in the final result.

In assessing Ricci's fundamental study, it should be remembered that the calculated errors depend on the validity of assumptions made in the irradiation geometry illustrated in Fig. 6.4. However, the sign of any error calculated on the basis of a different geometry can be determined with certainty using principles described by Ricci; the final result has therefore benefited by always being closer to the correct value than the uncorrected value. In particular Ricci has indicated a strong basis for the disparity in reported cross-section data for 14-MeV neutron activation.

To achieve a high order of precision in activation analysis with a neutron generator, a measure of the "sample-sensitive" neutron flux is of primary importance. Two approaches for making this measurement are generally used: first, the detection of secondary radiation produced by the interaction of neutrons with a selected element, be it the activity from a simultaneously irradiated reference sample or from a flux monitor, and second, the measurement of the neutrons directly by a counter that is responsive to neutrons of the desired energy and that has been calibrated previously against a known standard. An example of the first case is that of the copper-foil monitor using the reaction $^{63}Cu(n, 2n)^{62}Cu$, as suggested by the Texas convention.[25] As long as the degradation in neutron energy does not significantly affect the activity ratio of standard to monitor or of sample to monitor, this method can be quite useful. The ideal case is when the neutron flux monitoring is carried out using the element to be analyzed, assuming, of course, that the matrices in each case are similar in terms of their capability to degrade the incident neutron beam. Furthermore, counting of the sample and monitor is carried out under the same conditions. If a reference system is being used, for example, a dual sample-biaxial rotating system or two samples side by side, it is important that both samples experience an identical neutron environment.

Several workers[14, 26, 27] have found it convenient to use the direct neutron-counting principle using organic scintillators. A BF_3 counter surrounded by about 8 cm of paraffin makes a good detector for fast neutrons that is almost insensitive to gamma radiation. Such a counter, as described by Ladu et al.,[28] has a flat response for neutrons of energies between 20 keV and 14 MeV. Some difficulties arise in flux monitoring with integral devices like the BF_3 and scintillation plastic beads when dealing with short-lived isotopes, due to irreproducibility in the rise and fall times of neutron production.

Using a simple RC integrating device Anders and Briden[14] and Fujii et al.[29] connected the RC circuit to a fast neutron detector. The RC value was adjusted to equal the mean life $(1/\lambda)$ of the radioactive nuclide being measured. Other flux-monitoring techniques, such as associated α-particle counting[30] and the measurement of ^{16}N activity in the generator target cooling water,[31] experience the same type of difficulties. Iddings[32] has made a detailed study of flux-monitoring practices applicable to instrumental neutron activation analysis. His evaluation indicates that the cooling-water monitoring method is most unreliable for good precision analysis, since it is subject to fluid flow conditions, such as irreproducibility of water flow rates. Associated α-particle counting is affected seriously by any nonuniformity in the loading of the tritium target and therefore causes error in analysis. Gijbels,[33] in a recent critique on activation analysis with neutron generators, has summarized the characteristics of the principal flux-monitoring systems used in 14-MeV neutron activation analysis. His observations have been abstracted from the critique and are presented in Table 6.8.

For purposes where high accuracy $(\leq \pm 5\%)$ is not required, the foil activation method of flux normalization or flux monitoring can be employed and the use of flux monitors such as copper foil, as suggested by the Texas convention, can be recommended. However, criticism of this method has led to the use of several flux-monitoring techniques. Partington et al.[34] suggest the use of the reaction ^{27}Al(n, p)^{27}Mg for monitoring neutron fluxes near 14 MeV because of the relatively flat cross-section energy relationship around that energy. The ^{63}Cu(n, 2n)^{62}Cu reaction has a rather steep (140 mb/MeV) slope at around 14-MeV neutron energy. In fact, Gijbels[33] recommends the ^{27}Al(n, p)^{27}Mg reaction for all (n, p) reactions and the ^{63}Cu(n, 2n)^{62}Cu reaction for all (n, 2n) reactions induced by 14-MeV neutrons.

Jessen and Pierce[35] have proposed the reaction ^{31}P(n, p)^{31}Si as a standard method for measurement of 2.6-MeV neutron yields from the D–D reaction. Some of the criteria used for their selection are based on the following considerations.

a. The selected reaction should exhibit a suitably located and sharp threshold so that backscattered neutrons do not complicate geometric calculations.

b. The neutron cross section should be large enough to give reasonable counting statistics for irradiation and counting times normally used in activation analysis with D–D neutrons.

c. The flux monitor material must be easily available in a pure form.

d. The D–D neutron cross section must be known with good certainty.

The reaction threshold used by Jessen and Pierce was 1.9 MeV, and the average cross section over the pertinent energy range was 68 mb. The nuclear reactions ^{27}Al(n, p)^{27}Mg ($\sigma = 1.5$ mb) and ^{89}Y(n, n'γ)^{89m}Y

II. EXISTENCE AND EVALUATION OF SYSTEMATIC ERROR

TABLE 6.8. Flux-Monitoring System for 14-MeV Neutron Activation[a]

System	Proportional to Total Neutron Yield	Correction for Local Flux Variations?	Correction for Instantaneous Flux Variations?	Correction for Neutron Attenuation in Sample	Remark: Additional Correction Required?
Plastic scintillator					
(Low geometry)	Yes	No	Yes if MS/RC[b]	No	DN, n/γ attenuation[c]
Shadow of sample	No	Yes	Yes if MS/RC	Yes	DN, γ attenuation
BF$_3$ (low geometry)	Yes	No	Yes if MS/RC	No	DN, n/γ attenuation
Associated particle	Yes	No	Yes if MS/RC	No	DN, n/γ attenuation
^{16}N in cooling water	Yes	No	—	No	DN, n/γ attenuation
Reference system[d]	Yes if no beam wandering				
No rotation, $T_{1/2} \neq$ [e]		—	No[f]	No	DN, n/γ attenuation
No rotation, $T_{1/2} =$ [e]		—	Yes	No	(DN), n/γ attenuation
Reference system[d]					
Biaxial rotation, $T_{1/2} \neq$	Yes	Yes	No[f]	No	DN, n/γ attenuation
Biaxial rotation, $T_{1/2} =$	Yes	Yes	Yes	No	(DN), n/γ attenuation
Internal standard					
$T_{1/2} \neq$	—	Yes	No[f]	Yes	DN, γ attenuation
Same $T_{1/2}$	—	Yes	Yes	Yes	(DN), γ attenuation
Reference system[g]					
$T_{1/2} \neq$	—	Yes	No[f]	Yes	DN
Same $T_{1/2}$	—	Yes	Yes	Yes	(DN)

[a] Cf. reference 33.

[b] If instantaneous fluctuations are recorded using a multiscaler (MS) or a suitable RC integrating circuit.

[c] Apart from limitations described in columns 2 to 5, one must be aware of the different sensitivity of the detector for degraded neutrons if indicated DN. If, in the reference system, the same element is used, (DN) is not applicable; n/γ attenuation means that correction for different neutrons and/or gamma attenuation is required.

[d] Reference system: two tubes beside each other in front of the target.

[e] If no rotation, the sample is not "homogeneously activated" and must therefore be counted in the same geometry (tubes of rectangular section), spun at the detector, or counted in a well-type detector. $T_{1/2} \neq$ means that the reference has a different half-life from that of the radionuclide of interest.

[f] Unless irradiation short compared to shortest half-life.

[g] Reference system: two tubes behind each other, that is, monitor in shadow of sample. In this system a correction for the flux gradient is required.

($\sigma = 150$ mb) were used by Nargolwalla et al.[36] to normalize D–D neutron activation data in the compilation of elemental sensitivities for 3-MeV neutron activation analysis. Neutron flux determinations using the above-mentioned reactions were within 5% of each other. Weber and Guillaume[37] have recommended the use of the ^{115}In(n, n′γ)^{115m}In reaction ($\sigma = 360$ mb). It was suggested that the indium foil be sandwiched between two cadmium foils to minimize interference from the ^{115}In(n, γ)^{116m}In reaction.

Before concluding the discussion of flux-monitoring techniques, we shall treat the internal standard method briefly. Twitty and Fritz[38] used this technique successfully for the determination of oxygen in magnesium chips. Dutina,[39] in the determination of O_2 in Zircaloy, performed flux calibrations using the matrix, zirconium, as an internal standard. The specific activity of ^{89m}Zr ($T_{1/2} = 4.4$ min) was measured to monitor the neutron flux. Counting of the radioactivity was done 4 min after termination of irradiation. All counts above 0.4 MeV were measured. At oxygen levels above 1%, precisions of the order of $\pm 4\%$ were obtained.

F. ERRORS FROM SAMPLE ATTENUATION

Prior to 1965, analytical errors obtained in activation analysis with neutron generators were generally in excess of $\pm 5\%$. This is not surprising since no serious attempts were made to correct for systematic errors resulting from sample attenuation. Since 1965 great emphasis has been placed on improving the analytical precision by systematic studies of sample attenuation effects on the analytical result. One of the first such exhaustive efforts in the search for better precision and accuracy in analysis with neutron generators was reported by Anders and Briden.[14] To isolate errors from sample attenuation of the incident neutron flux and sample absorption of the subsequently counted gamma radiation, these workers developed an experimental configuration which provided adequate precision from identical samples analyzed repetitively. Deviation of the final result from the true value was related to sample attenuation effects. Treating this problem empirically, Anders and Briden set up approximate relationships from which corrections for neutron and gamma-ray attenuation effects could be computed. They state that prior to their work, errors resulting from sample attenuation processes must have been inherent in all reported analytical results in which neutron flux normalization practices, such as associate particle counting and cooling-water monitoring for ^{16}N, were used. To meet the initial requirement of adequate precision, Anders and Briden used the *RC* principle of neutron flux normalization and introduced the concept of sample spinning both at the irradiation and the counting stations. To measure the effect of neutron attenuation, they carefully selected a large number of different samples containing various amounts of oxygen and encapsulated them in their standard 7-cm^3-volume polyethylene irradiation containers. The selected samples weighed 4 to 14 g. The chemical composition of the samples was such that attenuation of the counted gamma rays (4.5–7.5 MeV) was minimized. Any measured deviation of their normalized results would therefore be attributed to the neutron attenuation effect. To lend theoretical confirmation of their results, the *total* macroscopic cross section, Σ_T, was used to describe the sample attenuation process for 14-MeV neutrons, on the premise that even a single collision of a fast neutron with the sample would effectively remove the probability of further interaction with ^{16}O nuclei in the deeper layers of the sample. Their argument was that, if a collision results in a nuclear reaction, the neutron is absorbed. If inelastic or elastic scattering were the neutron attenuating modes; then the energy of the scattered neutron would, on the average, be below 12 MeV where the cross section for the ^{16}O(n, p)^{16}N reaction is only about 25% of that at 14 MeV. Furthermore, since scattering processes result in a directional change of the incident neutron through the sample, the probability of the scattered neutron reaching the deeper layers of the sample in the geometric shadow behind the sample is considerably diminished. Therefore, the average neutron flux inside the sample is a function of the *total* macroscopic cross section multiplied by the thickness, D, of the sample as measured in the direction of the neutron flux. From these considerations, Anders and Briden plotted their experimental data as shown in Fig. 6.6. The self-shielding parameter, $D \times \Sigma_T$, was expressed as

$$D \cdot \Sigma_T = D \cdot \frac{1}{V} \cdot \sum_i \left[\sigma_T(14 \text{ MeV}) \cdot \frac{[W_i]}{[M_i]} 6.02 \times 10^{23} \right] \qquad (6.3)$$

where V is the sample volume (~ 7 cm^3), i represents the element in a given sample, W_i is the weight of element i, M_i is the atomic weight, and $\sigma_T(14 \text{ MeV})$ is the *total* microscopic cross section of element i, for 14-MeV neutrons. Although the overall spread of experimental points in Fig. 6.6 does

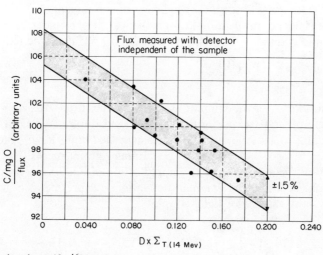

Fig. 6.6. Induced specific ^{16}N activity per flux as a function of the self-shielding parameter $D \times \Sigma_T$ (cf. reference 14).

not exceed $\pm 1.5\%$ from the average, a careful examination of this graph indicates an anomaly. For instance, the normalized activity in several cases shows an increase with increasing values of the self-shielding parameter, $D \times \Sigma_T$. This anomaly cannot be fully explained on the basis of the random statistical variation of the measurement. If the small gamma-ray attenuation differences between the samples had been considered, more systematic data could have been obtained; although Anders and Briden state that, owing to the "small" weight difference between the lightest (~ 4 g) and the heaviest (~ 14 g) samples, the correction for attenuation of the counted gamma rays would be within the statistical error limit of the method. It can also be contended that the use of Σ_T to describe the fast neutron attenuation process is not justified; this is discussed following the remaining description of the technique.

Using much heavier samples than those used to determine the neutron attenuation effect, Anders and Briden performed experiments to determine the sample attenuation effect of the counted gamma radiation. The selected samples varied in weight from about 8 to 30 g. Assuming an average sample thickness of 0.8 cm for the 1.5-cm-diameter sample, they applied the usual gamma-ray absorption law (Eq. 6.4),

$$\frac{I}{I_0} = \exp -\sum [\mu_i][\rho_i] \cdot d$$

$$= \exp -\sum [\mu_i] \cdot \frac{[W_i]}{V} \cdot d \tag{6.4}$$

Where μ_i is the mass absorption coefficient of component i, ρ_i is the partial density of element i in the sample, W_i is the weight of element i, and V is the total sample volume (~ 7 cm^3). The factor $([W_i]/V) \cdot d$ can be reduced to

$$\frac{[W_i]}{7} \times 0.8 = 0.114[W_i] \text{ g/cm}^2 \tag{6.5}$$

Calculation for μ_i for the samples used gives values from 0.025 cm^2/g for the lightest sample to 0.045 cm^2/g for the heaviest sample. To establish simple usable relationships, Anders and Briden suggested limits for $\bar{\mu}_L$, the average value of μ_i being 0.03 for samples containing light elements ($A < 53$) and 0.04 for heavy samples ($A > 53$). Simplification of Eq. 6.4 therefore yields a relationship in which the gamma attenuation is expressed as a function of sample weight, that is, for light samples,

$$\left(\frac{I}{I_0}\right)_{\text{average}} = \exp\left(-\frac{0.03 \times W \times 0.8}{7}\right)$$

$$\simeq 1 - 0.0034W \tag{6.6}$$

II. EXISTENCE AND EVALUATION OF SYSTEMATIC ERROR

and for heavy samples,

$$\left(\frac{I}{I_0}\right)_{\text{average}} \simeq 1 - 0.0045W \tag{6.7}$$

When Eqs. 6.6 and 6.7 were used for the series of samples, the corrected results agreed with the true oxygen content of the samples to within $\pm 2\%$. Although this method does not include any direct correction for differences in the neutron attenuation of samples used for the gamma-ray absorption work, Anders and Briden's final results are remarkably good. Above all, their experimental data indicate the very real nature of the gamma-ray attenuation in thick samples. For example, in the heaviest sample (31.7 g of Pb_3O_4), the uncorrected result showed a negative 14% difference from the true value of the oxygen content. Anders and Briden set the stage for further studies to improve the analytical accuracy with neutron generators, and their report describing their excellent pioneering efforts is highly recommended for a detailed study.[14]

It should be obvious from the study just described that, in comparative neutron activation analysis with neutron generators in which sample irradiations are performed either simultaneously (as in the case of the dual sample-biaxial rotational devices) or sequentially (sample in a single tube), both neutron and gamma-ray attenuation by the sample can introduce serious systematic errors. A theoretical treatment of these errors as given by Gijbels[33] is summarized below.

If the absorption of the 14-MeV neutron flux can be considered to be approximately exponential in behavior, it follows that

$$\frac{\phi}{\phi_0} = e^{-\Sigma \cdot d} \tag{6.8}$$

where ϕ represents neutron flux [neutrons/(cm² sec)] after penetration through a sample of thickness d cm, ϕ_0 is the neutron flux [neutrons/(cm² sec)] with no sample present, and Σ is the macroscopic cross section for 14-MeV neutrons (cm^{-1})*.

Consider a monodirectional and homogeneous neutron flux emanating from the tritium target of a neutron generator and incident on a sample. Then the average flux ϕ_{av} in the sample is given by

$$\phi_{\text{av}} = \frac{\phi_0[1 - e^{-\Sigma \cdot d}]}{\Sigma \cdot d} \tag{6.9}$$

* The use of Σ_R, the *removal* macroscopic cross section, in Eq. 6.8 instead of Σ_T, the *total* macroscopic cross section, as suggested by Anders and Briden,[14] was first proposed by Nargolwalla et al.[40] In the theoretical treatment by Gijbels,[33] justification for the use of Σ_R in place of Σ_T is given.

Also, for two samples of average thickness d_1 and d_2, each placed normal to the direction of the incident neutron flux, the average flux will be the same if $\Sigma_1 \times d_1 = \Sigma_2 \times d_2$. For a sample of thickness d, placed normal to the direction of the neutron flux, the average flux will be approximately equal to the local flux, ϕ_l, at some distance, $r \times d$, into the slab (where $0 \leq r \leq 0.5$). Therefore $\phi_{av} = \phi_l$ or

$$\frac{\phi_0[1 - e^{-\Sigma \cdot d}]}{\Sigma \cdot d} = \phi_0 \cdot e^{-\Sigma \cdot r \cdot d} \tag{6.10}$$

and

$$r = \frac{2.302}{\Sigma \cdot d}\left(\log \frac{\Sigma \cdot d}{1 - e^{-\Sigma \cdot d}}\right) \tag{6.11}$$

For $\Sigma \cdot d < 1$, $r \simeq 0.5$, and

$$\phi_{av} = \phi_0 e^{-\Sigma \cdot d/2} \tag{6.12}$$

This implies that for a homogeneous and unidirectional neutron flux, the average flux $\phi_{av} \simeq \phi_l$, where ϕ_l is the local neutron flux at a depth r, which is one-half of the total thickness d of the sample. The effective sample thickness is then $0.5d$. Because the macroscopic cross sections for 14-MeV neutrons are generally small, the assumption of the condition of small samples is not unreasonable, and one can compare the average flux in two samples of equivalent diameter as

$$\frac{\phi_{av_1}}{\phi_{av_2}} = \exp\left[(\Sigma_1 - \Sigma_2) \cdot \frac{d}{2}\right] \tag{6.13}$$

An experimental plot of ϕ_{av_1}/ϕ_{av_2} versus $\Delta\Sigma$ should approximate a linear relationship on semilogarithmic coordinates, the slope of which is the effective attenuation thickness, and therefore implies the very empirical nature of the parameters Σ and d.

In most experimental situations, the flux ratio, as expressed in Eq. 6.13, is of the order of 0.9 to 1.1 and therefore a simple linear relationship on linear coordinates is just as good an approximation.

For cylindrical samples (diameter $D = 2R$) under the same irradiation conditions the average flux,

$$\phi_{av} = \phi_0 \frac{\iint e^{-x} \cdot dx \cdot dy}{\pi R^2}$$

$$\simeq \phi_0 \exp\left(-\Sigma \cdot R\left\{1 + \frac{1}{8}[(R \cdot \Sigma)^2 + R \cdot \Sigma] + \cdots\right\}\right) \tag{6.14}$$

II. EXISTENCE AND EVALUATION OF SYSTEMATIC ERROR

From Eq. 6.14, the average flux ϕ_{av_1} and ϕ_{av_2} in each sample can be equated if $\Sigma_1 R_1 = \Sigma_2 R_2$. Again, if $\Sigma_1 \neq \Sigma_2$, then the average flux ratio, ϕ_{av_1}/ϕ_{av_2}, is a function of $\Delta\Sigma$, and for cylindrical samples of equal diameter a linear relationship between $\log[\phi_{av_1}/\phi_{av_2}]$ versus $\Delta\Sigma$ can be obtained. It may be noted that if the irradiation geometry for both samples is identical, the effect on the average sample flux from flux gradients due to a point or a disk neutron source need not enter into the discussion above.

In Gijbels' treatment[33] summarized thus far, no attempt has been made to qualify or define Σ as related to any particular type of nuclear interaction probability. The justification for the use of Σ_R, σ_R, the removal cross-section parameters, instead of Σ_T, σ_T, the total cross-section designations for Σ occuring in all equations from 6.8 to 6.14, is given by Gijbels.[33] By definition, Σ_T and σ_T, the macroscopic and microscopic *total* cross sections, respectively, for 14-MeV neutron interactions, represent the sum of all possible interaction probabilities, that is,

$$\sigma_T = \sigma_n + \sigma_{n'\gamma} + \sigma_{2n} + \sigma_f + \sigma_{n'\alpha} + \sigma_\gamma + \sigma_p + \sigma_\alpha \tag{6.15}$$

Where σ_n is the elastic collision cross section, and the sum of the remaining cross sections, that is, the nonelastic cross section, includes such interaction probabilities as the inelastic scattering (n, n'γ) and absorption probabilities such as (n, 2n), (n, fission), (n, n'α), (n, γ), (n, p), and (n, α). The possibility of even a single interaction with the sample effectively removing the neutron from initiating any further interactions, as assumed by Anders and Briden,[14] is not completely true. No doubt any absorption process will eliminate a neutron from the incident flux, and on the average, any inelastic scattering interaction will result in a decrease in the neutron energy such that, for most reactions with 14-MeV neutrons, the cross section for the inelastically scattered neutron will, in general, be smaller than at 14 MeV. For example, for the ^{16}O(n, p)^{16}N reaction, the effective reaction threshold is about 11 MeV. Inelastic collisions of 14-MeV neutrons with the sample nuclei will effectively reduce the neutron energy below this threshold, or at least to a value where the cross section for the ^{16}O(n, p)^{16}N reaction is negligible. Rosen and Stewart[41] indicate that for inelastic scattering of 14-MeV neutrons by bismuth, the probability of degraded neutrons being between 4 and 12 MeV is an order of magnitude lower than for those in the energy group 0.5 to 4 MeV. Furthermore, for scattering from light materials ($A < 30$), the angular distribution of inelastically scattered neutrons is almost symmetrical about 90°, whereas for heavy scattering materials like Sn, Pb, and Bi, the angular distribution is peaked in the forward direction, that is, in the direction of the incident neutron flux.

The properties of elastically scattered 14-MeV neutrons, on the other hand, are vastly different from those for the inelastic scattering process. For heavy elements, only a small change in the direction of initial neutron trajectory occurs; the elastically scattered neutrons are strongly peaked in the forward direction and the neutrons act as if no scattering had occurred. With lower and lower atomic number scatterers, the fast neutrons lose more and more of their incident energy by elastic scattering. From basic reactor theory dealing with the slowing down of fast neutrons, the number of elastic collisions required to decrease the neutron energy to some effective threshold energy can be calculated from the expression[42]

$$N = \frac{1}{\xi} \ln \frac{E_0}{E_{\text{eff}}} \qquad (6.16)$$

where E_0 is the 14-MeV neutron energy, E_{eff} is the effective threshold energy (e.g., 11 MeV), and ξ is the average logarithmic energy decrement per collision and is calculated from the expression[42]

$$\xi = 1 + \frac{(A-1)^2}{2A} \ln \frac{A-1}{A+1} \qquad (6.17)$$

where A is the mass number of the scattering material. For elements with $A > 10$ a good approximation for the value of ξ is given[42] by

$$\xi \simeq \frac{2}{A + 2/3} \qquad (6.18)$$

It can be calculated therefore that for aluminum ($\xi = 0.022$) and Plexiglas ($\xi = 0.054$) the number of elastic collisions required to reduce the neutron energy from 14 MeV to 11 MeV are 10 and 4, respectively. Thus it is obvious that multiple elastic collisions are required to degrade 14-MeV neutrons to below the reaction threshold. It follows that the exponential term $e^{-\Sigma_T \cdot d}$ represents the probability that a neutron will penetrate to a distance d in the sample without being affected by any absorption or scattering process, whereas the exponential $e^{-\Sigma_R \cdot d}$ represents the probability that a neutron will penetrate to a distance d in the sample without any nonelastic interactions. Here the nonelastic cross section σ_x is equal to the sum of all cross sections, less the elastic contribution. Therefore, the identity $\phi = \phi_0 e^{-\Sigma_x \cdot d}$ represents the 14-MeV neutron flux after traversing a thickness "d" in the sample. As a first approximation Σ_x, the macroscopic nonelastic cross section, is equivalent to $0.5 \times \Sigma_T$.

In developing the concept of removal cross sections it should be remembered that σ_T data, normally compiled, were obtained from simple transmission experiments and corrected for neutrons that were scattered in the sample and registered in the detection system. As was pointed out earlier,

II. EXISTENCE AND EVALUATION OF SYSTEMATIC ERROR

this in-scattering can be significant for 14-MeV neutrons elastically scattered in the predominantly forward direction. Consequently, without this correction for forward scattering, the transmission of fast neutrons would be higher and the measured cross section lower. In this case where the neutron flux is being measured in the shadow of the sample,[14] the measured value would be geometry dependent, and, as pointed out by Nargolwalla et al.[40] and Gijbels,[33] the experimental geometry used in the earlier study by Anders and Briden[14] appears to approximate the measurement of σ_x rather than σ_T. The application of the removal cross-section (σ_R, Σ_R) concept, as proposed by Nargolwalla et al.,[40] is based on a rigorous treatment developed for calculating shielding parameters for the design of biological shields for power reactors.[43] This concept has also been recommended in the recent critique by Gijbels[33] and in some of his earlier studies.[44-46]

The semiempirical removal cross-section technique is found to be adequate for calculating attenuation of neutrons by biological shields either rich or deficient in hydrogen. For materials rich in hydrogen, the removal cross section is almost independent of the incident neutron energy between 0.1 and 14 MeV. For substances deficient in hydrogen this is not the case. The treatment[43] of removal cross sections takes into account the first collision density, which includes a contribution from neutrons scattered through small angles. As defined,[47] the removal cross section

$$\sigma_R(E) = \sigma_x(E) + \sigma_n(E)[1 - \bar{\mu}(E)] \qquad (6.19)$$

where σ_R is the microscopic removal cross section; σ_x is the microscopic nonelastic cross section; σ_n is the microscopic elastic cross section; and $\bar{\mu}(E)$ is the average cosine of the elastic scattering angle in the laboratory system at energy E. Since the elastic cross section is predominantly in the forward direction, the last term in Eq. 6.19 is negligible for a first-order approximation. Therefore, the attenuation of fast neutrons can be represented essentially by absorption processes plus inelastic scattering, or

$$\sigma_R(E) \geq \sigma_x(E) \qquad (6.20)$$

For hydrogen, however, $\sigma_R = \sigma_T \simeq 0.7$ barns for 14-MeV neutrons. For other neutron energies Gijbels[33] suggests the relationship

$$\sigma_R(\text{H}) = \frac{5.13}{E^{0.75}} \qquad (6.21)$$

Experimentally[44-46, 48] it was shown that the neutron attenuation calculated using Σ_R gave results in good agreement with the measured attenuation, and the use of Σ_T gave poorer comparisons.

260 SOURCES AND REDUCTION OF SYSTEMATIC ERROR

To facilitate calculation of Σ_R, Nargolwalla et al.[26,49] have provided tabulations for both 15- and 3-MeV neutrons (see Table 4.13, Chapter 4)*. These tables can be applied to sample thicknesses small compared to the removal mean free path λ_R ($\lambda_R = 1/\Sigma_R$).

In the case of gamma-ray absorption in thick samples, the usual exponential absorption law used by Anders and Briden[14] is valid, that is,

$$\frac{I}{I_0} = e^{-(\mu_0 \cdot d)} \tag{6.22}$$

where μ_0 is the total linear absorption coefficient for the absorption of a given energy gamma ray. Values for μ_0 are available for most elements in a recent tabulation by Storm and Israel.[50] The average transmitted gamma-ray intensity is seen by the detector in a manner analogous to that discussed under the neutron attenuation process. Therefore, for a homogeneously activated sample slab,

$$I \simeq I_0 e^{-\mu_0 \cdot d/2} \tag{6.23}$$

where $d/2$ is the effective sample attenuation thickness for gamma-ray absorption. As with the treatment of the neutron attenuation process, it is possible to derive a relationship between the ratio of specific activities generated in two samples and $\Delta\mu_0$. A highly accurate and detailed study of the two attenuation effects was carried out at the National Bureau of Standards Laboratory[40,49]; the techniques developed and results obtained are presented in the following discussion.

To permit accurate measurement of neutron and gamma-ray attenuation correction factors applicable to the analysis of thick samples, it was first necessary to develop a highly reproducible and accurate analytical technique so that the effect of random error could be reduced to negligible proportions compared to the magnitude of the correction error itself. The analytical system developed and described in a report by Lundgren and Nargolwalla[51] included a dual sample-biaxial rotating irradiation assembly and used the principle of simultaneous irradiation, followed by sequential counting of the sample and standard in the same counting system. Experimental data from the irradiation of two identical samples (no attenuation correction) indicated that the precision of this analytical technique was almost entirely controlled by random statistics of radioactive nuclide counting. Having obtained this optimum precision consistently, Nargolwalla et al.[40] then developed a technique for the evaluation of systematic errors from

---

* The values given in Table 4.13 (Chapter 4) for 15-MeV neutrons differ from those given in reference 49, due to the small change in neutron energy.

II. EXISTENCE AND EVALUATION OF SYSTEMATIC ERROR

the two attenuation processes as experienced in the case of oxygen analysis, and extended their work by concluding a subsequent study[49] which could be applied to the general analytical case.

Consider the attenuation of 14-MeV neutrons and 6.1-MeV gamma radiation from the decay of ^{16}N in O_2 determinations. In the general case, the attenuation process can be expressed as

$$\frac{f}{f_0} = e^{-l \cdot X} \tag{6.24}$$

where f is the neutron flux [neutrons/(cm² sec)] or gamma-ray intensity (photons/sec) after traversing the effective attenuation sample thickness X (cm); f_0 represents the neutron flux [neutrons/(cm² sec)] or gamma-ray intensity (photons/sec) with no sample attenuation; and l is a proportionality constant in the differential expression $df/dX = -lf$ for example, absorption coefficient (cm^{-1}).

To differentiate between the two attenuation effects being considered, the constant l will have a different meaning to represent neutron attenuation and gamma-ray absorption. Suitable correction factors can be calculated or experimentally generated with X taking the form of an empirical constant describing some effective attenuating thickness of the sample. For 14.5-MeV neutrons the constant l is taken to be the total macroscopic removal cross section Σ_R, which can be calculated from the expression

$$\Sigma_R = \frac{1}{V} \sum_{i=1}^{n} \left(\frac{\sigma_{R(i)} \cdot W_i N_0}{M_i} \right) \tag{6.25}$$

where V is the sample volume (cm³); $\sigma_{R(i)}$ is the microscopic removal cross section of element i (cm²/atom) (the values of $\sigma_{R(i)}$ can be obtained from Table 4.13, Chapter 4); W_i is the weight of element i in the sample (g); N_0 is Avogadro's number, 6.023×10^{23} (atoms/g-atom); Σ_R is the total macroscopic removal cross section of the sample for 14.5-MeV neutrons (cm^{-1}); and M_i is the atomic weight of element i (g/g-atom). For the case of gamma-ray attenuation, the constant l in Eq. 6.24 takes the usual form of a total linear absorption coefficient, μ_0 (cm^{-1}), of the gamma energy of interest, and

$$\mu_0 = \sum_{i=1}^{n} \left[\left(\frac{\mu}{\rho}\right)_i \cdot \frac{W_i}{V} \right] \tag{6.26}$$

where μ/ρ_i is the mass attenuation coefficient for element i (cm²/g) (values for μ/ρ for gamma energies of interest are available in the literature[49]) and W_i/V is the partial density of element i (g/cm³).

For the experimental evaluation of Eq. 6.24, the example of oxygen determination was used. The main consideration is the measurement of the oxygen content, W_{ox}, in a sample by comparing its ^{16}N-induced radioactivity, A_x, with the oxygen content, W_o, and induced ^{16}N radioactivity, A_s, of a standard, that is,

$$W_{ox} = \frac{W_o}{A_s} \times A_x \tag{6.27}$$

Two standards of identical macroconstituent composition and weight (S_1 and S_2) were irradiated in the dual sample-biaxial rotating device.[51] It is understood that the ultimate sample to be analyzed is placed in position 1, and the standard in position 2 of the rotator; that is, the sample is counted first, and then the standard. The activity ratios are defined as $R_1 = A_{s1}/A_{s2}$ and $R_2 = A_x/A_s$, where A_{s1}, and A_{s2} refer to the induced radioactivity in samples S_1 and S_2, respectively, and Eq. 6.27 reduces to

$$W_{ox} = \frac{R_2}{R_1} \cdot W_o \tag{6.28}$$

where R_1 is the flux normalization factor to account for the two counting intervals and the time difference between the end of the first and the start of the second sample count. This ratio is determined by irradiating and counting two identical standards under prescribed conditions to be used for subsequent analysis. The experimentally measured radioactivity terms are corrected by multiplying them by an attenuation correction factor γ which is equal to f_0/f (Eq. 6.24). Since in the determination of the normalization constant the two standards are identical, $\gamma_{s1} = \gamma_{s2}$ and the corrected oxygen content, W'_{ox}, can be calculated from

$$W'_{ox} = W_o \cdot \frac{R_2}{R_1} \cdot \frac{\gamma_x}{\gamma_s} \tag{6.29}$$

Substituting Eq. 6.28 into Eq. 6.29 yields

$$\frac{W_{ox}}{W'_{ox}} = \frac{\gamma_s}{\gamma_x} \tag{6.30}$$

From Eq. 6.24

$$\frac{W_{ox}}{W'_{ox}} = \exp[(l_s - l_x) \cdot X] \tag{6.31}$$

the subscripts on the value of l have the same meaning as those on the radioactivity terms A_s and A_x. The observed oxygen content, W_{ox}, is measured by experimentally determining the quantities in Eq. 6.31. This result is compared to the true value to give the correction factor. For a given diameter, volume, and weight of the sample whose macroconstituent analysis is known, the appropriate value of l is calculated from Eqs. 6.25 and 6.26. If the difference in the calculated attenuation factors is plotted against W_{ox}/W'_{ox}, a linear relationship on semilogarithmic coordinates would confirm Eq. 6.31. However, it should be realized that in the above-mentioned experiment a product of the two attenuation factors is obtained and the proper generalized equation is

$$\frac{W_{ox}}{W'_{ox}} = \left(\frac{\gamma_s}{\gamma_x}\right)_{neutrons} \cdot \left(\frac{\gamma_s}{\gamma_x}\right)_{gamma}$$

$$= \exp[(\Sigma_{R,s} - \Sigma_{R,x})X_n + (\mu_{0,s} - \mu_{0,x})X_\gamma] \tag{6.32}$$

To determine the individual attenuation factors, a series of materials were selected so that one of the correction factor ratios approximates unity, for example, $\mu_0, x \simeq \mu_0, s$. After X_n is evaluated and a correction curve for neutron attenuation is established, the gamma attenuation factors and X_γ can be determined easily from Eq. 6.31 for a selection of materials having a wide range of μ_0 values. In the experiments carried out, a 500-μA beam current neutron generator was used. In this accelerator no provision for control of the beam size was available. The estimated fixed beam spot diameter was about 1 cm. Several samples, such as oxalic acid, calcium phosphate, titanium dioxide, and benzoic acid, were used for the determination of the 14.5-MeV neutron attenuation correction factor curve. Oxalic acid served as the primary standard. For these samples the calculated gamma attenuation factors were within $\pm 1\%$ of one another and suitable corrections were applied. All samples were encapsulated in standard 8-cm^3-volume polyethylene snap-cap cylindrical containers. Neutron attenuation factors for sample diameters of 1.45, 0.95, and 0.64 cm were determined experimentally. The two smaller diameters were achieved by filling polyethylene tubes of appropriate diameters with sample material and locating them centrally inside the 1.45-cm-diameter polyethylene containers.

After the neutron attenuation correction curve had been established for each of the three sample diameters, a selection of samples of iron oxide, lead oxide, and bismuth oxide was made. These samples were irradiated and counted under the prescribed irradiation and counting conditions using the same sample of oxalic acid as the standard. The observed oxygen content of these samples was corrected for neutron attenuation as predicted from the curve established previously. This corrected value was compared with the true value and the ratio gave the gamma attenuation correction factor. These factors were then expressed as a function of $\Delta\mu_0$. The two correction factor curves were used in the determination of oxygen in several NBS Standard Reference Materials. The 14-MeV neutron activation results showed good agreement with the certified values obtained by other analytical techniques. It is important to realize that a fairly good knowledge of the major constituents of the unknown sample is necessary for the calculation of $\Delta\Sigma_R$ and $\Delta\mu_0$. It is estimated that major constituents known to $\pm 5\%$ relative standard deviation of a single determination are adequate for this purpose. More important is the fact that the correction factors determined are only valid for the irradiation and counting geometries used to determine them and for the sample size and the configuration used. Therefore, the effective attenuation sample thickness as given by the slope of the respective correction curves must be reevaluated if any of the initial geometrical considerations are changed.

The study described above was extended to include the general case of monoenergetic gamma-ray counting. Because of a radical change in the irradiation geometry caused by the replacement of the 500-μA generator with a larger capacity, 2500-μA machine, a redetermination of the 14.5-MeV neutron and 6.1-MeV gamma-ray attenuation correction factors was necessary.[49] The effective sample thicknesses determined varied considerably from those measured in the earlier work.[40] We have explained the cause of this difference above; it was apparently overlooked by Gijbels[33] in his discussion of these studies.

The techniques described provided analyses of a high order of accuracy and precision. It was logical to extend the use of these techniques to study gamma-ray attenuation in the general case where photopeak counting is usually performed. Correction factors evaluated from this investigation[49] were applied in the certification of several elements in NBS Standard Reference Materials. An understanding of perturbations caused by thick samples permitted the use of a single standard for the determination of an element of interest in any matrix whose composition is reasonably well known.

In general, gamma spectrometry associated with activation analysis involves the measurement of photopeaks resulting from direct nuclear level transitions or from the annihilation of positrons. One of the conclusions drawn from the earlier study[40] was that the thickness of the attenuated sample would be independent of the primary gamma energy and dependent only on the efficiency of the detection system for a degraded gamma-ray contribution in the region of analysis (e.g., the Compton contribution in the analysis of gamma rays between 4.8 and 8.0 MeV for measurement of 6.1-MeV ^{16}N gamma rays). Therefore, if only the photopeak were counted, the slope of the gamma-ray correction factor curve would be greater than that obtained for integral counting processes like the one used for the analysis of oxygen, and would approximate a value one-half of the geometrical sample diameter. It was also surmised that the gamma-ray attenuation line from the annihilation process would show a small change in slope due to the bremsstrahlung contributions under the annihilation photopeak. On the other hand, the effective attenuation thickness for neutron absorption would remain unaffected as long as the whole geometry of the analytical system was not disturbed. Furthermore, this curve would be independent of the reaction threshold energies. In this investigation[49] all of the parameters mentioned above were verified. Judicious selection of samples permitted evaluation of the effect of different reaction threshold energies on the attenuation of 14.5-MeV neutrons, and the gamma-ray and positron energy dependence on the photopeak attenuation correction factors. Additional variables, such as the effect of sample density and bremsstrahlung contribution in the photopeak region, were also evaluated.

Correction factor correlations taken from this study[49] are shown in Figs. 6.7 to 6.9 and these curves were used to analyze several NBS Standard Reference Materials listed in Table 6.9.

TABLE 6.9. Analysis of NBS Standard Reference Materials[a]

Standard Reference Material	Element Determined by 14 MeV Neutron Activation		Concentration (%) Certified[c]
Cobalt cyclohexanebutyrate[1055a]	Co	17.49 ± 0.11[b]	17.4 ± 0.17
Silver 2-ethylhexanoate[1077]	Ag	42.49 ± 0.42[b]	42.4 ± 0.42
Magnesium cyclohexanebutyrate[1061a]	Mg	6.69 ± 0.04[b]	6.8 ± 0.14
Octaphenylcyclotetrasiloxane[1066]	Si	14.16 ± 0.11[b]	14.1 ± 0.14

[a] Cf. reference 49.
[b] Weighted standard error of the weighted mean (four determinations).
[c] National Bureau of Standards Certificate.

In these analyses the standards used were such that correction factors as high as 20% were applied to the experimental results. The accurate determination of the correction factor curves is reflected in the good agreement of the data shown in Table 6.9. To summarize these extensive studies carried out at the National Bureau of Standards, we shall mention several important conclusions.

a. Attenuation correction factor curves are only significant if identical analytical systems are used, that is, identical irradiation source, irradiation and counting geometries, and sample size and shape.

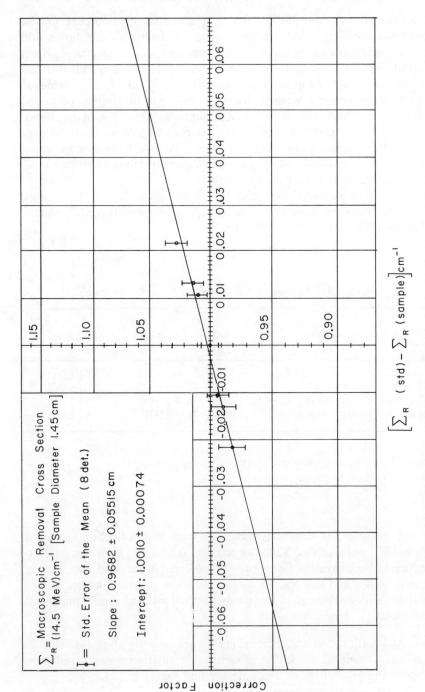

Fig. 6.7. The 14.5-MeV neutron attenuation correction factor curve (cf. reference 49).

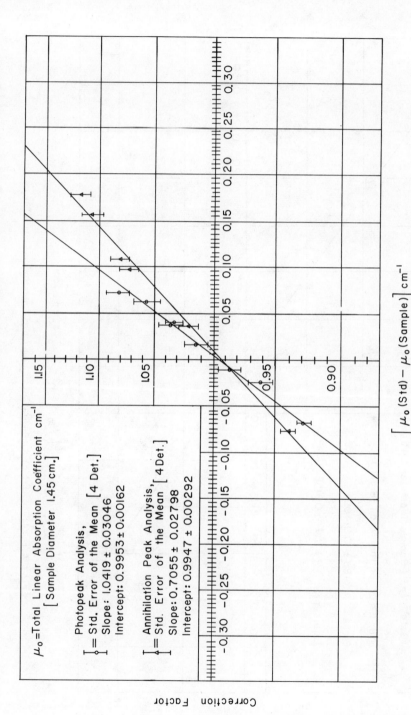

Fig. 6.8. Photopeak and annihilation peak gamma-ray attenuation curve (cf. reference 49).

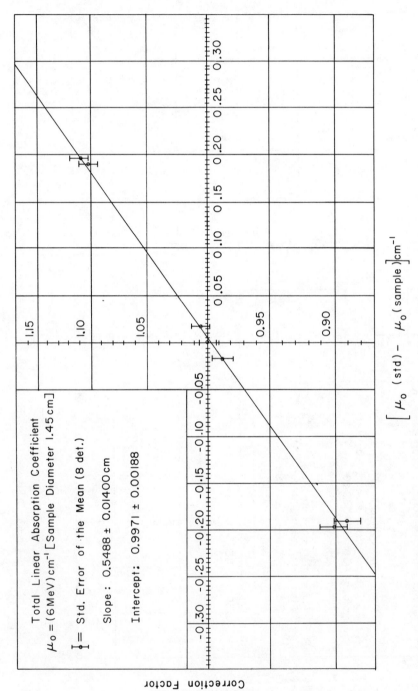

Fig. 6.9. The 4.8- to 8.0-MeV gamma-ray attenuation correction factor curve (cf. reference 49).

II. EXISTENCE AND EVALUATION OF SYSTEMATIC ERROR

b. Fast neutron attenuation characteristics are independent of the reaction threshold energies, and depend only on the calculated macroscopic removal cross-section differences between the sample and standard.

c. Gamma-ray attenuation is dependent on the counting efficiency of the detector system. The higher the counting efficiency, the lower is the slope of the gamma-ray attenuation correction curve.

Several techniques to study attenuation effects have been reported. Brune and Jirlow[52] studied this effect in the analysis of trace oxygen levels in aluminum using 14-MeV neutron activation analysis. The effects of neutron attenuation in the sample from scattering and from other slowing-down processes were studied.[52] In their compilation of elemental sensitivities for 3-MeV neutron activation analysis, Nargolwalla et al.[36] used the techniques described earlier[40,49] to correct for 3-MeV neutron and gamma-ray absorption by thick samples and for normalization of the 3-MeV neutron flux. Anderson and Algote[53] investigated the effect of sample bulk density on the determination of nitrogen by 14-MeV neutron activation analysis. Dibbs and McMahon[54] found that, in the analysis for silicon in iron ore by the reaction ^{28}Si(n, p)^{28}Al, a correction factor of 10% was necessary to correct for the absorption of the 1.78-MeV ^{28}Al gamma radiation in the sample. In this case the standard was 8.4 g of silica and the sample weighed 21 g.

For the case of sample and standard not independent during irradiation, Gijbels[33] has described a method for the measurement of sample attenuation. The system used involves irradiating small disks of sample and standard, one behind the other. The standard serves as a flux monitor for oxygen analysis. To correct for the flux gradient experienced in this irradiation geometry, two identical oxygen standards were irradiated and the ratio of their induced ^{16}N radioactivity was measured. For reasons given below, this method can lead to systematic errors in analysis. For example, in the determination of this ratio, the standard is shielded by a different material from that used in the actual analysis of an unknown sample. Gijbels reports a detailed analytical expression that permits the analysis for oxygen in samples with the proper correction factors applied.

G. INSTRUMENTAL ERRORS

In a presentation on some of the problems in precision activation analysis with 14-MeV neutron generators, Wood[55] stressed the uncertainty of analysis due to intrumental factors such as inadequate timing apparatus, lack of reproducibility in the intercalibration of two detector systems, and dead-time corrections. This discussion points out that very good precision and accuracy can be obtained only if carefully designed systems and techniques, proper flux-monitoring practices, accurate timing of critical measurements, and good sample preparation methods are used.

Perhaps one of the most important causes of systematic error from instrumental effects occurs in the analysis of oxygen in certain matrices. In this case, the high activity of the matrix can cause pulse pile-up effects in the energy region used normally for counting ^{16}N gamma radiation. In a study[48] of the pulse pile-up effect, it was observed that errors occur in the determination of oxygen in copper and aluminum matrices.

In a later study[56] a comprehensive investigation of this effect in 22 different matrices was carried out. The matrix elements were selected carefully on the basis of their importance to industry, their high 14-MeV neutron activation cross section, and unfavorable decay characteristics in terms of short half-life, emission of high-energy gamma rays, and bremsstrahlung production. The results obtained are shown in Table 6.10. As expected, the elements boron, fluorine, and uranium present severe problems even at oxygen levels greater than 1%. Interference from pulse pile-up was observed in the case of aluminum and copper. The data indicate that below 100 ppm, oxygen determination should be corrected for the pulse pile-up effect. Although one can also expect interferences from the 0.8-sec isomers ^{90m}Zr and ^{207m}Pb, these were not observed owing to the 4-sec delay before the first count could be measured. It was also expected that a pulse pile-up contribution from the bremsstrahlung from the ^{28}Si(n, p)^{28}Al reaction would be apparent. However, because of the effective beta absorber used between the detector and sample, contributions, if any, could not be detected above background. Again, it is emphasized that pulse pile-up

TABLE 6.10. Results of Interferences in Trace Oxygen Determination[a]

Matrix Element	Interference[b] (ppm equivalent oxygen/g matrix element)	
	Nuclear	Pulse Pile-up
Boron	1.3×10^4	—
Fluorine	4.9×10^5	—
Aluminum	—	10
Copper	—	20
Uranium	2.2×10^4	—

[a] Cf. reference 56.
[b] For elements Be, N, Na, Si, P, Ti, V, Cr, Mn, Fe, Ni, Zn, Se, Zr, Mo, Ba, and Pb, interference, if any, was not detectable above background.

effects evaluated for a specific system cannot be applied directly to other analytical systems.

In this treatment of possible sources of systematic errors and the description of various solutions to these problems offered in this chapter, it is concluded that activation analysis with neutron generators can provide high accuracy and precision. In fact, this analytical technique rightfully belongs with other sophisticated nonnuclear techniques in the analytical field. As pointed out in a recent review,[2] the improvement in the analytical accuracy and precision obtained with a neutron generator is a direct result of the great effort that has been expended in the study of systematic error.

REFERENCES

1. L. A. Currie, *Nucl. Instrum. Methods*, **100**, 387 (1972).
2. S. S. Nargolwalla, "Application of Neutron Generators to Activation Analysis," Proceedings, Second Oak Ridge Conference on the Use of Small Accelerators for Teaching and Research, Oak Ridge, Tennessee, March 23–25, 1970, USAEC Report CONF-700322.
3. J. Csikai, M. Buczkó, Z. Böcy, and A. Demény, *Atomic Energy Review*, Vol. VII, No. 4, IAEA, Vienna, 1969, pp. 93–128.
4. E. Ricci and F. F. Dyer, *Nucleonics*, **22** (6), 45 (1964).
5. S. C. Mathur and G. Oldham, *Nucl. Energy*, September–October 1967, pp. 136–141.
6. J. T. Gilmore and D. E. Hull, *Anal. Chem.*, **34**, 187 (1962).
7. J. T. Prud'homme, "Texas Nuclear Neutron Generators," Texas Nuclear Corporation, Austin, Texas, 1964.
8. J. M. A. Lenihan and S. J. Thomson, Eds., *Activation Analysis, Principles and Applications*, Academic Press, New York, 1965.
9. D. J. Veal and C. F. Cook, *Anal. Chem.*, **34**, 178 (1962).
10. E. L. Steele and W. W. Meinke, *Anal. Chem.*, **34**, 185 (1962).
11. J. Csikai, P. Bornemisza, and I. Hunyadi, *Nucl. Instrum. Methods*, **24**, 227 (1963).
12. O. U. Anders and D. W. Briden, *Anal. Chem.*, **37**, 530 (1965).
13. S. S. Nargolwalla, E. P. Przybylowicz, J. E. Suddueth, and S. L. Birkhead, *Anal. Chem.*, **41**, 168 (1969).
14. O. U. Anders and D. W. Briden, *Anal. Chem.*, **36**, 287 (1964).
15. R. F. Coleman, *Iron Steel Inst. (London), Spec. Rep.*, **68**, 1960.
16. K. G. Broadhead and H. H. Heady, *Anal. Chem.*, **37**, 759 (1965).
17. H. F. Priest, U.S. Army Materials Research Agency, Watertown, Massachusetts, private communication, June 1968.
18. R. L. Heath, "Scintillation Spectrometry Gamma-Ray Spectrum Catalogue," USAEC Report IDO-16880, 1964.
19. E. Ricci, *J. Inorg. Nucl. Chem.*, **27**, 41 (1965).
20. E. Ricci, *Trans. Am. Nucl. Soc.*, **7**, 203 (1964).
21. J. L. Fowler and J. E. Brolley, Jr., *Rev. Mod. Phys.*, **28**, 103 (1956).

22. S. Glasstone and M. C. Edlund, *The Elements of Nuclear Reactor Theory*, Van Nostrand, Princeton, New Jersey, 1958.
23. M. D. Goldberg, V. M. May, and J. R. Stehn, USAEC Report BNL-400, 1962.
24. D. J. Hughes and R. B. Schwartz, USAEC Report BNL-325, 1958.
25. Proceedings, 1965 International Conference on Modern Trends in Activation Analysis, College Station, Texas, April 19–22, 1965, pp. 388–390.
26. F. Girardi, J. Pauly, and E. Sabbioni, "Dosage de l'Oxygéne dans les Produits Organique et les Metauxpar Activation aux Neutrons de 14 MeV," Euratom Report EUR-2290, Brussels, 1965.
27. J. R. Vogt and W. D. Ehmann, *Radiochim. Acta*, **4**, 24 (1965).
28. M. Ladu, M. Pellicioni, and E. Totondi, *Nucl. Instrum. Methods*, **32**, 173 (1965).
29. I. Fujii, K. Miyoshi, H. Muto, and Y. Maeba, *Toshiba Rev.*, Autumn 1963.
30. J. Wing, *Anal. Chem.*, **36**, 559 (1964).
31. W. W. Meinke and R. W. Shideler, *Nucleonics*, **20** (3), 60 (1962).
32. F. A. Iddings, *Anal. Chim. Acta*, **31**, 206 (1964).
33. R. Gijbels, "Activation Analysis with Neutron Generators," in *Instrumental and Radiochemical Activation Analysis*, Monotopic Series, F. Adams, J. P. Op de Beeck, P. Van den Winkel, R. Gijbels, D. de Soete, and J. Hoste (Eds.), Chemical Rubber Company, Cleveland, Ohio.
34. D. Partington, D. Crupton, and S. E. Hunt, *The Analyst*, **95**, 257 (1970).
35. P. J. Jessen and K. C. Pierce, *Trans. Am. Nucl. Soc.*, **10** (1), 85 (1967).
36. S. S. Nargolwalla, J. Niewodniczanski, and J. E. Suddueth, *J. Radioanal. Chem.*, **5**, 403 (1970).
37. G. Weber and M. Guillaume, *Radiochem. Radioanal. Lett.*, **3**, 97 (1970).
38. B. L. Twitty and K. M. Fritz, "A Rapid Determination of Oxygen in High-Purity Magnesium Chips by Neutron Activation Analysis," National Lead Company of Ohio, Cincinnati, Ohio, Report NLCO-973, May 1966.
39. D. Dutina, "Determination of Oxygen in Zircaloy by Fast Neutron Activation," Knolls Atomic Power Laboratory, Schnectady, New York, Report KAPL-2000-19, I.14–I.23, 1962.
40. S. S. Nargolwalla, M. R. Crambes, and J. R. DeVoe, *Anal. Chem.*, **40**, 666 (1968).
41. L. Rosen and L. Stewart, *Phys. Rev.*, **99**, 1052 (1955).
42. S. Glasstone, *Principles of Nuclear Reactor Engineering*, Van Nostrand, Princeton, New Jersey, 1957, p. 152.
43. A. F. Avery, D. E. Bendall, J. Butler, and K. T. Spinney, "Methods of Calculation for Use in the Design of Shields for Power Reactors," UKAE, Report AERE-R-3216, 1960.
44. R. Gijbels, A. Speecke, and J. Hoste, *Anal. Chim. Acta*, **43**, 183 (1968).
45. R. Gijbels, A. Speecke, and J. Hoste, Proceedings, 1968 International Conference on Modern Trends in Activation Analysis, National Bureau of Standards, Gaithersburg, Maryland, NBS Special Publication 312, Vol. II, 1969, p. 1298.
46. R. Gijbels, J. Hoste, and A. Speecke, "The Industrialization of 14-MeV Neutron Activation Analysis for Oxygen in Steel," Euratom Report EUR-4297, Luxemburg, September 1969.
47. B. T. Price, C. C. Horton, and K. T. Spinney, *Radiation Shielding*, Pergamon Press, London, 1957.

48. J. Hoste, D. De Soete, and A. Speecke, "The Determination of Oxygen in Metals by 14-MeV Neutron Activation Analysis," Euratom Report EUR-3565e, Brussels, September 1967.
49. S. S. Nargolwalla, M. R. Crambes, and J. E. Suddueth, *Anal. Chim. Acta*, **49**, 425 (1970).
50. E. Storm and H. I. Isreal, Los Alamos Scientific Laboratory Report LA-3753 1967.
51. F. A. Lundgren and S. S. Nargolwalla, *Anal. Chem.*, **40**, 672 (1968).
52. D. Brune and K. Jirlow, *J. Radioanal. Chem.*, **2**, 49 (1969).
53. G. H. Anderson and J. M. Algote, *J. Radioanal. Chem.*, **3**, 261 (1969).
54. H. P. Dibbs and C. McMahon, Mines Branch Investigation Report IR64-82, Ottawa, Canada, 1964.
55. D. E. Wood, "Problems in Precision Activation Analysis with Fast Neutrons," Proceedings, Conference on the Use of Small Accelerators for Teaching and Research, Oak Ridge Associated Universities, Oak Ridge, Tennessee, USAEC Report CONF-680411, April 8–10, 1968.
56. S. S. Nargolwalla, J. E. Suddueth, and H. L. Rook, *Trans. Am. Nucl. Soc.*, **13** (1), 78 (1970).

CHAPTER

7

APPLICATIONS

I. INTRODUCTION

Since the purpose of this treatise on neutron generator activation analysis is to stimulate a broader use of the technique, this chapter on applications is perhaps one of the most important insofar as the reader's initial interest is concerned. The intent in this chapter is to point out the breadth of applications for this technique and to cite representative applications. For some elements, applications of the neutron generator have been fully developed because of the lack of suitable alternate chemical or instrumental methods of analysis. For other elements this technique has not been explored because older, established methods readily fill present needs. For elements in this latter category, the potentialities of neutron activation with the generator are pointed out since, if the technique is available in a laboratory, it may be feasible to determine these elements more readily than with the older established methods, and multielement analyses using a single procedure may be possible in a specific matrix.

In Chapter 1, a number of advantages were identified as important reasons for utilizing neutron generator activation analysis in the analytical laboratory. It is relatively easy to contrast this technique with those alternate methods of analysis which require solution chemistry. Where this technique is capable of measuring the element with sufficient sensitivity and there are no direct interfering reactions, there is no question as to its superiority. The usual problems with sample dissolution are not present and the time per analysis is likely to be shorter. However, the technique competes more directly with atomic absorption spectrometry and x-ray fluorescence spectrometry in speed of analysis, sensitivity, precision, and accuracy. All are capable of measuring elements from the parts per million to the macroconstituent range with high precision and accuracy. The fact that they are not equally sensitive to all elements and that each technique has certain "blind spots" and weaknesses makes the choice of a method dependent on the analysis to be carried out. Thus discussion of the strengths and weaknesses of these three techniques for all elements would be a tedious undertaking and not particularly useful to the reader with a problem in analysis that did not fit any of the categories. On the other hand, it is appropriate to point out some general characteristics of these three competing methods which may serve as guidelines for the worker to select the best method for his particular analysis.

First, it should be emphasized that each technique is based on different properties of the element being measured. Atomic absorption processes depend on the properties of the outer-shell electrons, x-ray fluorescence is a function of the binding energies of the inner-shell electrons, and neutron activation analysis is a function of the nuclear properties of the element. These different dependencies enhance the possibility that one of the three methods will provide the required selectivity necessary to measure the element in a mixture. Moreover, the response intensity is also related to the different physical properties of each element and thus the sensitivity of the techniques would be expected to be quite different. Based on these considerations, one finds the techniques to be generally complementary.

Broadly speaking, the following considerations should be borne in mind when choosing between these techniques. The atomic absorption technique requires that the sample be soluble in a solvent which will support combustion but not be too flammable. This can prove to be a limitation for refractory materials or when a large sample must be taken to ensure sufficient sensitivity for a trace constituent. For such samples, the fluorescence and neutron generator activation techniques have a decided advantage since no dissolution of the sample is required. While spectral interferences are infrequent in atomic absorption work, matrix effects are present and must be recognized and corrected for in the analysis. This type of interference is difficult to predict and can only be properly taken into account by calibrating the element in the sample matrix or by using a standard addition procedure.

X-ray fluorescence analysis is, in general, a good technique for the determination of elements above atomic number 20. Below this atomic number, the fluorescence x-ray intensity diminishes due to an increased yield of Auger electron emission, and thus the sensitivity for the elements decreases for elements of low atomic number. Because x-rays are readily attenuated, x-ray emission calibration curves must be made using a matrix as similar to the sample as possible. In addition, the surface roughness and particle size of powdered samples must be carefully controlled if precise and accurate analyses are to be obtained. For samples in which these parameters can be controlled sufficiently, the technique provides one of the most rapid methods of analysis available.

Neutron activation analysis with fast neutrons is complementary to x-ray fluorescence analysis in element sensitivity. It has excellent sensitivity for the light elements where the x-ray technique is weak. The classic example, of course, is the determination of oxygen, which can be measured with exceedingly high sensitivity and selectivity by neutron activation. There are a few light elements, such as lithium, boron, and carbon, for which neither technique provides sufficient measurement sensitivity. Fast neutron activation suffers far less from sample attenuation problems than x-ray fluorescence because of the higher energy of the gamma radiation emitted. On the other hand, when large samples are used (gram quantities), neutron and gamma attenuation may occur. Both attenuation processes have been studied quantitatively, and procedures are now available which will provide accurate corrections (cf. Chapter 6). The major limitations of neutron generator activation are its insensitivity for certain elements, the lack of suitable nuclear reactions for others, and the sizable number of positron-emitting nuclides which are produced with fast neutron reactions. Where more than one positron emitter is produced in a matrix upon irradiation, and the half-lives of the positron-emitting isotopes are similar, a precise analysis is precluded. Decay curve analysis is useful where a substantial difference exists between the half-lives of positron emitters, but it is not capable of high precision under less ideal conditions.

In the following sections, each element which can be measured with reasonable sensitivity by neutron activation analysis with a neutron generator is mentioned along with the nuclear data pertinent to selecting conditions for its determination. The likely interferences are also discussed. Significant literature describing applications of this technique for the determination of the element is discussed along with information on the reported sensitivities and any other unique aspects of the analysis mentioned in the literature.

With respect to reported detection limits or sensitivities for the element, the reader is cautioned to use these figures of merit only as an approximate indication of useful sensitivity. Uncontrolled factors such as the energy distribution of the particular generator used as well as the worker's choice of irradiation, decay, and counting conditions introduce significant enough differences in sensitivity that the figures cannot be compared readily. In addition, there are several definitions of sensitivity extant which make comparison even more difficult. Some authors believe the best criterion of the detection limit to be the amount of element which results in a signal that is three times the standard deviation of the background count; others determine the amount of element which will give a fixed number of counts, such as 100 counts above the background level. It is not the authors' intent to resolve this issue but merely to alert the reader that where significant differences occur in reported detection limits, in most cases they can be attributed to one or more of the factors just discussed.

Two tables from review papers on applications are reproduced with the authors' permission in the Appendices. Readers will find these useful to supplement information about each element given in this chapter. Found in the Appendices are the following tabular summaries:

Appendix III. Experimental Sensitivities for 14-MeV and Thermal Activation with a Neutron Generator, from H. P. Dibbs, "Activation Analysis with a Neutron Generator," Department of Mines and Technical Surveys, Research Report R155, Ottawa, Canada, February 1965.

Appendix IV. Experimental Sensitivities for 3-MeV Neutron Activation Analysis, from S. S. Nargolwalla, J. Niewodniczanski, and J. E. Suddueth, *J. Radioanal. Chem.*, **5**, 403 (1970).

Unfortunately, the useful compendium by Cuypers and Cuypers[1] is too extensive to reprint here, although it is one of the largest summaries of experimental activation analysis conditions for use with the neutron generator. Information from this source is quoted liberally throughout this chapter.

II. PREDICTION OF ANALYSIS CONDITIONS

To assist the analyst in setting up optimum conditions for his analysis, pertinent nuclear data have been assembled and put into a form for convenient use. In the section on each element, activation, decay, and counting integral curves have been provided for both the element and its common interferences. The interferences listed for each element are those which produce the same nuclide as that obtained from the element being analyzed, and also those elements which produce nuclides that have gamma emissions in the same region as the nuclide being analyzed. By combining information from these curves for a given element, the analyst can arrive at optimum conditions for a particular analysis. The curves permit an estimate of actual measured activity for a particular neutron activation system once the system has been cross-calibrated with the procedure described here. The actual measured activity, A_m, is related to the other parameters of the analysis system and the nuclear data in the following manner:

$$A_m = I_g \times D_g \times F_c \times A_c \times D_f \times C_i \qquad (7.1)$$

where,

- I_g = a geometric correction factor to correct for the difference between the irradiation geometry actually used and that assumed in these calculations.
- D_g = a geometric correction factor to account for differences between the counting geometry actually used and that assumed for these calculations, a point source, 0 cm from a 3 × 3-in. sodium iodide crystal placed on the axis of that crystal (i.e., 2π geometry).
- F_c = a flux correction factor which represents the ratio of the actual flux used to a flux of 10^9 neutrons/(cm² sec) which was used in the calculation of the activation curves.

A_c = the measurable activity for a given irradiation time if the flux is 10^9 neutrons/(cm² sec) and all geometric and decay factors are unity. This value can be read from the "activation" curves provided for each element.

D_f = the decay factor for a selected decay time and is read from the decay curves provided for each element.

C_i = the counting integral factor which can be read from the curves supplied with each element for a desired counting interval.

If the curves supplied in this chapter are to be used on a continuing basis with a given instrumental arrangement, the product of the factors I_g and D_g can be determined by a known amount of an element in the system and comparing the values obtained with values predicted from the other factors. Thereafter the only adjustment necessary will be made for the variation in flux. On a relative basis, the product of A_c, D_f, and C_i may be used to select optimum conditions for the measurement of an element in the presence of an interference, and possible interference may be minimized through proper selection of irradiation, decay, and counting conditions. The method of calculation of each of these working curves is described below.

A. ACTIVATION CURVES

The activation curves were calculated from the activation equation 7.2, taking into account the gamma-ray branching ratio, B_r, the absolute detector efficiency, D_e, and the peak-to-total ratio, P, for the particular photopeak.

$$A_c = \Phi \cdot N \cdot \sigma(1 - e^{-\lambda t}) \cdot B_r \cdot D_e \cdot P \qquad (7.2)$$

where

Φ = the neutron flux, neutrons/(cm² sec)
N = the number of target atoms
σ = the cross section, cm²
t = the irradiation time
λ = the decay constant.

Computer techniques were used to generate the activation curves. The reference sources for the physical constants used in these calculations are listed below.

Cross-Section Values
1. B. T. Kenna and F. J. Conrad, "Tabulation of Cross Sections, Q-Values, and Sensitivities for Nuclear Reactions of Nuclides with 14 MeV Neutrons," SC-RR-66-229, Sandia Laboratory, Albuquerque, New Mexico, June 1966.

2. M. D. Goldberg, S. F. Mughabaghad, S. N. Purohit, B. A. Magurno, and V. M. May, "Neutron Cross Sections," BNL 325, 2nd ed., Supplement No. 2, August 1966.
3. J. Csikai, M. Buczkó, Z. Bödy, and A. Demény, "Nuclear Data for Neutron Activation Analysis," *Atomic Energy Review*, Vol. VII, No. 4, IAEA, Vienna, 1969.
4. M. Cuypers and J. Cuypers, "Gamma-Ray Spectra and Sensitivities for 14-MeV Neutron Activation Analysis," Texas A and M, College Station, Texas, April 12, 1966.

Natural Abundance, Half-Life, and Gamma-Ray Yield Values
1. C. M. Lederer, J. M. Hollander, and I. Perlman, *Table of Isotopes*, 6th ed., Wiley, New York 1967.

Scintillation Counting Factors
1. R. L. Heath, "Scintillation Spectrometry Gamma Ray Spectrum Catalogue," Vol. 1, 2nd ed., USAEC Report IDO-16880, 1964.

The absolute detection efficiency, D_e, was taken from the curves provided for a point source on a 3 × 3-in. sodium iodide crystal for a distance of 0 cm from the detector by Heath. This configuration was chosen arbitrarily to permit relative comparisons of the activity produced by different elements in the sample. For more exact comparisons, geometric corrections for the actual counting system used should be introduced into the calculation to relate the system to the reference point chosen for these calculations.

B. DECAY CURVES

The decay factor curves shown for each element are calculated on the basis of the decay equation

$$D_f = \frac{A_t}{A_0} = e^{-\lambda t_d} \tag{7.3}$$

where
D_f = the decay factor previously mentioned
A_t = the activity after a decay time t_d
A_0 = the activity at the end of the irradiation
t_d = the decay period.

The factor indicates what portion of the activity present at the termination of irradiation is present after the indicated decay time.

C. COUNTING INTEGRAL FACTOR

The counting integral factor, C_i, is a convenient way to estimate the integrated number of counts accumulated during a counting interval. If the counting rate of the sample at the beginning of the counting interval is known, multiplication by this factor will give an estimate of the total accumulated counts. No attempt has been made to include corrections for dead time since counting under a condition where significant dead-time losses occur will not lead to the highest precision attainable in a system. This factor is represented in the following equation:

$$C_i = \frac{1}{\lambda}(1 - e^{-\lambda t_c}) = \frac{T_{1/2}(1 - e^{-\lambda t_c})}{0.693} \tag{7.4}$$

where t_c is the counting time interval.

D. USE OF WORKING CURVES

Two specific examples of the use of these curves may serve to illustrate their value in predicting conditions for an analysis. The two examples cover the interference-free determination of an element and their use in a situation in which two elements give the same photopeak from two nuclides of differing half-life. In this latter situation the analyst must choose conditions of irradiation, decay, and counting to optimize the response for the element of interest.

In the interference-free determination of an element, a qualitative examination of these curves will permit the selection of the most sensitive reaction and irradiation, decay, and counting times which will maximize the response for that element. For example, in the determination of aluminum an examination of Fig. 7.1 shows that the (n, p) reaction on ^{27}Al is the most sensitive reaction to use for the determination of aluminum. Moreover, it is clear that an irradiation time of 10 min is more favorable than 5 min but that beyond 10 min the small gain in activation does not warrant longer irradiation times because "saturation" is being approached. An examination of the decay and counting curves (Fig. 7.2 and 7.3) for aluminum further indicates that if the (n, p) reaction is used for the analysis, it is desirable to minimize the decay time to permit the complete decay of any activity such as that due to oxygen or other short-lived isotopes which may have photopeaks at energies higher than 0.84 MeV. The decay curve also indicates that should longer decay times be desirable to permit other activities to decay and if sensitivity for the determination of aluminum is not of concern, perhaps the (n, α) reaction on aluminum producing longer-lived ^{24}Na would be a more desirable reaction to use for analytical purposes. However, assuming the (n, p) reaction on aluminum can be used for its measurement, an examination of the counting integral curve will indicate the length of counting time to maximize the total number of counts in the analytical measurement. If sensitivity is a problem, one would want to count the sample as long as 20 min. Beyond 20 min the net gain in accumulated counts begins to diminish significantly. Thus optimum conditions for a high-sensitivity determination of aluminum would be to irradiate the sample for 10 min, use a 1-min decay to permit the decay of oxygen, which is likely to be present, and, finally, to count the sample for 20 min. The working activation, decay, and counting integral curves permit a rapid determination of these conditions. If it is desirable to treat these conditions quantitatively to establish the levels of activity expected more exactly, the curves would have to be calibrated with known standards so that geometrical factors given in Eq. 7.1 could be evaluated.

The second example which illustrates the use of these curves is one in which two elements give photopeaks of approximately the same energy and thus discrimination of the elements must be made on the basis of difference in half-lives. If aluminum is to be measured in the presence of an equivalent weight of iron, these working curves could be used in the following manner. An examination of the working curves for aluminum shows that the 0.84-MeV photopeak is the most sensitive. Unfortunately, iron interferes in this determination since it gives a photopeak at the same energy. Thus the two elements must be separated on the basis of difference in half-lives of the ^{27}Mg ($T_{1/2} = 9.50$ min) and ^{56}Mn ($T_{1/2} = 154.56$ min). Since the nuclide produced from the element of interest is of shorter half-life than the matrix element which interferes, it is clear that discrimination must be made in the irradiation phase of the analysis rather than the decay or counting. The degree to which this can be done will depend on the total amount of aluminum which is present in the sample since one trades off the sensitivity of the analysis versus discrimination between aluminum and iron. Using the activation curves for aluminum and iron, it can be shown that for equivalent amounts of the two elements the iron will contribute 5.4% of the total activity at 0.84 MeV after 1 min of irradiation. On the other hand, a 10-min irradiation increases to 9.1% the contribution of iron to the total activity at 0.84 MeV. Thus, based on the amounts of iron and aluminum present and the acceptable degree of interference, one can decide upon irradiation conditions which will give sufficient sensitivity and discrimination of aluminum from the iron. Having activated the samples it is clear from an examination of the decay curves that a minimum decay time should be used to keep the aluminum activity as high as possible relative to the iron. The longer the decay time, the greater will be the contribution of iron to the total measurement. This is also true in choosing the counting interval. The counting interval should be chosen to get sufficient counts so that the aluminum statistics are favorable; however, the longer the counting interval, the greater will be the contribution from the iron in the matrix to the measured counts.

If the working curves are to be used in connection with a Ge(Li) detector instead of the sodium iodide crystal, a conversion for the difference in detector efficiencies must be made. Since the two detectors differ in their efficiency as a function of photon energy, a simple conversion factor cannot be applied. Moreover, peak-to-total ratios for both detectors are significantly different. It should be possible to relate a given detector system to the working curves for each element by "calibrating" the system throughout the energy region of interest.

III. ACTIVATION ANALYSIS CONDITIONS FOR THE ELEMENTS

ALUMINUM

Aluminum has some ideal characteristics for activation analysis with neutrons. In nature, aluminum occurs as one isotope of mass 27 which has moderately good reaction cross sections for both fast and thermal neutron reactions. In addition, certain of the isotopes produced from these reactions have half-lives of several minutes and emit gamma rays with energies above 0.5 MeV. The combination of these properties makes the neutron activation technique particularly attractive for the measurement of aluminum.

When compared with other analytical methods for aluminum, the neutron activation technique looks very good indeed. Aluminum is not easily determined in complex matrices by wet chemical or other instrumental methods of analysis. Its measurement by atomic absorption spectroscopy, for example, is complicated by oxide formation in the flame. When excited with x-rays, aluminum emits soft x-ray emission which is readily subject to attenuation by the sample matrix.

14-MeV NEUTRONS

The pertinent nuclear data for the activation of aluminum with 14-MeV neutrons are given in Table 7.1. The relative intensities of the major gamma-ray emissions produced in the activation of aluminum with 14-MeV neutrons as a function of the irradiation time are shown in Fig. 7.1. For short

TABLE 7.1. Nuclear Data for Reactions with 14-MeV Neutrons—Aluminum

Nuclear Reaction	Target Isotope Abundance (%)	Cross Section (mb)	Half-Life (min)	Gamma Energy (MeV)	Gamma Yield (%)	Detector Efficiency (%)	Peak/Total Ratio
1 = ^{27}Al(n, α)^{24}Na	100.00	120	900.0	1.37	100	30	0.35
2 = ^{27}Al(n, α)^{24}Na	100.00	120	900.0	2.75	100	24	0.22
3 = ^{27}Al(n, γ)^{28}Al	100.00	0.5	2.31	1.78	100	27	0.29
4 = ^{27}Al(n, p)^{27}Mg	100.00	81	9.50	0.84	70	34	0.44
5 = ^{27}Al(n, p)^{27}Mg	100.00	81	9.50	1.01	30	32	0.41

III. ACTIVATION ANALYSIS CONDITIONS FOR THE ELEMENTS

irradiation times, the most sensitive photopeak for measuring aluminum is at 0.84 MeV from ^{27}Mg. Where other considerations, such as the freedom from interference, are controlling, the higher energy photopeaks at 1.01 MeV (^{27}Mg) or 1.37 and 2.75 MeV (^{24}Na) may be preferred, although these are used with some sacrifice in sensitivity. The decay and counting integral curves for aluminum are given in Figs. 7.2 and 7.3.

Cuypers and Cuypers[1] report a detection limit of 140 μg for aluminum using the 0.84-MeV peak with a 5-min irradiation time at a flux of 5×10^8 neutrons/(cm^2 sec). D'Agostino and Kuehne[2] indicated a 2-μg minimum detectable quantity using a 1-hr irradiation at a flux of 10^9 neutrons/(cm^2 sec). These values are in general agreement with those calculated from reported cross sections (Fig. 7.1) for the 0.84-MeV photopeak.

There are four elements which may interfere with the determination of aluminum with 14-MeV neutrons if they are present in sufficient quantities. They are iron, magnesium, silicon, and strontium (Table 7.2). Magnesium interferes with the (n, α) reaction for aluminum by producing ^{24}Na,

TABLE 7.2. Nuclear Data for Reactions with 14-MeV Neutrons—Aluminum Interference

Nuclear Reaction	Target Isotope Abundance (%)	Cross Section (mb)	Half-Life (min)	Gamma Energy (MeV)	Gamma Yield (%)	Detector Efficiency (%)	Peak/Total Ratio
1 = ^{30}Si(n, α)^{27}Mg	3.05	45.9	9.50	0.84	70	34	0.44
2 = ^{30}Si(n, α)^{27}Mg	3.05	45.9	9.50	1.01	30	32	0.41
3 = ^{24}Mg(n, p)^{24}Na	78.60	186	900.0	1.37	100	30	0.35
4 = ^{24}Mg(n, p)^{24}Na	78.60	186	900.0	2.75	100	25	0.22
5 = ^{56}Fe(n, p)^{56}Mn	91.68	103	154.56	0.84	99	34	0.44
6 = ^{88}Sr(n, p)^{88}Rb	82.56	17.7	17.80	0.90	13	33	0.43
7 = ^{88}Sr(n, p)^{88}Rb	82.56	17.7	17.80	1.86	21	27	0.28

whereas silicon interferes with the (n, p) reaction by producing ^{27}Mg. The activation, decay, and counting integral curves for the interfering reactions are shown in Figs. 7.4 to 7.6. It is apparent from comparing the sensitivity of these activations with those for aluminum that the cross sections are smaller. Thus the relative ratios of aluminum to silicon or magnesium in the sample will determine the usefulness of the 0.84- and 1.37-MeV gamma photopeaks for the analysis; where this ratio is unfavorable, it may be necessary to utilize some of the higher energy photopeaks for the determination of aluminum. Iron interferes in the determination of aluminum by producing ^{56}Mn which has major gamma emissions at 0.85, 1.81, and 2.11 MeV. The sensitivity of the interfering photopeaks is plotted in Fig. 7.4 as a function of activation

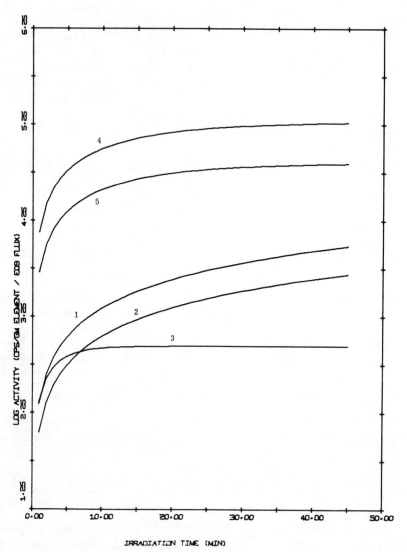

Fig. 7.1. Aluminum—activation curves (see Table 7.1).

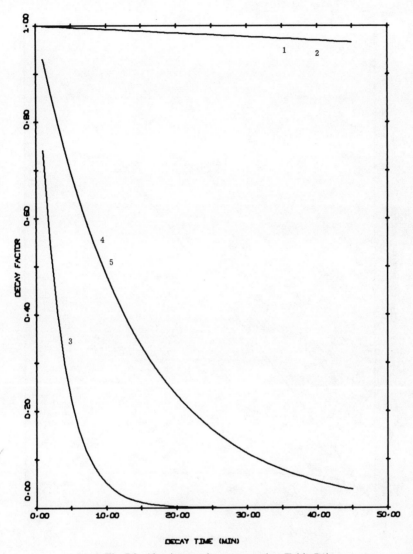

Fig. 7.2. Aluminum—decay curves (see Table 7.1).

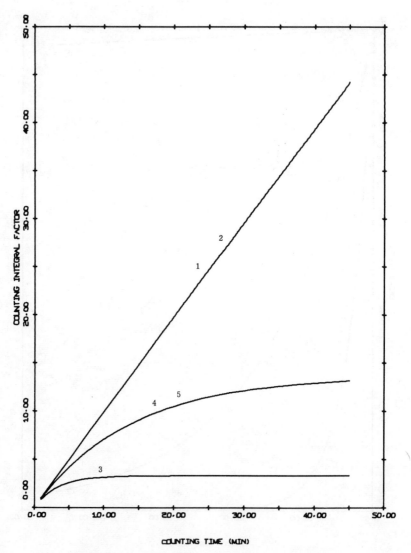

Fig. 7.3. Aluminum—integral counting curves (see Table 7.1).

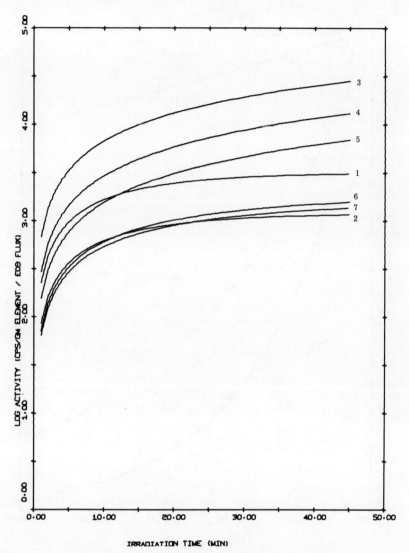

Fig. 7.4. Aluminum interferences—activation curves (see Table 7.2).

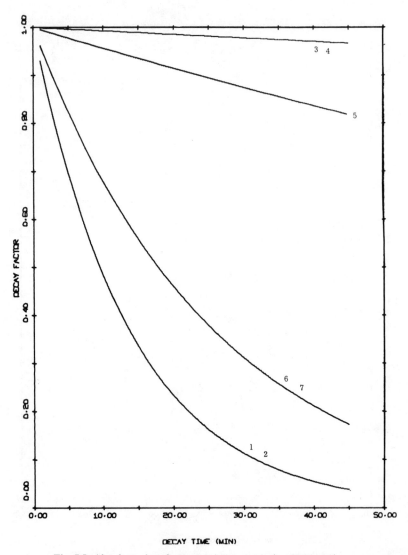

Fig. 7.5. Aluminum interferences—decay curves (see Table 7.2).

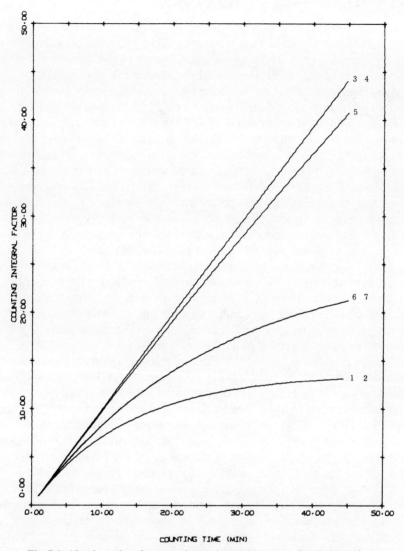

Fig. 7.6. Aluminum interferences—integral counting curves (see Table 7.2).

time. Under circumstances where the ratio of iron to aluminum is unfavorable, it may be necessary to use the less sensitive 2.75-MeV peak from ^{24}Na for the determination of aluminum. Suitable corrective procedures might also be worked out using some of the lower energy photopeaks for aluminum if the iron content is not too high. Large amounts of strontium will also interfere with aluminum analysis through the production of ^{88}Rb.

A large number of applications of 14-MeV neutron activation have been reported for the determination of aluminum. Most of them make use of the very sensitive 0.84-MeV photopeak of ^{27}Mg. An excellent example of the use of this technique for the determination of aluminum in complex matrices was reported by D'Agostino and Kuehne[2] who used this procedure to determine metallic wear in aircraft hydraulic systems. Richardson and Harrison[3] report the use of the 0.84- and 1.01-MeV photopeaks for determining aluminum in composite propellants containing ammonium perchlorate. Using a mixed source of neutrons (14-MeV and thermal) they measured aluminum and chlorine simultaneously. They show some interesting data comparing the activation analysis method with x-ray fluorescence which indicate larger matrix problems with the x-ray technique. A number of applications have been reported for the determination of aluminum in bauxite ores,[4] coal,[5-9] meteorites,[10] and other minerals.[7,11,12] A number of feasibility studies have been reported for remote analysis of aluminum on lunar surfaces.[13-16] Remote analysis has also been suggested for aluminum on the ocean bottom.[17] Several applications have been reported for the determination of aluminum for use in process control.[18] The determination of aluminum in copper alloys[19] and in steel[20] has been reported. Gray has also reported determination of aluminum along with iron and copper in engine oil.[21] Golanski[22] demonstrated the use of pulsing techniques for the determination of aluminum via the ^{27}Al(n, α)^{24m}Na reaction. ^{24m}Na has a 19.3-msec half-life with a gamma emission at 0.475 MeV. With pulsed fluxes of 10^8 neutrons/(cm^2 sec), the technique might be sensitive enough for semimicroanalysis. This approach may have some real value for the measurement of aluminum in matrices containing silicon or magnesium. It is clear from this list that the application of 14-MeV neutrons for the determination of aluminum in a variety of matrices is useful and indeed superior to alternate analytical methods for certain types of materials.

3-MeV AND THERMAL NEUTRONS

Aluminum is also activated using 3-MeV neutrons. Because the available neutron flux is substantially lower from the D–D reaction than from the D–T reaction, the sensitivity for aluminum using this technique is lower by a

III. ACTIVATION ANALYSIS CONDITIONS FOR THE ELEMENTS

factor of 100 than with 14-MeV neutrons. Nargolwalla et al.[23] report a sensitivity of 6.3×10^2 counts/g for the (n, γ) reaction and 4.7×10^2 counts/g for the (n, p) reaction using a flux of 10^6 neutrons/(cm² sec). This type of activation can be used when the amount of aluminum present is high and/or when matrix interferences, such as magnesium or phosphorus, prevent the use of 14-MeV neutrons. The only known interference for aluminum under these conditions is manganese which produces photopeaks from ^{56}Mn in the same regions as ^{27}Mg and ^{28}Al.

The thermal activation cross section of aluminum is substantially higher than that with 3-MeV neutrons. When interference from other elements which are activated strongly with thermal neutrons is not anticipated, this procedure might be preferred to activation with 3-MeV neutrons or fast neutrons. Dibbs[24] reports a predicted sensitivity of 50 μg using a thermal flux of 1×10^8 neutrons/(cm² sec).

ANTIMONY

Antimony occurs naturally in two isotopic forms of mass 121 and 123. The isotope of mass 121 can be activated with fast neutrons. The nuclear reactions of antimony which are important in fast neutron activation are listed in Table 7.3. The predominant activity produced with

TABLE 7.3. Nuclear Data for Reactions with 14-MeV Neutrons—Antimony

Nuclear Reaction	Target Isotope Abundance (%)	Cross Section (mb)	Half-Life (min)	Gamma Energy (MeV)	Gamma Yield (%)	Detector Efficiency (%)	Peak/Total Ratio
1 = ^{121}Sb(n, 2n)^{120m}Sb	57.25	1310	8352	0.09	81	50	1.00
2 = ^{121}Sb(n, 2n)^{120m}Sb	57.25	1310	8352	0.20	88	50	0.93
3 = ^{121}Sb(n, 2n)^{120m}Sb	57.25	1310	8352	1.03	99	32	0.41
4 = ^{121}Sb(n, 2n)^{120m}Sb	57.25	1310	8352	1.17	100	31	0.38
5 = ^{121}Sb(n, 2n)^{120}Sb	57.25	750	15.90	0.51	87	40	0.63
6 = ^{121}Sb(n, 2n)^{120}Sb	57.25	750	15.90	1.17	1	31	0.38

fast neutron activation is from the positron emitter ^{120}Sb; thus the determination of antimony in complex matrices is limited to those samples which do not yield other positron emitters. The cross-section–half-life relationship is favorable, however, for the determination of fairly small amounts of antimony in systems containing no other positron emitters.

14-MeV NEUTRONS

Figure 7.7 shows the relative measurable activities from the major gamma emissions of nuclides produced in the activation of antimony with 14-MeV neutrons. The ^{121}Sb isotope activates most readily to produce ^{120}Sb and ^{120m}Sb nuclides. While the most sensitive γ-emission is at 0.51 MeV, from ^{120}Sb, there may be some cases in which the less sensitive 1.17-MeV photopeak may be useful in matrices which also contain other positron emitters. Decay and counting integral curves for antimony are given in Figs. 7.8 and 7.9. Cuypers and Cuypers[1] list a detection limit of 0.17 mg for antimony using the 0.51-MeV annihilation peak, whereas Dibbs[24] reported an experimental sensitivity of 18 μg. Crambes[25] describes the determination of antimony by activation with 14-MeV neutrons followed by rapid radiochemical separation of the antimony prior to its measurement. This author reported a sensitivity for the measurement of antimony of approximately 0.1 mg, which is consistent with the work of Cuypers and Cuypers.[1] These data are consistent with the activation curve shown in Fig. 7.5. There are no other reported applications of this technique for the determination of antimony. Tables 7.76, 7.77, and 7.78 should be consulted for possible interfering positron emitters.

3-MeV AND THERMAL NEUTRONS

Work by Nargolwalla et al.[23] in examining the utility of 3-MeV neutrons for the determination of antimony indicate that ^{123}Sb undergoes an (n, γ) reaction to produce ^{124m}Sb. The activation, however, has a relatively low cross section so that only gram quantities of material could be determined in this fashion.

The (n, γ) reactions described for 3-MeV neutrons occur with somewhat better facility using thermal neutrons. The sensitivity, however, is not sufficiently better to recommend this as a useful analytical approach utilizing the thermal fluxes available by moderating the fast flux from a neutron generator.

ARSENIC

The sensitivity for the neutron activation analysis of arsenic with fast neutrons is not high. Arsenic is monoisotopic and undergoes a variety of reactions with fast neutrons, but only with small cross sections. Because of its importance in pesticides and toxicology, a number of conventional analytical methods have been developed for its trace determination. Thus neutron activation has only been used for the measurement of arsenic when the high sensitivity obtained from a reactor flux was needed.

The useful nuclear reactions for the determination of arsenic are listed in Table 7.4.

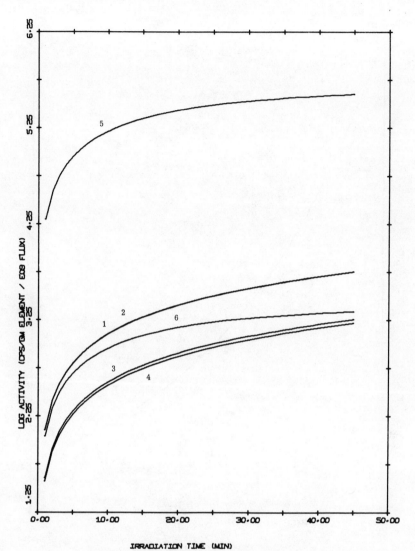

Fig. 7.7. Antimony—activation curves (see Table 7.3).

293

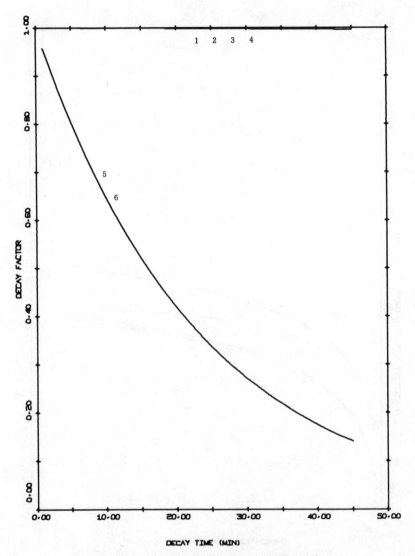

Fig. 7.8. Antimony—decay curves (see Table 7.3).

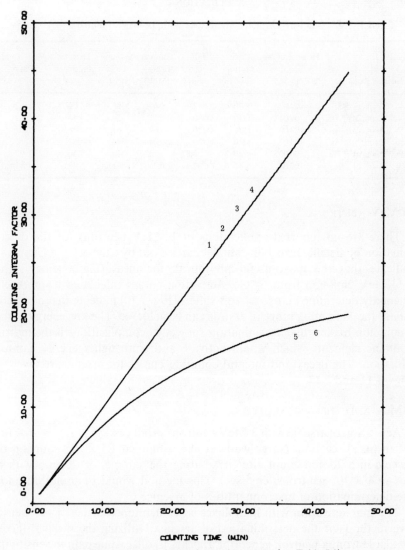

Fig. 7.9. Antimony—integral counting curves (see Table 7.3).

TABLE 7.4. Nuclear Data for Reactions with 14-MeV Neutrons—Arsenic

Nuclear Reaction	Target Isotope Abundance (%)	Cross Section (mb)	Half-life (min)	Gamma Energy (MeV)	Gamma Yield (%)	Detector Efficiency (%)	Peak/Total Ratio
1 = ^{75}As(n, α)^{72}Ga	100.00	10.4	852.00	0.84	96	34	0.44
2 = ^{75}As(n, 2n)^{74}As	100.00	1110	24480	0.51	59	40	0.63
3 = ^{75}As(n, 2n)^{74}As	100.00	1110	24480	0.60	61	38	0.57
4 = ^{75}As(n, 2n)^{74}As	100.00	1110	24480	0.64	14	37	0.55
5 = ^{75}As(n, p)^{75m}Ge	100.00	10	0.82	0.14	34	50	0.97
6 = ^{75}As(n, p)^{75}Ge	100.00	18	82.00	0.26	11	48	0.87

14-MeV NEUTRONS

There are no reported applications of 14-MeV neutrons for the determination of arsenic. Feasibility studies carried out by Cuypers and Cuypers[1] indicate that the most useful photopeak for measuring arsenic is the 0.14-MeV emission from ^{75m}Ge. Activation curves calculated from experimentally determined cross-section values (Fig. 7.10) indicate that a sensitivity in the 1- to 5-mg range of arsenic can be achieved. The low energy of the γ-emission makes this determination subject to complications if the matrix contains elements which would activate and emit higher energy gamma radiation. The decay and integral counting curves for arsenic are given in Figs. 7.11 and 7.12.

3-MeV AND THERMAL NEUTRONS

Activation of arsenic with 3-MeV neutrons produces ^{76}As ($T_{1/2}$ = 26.4 hr) by an (n, γ) reaction. Nargolwalla et al.[23] obtained 1.3×10^3 counts/g of arsenic in a 20-min count after irradiating the sample for 20 min with a flux of 1×10^6 neutrons/(cm^2 sec). This method would be useful only for the determination of macroquantities of arsenic.

Thermal activation with high fluxes is, as mentioned earlier, a very sensitive method for the determination of arsenic. Utilizing the thermal fluxes available from a neutron generator appears to offer some gain in sensitivity for the measurement of arsenic over the sensitivity obtained with the 14-MeV technique. Dibbs[24] reported an experimental sensitivity of 200 μg using the 0.56-MeV photopeak from ^{76}As which was activated with the thermalized flux from a neutron generator.

BARIUM

Barium occurs in nature in seven isotopic forms of mass 130, 132, and 134 through 138. The nuclides produced by activation of barium have desirable

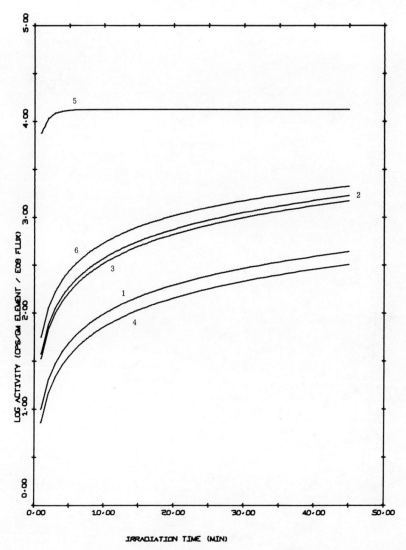

Fig. 7.10. Arsenic—activation curves (see Table 7.4).

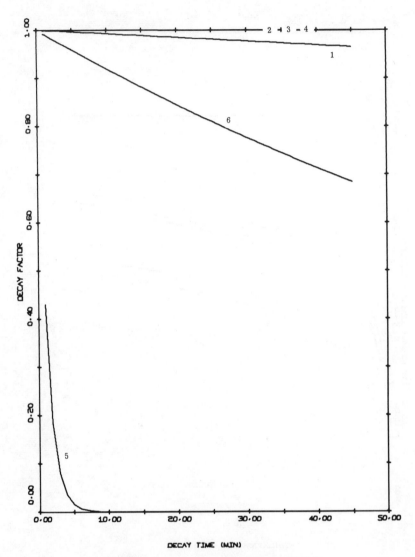

Fig. 7.11. Arsenic—decay curves (see Table 7.4).

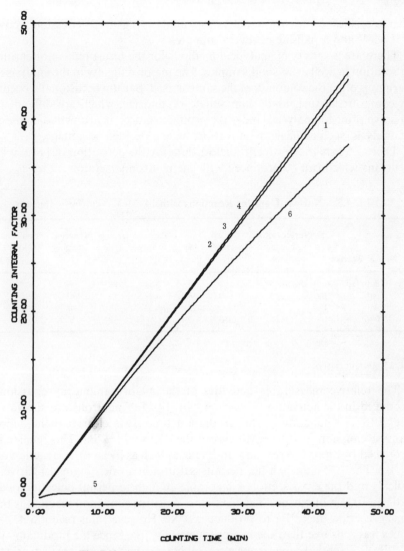

Fig. 7.12. Arsenic—integral counting curves (see Table 7.4).

characteristics for neutron activation analysis—gamma emission above 0.51 MeV and a half-life of several minutes.

There are a variety of analytical methods for the measurement of barium in solution as well as in solid samples. The major difficulty in the analysis of barium is often the solubility of the sample itself. Barium occasionally occurs in compounds, such as barium sulfate or fluoride, which are difficult to dissolve prior to analysis. Under these circumstances, instrumental methods of analysis, such as neutron activation, have a decided advantage.

Table 7.5 lists the pertinent nuclear data for the activation of barium by neutrons which can be produced with the neutron generator.

TABLE 7.5. Nuclear Data for Reactions with 14-MeV Neutrons—Barium

Nuclear Reaction	Target Isotope Abundance (%)	Cross Section (mb)	Half-life (min)	Gamma Energy (MeV)	Gamma Yield (%)	Detector Efficiency (%)	Peak/Total Ratio
1 = ^{138}Ba(n, α)^{135m}Xe	71.66	13	15.60	0.53	80	40	0.61
2 = ^{134}Ba(n, 2n)^{133m}Ba	2.42	940	2334	0.28	17	48	0.86
3 = ^{136}Ba(n, 2n)^{135m}Ba	7.81	700	1722	0.27	16	48	0.87
4 = ^{138}Ba(n, 2n)^{137m}Ba	71.66	1250	2.55	0.66	89	37	0.54

14-MeV NEUTRONS

The relative measurable activities of the major gamma-ray emissions produced in the activation of barium with 14-MeV neutrons are shown in Fig. 7.13 as a function of the irradiation time. It is clear that the major gamma emission is at 0.66 MeV from the decay of ^{137m}Ba. This nuclide is produced by an (n, 2n) reaction on ^{138}Ba, as well as by an (n, n') reaction on ^{137}Ba. The 2.55-min half-life permits saturation to be obtained relatively quickly and the cross section is favorable enough to permit detection of as little as 100 μg of barium. The next most sensitive activation reaction is the (n, α) reaction on ^{138}Ba to produce ^{135m}Xe. However, this reaction is 100 times less sensitive than the one producing ^{137m}Ba. Since the production of ^{137m}Ba is essentially interference free, there are few conceivable situations wherein this reaction cannot be used directly.

The sensitivity for the determination of barium is quite high. Figure 7.13 indicates a saturation sensitivity of approximately 1×10^6 dis/(sec g) of barium utilizing the (n, 2n) reaction on ^{138}Ba. This sensitivity is confirmed by experimental observations reported in the literature [26,-27]; Dibbs [24] reported an experimental sensitivity of 27 μg of barium.

There are no known direct interferences for barium. The only problem might arise from a matrix which, upon irradiation, would produce nuclides

III. ACTIVATION ANALYSIS CONDITIONS FOR THE ELEMENTS

having gamma energies above 0.66 MeV. In this case, this photopeak would have some Compton continuum beneath it.

There are relatively few reported applications of the neutron generator to the determination of barium in practical samples. Three applications of this technique have been made to estimate barium in various hydrocarbon matrices.[26-28] In these applications, barium was determined in the presence of other trace elements, such as phosphorus and silicon, by direct activation. The concentration range of interest was between 100 and 500 ppm. Chiba[29] reported the use of 14.7-MeV neutrons for the measurement of barium-to-iron ratios in barium ferrite, a compound used for magnetic materials. Samples of 70 to 130 mg of the compound were used and a precision of 2% was reported.

Despite the rather meager number of reported applications of this technique to the measurement of barium, it would appear to offer substantial promise from the standpoint of being interference free and of reasonably high sensitivity. Undoubtedly, competition from other nondestructive instrumental methods of analysis, such as x-ray fluorescence, which gives excellent results for barium, results in the low level of applications for this technique for the measurement of barium.

The decay and integral counting curves for barium are shown in Figs. 7.14 and 7.15.

3-MeV AND THERMAL NEUTRONS

Barium can also be activated using 3-MeV neutrons.[23] Reactions of barium isotopes with 3-MeV neutrons produce the same nuclide, ^{137m}Ba, which is produced in predominance with 14-MeV neutrons. The difference lies in the isotope from which this radioactive nuclide is produced. The ^{136}Ba and ^{137}Ba isotopes are utilized in this activation by an (n, γ) and (n, n'γ) reaction, respectively. The experimental sensitivity for the determination of barium by the use of 3-MeV neutrons is approximately 300 times less than with 14-MeV neutrons. This reaction, like the one with 14-MeV neutrons, appears to be essentially interference free. Thus, where high sensitivity is not a primary requirement and target lifetime may be, this mode of activation for barium may have some decided advantages. Barium can also be activated using the moderated flux from the neutron generator to produce the following reactions: ^{136}Ba(n, γ)^{137}Ba and ^{138}Ba(n, γ)^{139}Ba.

While these reactions occur with substantially higher cross sections than reactions with either 3- or 14-MeV neutrons, the lower thermal flux yield available from the neutron generator results in a relatively low detection sensitivity for barium using thermal neutrons. Dibbs[24] shows that the predicted sensitivity for the (n, γ) reaction on ^{138}Ba yielding 85-min ^{139}Ba (0.167-MeV photopeak) has a sensitivity of 190 μg of barium.

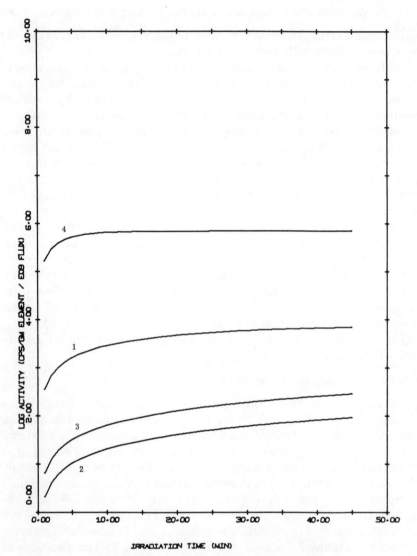

Fig. 7.13. Barium—activation curves (see Table 7.5).

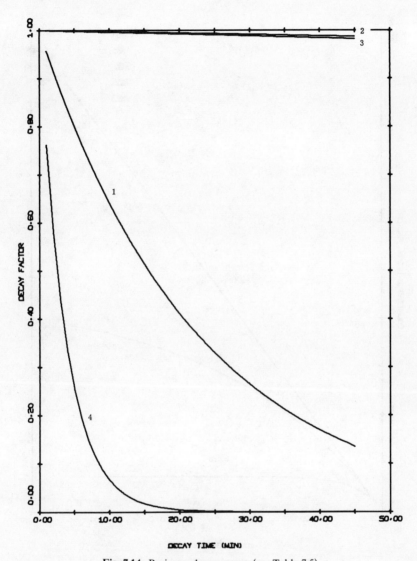

Fig. 7.14. Barium—decay curves (see Table 7.5).

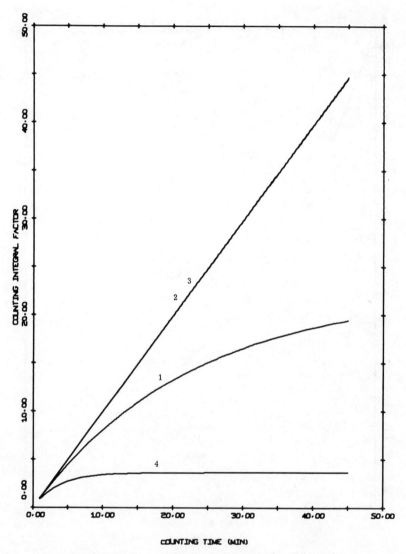

Fig. 7.15. Barium—integral counting curves (see Table 7.5).

BORON

Boron occurs in nature in two isotopic forms of mass 10 and 11. It is one of the more difficult elements to determine with high sensitivity, since the variety of solution chemical reactions is limited. Boron undergoes reactions with neutrons very readily but the nuclides produced are short-lived and therefore require special techniques if they are to be used for analytical purposes. Table 7.6 shows the nuclear reactions of interest for the activation of boron. The methods for measuring boron using the neutron generator cannot be highly recommended either in terms of sensitivity or of freedom from interference.

TABLE 7.6. Nuclear Data for Reactions with 14-MeV Neutrons—Boron

Nuclear Reaction	Target Isotope Abundance (%)	Cross Section (mb)	Half-life (min)	Gamma Energy (MeV)	Gamma Yield (%)	Detector Efficiency (%)	Peak/Total Ratio
$1 = {}^{11}B(n,p){}^{11}Be$	81.50	3.00	0.23	2.14	32	26	0.26
Boron Interference							
$2 = {}^{16}O(n,p){}^{16}N$	99.76	42.0	0.12	6.13	69	23	0.14

14-MeV NEUTRONS

Activation of boron with 14-MeV neutrons produces two nuclides which may be used for the measurement of boron. The activation curves for the production of ^8Li and ^{11}Be are shown in Fig. 7.16. Decay and integral counting curves are shown in Figs. 7.17 and 7.18. Because both of these activities are relatively short-lived they must be used under highly specialized conditions. If one chooses to use ^8Li ($T_{1/2} = 0.83$ sec), the system has to be reasonably free of high-energy γ-emissions since ^8Li is only a β-emitter, and β-counting or the counting of high-energy bremsstrahlung must be used. In addition, transit times from the generator to the counting site must be very rapid to achieve sufficient sensitivity. Oxygen in the sample, of course, would be a complication because of the production of ^{16}N. Although less sensitive, the (n, p) reaction on ^{11}B is somewhat easier to use because of the longer-lived nuclide which is produced. Interference from oxygen, however, with the 2.14-MeV emission may be expected unless decay curve corrections are made.

There have been several applications of neutron generators for the determination of boron. Broadhead et al.[12] determined boron in vanadium alloys by measuring the high-energy bremsstrahlung from ^{11}Be. These materials contained very low oxygen, which was determined by vacuum fusion, and

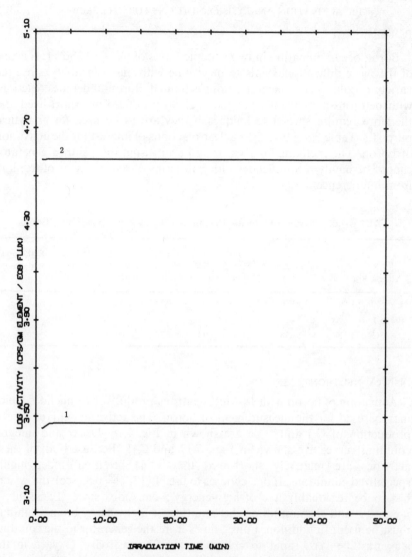

Fig. 7.16. Boron—activation curves (see Table 7.6).

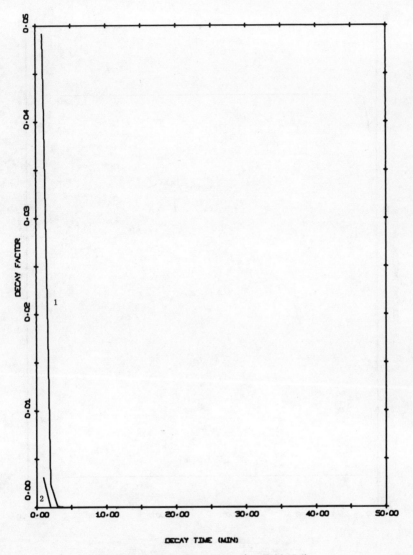

Fig. 7.17. Boron—decay curves (see Table 7.6).

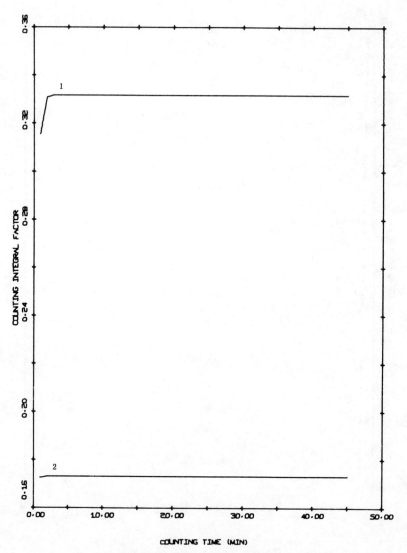

Fig. 7.18. Boron—integral counting curves (see Table 7.6).

appropriate corrections were made in the boron analysis. Using the 2.14-MeV emission from ^{11}Be, workers at the Mound Laboratory[30] obtained sensitivity of 400 counts/mg of boron using a fast neutron flux of 1×10^9 neutrons/(cm^2 sec).

Rasmussen and Thompson[31] devised a way of measuring the prompt α-emission from the (n, α) reaction to obtain better sensitivity for the measurement of boron. A technique which is gaining favor for the measurement of trace levels of boron is to record the α-tracks emitted upon irradiation of boron-containing materials using cellulose acetate-butyrate films. The α-tracks are visualized by subsequent processing of the film in alkali. Using reactor fluxes, sensitivities as low as 10^{-12} g of boron track have been reported.[32]

3-MeV AND THERMAL NEUTRONS

The (n, α) reaction with ^{10}B occurs with low-energy neutrons at somewhat higher efficiencies than with 14-MeV neutrons. However, the (n, p) reaction described above cannot occur because the threshold for the reaction is not exceeded.

BROMINE

The two naturally occurring bromine isotopes are ^{81}Br and ^{79}Br. Neutron reactions with ^{79}Br are the more sensitive for activation analysis. While the activation cross sections and half-lives for nuclides produced from bromine are favorable to obtain high sensitivity, the γ-emissions are at relatively low energies, thus posing potential problems of interference from other matrix elements.

Because of their wide occurrence in nature, the halogens as a group have received very extensive study, and today a wide variety of analytical methods exist for their determination. The technique of x-ray fluorescence analysis, for example, can compete very favorably with neutron activation analysis for the determination of semimicrolevels of bromine in many matrices. Yet the significantly larger attenuation effects on the soft fluorescent x-ray radiation, as compared with the harder γ-radiation from nuclides produced by neutron activation, favor the use of the nuclear method for matrices where substantial attenuation may be expected. Halogens commonly occur together in nature, and chemical methods for their simultaneous determinations suffer from the similarities in their chemical properties. Methods based on their nuclear properties can provide a way of discriminating between them. Specifically, neutron activation can provide a very rapid method for the determination of bromine at very low levels in matrices where other interfering elements are absent. Here the method can compete favorably in sensitivity with other instrumental methods such as x-ray fluorescence.

Table 7.7 lists the pertinent nuclear data for the activation of bromine with neutrons that can be produced with a neutron generator.

TABLE 7.7. Nuclear Data for Reactions with 14-MeV Neutrons—Bromine

Nuclear Reaction	Target Isotope Abundance (%)	Cross Section (mb)	Half-life (min)	Gamma Energy (MeV)	Gamma Yield (%)	Detector Efficiency (%)	Peak/Total Ratio
1 = ^{79}Br(n, n'γ)^{79m}Br	50.52	94	0.08	0.21	100	50	0.92
2 = ^{79}Br(n, 2n)^{78}Br	50.52	1141	6.50	0.51	184	40	0.63
3 = ^{79}Br(n, 2n)^{78}Br	50.52	1141	6.50	0.61	14	38	0.56
4 = ^{81}Br(n, 2n)^{80g}Br	49.48	440	17.60	0.51	5	40	0.63
5 = ^{81}Br(n, 2n)^{80g}Br	49.48	440	17.60	0.62	7	38	0.56

14-MeV NEUTRONS

The relative intensities of the major γ-ray emissions produced in the activation of bromine with 14-MeV neutrons are shown in Fig. 7.19. The most sensitive reaction yields ^{78}Br, a positron emitter with γ-radiation at 0.51 MeV. Unfortunately, the 6.5-min half-life of this nuclide puts it into an time region where approximately nine other positron emitters may interfere with it (see Tables 7.76 and 7.77). Decay curve analysis may be used to differentiate bromine from some of the longer lived positron emitters. To circumvent the interference from matrix elements which may also produce positron emitters, very short irradiation times may be used and the bromine determined via ^{79m}Br ($T_{1/2} = 4.8$ sec; cf. Table 7.7). Although the γ-emission at 0.21 MeV may have to be corrected for the contribution from the Compton and bremsstrahlung contributions of other higher energy γ-emissions, the very short half-life of this isotope can provide a means for discrimination by suitable adjustment of the radiation and decay times. The decay and counting integral curves are shown in Figs. 7.20 and 7.21.

Utilizing the 0.51-MeV radiation from ^{78}Br, as little as 50 μg of bromine may be detected in a favorable matrix. The use of ^{79m}Br provides a method which is approximately 20 times less sensitive.

The number of reported applications for 14-MeV activation of bromine is small. It has been used for the determination of bromine in river water, using the 0.51-MeV γ-emission from ^{78}Br.[33] In this application, interference from the proton recoil reaction on oxygen producing ^{13}N was handled by the decay curve type of correction. Oldham and Darrall[34] demonstrated the utility of the short half-life of ^{79m}Br ($T_{1/2} = 4.8$ sec) for the determination of bromine in the presence of other interfering positron emitters. They demonstrated the utility of this peak for the measurement of bromine in the

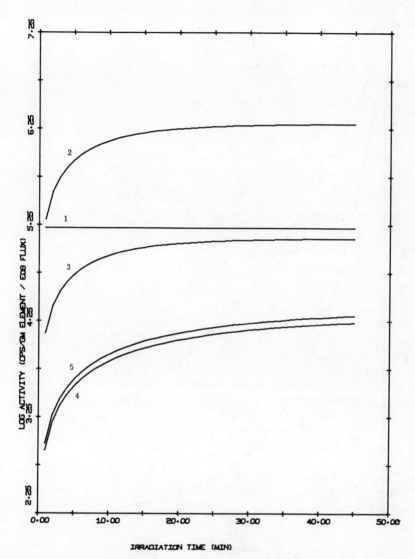

Fig. 7.19. Bromine—activation curves (see Table 7.7).

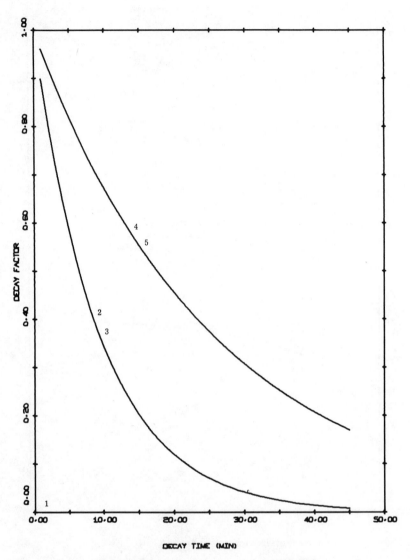

Fig. 7.20. Bromine—decay curves (see Table 7.7).

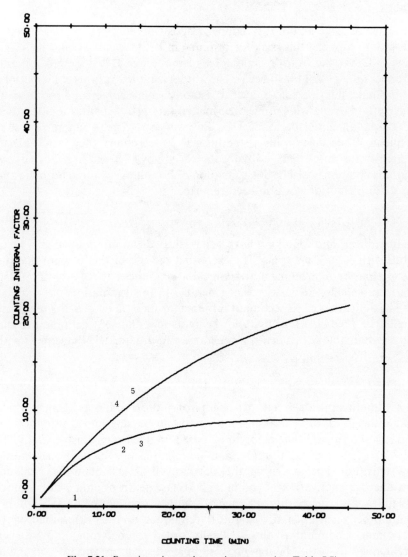

Fig. 7.21. Bromine—integral counting curves (see Table 7.7).

presence of large amounts of copper (a positron emitter) and developed a procedure utilizing this peak for bromine in the presence of fluorine which gave ^{19}O via the (n, p) reaction. This isotope has a 29.4-sec half-life and produces a γ-peak at 0.20-MeV. The authors studied various combinations of bromine, fluorine, and copper to establish the accuracy and precision of the method over a wide range of elemental ratios. Using an irradiation time of 10 sec, an initial decay of 3 sec, followed by a 5-sec count, and a subsequent 40-sec delay with an additional 30-sec count, they were able to demonstrate better than $\pm 5\%$ absolute accuracy over extreme ratios of the various components. The determination of bromine in sodium bromide and in sodium chloride has also been reported.[35]

3-MeV AND THERMAL NEUTRONS

Broadhead and Shanks[36] have demonstrated the utility of (n, n'γ) reactions with 3-MeV neutrons. They reported the use of this reaction for the determination of bromine in sodium chloride in the range of 0.2 to 25%. No measurable chlorine activity was generated in this irradiation.

The activation cross section of bromine for thermal neutrons is relatively high. The predicted sensitivity for bromide measurement with the thermal fluxes attainable with a neutron generator is approximately the same as with the 14-MeV technique.[24]

CADMIUM

Cadmium appears in nature in eight isotopic forms. It is an element which has rather ideal analytical chemistry and can be determined using a wide variety of techniques including those based on nuclear chemistry. Cadmium has favorable cross-section and half-life relationships for its determination by activation analysis using thermal as well as fast neutrons. Nuclides produced by activation of cadmium with fast neutrons emit γ-radiation of relatively low energy; thus the nuclear method does not compete favorably with other instrumental analytical techniques for the determination of cadmium. Cadmium is almost ideally suited for determination by such techniques as x-ray fluorescence, atomic absorption spectroscopy, and polarography. This accounts for the fact that there are few methods reported in the literature for its determination using neutrons produced by a neutron generator.

14-MeV NEUTRONS

Figure 7.22 shows the relative intensities of the major γ-emissions produced by neutron activation with 14-MeV neutrons and Table 7.8 lists the pertinent nuclear data. Decay and counting integral curves for cadmium

TABLE 7.8. Nuclear Data for Reactions with 14-MeV Neutrons—Cadmium

Nuclear Reaction	Target Isotope Abundance (%)	Cross Section (mb)	Half-life (min)	Gamma Energy (MeV)	Gamma Yield (%)	Detector Efficiency (%)	Peak/Total Ratio
1 = ^{106}Cd(n, 2n)^{105}Cd	1.21	827	55.00	0.51	100	40	0.63
2 = ^{112}Cd(n, 2n)^{111m}Cd	12.39	1390	48.60	0.15	30	50	0.97
3 = ^{112}Cd(n, 2n)^{111m}Cd	12.39	1390	48.60	0.25	94	50	0.88
4 = ^{113}Cd(n, p)^{113}Ag	12.26	40	1.20	0.30	67	47	0.84

are shown in Figs. 7.23 and 7.24. It is apparent that the method is suitable for the determination of as little as 0.1 of a milligram of cadmium. Some choice is available in the isotopes to be used for the analysis, depending on the presence of other matrix elements. While there are no direct interferences with the measurement of cadmium, the usual problem with higher energy γ-emissions from other matrix elements does affect the cadmium determination because it produces nuclides having γ-emissions primarily at low energies. To date there have been no reported applications of 14-MeV neutrons for the activation analysis of cadmium.

3-MeV AND THERMAL NEUTRONS

Broadhead and Shanks[36] have reported experimental data for the activation of cadmium with 3-MeV neutrons. When compared with the activation of cadmium with fast neutrons at 10^9 neutrons/(cm^2 sec), the 3-MeV technique is tenfold less sensitive. Under certain matrix conditions, this sensitivity might be sacrificed to achieve an interference-free analysis with the lower energy neutrons.

The activation of cadmium with thermal neutrons produced by a neutron generator is about a factor of 50 lower in sensitivity over the 3-MeV technique. Dibbs[24] found that thermal neutron activation of cadmium gave a sensitization of 100 μg from ^{111m}Cd. Thus thermal activation analysis for cadmium using a neutron generator has no practical advantages over activation with 14-MeV neutrons.

CALCIUM

Calcium is not readily activated with the fast or thermal neutron fluxes available from a neutron generator. Although the nuclides produced in the activation of calcium have γ-emissions of high energy, which makes them ideal for measurement in complex matrices, low cross sections and somewhat long half-lives contribute to a very low detection sensitivity for calcium

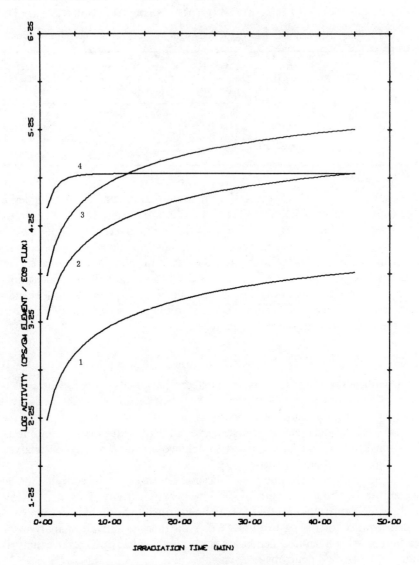

Fig. 7.22. Cadmium—activation curves (see Table 7.8).

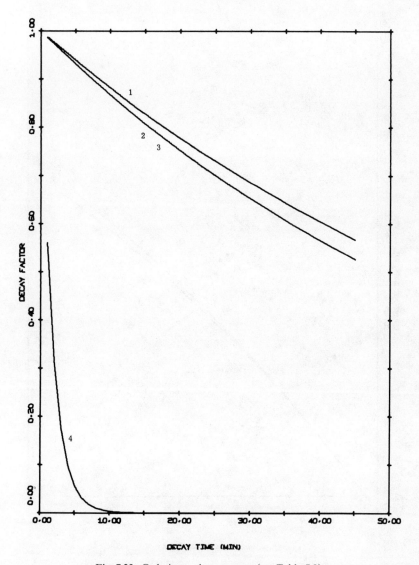

Fig. 7.23. Cadmium—decay curves (see Table 7.8).

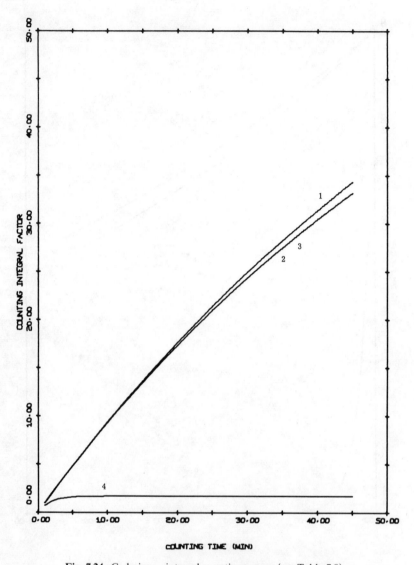

Fig. 7.24. Cadmium—integral counting curves (see Table 7.8).

III. ACTIVATION ANALYSIS CONDITIONS FOR THE ELEMENTS

by this technique. These activation features, of course, aid in the determination of other elements in a matrix containing large amounts of calcium.

The determination of calcium by other methods including atomic absorption, x-ray fluorescence, and flame emission techniques makes its determination by neutron activation unattractive and unnecessary. Only unusual circumstances would dictate the use of this technique for the determination of calcium.

14-MeV NEUTRON ACTIVATION

The primary nuclear reactions of calcium with neutrons are listed in Table 7.9. The activation curves for calcium with 14-MeV neutrons are shown in Fig. 7.25 and the decay and counting integral curves are given in Figs. 7.26 and 7.27. There have been a few applications reported for the determination of calcium. One such application involved *in vivo* activation of humans,

TABLE 7.9. Nuclear Data for Reactions with 14-MeV Neutrons—Calcium

Nuclear Reaction	Target Isotope Abundance (%)	Cross Section (mb)	Half-life (min)	Gamma Energy (MeV)	Gamma Yield (%)	Detector Efficiency (%)	Peak/Total Ratio
1 = ^{44}Ca(n, α)^{41}Ar	2.06	30	109.80	1.29	99	30	0.36
2 = ^{44}Ca(n, p)^{44}K	2.06	25	22.00	1.16	61	31	0.39

using radiation doses small enough to be comparable to those used in diagnostic radiology. The induced radioactivity enabled an estimate to be made of a total body calcium.[37] Hull and Gilmore[28] investigated the application of 14-MeV neutron activation for the determination of calcium in petroleum products. They concluded that the technique was not particularly promising for the determination of calcium at the normal levels found in oil. Schramel[38] reported the determination of calcium by fast neutron activation in biological materials. Using 10-min irradiations, they reported detection sensitivity for calcium of 400 μg/g. Dibbs[24] reported an experimental sensitivity of 900 μg for calcium.

3-MeV AND THERMAL NEUTRONS

Because of the low sensitivities obtained by activation with 3-MeV and thermal neutrons, these methods are not recommended for the determination of calcium.

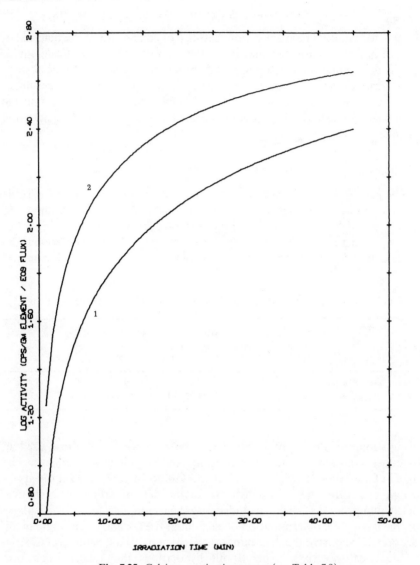

Fig. 7.25. Calcium—activation curves (see Table 7.9).

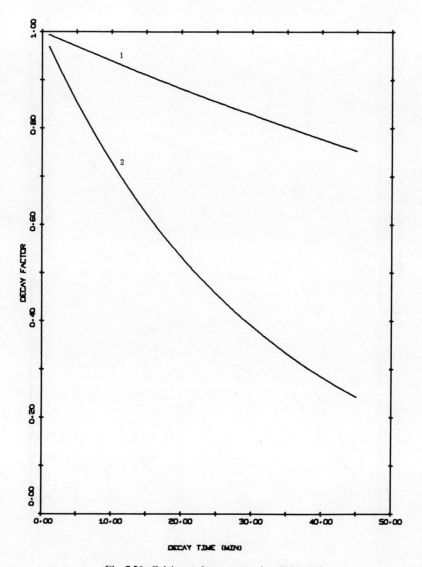

Fig. 7.26. Calcium—decay curves (see Table 7.9).

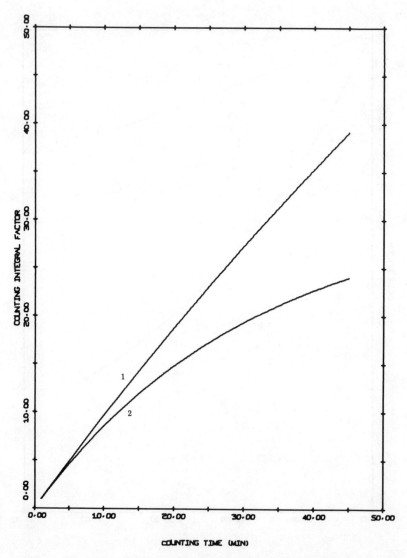

Fig. 7.27. Calcium—integral counting curves (see Table 7.9).

CERIUM

Neutron activation parameters for the determination of cerium are quite favorable. Cerium exists in nature in four isotopic forms of mass numbers 136, 138, 140, and 142. The isotope with mass of 140 is the most important from the standpoint of neutron activation analysis with a neutron generator.

TABLE 7.10. Nuclear Data for Reactions with 14-MeV Neutrons—Cerium

Nuclear Reaction	Target Isotope Abundance (%)	Cross Section (mb)	Half-life (min)	Gamma Energy (MeV)	Gamma Yield (%)	Detector Efficiency (%)	Peak/Total Ratio
1 = ^{140}Ce(n, α)^{137m}Ba	88.48	12.1	2.55	0.66	89	36	0.54
2 = ^{140}Ce(n, 2n)^{139m}Ce	88.48	1200	0.92	0.75	93	35	0.50

Neutron activation analysis can compare favorably with other analytical methods for the determination of cerium. Similarity of the rare earth elements in chemical properties makes the determination of individual rare earths somewhat difficult by wet chemical methods. Instrumental methods such as x-ray fluorescence or neutron activation can provide a superior way for determining certain rare earth elements, including cerium.

14-MeV NEUTRONS

The activation curves for cerium with 14-MeV neutrons are shown in Fig. 7.28; the decay and integral counting curves are shown in Figs. 7.29 and 7.30. The most sensitive reaction for the measurement of cerium involves the (n, 2n) reaction on ^{140}Ce to produce ^{139m}Ce which has a photopeak at 0.75 MeV and a half-life of 55 sec. From nuclear parameters, this reaction has the sensitivity to measure as little as 10 µg of cerium. Neodymium and samarium may interfere directly since they produce nuclides with emissions at 0.75 MeV. The activation curves for these interferences are shown in Fig. 7.31, and their decay and counting curves are given in Figs. 7.32 and 7.33 (see Table 7.11). Cuypers and Menon[39,40] have described the application of 14.7-MeV neutron activation for the determination of cerium in a variety of rare-earth-bearing minerals. Using a flux of 3×10^7 neutrons/(cm² sec), they report a sensitivity of 300 µg for the determination of cerium. This reported sensitivity is consistent with that shown in the activation curve (Fig. 7.28). Dibbs[24] reported an experimental sensitivity of 70 µg using a flux of 5×10^8 neutrons/(cm² sec). Broadhead and co-workers[12] have described the use of this technique for the assay of rare earth fluorides for their purity.

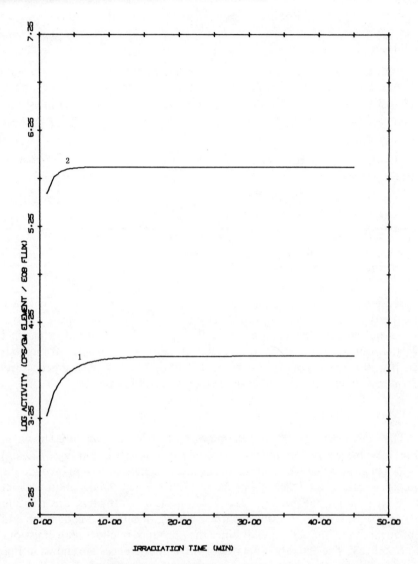

Fig. 7.28. Cerium—activation curves (see Table 7.10).

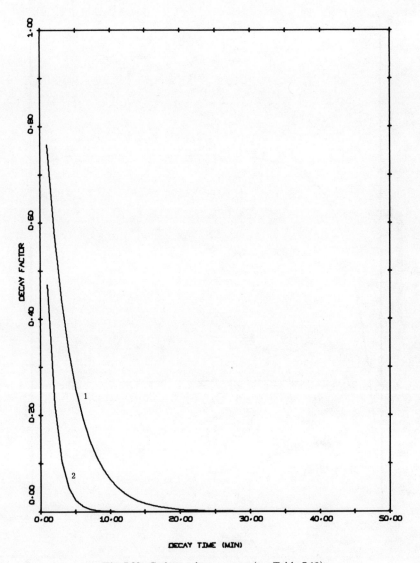

Fig. 7.29. Cerium—decay curves (see Table 7.10).

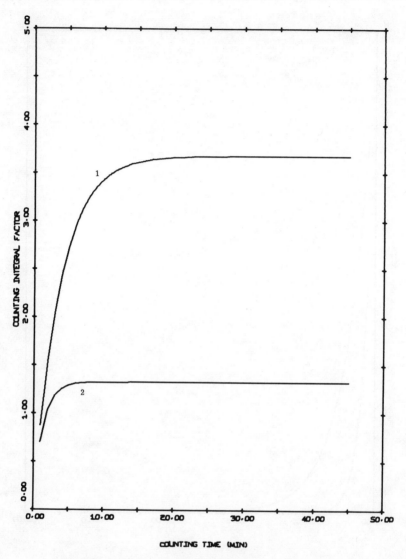

Fig. 7.30. Cerium—integral counting curves (see Table 7.10).

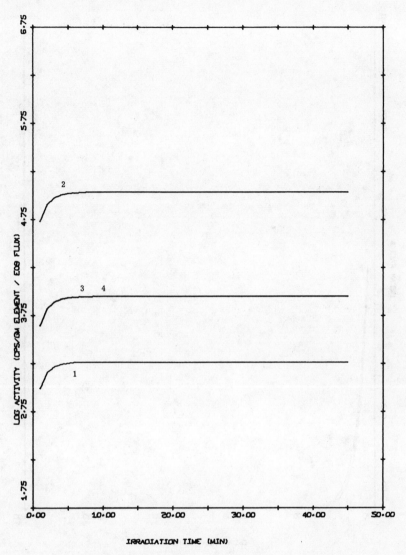

Fig. 7.31. Cerium interferences—activation curves (see Table 7.11).

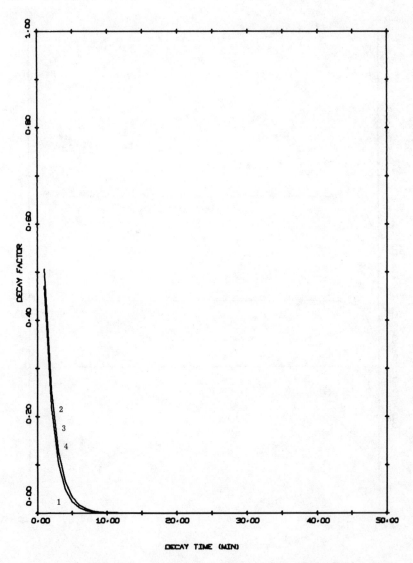

Fig. 7.32. Cerium interferences—decay curves (see Table 7.11).

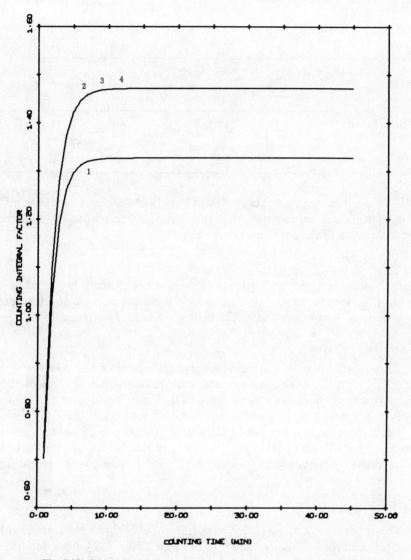

Fig. 7.33. Cerium interferences—integral counting curves (see Table 7.11).

TABLE 7.11. Nuclear Data for Reactions with 14-MeV Neutrons—Cerium Interferences

Nuclear Reaction	Target Isotope Abundance (%)	Cross Section (mb)	Half-life (min)	Gamma Energy (MeV)	Gamma Yield (%)	Detector Efficiency (%)	Peak/Total Ratio
1 = ^{142}Nd(n, α)^{139m}Ce	27.13	10	0.92	0.75	93	35	0.50
2 = ^{142}Nd(n, 2n)^{141m}Nd	27.13	545	1.02	0.75	100	35	0.50
3 = ^{144}Sm(n, α)^{141m}Nd	3.16	400	1.02	0.75	100	35	0.50
4 = ^{144}Sm(n, 2n)^{143m}Sm	3.16	400	1.02	0.75	100	35	0.50

These reports, along with data given by Cuypers and Cuypers,[1] indicate that this procedure is satisfactory in both sensitivity and selectivity for the determination of cerium in a variety of matrices.

3-MeV AND THERMAL NEUTRONS

There have been no applications of lower energy neutrons produced by a neutron generator for the determination of cerium because the inherent sensitivity of these reactions is at least two orders of magnitude less sensitive.

CESIUM

Cesium exists in nature as one isotope of mass number 133. As a member of the alkali metals group, its analytical reactions in solution are limited. It is more commonly measured using flame emission or absorption techniques although it does give an analytically useful response with the x-ray fluoresence technique. In comparison, neutron activation procedures for cesium are not highly sensitive. This low sensitivity probably accounts for the lack of reported investigations of neutron activation methods for cesium using the neutron generator.

14-MeV NEUTRONS

The only reaction of use for the measurement of cesium by activation with 14-MeV neutrons is the (n, 2n) reaction (see Table 7.12). This produces 6.54-day ^{132}Cs with a characteristic photopeak at 0.67 MeV. Cuypers and

TABLE 7.12. Nuclear Data for Reactions with 14-MeV Neutrons—Cesium

Nuclear Reaction	Target Isotope Abundance (%)	Cross Section (mb)	Half-life (min)	Gamma Energy (MeV)	Gamma Yield (%)	Detector Efficiency (%)	Peak/Total Ratio
1 = ^{133}Cs(n, 2n)^{132}Cs	100.00	1550	9360	0.67	99	36	0.53
2 = ^{133}Cs(n, p)^{133}Xe	100.00	10.5	7588	0.08	37	50	1.00

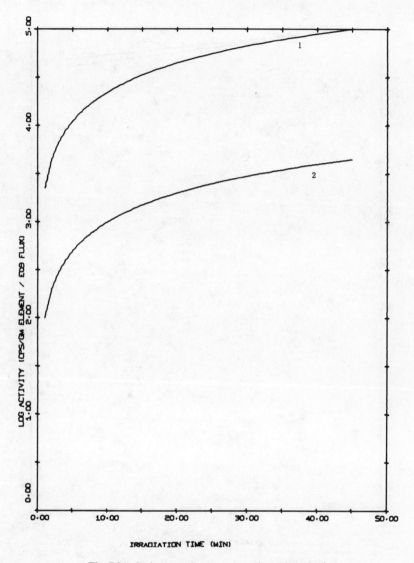

Fig. 7.34. Cesium—activation curves (see Table 7.12).

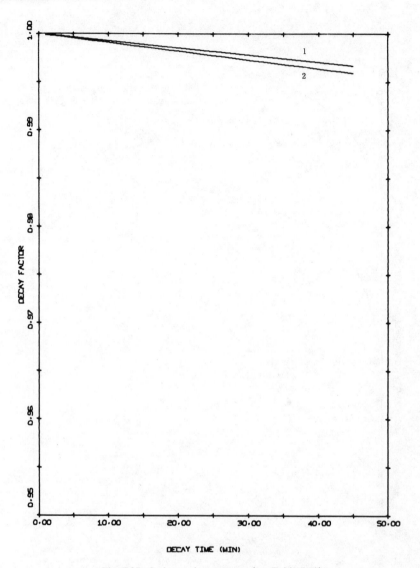

Fig. 7.35. Cesium—decay curves (see Table 7.12).

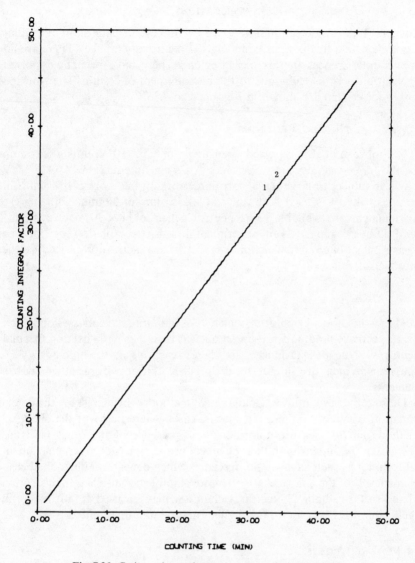

Fig. 7.36. Cesium—integral counting curves (see Table 7.12).

Cuypers[1] report a detection limit of 8.5 mg for cesium using a 5-min irradiation time with a flux of approximately 10^8 neutrons/(cm^2 sec). There are no known interferences in this measurement. There have been no reported applications of this technique to the measurement of cesium. The working curves for cesium are shown in Figs. 7.34 to 7.36.

3-MeV AND THERMAL NEUTRONS

Nargolwalla et al.[23] reported a sensitivity of 1.5×10^4 counts/g of cesium for the (n, γ) reaction. However, the ^{134m}Cs has an emission at 0.128 MeV which is subject to large self-absorption errors. In this work a 10^6-neutron/(cm^2 sec) flux was used with an irradiation time of 20 min. Utilizing the thermal neutrons available from a neutron generator [1×10^8 neutrons/(cm^2 sec)] Dibbs[24] reported an experimental sensitivity of 150 μg for the measurement of cesium, which makes the thermal activation procedure the most sensitive.

CHLORINE

The two isotopes of chlorine which exist in nature have masses of 35 and 37, respectively, and undergo useful nuclear reactions with fast and thermal neutrons. Nuclides produced in these reactions have high-energy γ-emissions which are useful to distinguish chlorine from other matrix elements.

Despite the large number of analytical methods which exist for the determination of chlorine, there are specific cases where many of the chemical methods fail to measure chlorine in the presence of the other halogens because of the similarity in their chemical properties. Moreover, instrumental techniques such as x-ray fluorescence, which depend on inner-shell electronic levels of the chlorine, are often not suitable because of self-absorption of the soft x-radiation. Nuclear techniques may be used to advantage in these situations because of the higher energy radiation emitted.

14-MeV NEUTRONS

Activation of chlorine with 14-MeV neutrons produces three nuclides in abundance. The nuclear reactions of interest for the determination of chlorine with 14-MeV neutrons are given in Table 7.13. The activation curves for these nuclides are shown in Fig. 7.37, and the decay and counting curves are given in Figs. 7.38 and 7.39. Although the most sensitive method is based on the annihilation radiation from ^{34m}Cl, perhaps the most useful from the standpoint of freedom from interference is the 3.09-MeV γ-emission from ^{37}S. This latter photopeak is at a high energy and thus is free from interference, whereas the annihilation radiation is subject to possible interference

III. ACTIVATION ANALYSIS CONDITIONS FOR THE ELEMENTS

TABLE 7.13. Nuclear Data for Reactions with 14-MeV Neutrons—Chlorine

Nuclear Reaction	Target Isotope Abundance (%)	Cross Section (mb)	Half-life (min)	Gamma Energy (MeV)	Gamma Yield (%)	Detector Efficiency (%)	Peak/Total Ratio
$1 = {}^{35}Cl(n, 2n){}^{34m}Cl$	75.40	3.5	32.00	0.15	45	49	0.97
$2 = {}^{35}Cl(n, 2n){}^{34m}Cl$	75.40	3.5	32.00	0.51	100	40	0.63
$3 = {}^{35}Cl(n, 2n){}^{34m}Cl$	75.40	3.5	32.00	1.17	12	31	0.38
$4 = {}^{35}Cl(n, 2n){}^{34m}Cl$	75.40	3.5	32.00	2.12	38	36	0.26
$5 = {}^{35}Cl(n, 2n){}^{34m}Cl$	75.40	3.5	32.00	3.30	12	24	0.20
$6 = {}^{37}Cl(n, \alpha){}^{34}P$	24.60	52.4	0.21	2.13	25	28	0.26
$7 = {}^{37}Cl(n, p){}^{37}S$	24.60	33.4	5.06	3.09	90	24	0.20

from other positron emitters. In addition, the short-lived ^{34}P is useful only in isolated cases where oxygen-containing materials are absent and cannot interfere. Hull and Gilmore[41, 28] have measured the chlorine content of lubricating oils using the ^{37}S isotope. The analyses for chlorine were carried out in the presence of calcium, barium, zinc, nitrogen, and phosphorus, as well as oxygen. Chlorine levels from 0.01 to 0.59% were analyzed using this technique. A computer routine was used to handle the complex spectra for calculation of individual element concentrations. Hull[27] has also reported the application of this technique to petroleum products. Early work on the activation of chlorine was carried out using the (n, α) reaction on ^{35}Cl to produce ^{32}P. This reaction was used to measure chlorine in methacrylates and paraffin, and in graphite using a 12- to 14-hr irradiation at 5×10^8 neutrons/(cm² sec). Radiochemical separations were carried out to isolate the phosphorus and precipitate it as magnesium ammonium phosphate. There appears to be no advantage to this approach in view of the more convenient (n, p) reactions giving γ-emitters. Przybylowicz et al.[42] have utilized the ^{37}S nuclide for the measurement of chlorine in photographic emulsions containing silver and the other halides. The method provides a direct rapid measure of chlorine in bulk photographic emulsions. The sensitivity of this technique appears to be excellent with as little as 10 μg of chlorine being detectable. If the annihilation radiation from the decay of ^{34m}Cl is to be used, interference from other possible positron emitters, listed in Tables 7.76, 7.77, and 7.78, should be considered.

3-MeV AND THERMAL NEUTRONS

The sensitivity for the determination of chlorine with 3-MeV neutrons is very low.[23] At this neutron energy, the (n, γ) neutron reaction has a relatively low cross section, which is not useful for the analytical determination of chlorine.

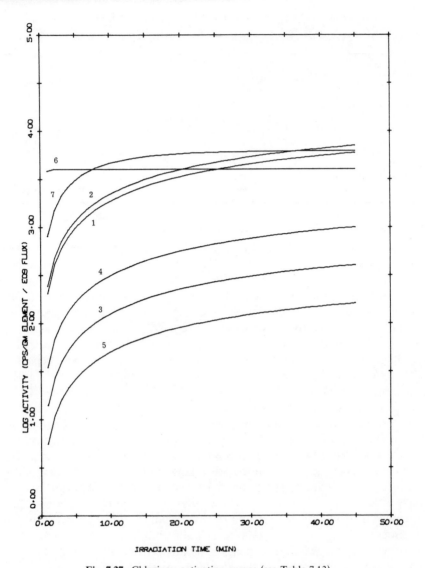

Fig. 7.37. Chlorine—activation curves (see Table 7.13).

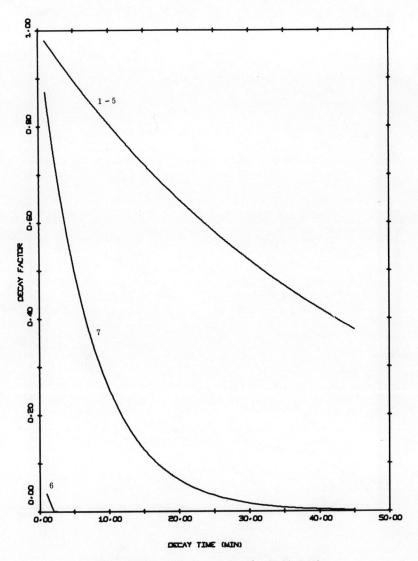

Fig. 7.38. Chlorine—decay curves (see Table 7.13).

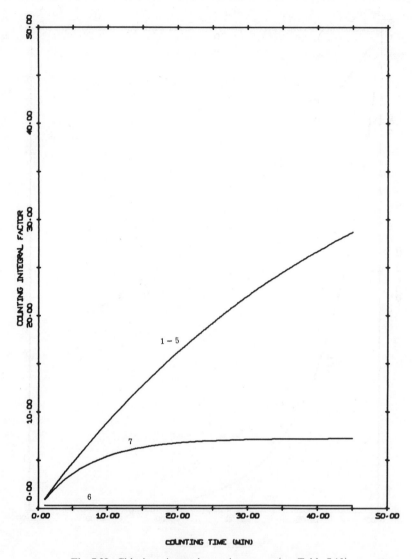

Fig. 7.39. Chlorine—integral counting curves (see Table 7.13).

III. ACTIVATION ANALYSIS CONDITIONS FOR THE ELEMENTS   339

Thermalizing the neutron flux from a generator can provide a useful means of measuring macroscopic quantities of chlorine since the reaction is less sensitive than with 14-MeV neutrons. The method has found particular use in whole-body irradiation of humans with neutrons for the estimation of total chlorine content. Several workers[37,43] have demonstrated the utility of this technique and note that the results obtained by this method are the first direct results for whole-body chlorine content that have been reported. Ingenious methods were used to calibrate the system in order that reasonably good estimates could be obtained. The authors note that the radiation dose to the subject was no greater than that received in diagnostic radiology with x-rays.

CHROMIUM

Chromium occurs in nature in four isotopic forms. The isotopes with mass 52 and 53 are the most useful for neutron activation analysis with fast neutrons. The isotopes with mass 50 and 54 do not undergo nuclear reactions with sufficient sensitivity to be useful. Chromium is one of the elements which can be determined with very high sensitivity using neutron generator activation analysis.

There has been substantial interest in chromium as a minor constituent in various ferrous alloys. Thus its analytical chemistry in complex matrices has been thoroughly studied with a wide variety of instrumental and wet chemical methods including such nondestructive methods as x-ray fluorescence analysis. The major advantage of neutron activation analysis for chromium would be speed and reduction of sample attenuation problems which plague the x-ray fluorescence technique for high-density samples.

14-MeV NEUTRONS

Table 7.14 gives the pertinent nuclear data for the activation of chromium using 14-MeV neutrons, and the activation, decay, and integral counting curves for the various activities are shown in Figs. 7.40 to 7.42. The (n, p)

TABLE 7.14. Nuclear Data for Reactions with 14-MeV Neutrons—Chromium

Nuclear Reaction	Target Isotope Abundance (%)	Cross Section (mb)	Half-life (min)	Gamma Energy (MeV)	Gamma Yield (%)	Detector Efficiency (%)	Peak/Total Ratio
1 = ^{50}Cr(n, 2n)^{49}Cr	4.31	27	41.90	0.15	13	49	0.97
2 = ^{50}Cr(n, 2n)^{49}Cr	4.31	27	41.90	0.51	186	40	0.63
3 = ^{52}Cr(n, 2n)^{51}Cr	83.76	285	40032	0.32	9	47	0.81
4 = ^{52}Cr(n, p)^{52}V	83.76	78	3.76	1.43	100	29	0.34
5 = ^{53}Cr(n, p)^{53}V	9.55	48	2.00	1.00	100	32	0.41

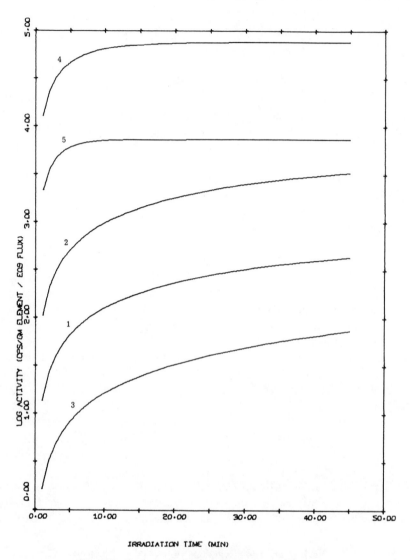

Fig. 7.40. Chromium—activation curves (see Table 7.14).

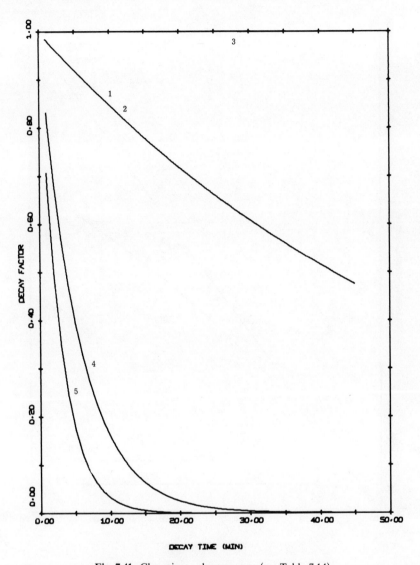

Fig. 7.41. Chromium—decay curves (see Table 7.14).

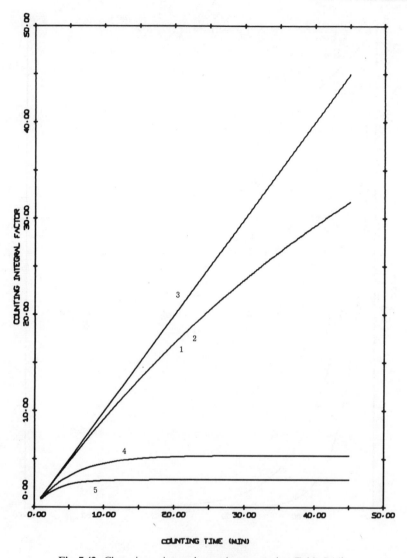

Fig. 7.42. Chromium—integral counting curves (see Table 7.14).

III. ACTIVATION ANALYSIS CONDITIONS FOR THE ELEMENTS 343

reaction on ^{52}Cr producing ^{52}V is the most sensitive and yields a 1.43-MeV gamma ray.

There is relatively little reported literature on the use of a neutron generator for the determination of chromium. Steele and Meinke[44] measured the sensitivity of the (n, p) reaction yielding ^{52}V. Utilizing a flux of 10^8 neutrons/(cm^2 sec) and an irradiation time equivalent to the half-life (3.76 min), these authors reported a specific activity of 139,810 counts/(min g) of chromium. The lower limit of detection based on these specific activity measurements was 8 ppm of chromium. Dibbs[24] has demonstrated a sensitivity of 300 μg using this same reaction. Broadhead et al.[12] reported the analysis of metal phosphides, including chromium phosphides, using the same activated isotope and γ-emission.

The major interferences in the determination of chromium using the (n, p) reaction are manganese and vanadium. Both give the same isotope upon irradiation with 14-MeV neutrons, although the level of interference by each element differs substantially. Interference by manganese is produced by an (n, α) reaction on ^{55}Mn for which the sensitivity is approximately the same as that for chromium. Vanadium interferes through an (n, γ) reaction on ^{51}V. This reaction is, however, almost 15 times less sensitive than the reaction for the activation of chromium. The activation, decay, and integral counting curves for these interfering reactions are shown in Figs. 7.43 to 7.45. Table 7.15 gives the important parameters for the production of these interferences. Possible interferences with the reaction yielding the positron-emitting ^{49}Cr are listed in Tables 7.76 to 7.78.

TABLE 7.15. Nuclear Data for Reactions with 14-MeV Neutrons—Chromium Interferences

Nuclear Reaction	Target Isotope Abundance (%)	Cross Section (mb)	Half-life (min)	Gamma Energy (MeV)	Gamma Yield (%)	Detector Efficiency (%)	Peak/Total Ratio
1 = ^{51}V(n, γ)^{52}V	99.76	0.80	3.76	1.43	100	29	0.34
2 = ^{55}Mn(n, α)^{52}V	100.00	52.5	3.76	1.43	100	29	0.34

3-MeV AND THERMAL NEUTRONS

There are no useful reactions for chromium using these lower energy neutrons with the fluxes available from a neutron generator.

COBALT

Cobalt exists in nature in one isotopic form of mass number 59. While it undergoes several nuclear reactions with both thermal and fast neutrons, it

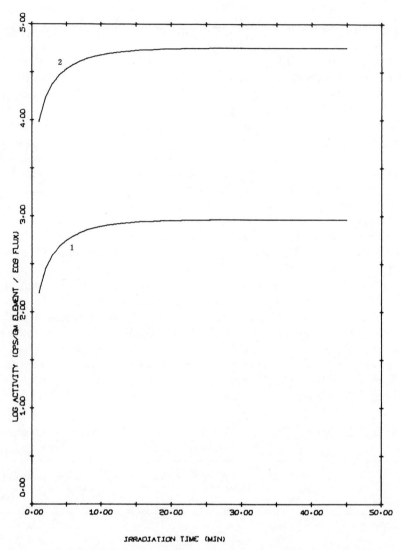

Fig. 7.43. Chromium interferences—activation curves (see Table 7.15).

344

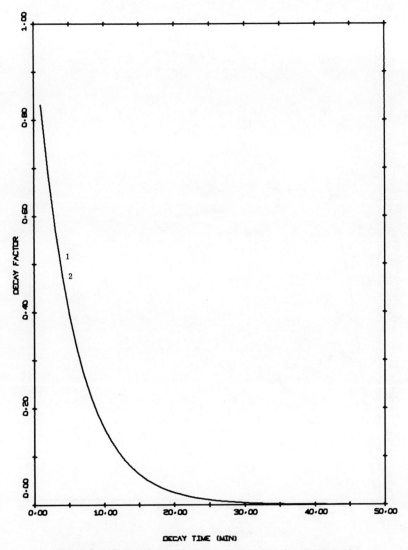

Fig. 7.44. Chromium interferences—decay curves (see Table 7.15).

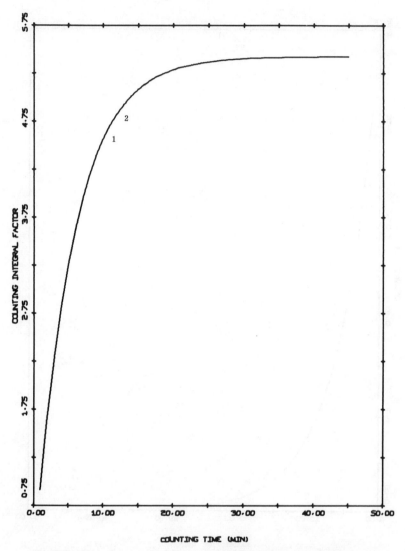

Fig. 7.45. Chromium interferences—integral counting curves (see Table 7.15).

III. ACTIVATION ANALYSIS CONDITIONS FOR THE ELEMENTS 347

cannot be determined with high sensitivity using the fluxes available from a neutron generator. However, the reaction parameters are good enough to permit the determination of milligram quantities of cobalt in some circumstances.

Cobalt can readily be determined by a wide variety of analytical methods, both instrumental and wet chemical. Thus the need to develop neutron activation methods for semimicroquantities of cobalt has not been pressing. The neutron activation method should be seriously considered in circumstances where milligram amounts of cobalt are to be determined in matrices which are difficult to solubilize for wet chemical methods or where sample attenuation problems for instrumental methods cannot be easily overcome.

14-MeV NEUTRONS

The pertinent nuclear parameters for the determination of cobalt are given in Table 7.16. The most sensitive reaction for the determination of cobalt is the (n, α) reaction to produce ^{56}Mn. Cuypers and Cuypers[1] report

TABLE 7.16. Nuclear Data for Reactions with 14-MeV Neutrons—Cobalt

Nuclear Reaction	Target Isotope Abundance (%)	Cross Section (mb)	Half-life (min)	Gamma Energy (MeV)	Gamma Yield (%)	Detector Efficiency (%)	Peak/Total Ratio
$1 = {}^{59}\text{Co}(n, \alpha){}^{56}\text{Mn}$	100.00	39.1	154.56	0.85	99	34	0.44
$2 = {}^{59}\text{Co}(n, \alpha){}^{56}\text{Mn}$	100.00	39.1	154.56	1.81	29	32	0.28
$3 = {}^{59}\text{Co}(n, \alpha){}^{56}\text{Mn}$	100.00	39.1	154.56	2.11	15	27	0.26
$4 = {}^{59}\text{Co}(n, 2n){}^{58m}\text{Co}$	100.00	4.0	540.00	0.03	100	50	1.00

a detection limit of 4.3 mg for cobalt using a 5-min radiation and a flux of approximately 10^8 neutrons/(cm^2 sec). The activation curves shown in Fig. 7.46, which are based on literature cross-section values, indicate a similar sensitivity level. Figures 7.47 and 7.48 show the decay and counting integral curves for these nuclides. The most favorable photopeak to use for this measurement is at 0.85 MeV. The next most sensitive peak is from this same isotope, at 1.81 MeV, and is approximately a factor of 5 lower in sensitivity.

There are two interfering reactions in the determination of cobalt (Table 7.17), and the activation, decay, and counting integral curves for these are given in Figs. 7.49 to 7.51. The most serious interference is from iron, via an (n, p) reaction, to produce ^{56}Mn which is formed with slightly higher sensitivity than the reaction for cobalt. Manganese also interferes via an (n, γ) reaction to produce the same isotope, although this interference is approximately two orders of magnitude less sensitive.

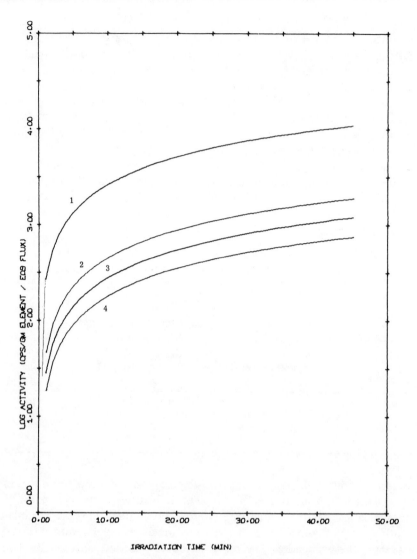

Fig. 7.46. Cobalt—activation curves (see Table 7.16).

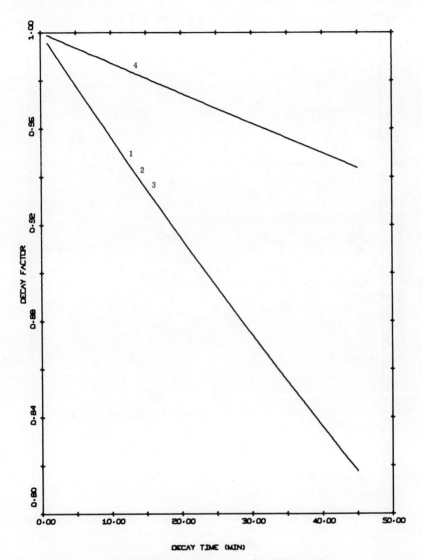

Fig. 7.47. Cobalt—decay curves (see Table 7.16).

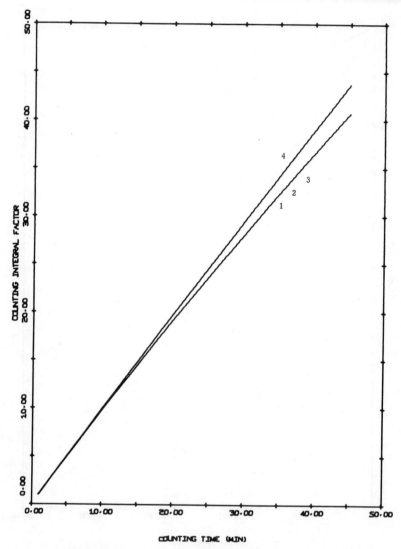

Fig. 7.48. Cobalt—integral counting curves (see Table 7.16).

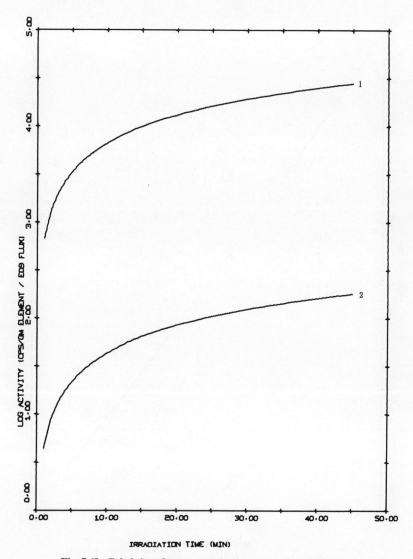

Fig. 7.49. Cobalt interferences—activation curves (see Table 7.17).

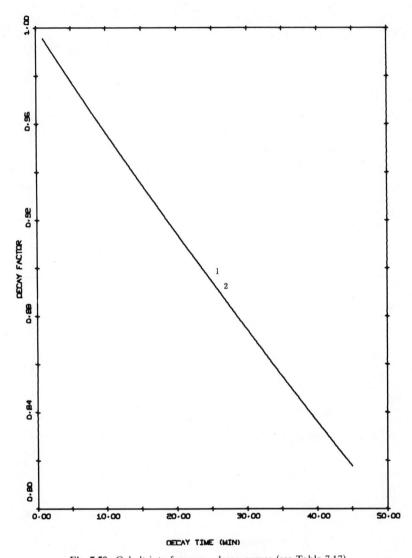

Fig. 7.50. Cobalt interferences—decay curves (see Table 7.17).

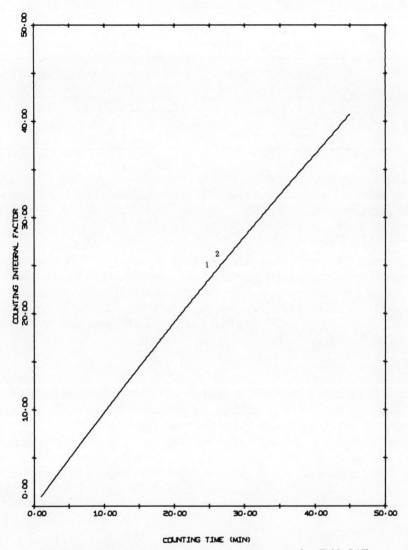

Fig. 7.51. Cobalt interferences—integral counting curves (see Table 7.17).

TABLE 7.17. Nuclear Data for Reactions with 14-MeV Neutrons—Cobalt Interferences

Nuclear Reaction	Target Isotope Abundance (%)	Cross Section (mb)	Half-life (min)	Gamma Energy (MeV)	Gamma Yield (%)	Detector Efficiency (%)	Peak/Total Ratio
1 = ^{56}Fe(n, p)^{56}Mn	91.68	103	154.80	0.85	99	34	0.44
2 = ^{55}Mn(n, γ)^{56}Mn	100.00	0.60	154.80	0.85	99	34	0.44

Other than in the activation spectra catalog of Cuypers and Cuypers,[1] there have been no reported applications of this technique for the determination of cobalt.

3-MeV AND THERMAL NEUTRONS

Although (n, γ) reactions occur with 3-MeV neutrons,[23] the detection sensitivity with the flux available from a neutron generator is too low to be seriously considered for purposes of an analytical method. However, Dibbs[24] has reported excellent sensitivity using the available thermal flux from a generator [1×10^8 neutrons/(cm^2 sec)]. His detection limit (100 photopeak counts) was 40 μg in a 5-min irradiation.

COPPER

The two naturally occurring copper isotopes have masses of 63 and 65, respectively. Both isotopes undergo useful nuclear reactions with the fast neutron fluxes available with the neutron generator.

On the basis of sensitivity, nuclear activation methods using neutrons can compete successfully with other instrumental and wet chemical methods for the determination of copper. However, the method does lack specificity because the most sensitive reaction with generator-produced neutrons yields a positron-emitting nuclide.

14-MeV NEUTRONS

The activation, decay, and integral counting curves for the reaction of 14-MeV neutrons with copper are shown in Figs. 7.52 to 7.54. These reactions are summarized in Table 7.18. These curves indicate that the (n, 2n) reaction on ^{63}Cu to produce the positron emitter ^{62}Cu is by far the most sensitive. Irradiation with 10^9 neutrons/(cm^2 sec) for 5 min produces sufficient activity to detect as little as 10 μg of copper. No other reaction of copper with this flux of 14-MeV neutrons has sufficient sensitivity to be observed. Data on other positron-emitting nuclides which may interfere are given in Tables 7.76, 7.77, and 7.78, and in Figs. 7.223 through 7.231.

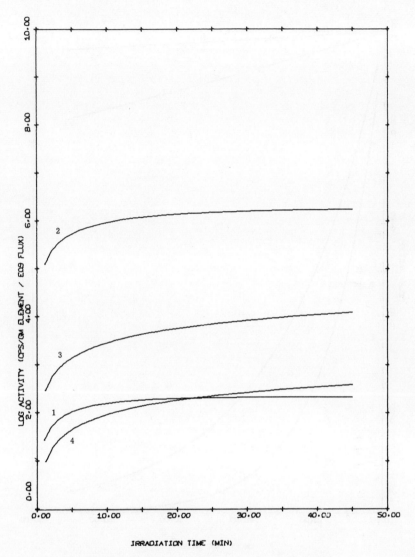

Fig. 7.52. Copper—activation curves (see Table 7.18).

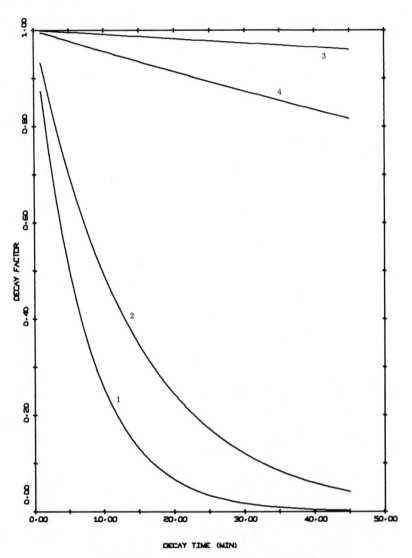

Fig. 7.53. Copper—decay curves (see Table 7.18).

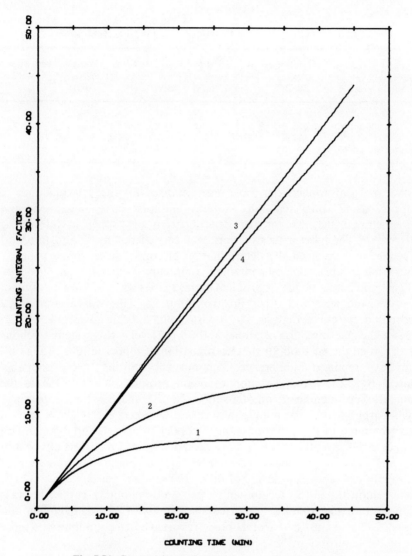

Fig. 7.54. Copper—integral counting curves (see Table 7.18).

TABLE 7.18. Nuclear Data for Reactions with 14-MeV Neutrons—Copper

Nuclear Reaction	Target Isotope Abundance (%)	Cross Section (mb)	Half-life (min)	Gamma Energy (MeV)	Gamma Yield (%)	Detector Efficiency (%)	Peak/Total Ratio
1 = ^{65}Cu(n, γ)^{66}Cu	30.90	6.2	5.1	1.04	9	32	0.41
2 = ^{63}Cu(n, 2n)^{62}Cu	69.10	550	9.8	0.51	195	40	0.63
3 = ^{65}Cu(n, 2n)^{64}Cu	30.90	1100	768.0	0.51	38	40	0.63
4 = ^{65}Cu(n, p)^{65}Ni	30.90	29	153.6	1.48	25	29	0.33

A variety of applications have been reported for the determination of copper using this reaction. Several authors have suggested and demonstrated the use for the determination of copper in lubricating oils.[2, 21, 46] The most extensive work was carried out by D'Agostino and Kuehne[2] and involved the determination of copper in a complex matrix consisting of lubricating oil containing aluminum, copper, iron, magnesium, silicon, and zinc. The major problem in this matrix was the differentiation of copper from zinc and other positron emitters. This was accomplished through decay curve analysis. The reported data indicate an accuracy of 3 to 5% for the determination of copper at the 0.1% level in this system. Gibbons et al.[45] used this method for the determination of copper in lubricating oils, but have reported interference from positron-emitting ^{13}N presumably caused by "knock-on" reaction of recoil protons on ^{13}C. Due to the generation of significant amounts of ^{13}N, the analysis for copper in a hydrocarbon matrix was not highly accurate. Gorski et al.[46, 47] have applied this technique to the determination of copper in ores. Using 8-g samples, copper levels from 0.01 to 12% have been determined requiring only 4 min per analysis. They propose the method for the routine determination of copper in ore samples at a rate of up to 100 samples per day. Daly et al.[48] have demonstrated this technique for the determination of copper in brass alloys. Steele and Meinke[44] reported a sensitivity of 25 ppm of copper at a flux of 10^8 neutrons/(cm^2 sec). Dibbs[24] reported a detection limit of 6 μg at a 5 × 10^8-neutron/(cm^2 sec) flux.

3-MeV AND THERMAL NEUTRONS

The (n, γ) reactions on ^{63}Cu and ^{65}Cu give low sensitivity with generator-produced neutrons of 3-MeV energy. Nargolwalla et al.[23] indicate that the production of ^{66}Cu is more sensitive by a factor of 2 than the production of ^{64}Cu, using irradiation times of 10 min at a flux of 3.5 × 10^5 neutrons/(cm^2 sec) with 3-MeV neutrons. The sensitivity for copper, using thermal neutron fluxes available from a neutron generator, is also about a factor of 20 lower than with 14-MeV neutrons.

DYSPROSIUM

Dysprosium is an element which can be determined with relatively high sensitivity using neutron generator activation analysis. It occurs in nature in the seven isotopic forms of mass 156, 158, and 160 through 164.

The rare earths as a class have been the subject of extensive analytical studies because of their importance in the metallurgical industry and the complexity of their analysis in the presence of one another. Today chemical methods exist for the determination of the individual rare earths, but they can be applied only after chemical separation.

Methods based on neutron activation analysis have been used in connection with radiochemical separations to determine rare earths after reactor activation of samples. For certain specific rare earths such as dysprosium, the neutron generator method provides a more specific, totally instrumental method.

14-MeV NEUTRONS

The pertinent nuclear data for fast neutron reactions with dysprosium are given in Table 7.19. The activation, decay, and integral counting curves for

TABLE 7.19. Nuclear Data for Reactions with 14-MeV Neutrons—Dysprosium

Nuclear Reaction	Target Isotope Abundance (%)	Cross Section (mb)	Half-life (min)	Gamma Energy (MeV)	Gamma Yield (%)	Detector Efficiency (%)	Peak/Total Ratio
1 = ^{162}Dy(n, p)^{162}Tb	25.53	4	7.50	0.26	40	48	0.87
2 = ^{163}Dy(n, p)^{163}Tb	24.97	3	390.00	0.51	100	40	0.63
3 = ^{164}Dy(n, γ)^{165}Dy	28.18	81.5	139.20	0.09	4	50	1.00
4 = ^{164}Dy(n, γ)^{165m}Dy	28.18	8	1.26	0.11	3	50	1.00
5 = ^{164}Dy(n, α)^{161}Gd	28.18	4.5	3.70	0.36	66	44	0.77

the reaction of 14-MeV neutrons with dysprosium are shown in Figs. 7.55 to 7.57. These curves indicate that the most sensitive reaction for dysprosium with 14-MeV neutrons is the (n, γ) reaction to produce ^{165m}Dy. The primary emissions from this nuclide at 0.05 and 0.11 MeV can be used to detect as little as 0.1 mg of dysprosium, as reported by Cuypers and Cuypers.[1] There have been no literature applications of the neutron generator for the determination of dysprosium except for the catalog data by Cuypers.

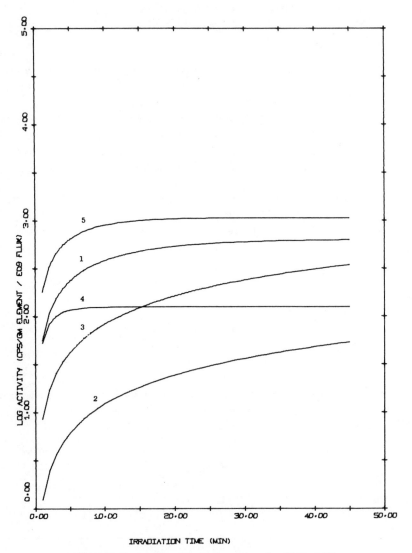

Fig. 7.55. Dysprosium—activation curves (see Table 7.19).

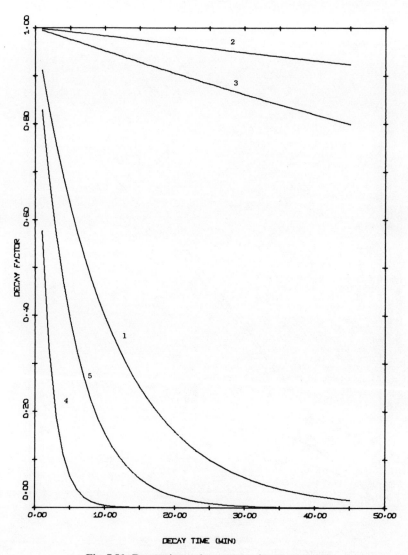

Fig. 7.56. Dysprosium—decay curves (see Table 7.19).

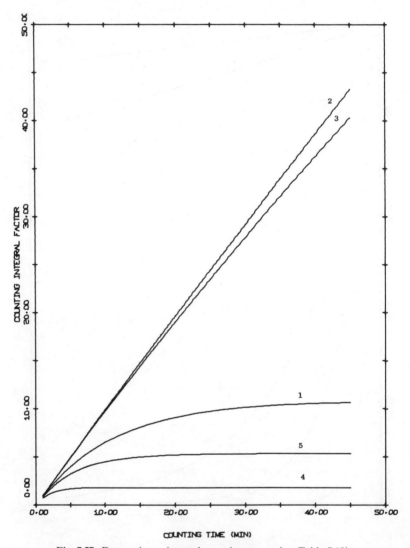

Fig. 7.57. Dysprosium—integral counting curves (see Table 7.19).

3-MeV AND THERMAL NEUTRONS

Nargolwalla et al.[23] measured the sensitivity of the (n, γ) reaction to produce ^{165m}Dy using a 10^6-neutron/(cm² sec) flux of 3-MeV neutrons. The cross section for the (n, γ) reaction is significantly higher at these neutron energies than with 14-MeV neutrons; however, the lower available flux results in a factor of 10 decrease in sensitivity. Nonetheless, these conditions may be attractive for macrodetermination of dysprosium if other matrix elements which would produce interferences with 14-MeV neutrons are present. Thermalization of 14-MeV neutrons results in a method of higher sensitivity for dysprosium since the cross section for reaction with thermal neutrons is even more favorable than with the 3-MeV neutrons. Dibbs[24] reports a sensitivity (100 counts over background) of 20 μg via ^{165m}Dy with a flux of 10^8 neutrons/(cm² sec).

ERBIUM

Erbium exists in nature in six isotopic forms which range from mass 162 to 170. While erbium undergoes reactions with neutrons of various energies, the most useful reactions are those occurring with low-energy neutrons.

The analytical chemistry of erbium has not been extensively explored except in the context of analyzing rare earth mixtures by various ion-exchange separation procedures. Because these are long, involved procedures, totally instrumental methods, such as x-ray fluorescence and neutron activation analysis, have attractive features for the analysis of erbium.

14-MeV NEUTRONS

Table 7.20 lists the pertinent nuclear reactions for the activation of erbium with 14-MeV neutrons. The predominant reaction produces ^{167m}Er, a nuclide having a 2.5-sec half-life. This short half-life, coupled with the low-energy gamma emission at 0.049 and 0.208 MeV, makes the determina-

TABLE 7.20. Nuclear Data for Reactions with 14.7–MeV Neutrons—Erbium

Nuclear Reaction	Target Isotope Abundance (%)	Cross Section (mb)	Half-Life (min)	Gamma Energy (MeV)	Gamma Yield (%)	Detector Efficiency (%)	Peak/Total Ratio
1 = ^{164}Er(n, p)^{164}Ho	1.56	44	37.00	0.09	100	50	1.00
2 = ^{168}Er(n, 2n)^{167m}Er	27.07	190	0.04	0.21	43	49	0.92
3 = ^{168}Er(n, p)^{168}Ho	27.07	2.5	3.30	0.85	100	34	0.44
4 = ^{170}Er(n, p)^{170}Ho	14.88	1.8	0.75	0.43	100	43	0.71

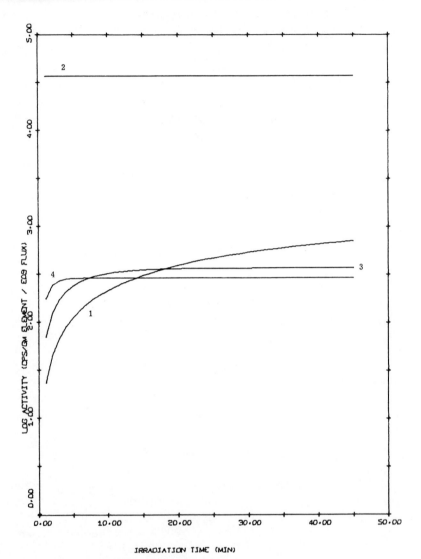

Fig. 7.58. Erbium—activation curves (see Table 7.20).

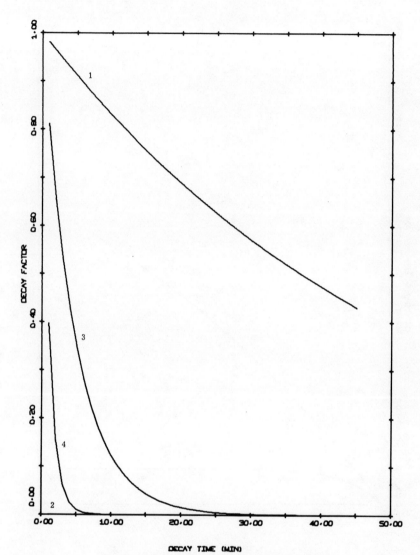

Fig. 7.59. Erbium—decay curves (see Table 7.20).

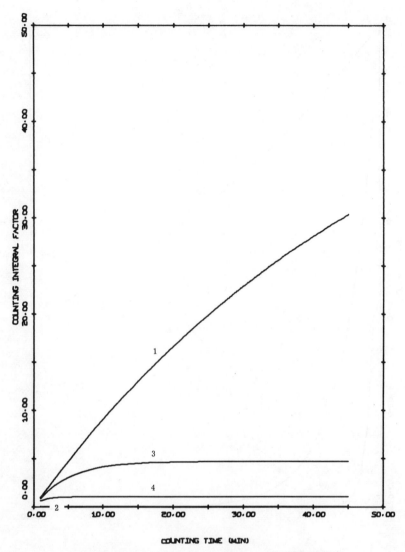

Fig. 7.60. Erbium—integral counting curves (see Table 7.20).

III. ACTIVATION ANALYSIS CONDITIONS FOR THE ELEMENTS   367

tion of erbium with 14-MeV neutrons unattractive in most matrices. Obviously the presence of trace amounts of oxygen will grossly interfere with the determination due to the high Compton and bremsstrahlung which is produced at low gamma energies. There have been no reported methods for the determination of erbium using 14-MeV neutron activation. The activation, decay, and integral working curves for erbium are shown in Figs. 7.58 through 7.60.

3-MeV AND THERMAL NEUTRONS

The production of the ^{167m}Er isotope can be achieved using 3-MeV neutrons which eliminates the interference of oxygen in the previously described determination. Broadhead and Shanks[36] have studied this reaction and reported cross-section values for it with 3-MeV neutrons. Nargolwalla et al.[23] reported an experimental sensitivity for this reaction for 3-MeV neutrons of 5.7×10^3 counts in the photopeak region per gram of erbium when irradiated with a flux of 10^6 neutrons/(cm^2 sec). A second, perhaps more useful, reaction occurs with 3-MeV neutrons involving the neutron capture by the ^{170}Er isotope to produce ^{171}Er($T_{1/2} = 7.52$ hr). These gamma emissions are at somewhat higher energies (0.30 MeV), but the sensitivity is approximately a factor of 4 lower. Nargolwalla et al.[23] report an experimental sensitivity for the capture reaction of 1.1×10^3 total counts/g of erbium with a flux of 10^6 neutrons/(cm^2 sec) of 3-MeV neutrons. Dibbs[24] reported a sensitivity of 260 μg for erbium (100 counts above background) using the ^{171}Er nuclide and a thermal neutron flux of 10^8 neutrons/(cm^2 sec).

EUROPIUM

Europium exists in nature as two isotopes of mass 151 and 153. As one of the rare earths, its chemistry is similar to that of a number of elements in the periodic table with which it is found in nature. Thus neutron activation analysis techniques could provide a means of differentiating europium from some of the chemically similar elements in the rare earth group. While europium does activate to produce unique products, there are several factors which may limit the use of this technique.

14-MeV NEUTRONS

The most sensitive reaction for the measurement of europium is ^{153}Eu(n, 2n)^{152m}Eu. Cuypers and Cuypers[1] list a detection limit of 2.9 mg for this reaction which uses the low-energy photopeak at 0.09 MeV for the measurement. The use of this peak may severely limit the applicability of this

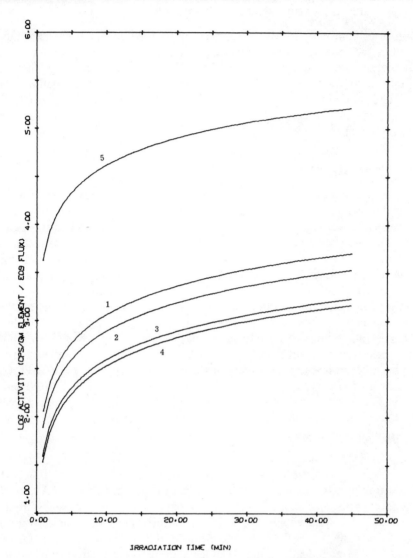

Fig. 7.61. Europium—activation curves (see Table 7.21).

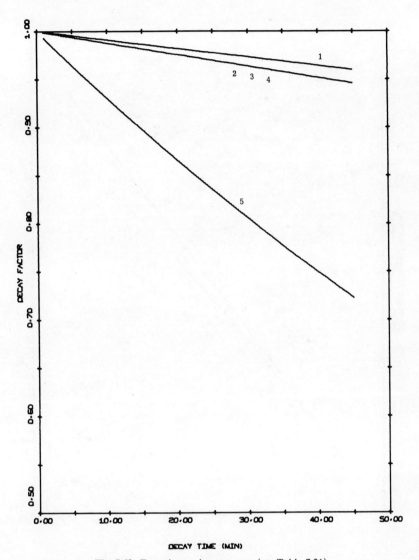

Fig. 7.62. Europium—decay curves (see Table 7.21).

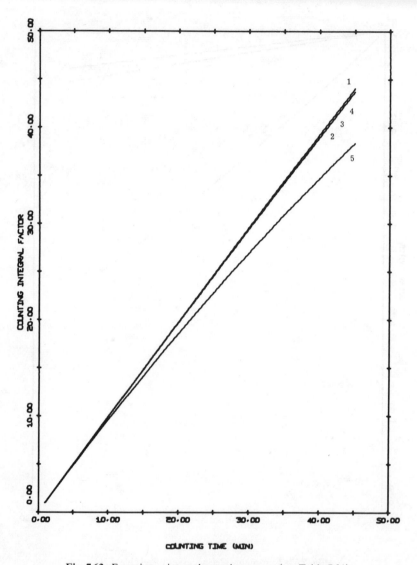

Fig. 7.63. Europium—integral counting curves (see Table 7.21).

III. ACTIVATION ANALYSIS CONDITIONS FOR THE ELEMENTS

technique to complex matrices in which other elements may activate to produce a high counting background in this low-energy region. One possible advantage in the use of this reaction is the long half-life of the product isotope which permits the decay of other activated matrix elements prior to counting.

Table 7.21 lists the reactions of possible interest using the 14-MeV technique. Figures 7.61 through 7.63 show the activation, decay, and counting integral curves for the activation of europium.

TABLE 7.21. Nuclear Data for Reactions with 14-MeV Neutrons—Europium

Nuclear Reaction	Target Isotope Abundance (%)	Cross Section (mb)	Half-life (min)	Gamma Energy (MeV)	Gamma Yield (%)	Detector Efficiency (%)	Peak/Total Ratio
1 = ^{151}Eu(n, 2n)^{150}Eu	47.77	500	756.0	0.58	60	38	0.58
2 = ^{153}Eu(n, 2n)^{152m}Eu	52.23	750	558.0	0.12	8	50	0.99
3 = ^{153}Eu(n, 2n)^{152m}Eu	52.23	750	558.0	0.84	13	34	0.44
4 = ^{153}Eu(n, 2n)^{152m}Eu	52.23	750	558.0	0.96	12	33	0.42
5 = ^{153}Eu(n, 2n)^{152m}Eu	52.23	750	96.0	0.09	74	50	1.00

There have been no reported applications of this technique to the measurement of europium.

3-MeV AND THERMAL NEUTRONS

Nargolwalla et al.[23] reported that no unique photopeaks were evident in the activation of europium with 3-MeV neutrons. Yet, using a counting energy window from 0.025 to 1.65 MeV, they report a sensitivity of 5×10^4 counts/g of europium [10^6-neutron/(cm^2 sec) flux]. The most sensitive reaction for the measurement of europium with a neutron generator is the (n, γ) reaction on ^{151}Eu which has a capture cross section of 2800 barns. Dibbs[24] reported an experimental sensitivity of 3.5 μg using a thermal flux of 10^8 neutrons/(cm^2 sec) utilizing the photopeak at 0.12 MeV. This amount of europium was found to give a net of 100 counts above background.

FLUORINE

The only naturally occurring isotope of fluorine is that of mass 19. It undergoes very useful reactions with 14-MeV neutrons, thus providing a rapid, sensitive method of measuring fluorine in a wide variety of matrices. The ideal activation parameters and the difficulty of determining fluorine by most other analytical methods make this approach one of the methods of choice for fluorine.

The importance of fluorine as a trace element in water supplies, as a significant constituent of minerals in the earth's crust, and as an important constituent of synthetic polymers, to mention a few of its occurrences, has focused a great deal of attention on the analytical chemistry of fluorine. Most methods for the determination of fluorine require conversion of the element to the fluoride ion and its subsequent measurement, usually by photometric, titrimetric, or potentiometric methods. Except for neutron activation analysis, there are no other nondestructive instrumental methods for the analysis of this element. This fact points up the uniqueness of this procedure for fluorine and emphasizes its importance.

14-MeV NEUTRONS

The important nuclear reactions of fluorine with 14-MeV neutrons are tabulated in Table 7.22.

TABLE 7.22. Nuclear Data for Reactions with 14-MeV Neutrons—Fluorine

Nuclear Reaction	Target Isotope Abundance (%)	Cross Section (mb)	Half-life (min)	Gamma Energy (MeV)	Gamma Yield (%)	Detector Efficiency (%)	Peak/Total Ratio
1 = $^{19}F(n, \alpha)^{16}N$	100.00	50.0	0.12	6.13	69	23	0.14
2 = $^{19}F(n, \alpha)^{16}N$	100.00	50.0	0.12	7.12	5	23	0.13
3 = $^{19}F(n, 2n)^{18}F$	100.00	60.6	112.00	0.51	97	40	0.63
4 = $^{19}F(n, p)^{19}O$	100.00	14.3	0.48	0.20	64	50	0.93
5 = $^{19}F(n, p)^{19}O$	100.00	14.3	0.48	1.36	39	30	0.35

There are three possible reactions of neutrons with fluorine which may be used for analytical determination. The activation, decay, and integral counting curves for the three possible reactions of fluorine are listed in Figs. 7.64 to 7.66. By far the most sensitive reaction is the (n, α) reaction resulting in the formation of ^{16}N. Unfortunately, this reaction produces the same nuclide as the (n, p) reaction with oxygen and can be used for the determination of fluorine only if oxygen is absent or can be corrected for in some manner. Oxygen is a major interference since its activation parameters for the production of this nuclide are more favorable than for fluorine. The activation, decay, and integral counting curves for the oxygen interference are given in Figs. 7.67 to 7.69, with the pertinent activation parameters listed in Table 7.23. In oxygen-containing matrices the (n, p) reaction on fluorine to produce ^{19}O may be adapted for the measurement of fluorine. The ^{19}O nuclide has characteristic gamma emissions at 0.2 and 1.37 MeV. Ordinarily the measurement of either of the gamma rays from ^{19}O would be satisfactory but the large Compton scatter from the high-energy gamma rays of ^{16}N,

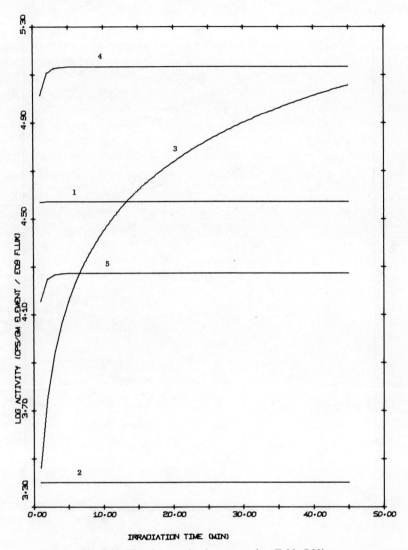

Fig. 7.64. Fluorine—activation curves (see Table 7.22).

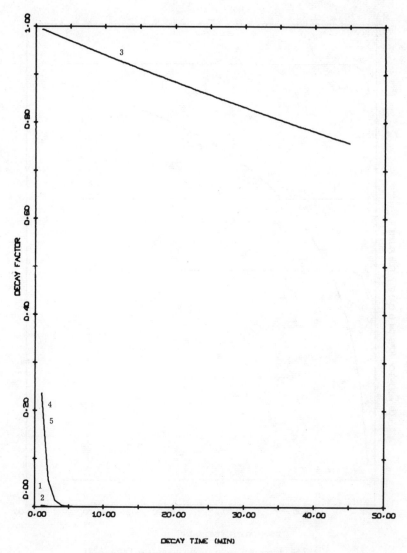

Fig. 7.65. Fluorine—decay curves (see Table 7.22).

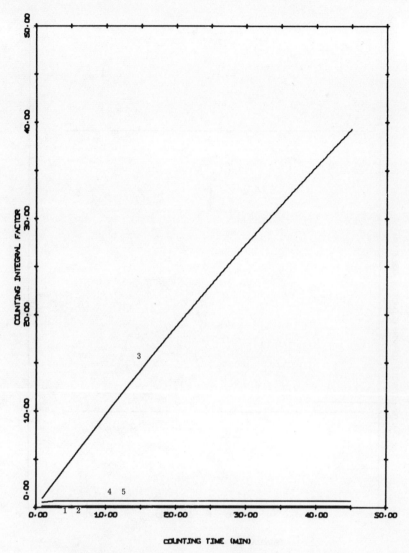

Fig. 7.66. Fluorine—integral counting curves (see Table 7.22).

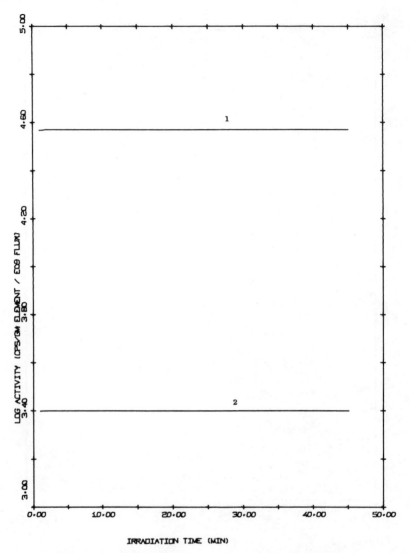

Fig. 7.67. Fluorine interferences—activation curves (see Table 7.23).

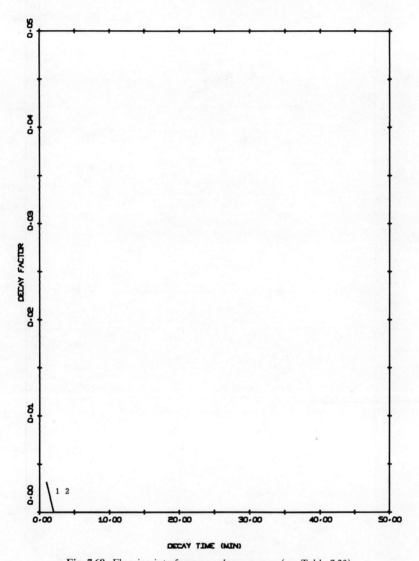

Fig. 7.68. Fluorine interferences—decay curves (see Table 7.23).

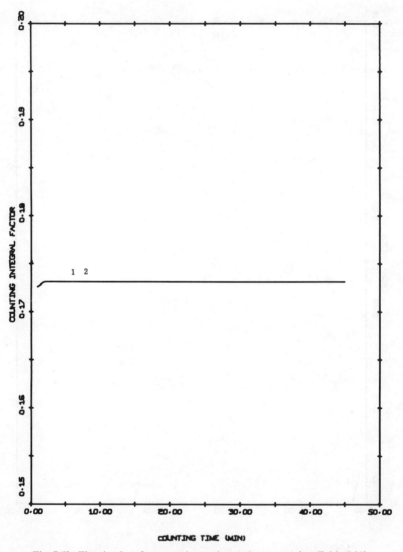

Fig. 7.69. Fluorine interferences—integral counting curves (see Table 7.23).

TABLE 7.23. Nuclear Data for Reactions with 14-MeV Neutrons—Fluorine Interferences

Nuclear Reaction	Target Isotope Abundance (%)	Cross Section (mb)	Half-life (min)	Gamma Energy (MeV)	Gamma Yield (%)	Detector Efficiency (%)	Peak/Total Ratio
$1 = {}^{16}O(n,p){}^{16}N$	99.76	42.00	0.12	6.13	69	23	0.14
$2 = {}^{16}O(n,p){}^{16}N$	99.76	42.00	0.12	7.12	5	23	0.13

produced in an oxygen-containing sample, results in a high background under the 0.20-MeV photopeak. Gamma-gamma coincidence techniques have been used to reduce the Compton scatter background and effectively allow the determination of fluorine without interference from oxygen or other nuclides which emit higher energy particles. Although it is possible for a gamma ray of ^{16}N to undergo multiple scatter between the two gamma detectors, giving up just the correct energies for each coincidence channel, the probability is insignificant. Several reported applications of the coincidence counting technique for the determination of fluorine via the ^{19}O nuclide have demonstrated the accuracy and the sensitivity of this approach. Nelson and Bussell[49] describe a successful method for determining fluorine in the presence of large quantities of oxygen using the coincidence approach. Samples and standards were activated for 1 min, and after a 20-sec delay, they were counted for 100 sec. This work demonstrated the applicability of the technique over a range of fluorine levels from 50 to 400 mg in the presence of 450 mg of oxygen. The method was applied for the determination of fluorine in boron samples. Perry[50] utilized Nelson and Bussell's approach for the determination of trace amounts of fluorine in zinc electrorefining solutions. These authors compared the detection limit for fluorine by direct photopeak analysis of the 0.2-MeV peak and by using the coincidence counting of the 0.2- and 1.37-MeV gamma rays. In the first case, the detection limit was 325 ppm, whereas the coincidence technique lowered this limit to 275 ppm. In the latter case, the sensitivity limit was still too high for successful analysis of the normal levels of fluorine found in zinc electrorefining solutions. The work, however, demonstrates the applicability of this procedure for eliminating the interference from oxygen. Other authors have found it possible to minimize the interference from oxygen by suitably adjusting decay times before counting. Nargolwalla and Jervis[51, 52] describe the continuous analysis of trace fluorine in water streams by measuring the gamma emission at 0.20 MeV from ^{19}O. A delay of 135 sec in the flow stream was used to minimize the contribution from ^{16}N produced from the oxygen. It was found, however, that the trace elements found in drinking water contribute approximately 20% of the counting rate equivalent to that registered with fluorine activated at 1 ppm. Assuming that this background was constant, suitable corrections could be made. These authors found that 1 ppm of fluorine could be determined with the precision of 10%, utilizing a flux of 10^9 neutrons/(cm^2 sec). A third possible reaction which has been used for the determination of fluorine is that resulting from the (n, 2n) reaction which produces ^{18}F. This reaction has similar sensitivity to the other two. However, it produces a positron emitter of 110-min half-life. With this long half-life it is possible to let other shorter lived positron emitters decay, if they happen to be present, prior to the counting of the fluorine. Broadhead et al.[12] used this reaction to measure fluorine in electrolytes consisting of lithium, barium, and rare earth fluorides. Because of the long half-life of the nuclide produced, irradiation times need to be somewhat longer if this reaction is to be utilized. Blackburn[53] described the determination of fluorine in organic compounds using this reaction. He used irradiation times of 30 min at a flux of 10^7 neutrons/(cm^2 sec) for compounds containing more than 10 mg of fluorine. Blackburn reported relative errors of less than 1.4% for a variety of organic fluorine compounds which ranged in fluorine content from 17 to 76%. These analyses were successful on samples smaller than 200 mg. England et al.[54] demonstrated the determination of fluorine in a wide variety of organic and organometallic compounds utilizing the (n, 2n) reaction. The precision of their determinations was better than 0.3% for all the materials examined. Samples were irradiated for 20 min in a flux of 10^8 neutrons/(cm^2 sec) and allowed to decay for 2 hr to permit short half-life positron emitters and other activated species to decay. Possible positron emitters which could interfere with the measurement of fluorine using this reaction are listed in Tables 8.77 and 8.78.

Other applications of the 14-MeV technique to fluorine determination include measurements in plants[55] and inorganic silica-alumina gels.[56]

3-MeV AND THERMAL NEUTRONS

There are no useful reactions of fluorine with generator-produced neutrons of 3-MeV or thermal energies. Although the threshold energy for the (n, α) reaction of fluorine is low, no detectable amount of ^{16}N is produced by activating fluorine with 3-MeV neutrons. Californium-252 has recently been shown to provide neutrons in the correct energy region (7–10 MeV) to activate fluorine but not oxygen to any significant extent.[57]

GADOLINIUM

The seven naturally occurring isotopes of gadolinium have masses of 152, 154 through 158, and 160. None of these isotopes has ideal characteristics

III. ACTIVATION ANALYSIS CONDITIONS FOR THE ELEMENTS

for activation analysis using generator-produced neutrons. The primary reactions produced even with 14-MeV neutrons are (n, γ) reactions which occur with much higher efficiency at lower neutron energies. It is difficult to produce these with high flux using a neutron generator, and the methods utilizing a neutron generator generally have poor sensitivity.

14-MeV NEUTRONS

The pertinent nuclear data for the activation of gadolinium with 14-MeV neutrons are given in Table 7.24. The activation, decay, and

TABLE 7.24. Nuclear Data for Reactions with 14-MeV Neutrons—Gadolinium

Nuclear Reaction	Target Isotope Abundance (%)	Cross Section (mb)	Half-life (min)	Gamma Energy (MeV)	Gamma Yield (%)	Detector Efficiency (%)	Peak/Total Ratio
1 = ^{160}Gd(n, α)^{157}Sm	21.90	2	0.50	0.57	100	39	0.59
2 = ^{160}Gd(n, γ)^{161}Gd	21.90	19	3.70	0.36	66	44	0.77
3 = ^{160}Gd(n, γ)^{161}Gd	21.90	19	3.70	0.32	25	45	0.81
4 = ^{160}Gd(n, γ)^{161}Gd	21.90	19	3.70	0.10	11	50	1.00
5 = ^{160}Gd(n, 2n)^{159}Gd	21.90	1470	1080	0.36	9	44	0.77

integral counting curves for these reactions are shown in Figs. 7.70 to 7.72. The most sensitive reaction of gadolinium with 14-MeV neutrons is neutron capture with ^{160}Gd. The photopeak at 0.36 MeV provides a detection limit according to Cuypers and Cuypers[1] of approximately 14 mg of the element using a flux of 10^8 neutrons/(cm^2 sec). There appears to be no detectable interference with this photopeak from other elements including rare earths. There have been no reported applications of fast neutron activation for the determination of gadolinium.

3-MeV AND THERMAL NEUTRONS

Broadhead and Shanks[36] have studied the reaction of 2.8-MeV neutrons with gadolinium and reported a cross-section of 17 millibarns for the (n, γ) reaction. Nargolwalla et al.[23] reported a net photopeak count of 2.4×10^3 for a 10-min irradiation and 10-min count of 1 g of gadolinium at a 3-MeV neutron flux of 10^6 neutrons/(cm^2 sec). Activation with thermal neutrons from a neutron generator produces somewhat better sensitivity than with 14-MeV neutrons. Dibbs[24] reports a sensitivity of 1.7 mg using ^{161}Gd (100 counts above background) using a flux of 10^8 neutrons/(cm^2 sec). It appears that the neutron activation method for gadolinium is not sufficiently

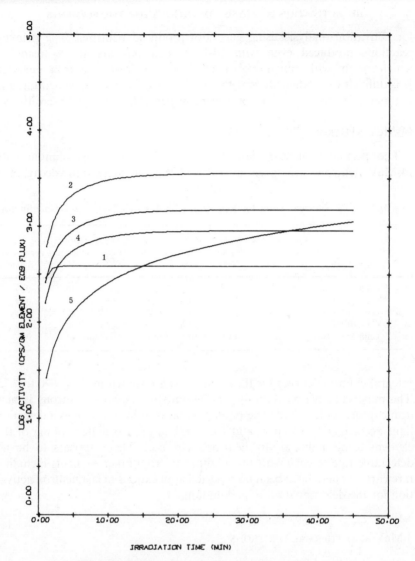

Fig. 7.70. Gadolinium—activation curves (see Table 7.24).

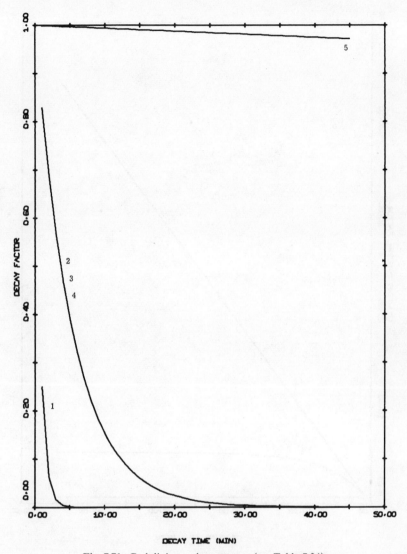

Fig. 7.71. Gadolinium—decay curves (see Table 7.24).

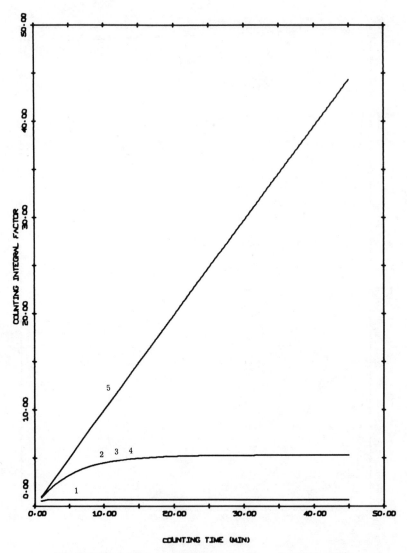

Fig. 7.72. Gadolinium—integral counting curves (see Table 7.24).

III. ACTIVATION ANALYSIS CONDITIONS FOR THE ELEMENTS 385

sensitive to provide a useful analytical method for trace analysis but it may, under certain circumstances, be sufficiently sensitive for macroconstituent analysis of gadolinium.

GALLIUM

Gallium exists in nature in two isotopic forms of mass 69 and 71. The mass 69 isotopes undergoes an analytically useful reaction. Recognizing that gallium is difficult to determine using chemical procedures, the neutron activation analysis method may offer an attractive alternative.

14-MeV NEUTRONS

The pertinent nuclear reactions for gallium with 14-MeV neutrons are tabulated in Table 7.25. The activation, decay, and counting integral curves

TABLE 7.25. Nuclear Data for Reactions with 14-MeV Neutrons—Gallium

Nuclear Reaction	Target Isotope Abundance (%)	Cross Section (mb)	Half-life (min)	Gamma Energy (MeV)	Gamma Yield (%)	Detector Efficiency (%)	Peak/Total Ratio
1 = ^{69}Ga(n, α)^{66}Cu	60.50	105	5.10	1.04	9	32	0.41
2 = ^{69}Ga(n, 2n)^{68}Ga	60.50	800	68.30	0.51	176	40	0.63
3 = ^{71}Ga(n, 2n)^{70}Ga	39.50	700	21.10	1.04	1	32	0.41
4 = ^{71}Ga(n, p)^{71}Zn	39.50	10	2.40	0.51	13	40	0.63

for 14-MeV neutrons with gallium are shown in Figs. 7.73 through 7.75. It is apparent that the most sensitive reaction is the (n, 2n) reaction with ^{69}Ga to produce ^{68}Ga ($T_{1/2}$ = 68.3 min). This is a positron emitter which can be well differentiated from most other positron emitters by decay curve analysis (cf. Tables 7.76–7.78). The primary interferences for gallium are chromium and zinc, positron emitters with half-lives of 42 and 38 min, respectively. The only reported application of the 14-MeV technique for the determination of gallium was by Crambes.[25] No investigation of the sensitivity of this determination was made by Crambes; Cuypers and Cuypers[1] report a detection limit of 0.18 mg using annihilation irradiation at 0.51 MeV. Using a similar flux [5×10^8 neutrons/(cm^2 sec)], Dibbs[24] reported a sensitivity of 8 μg (100 counts above background).

3-MeV AND THERMAL NEUTRONS

Nargolwalla et al.[23] reported an experimental sensitivity of 7×10^2 counts/g for a 10^6-neutron/(cm^2 sec) irradiation of 1 g of gallium for

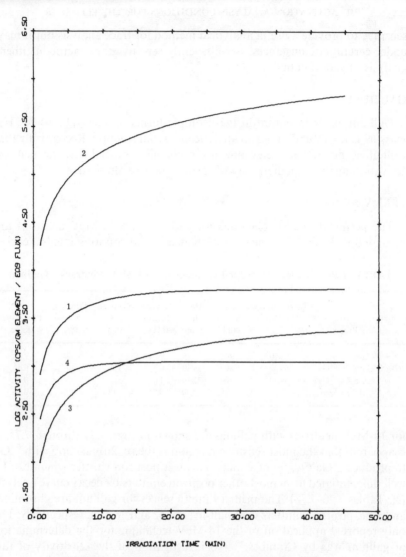

Fig. 7.73. Gallium—activation curves (see Table 7.25).

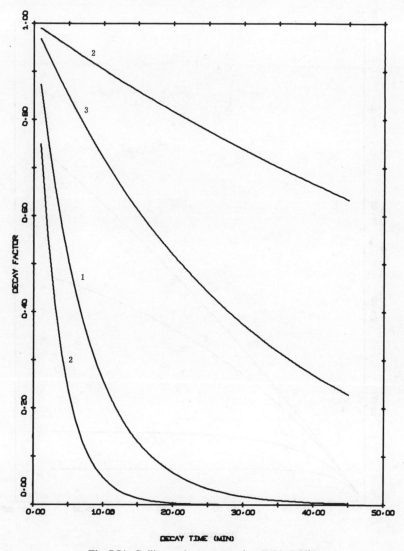

Fig. 7.74. Gallium—decay curves (see Table 7.25).

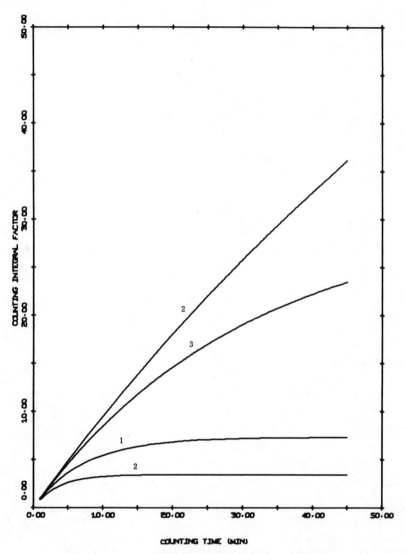

Fig. 7.75. Gallium—integral counting curves (see Table 7.25).

20 min to produce the (n, γ) product, ^{72}Ga. This sensitivity does not appear to be sufficient to make the method analytically useful.

GERMANIUM

Germanium exists in nature in five isotopic forms of mass 70, 72, 73, 74, and 76. Germanium is not easily determined using wet chemical methods of analysis because of its rather limited solution chemistry. Thus, instrumental analytical methods such as x-ray fluorescence are preferred for its determination. In this realm, the neutron activation analysis method would seem to offer an alternate approach. Unfortunately, the neutron activation methods are not highly sensitive. In addition, since the most useful nuclear reactions with germanium produced nuclides with gamma emissions at fairly low energy, the usual problem of a high background correction may occur for an analysis of germanium in a complex matrix.

14-MeV NEUTRONS

The nuclear parameters important in the activation analysis of germanium are listed in Table 7.26, and the activation, decay, and integral counting curves are given in Figs. 7.76 to 7.78.

TABLE 7.26. Nuclear Data for Reactions with 14-MeV Neutrons—Germanium

Nuclear Reaction	Target Isotope Abundance (%)	Cross Section (mb)	Half-life (min)	Gamma Energy (MeV)	Gamma Yield (%)	Detector Efficiency (%)	Peak/Total Ratio
1 = ^{74}Ge(n, p)^{74}Ga	36.74	6	7.90	0.60	100	38	0.57
2 = ^{76}Ge(n, 2n)^{75}Ge	7.67	1820	82.00	0.27	11	48	0.87
3 = ^{76}Ge(n, 2n)^{75m}Ge	7.67	726	0.82	0.14	34	50	0.97

The reaction of highest sensitivity for germanium is the (n, 2n) reaction on ^{75}Ge, yielding 49-sec ^{75m}Ge nuclide with an emission at 0.14 MeV (internal conversion). Cuypers and Cuypers[1] have indicated a detection limit of 0.38 mg of germanium using this γ-emission. The daughter product of ^{75m}Ge, ^{75}Ge, has a γ-emission of 0.27-MeV energy which may be used for the measurement of germanium, but the detection limit is an order of magnitude less sensitive. There is also a third nuclide, ^{74}Ga, produced by an (n, p) reaction which gives a γ-emission at 0.60 MeV. This reaction is even less sensitive than the (n, 2n) reaction with a listed detection limit of 5.8 mg. Mathur and Oldham[58] list the (n, α) reaction on ^{74}Ge yielding the positron emitter ^{71}Zn as the reaction of highest sensitivity for the determination of

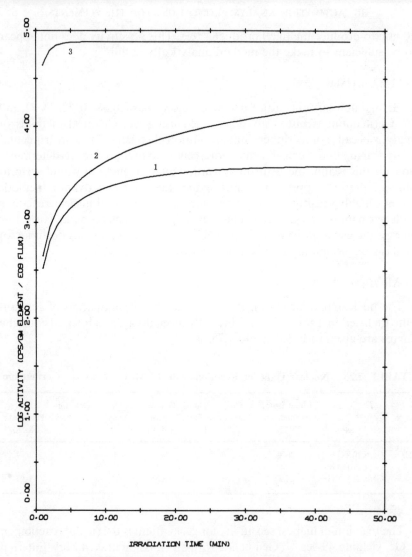

Fig. 7.76. Germanium—activation curves (see Table 7.26).

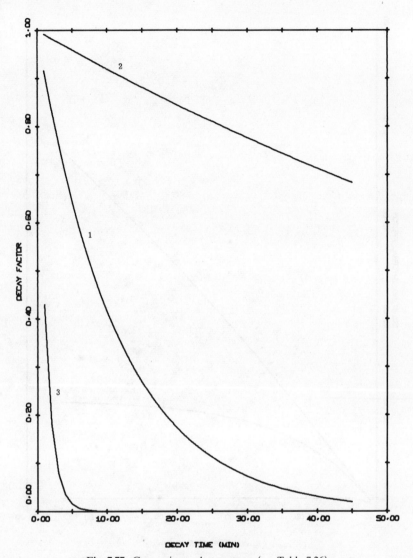

Fig. 7.77. Germanium—decay curves (see Table 7.26).

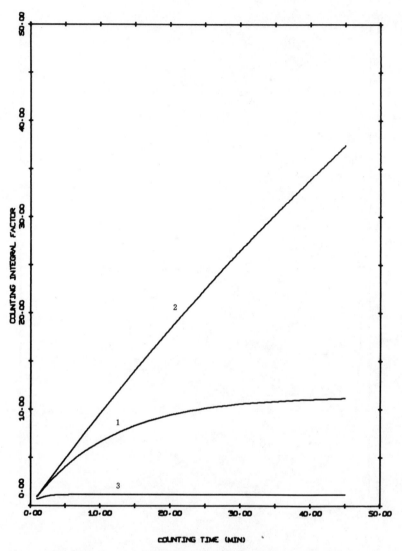

Fig. 7.78. Germanium—integral counting curves (see Table 7.26).

III. ACTIVATION ANALYSIS CONDITIONS FOR THE ELEMENTS

germanium. This is not borne out by actual experimental spectra published by Cuypers and Cuypers, nor is it indicated from calculations of induced activity based on published activation parameters. To date there have been no published applications of the neutron activation method for germanium; other analytical techniques appear to satisfy the analytical needs in this area for the present.

3-MeV AND THERMAL ACTIVATION

Germanium undergoes (n, γ) reactions with 3-MeV neutrons, and Nargolwalla et al.[23] report that the (n, γ) reactions on both ^{74}Ge and ^{76}Ge occur readily. Using a 10-min irradiation, a 4-sec delay, and a 10-min counting time, Nargolwalla et al. report a sensitivity of 9×10^2 counts/g of germanium for a neutron flux of 10^6 neutrons/(cm^2 sec). On an equal flux basis, sensitivity of the 3-MeV technique compares favorably with that of the 14-MeV procedure. Activation with the thermal neutron flux available from a neutron generator should offer no advantage in sensitivity for the measurement of germanium.

GOLD

Gold exists in nature in only one isotopic form, that of mass 197. Because it is a precious metal, analytical procedures for gold are very important in the mining and metallurgy of this material. A number of chemical and instrumental procedures exist for its determination. Most of the chemical methods are laborious and lengthy due to the difficulties in dissolving gold and its compounds and limited solution chemistry. The best instrumental methods for gold, other than neutron activation analysis, are those based on x-ray fluorescence. Here the usual limitations due to sample self-absorption problems are present. Neutron activation methods offer an attractive alternative for the determination of gold, depending on the matrix.

14-MeV NEUTRONS

Gold undergoes several analytically useful reactions with fast neutrons produced with a neutron generator. The important nuclear reactions for gold are tabulated in Table 7.27.

The most sensitive reaction for the activation of gold with 14-MeV neutrons is that based on the production of ^{197m}Au from an inelastic scattering reaction. The photopeak at 0.28 MeV is capable of providing a detection limit of about 0.60 mg of gold. Unfortunately, this reaction suffers severe interference from the presence of oxygen in the sample due to the Compton scattering from 7.2-sec ^{16}N.

TABLE 7.27. Nuclear Data for Reactions with 14-MeV Neutrons—Gold

Nuclear Reaction	Target Isotope Abundance (%)	Cross Section (mb)	Half-life (min)	Gamma Energy (MeV)	Gamma Yield (%)	Detector Efficiency (%)	Peak/Total Ratio
1 = ^{197}Au(n, n'γ)^{197m}Au	100.00	200	0.12	0.13	8	50	0.98
2 = ^{197}Au(n, n'γ)^{197m}Au	100.00	200	0.12	0.28	75	48	0.86
3 = ^{197}Au(n, 2n)^{196m}Au	100.00	142	582	0.19	32	50	0.93
4 = ^{197}Au(n, 2n)^{196}Au	100.00	1722	8899	0.33	25	46	0.80
5 = ^{197}Au(n, 2n)^{196}Au	100.00	1722	8899	0.36	94	45	0.77
6 = ^{197}Au(n, 2n)^{196}Au	100.00	1722	8899	0.43	6	42	0.71

The less sensitive reactions of gold involve production of ^{196}Au and the utilization of the photopeaks at 0.43 MeV. These reactions are only useful for the determination of large amounts of gold. Cuypers and Cuypers[1] report a detection limit for the 0.33- and 0.35-MeV γ-ray of 7.0 mg, whereas utilizing the photopeak at 0.3 MeV is reported to give a detection limit of 19 mg. Thus the neutron activation method for gold using 14-MeV neutrons cannot be considered a trace analysis procedure; however, it has useful sensitivity for the measurement of semimicroquantities of gold. The activation curves for the production of these nuclides from the irradiation of gold with 14-MeV neutrons are shown in Fig. 7.79. Decay and counting integral curves are given in Figs. 7.80 and 7.81.

3-MeV AND THERMAL NEUTRONS

The neutron inelastic scattering reaction on gold to produce ^{197m}Au is observed with 3-MeV neutrons. Using a flux of 10^6 neutrons/(cm^2 sec), Nargolwalla et al.[23] found a sensitivity of 2.4×10^4 total counts for a 1-g sample of gold using a 1-min irradiation time and a 1-min counting time. Under these conditions, oxygen is not activated, thus interference with the 0.12-MeV γ-emission would be expected to be low. In addition, they noted that significant amounts of ^{198}Au are produced with 3-MeV irradiation of gold using the photopeak at 0.41 MeV, and experimental sensitivity of 1.8×10^3 counts/g of element was obtained with a flux of 10^6 neutrons/(cm^2 sec). The utilization of 3-MeV neutrons, while not extremely sensitive, might be used for macroconstituent analysis of gold-containing materials. Dibbs[24] has reported a sensitivity of 8 μg (100 counts above background) using a thermal flux of 10^8 neutrons/(cm^2 sec). He utilized the 0.41-MeV photopeak from ^{198}Au for this determination.

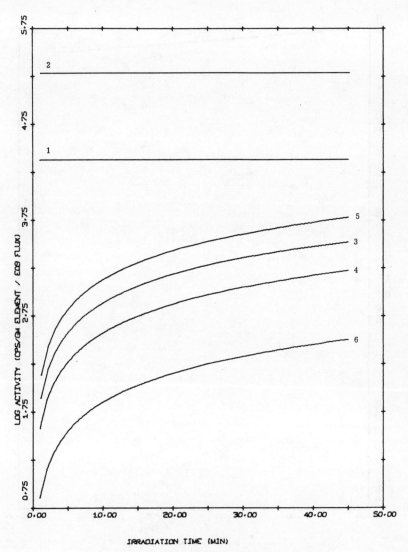

Fig. 7.79. Gold—activation curves (see Table 7.27).

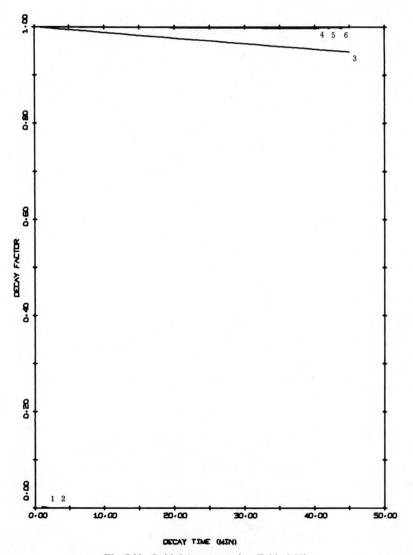

Fig. 7.80. Gold-decay curves (see Table 7.27).

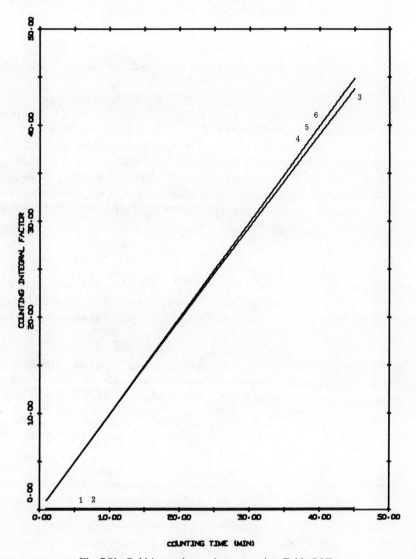

Fig. 7.81. Gold-integral counting curves (see Table 7.27).

HAFNIUM

Hafnium exists in nature in the six isotopic forms of mass 174 and 176 through 180.

The analytical chemistry of hafnium is not highly developed because of its rare occurrence in nature and so it is difficult to compare the neutron activation technique with other existing procedures for hafnium. A classical differentiation of hafnium from zirconium in ores where these two elements occur together can apparently be carried out using neutron activation techniques because of the difference in γ-ray spectra of isotopes produced upon irradiation. Thus the neutron activation method might recommend itself as a solution to this analytical problem which is not easily handled by other techniques.

14-MeV NEUTRONS

The activation parameters for the important reactions of hafnium with 14-MeV neutrons are listed in Table 7.28, and Figs. 7.82 through 7.84 give the working curves for the determination of hafnium using these reactions.

TABLE 7.28. Nuclear Data for Reactions with 14-MeV Neutrons—Hafnium

Nuclear Reaction	Target Isotope Abundance (%)	Cross Section (mb)	Half-life (min)	Gamma Energy (MeV)	Gamma Yield (%)	Detector Efficiency (%)	Peak/Total Ratio
$1 = {}^{179}\text{Hf}(n, \alpha){}^{176m}\text{Yb}$	13.75	2.2	0.19	0.29	100	50	0.85
$2 = {}^{178}\text{Hf}(n, p){}^{178}\text{Lu}$	27.10	5.5	20.00	0.43	100	42	0.71
$3 = {}^{180}\text{Hf}(n, 2n){}^{179m}\text{Hf}$	35.22	510	0.31	0.22	94	50	0.91

With 14-MeV neutrons the reactions of major importance occur with the isotopes of masses 179 and 180. The most sensitive reaction listed by Cuypers and Cuypers[1] is the production of ^{179m}Hf through either an inelastic scattering reaction on ^{179}Hf or an (n, 2n) reaction on ^{180}Hf. This isotope has a reported detection limit of 1 mg. Because of the short half-life of the ^{179m}Hf, care must be taken to minimize the contribution of any activity due to activation of oxygen present in the sample. Cuypers suggests a decay time of 19 sec as optimum. The second most sensitive peak is the emission of 0.29 MeV from ^{176m}Yb, and has a detection limit of 6.9 mg.[1] There have been no reports of applications of the 14-MeV technique to the determination of hafnium.

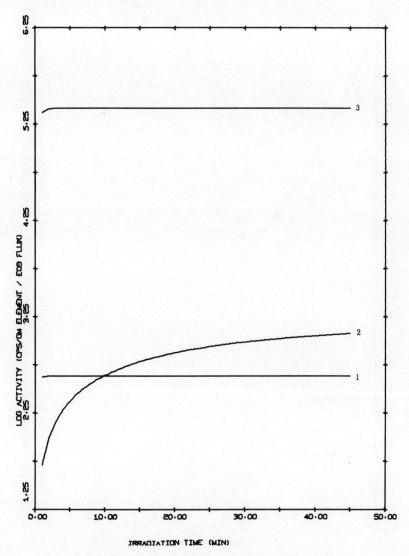

Fig. 7.82. Hafnium—activation curves (see Table 7.28).

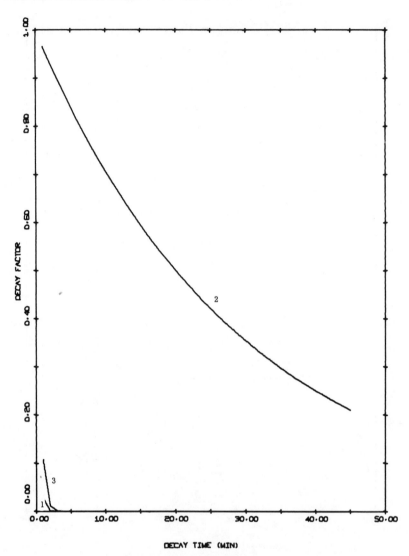

Fig. 7.83. Hafnium—decay curves (see Table 7.28).

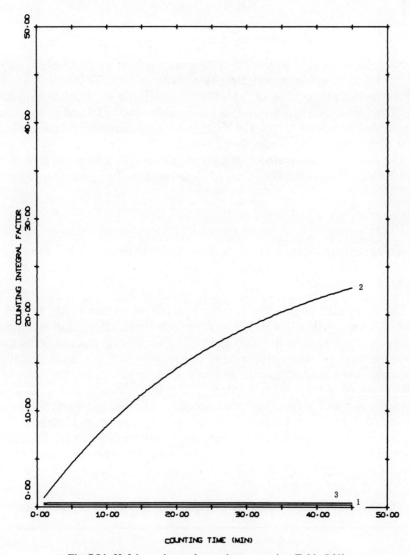

Fig. 7.84. Hafnium—integral counting curves (see Table 7.28).

3-MeV AND THERMAL NEUTRONS

Nargolwalla et al.[23] have utilized the emission from the 0.217-MeV γ-ray of ^{179m}Hf to obtain an experimental sensitivity of 7.0×10^2 counts/g for a 10^6-neutron/(cm^2 sec) flux of 3-MeV neutrons. In these reactions Nargolwalla and his co-workers utilized the (n, γ) reaction on ^{178}Hf and an inelastic scattering reaction of ^{179}Hf. The irradiation conditions were 3-min irradiation with a 3-sec decay and a 3-min count.

The most sensitive method for hafnium is activation with thermal neutrons. Dibbs[24] reports a sensitivity of 25 μg using the (n, γ) reaction to yield ^{179m}Hf.

INDIUM

Indium exists in nature in two isotopic forms of mass 113 and 115.

14-MeV NEUTRONS

The activation parameters for the analysis of indium with 14-MeV neutrons are listed in Table 7.29, and the activation, decay, and counting integral curves are given in Figs. 7.85 to 7.87. These curves represent only the reactions with 14-MeV neutrons and not those which are experimentally observed upon irradiating indium with a neutron generator. The neutron activation of indium with 14-MeV neutrons presents a rather unique situation in that the dominant reaction observed is the (n, γ) reaction producing ^{116m}In. This is found because of the very high cross section for the capture reaction, its low energy dependence, and the fact that the neutrons emitted from a tritium target are degraded in energy so that a continuum of lower energy neutrons is available.

TABLE 7.29. Nuclear Data for Reactions with 14-MeV Neutrons—Indium

Nuclear Reaction	Target Isotope Abundance (%)	Cross Section (mb)	Half-life (min)	Gamma Energy (MeV)	Gamma Yield (%)	Detector Efficiency (%)	Peak/Total Ratio
1 = ^{113}In(n, 2n)^{112}In	4.23	1500	14.00	0.51	44	40	0.63
2 = ^{115}In(n, 2n)^{114m}In	95.77	1540	20.70	0.19	17	50	0.93

The γ-ray spectrum obtained from the irradiation of indium with 14-MeV neutrons is that of the ^{116m}In isotope. Cuypers and Cuypers[1] note a sensitivity of 0.3 mg of indium if the 0.406-MeV peak of ^{116m}In is used.

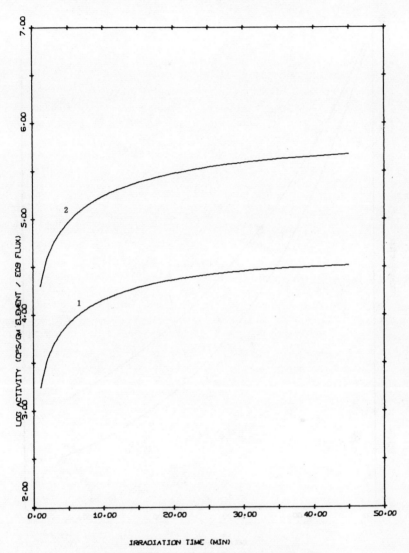

Fig. 7.85. Indium—activation curves (see Table 7.29).

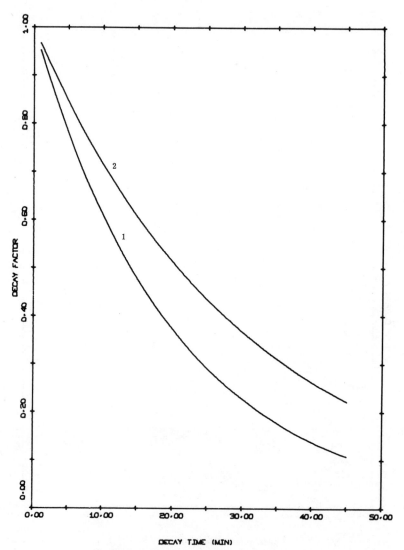

Fig. 7.86. Indium—decay curves (see Table 7.29).

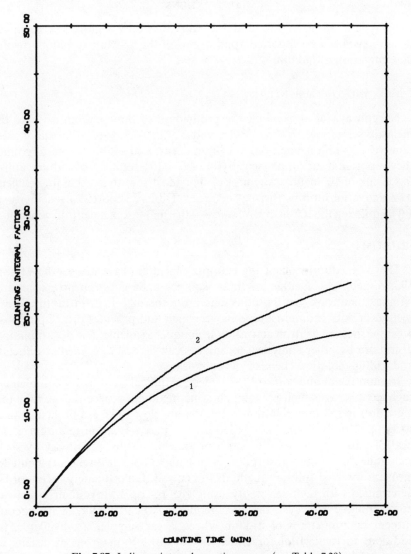

Fig. 7.87. Indium—integral counting curves (see Table 7.29).

Other peaks, those at 1.08 and 1.27-MeV, are only slightly less sensitive. There have been no reported applications of the 14-MeV technique to the determination of indium.

3-MeV AND THERMAL NEUTRONS

Nargolwalla et al.[23] noted the production of three indium isotopes by inelastic scattering, ^{113m}In, ^{115m}In, and ^{116m}In. Using experimental conditions of a 20-min irradiation, a 0.5-min decay, and a 20-min counting time, they reported a maximum sensitivity of 6×10^4 counts/g of indium utilizing a flux of 10^6 neutrons/(cm^2 sec). The most sensitive technique for neutron activation utilizes thermal neutrons. Dibbs[24] reported a sensitivity of 1.3 µg utilizing the 10^8-neutron/(cm^2 sec) thermal flux of a neutron generator.

IRIDIUM

The two naturally occurring isotopes of iridium have masses of 191 and 193, respectively. Neither of these isotopes undergoes sensitive reactions with fast neutrons. On the other hand, the thermal neutron capture cross section of both iridium isotopes is very high and provides one of the most sensitive thermal neutron activation techniques available. The determination of iridium by other analytical techniques can be best described as difficult, and iridium solution chemistry is complex.

Iridium metal and some of its oxides are difficult to dissolve completely; such techniques as high-pressure chlorine treatment is necessary to take the metal in the solution. Solution equilibria are slow to be established so that spectrophotometric methods based on complex formation require a great deal of patience and care to get reproducible results. Sensitivity by such techniques as atomic absorption is not high. Thus iridium is difficult to measure by most solution methods of analysis. Unfortunately it cannot be determined with high sensitivity using the 14- or 3-MeV neutron activation technique. However, these techniques may be useful for the macroconstituent determination of iridium where larger samples are available. In addition, thermalization of either 14- or 3-MeV neutrons is useful in determining even small amounts of iridium by the (n, γ) reaction.

14-MeV NEUTRONS

The activation parameters for 14-MeV neutron activation of iridium are given in Table 7.30, and the working curves are given in Figs. 7.88 to 7.90.

The most useful and sensitive reaction for the determination of iridium with 14-MeV neutrons is the production of ^{190m}Ir which has iridium x-ray emission at 0.063-MeV. Cuypers and Cuypers[1] reported a detection limit of

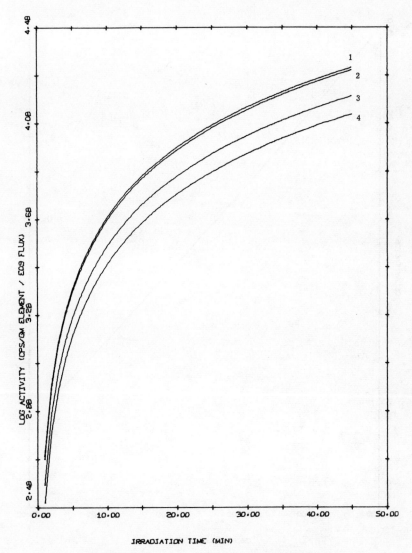

Fig. 7.88. Iridium—activation curves (see Table 7.30).

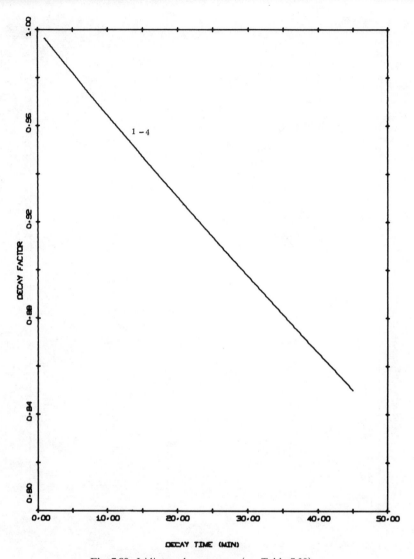

Fig. 7.89. Iridium—decay curves (see Table 7.30).

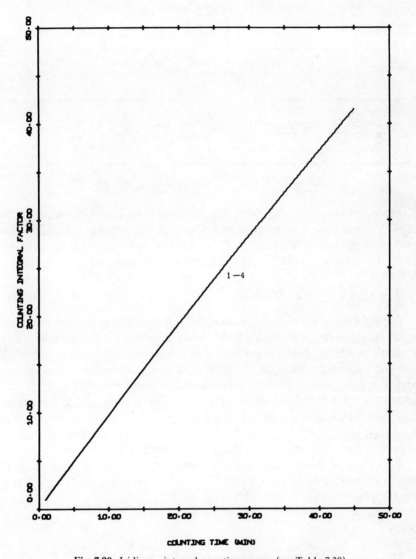

Fig. 7.90. Iridium—integral counting curves (see Table 7.30).

TABLE 7.30. Nuclear Data for Reactions with 14-MeV Neutrons—Iridium

Nuclear Reaction	Target Isotope Abundance (%)	Cross Section (mb)	Half-life (min)	Gamma Energy (MeV)	Gamma Yield (%)	Detector Efficiency (%)	Peak/Total Ratio
1 = ^{191}Ir(n, 2n)^{190m}Ir	38.50	367	192.00	0.19	66	50	0.93
2 = ^{191}Ir(n, 2n)^{190m}Ir	38.50	367	192.00	0.36	88	44	0.77
3 = ^{191}Ir(n, 2n)^{190m}Ir	38.50	367	192.00	0.50	92	40	0.63
4 = ^{191}Ir(n, 2n)^{190m}Ir	38.50	367	192.00	0.62	93	38	0.56

10 mg of iridium utilizing this peak. Shorter lived emissions from ^{191m}Ir provide somewhat lower sensitivity although the gamma energies of the emissions are at higher energies. There have been no reported applications of the 14-MeV technique to the determination of iridium.

3-MeV AND THERMAL NEUTRONS

The determination of iridium using 3-MeV neutrons is of approximately equivalent sensitivity to 14-MeV activation even with a 40-fold lower flux. Inelastic scattering reaction on the ^{191}Ir produces γ-emissions at 1.29 MeV and the iridium x-rays from ^{191m}Ir. With a 1-min activation, a 3-sec decay, and a 1-min counting time, Nargolwalla et al.[23] reported a maximum sensitivity of 5×10^3 counts/g of iridium for a flux of 10^6 neutrons/(cm^2 sec). Because the capture cross section of ^{191}Ir is 750 barns, it is suggested that thermalization of either the 3- or the 14-MeV flux would provide a very sensitive way of determining lower levels of iridium.

Activation of iridium with thermal neutrons provides the most sensitive means of measuring iridium. Dibbs[24] reports a sensitivity of 25 μg utilizing the 0.33-MeV emission from ^{194}Ir.

IRON

Iron exists in nature in the four isotopic forms of mass 54, 56, 57, and 58. While there has been a significant number of applications of neutron activation to the measurement of iron reported in the literature, one recognizes that this is but a minute fraction of the methods and applications which exist for the determination of iron. It is perhaps the most widely studied element from an analytical standpoint, and there are in existence today many excellent methods of analysis for iron which can compete very favorably with the neutron activation method. It would be difficult to make a strong case for the use of the neutron activation technique for the determination of iron in view of these other excellent methods. The prime reasons for its use would be in connection with specific materials in which other elements are to be measured by neutron activation or where unusually difficult sample preparation can be avoided through the use of this nondestructive method.

14-MeV NEUTRONS

The most abundant isotope of iron is mass 56, and it is the most useful for activation analysis with 14-MeV neutrons. It undergoes an (n, p) reaction to yield ^{56}Mn which gives a unique series of gamma rays of 2.56-hr half-life. The activation parameters for the isotopes of iron are given in Table 7.31. From these data and the activation working curves shown in Figs. 7.91 to 7.93, it can be seen that iron can be determined with relatively good sensitivity. Cuypers and Cuypers[1] report a detection limit of 1.3 mg for the 0.84-MeV peak. By doubling the irradiation and counting times, Dibbs[24] found a sensitivity of 0.14 mg.

TABLE 7.31. Nuclear Data for Reactions with 14-MeV Neutrons—Iron

Nuclear Reaction	Target Isotope Abundance (%)	Cross Section (mb)	Half-life (min)	Gamma Energy (MeV)	Gamma Yield (%)	Detector Efficiency (%)	Peak/Total Ratio
1 = ^{54}Fe(n, 2n)^{53}Fe	5.84	15	8.50	0.51	196	40	0.63
2 = ^{56}Fe(n, p)^{56}Mn	91.68	103	154.56	0.85	99	34	0.44
3 = ^{56}Fe(n, p)^{56}Mn	91.68	103	154.56	1.81	29	27	0.28
4 = ^{56}Fe(n, p)^{56}Mn	91.68	103	154.56	2.11	15	26	0.26
5 = ^{57}Fe(n, p)^{57}Mn	2.17	60	1.70	0.12	100	50	0.99
6 = ^{57}Fe(n, p)^{57}Mn	2.17	60	1.70	0.13	100	50	0.98

There are a number of reported applications of activation analysis to the determination of iron. Gibbons et al.[45] reported on the determination of iron in oil additives in the presence of sulfur and chlorine. These authors measured the photopeaks from these elements simultaneously; the Covell method of peak integration was applied. The 0.85-MeV photopeak was used for the measurement of iron, with a reported saturation activity of 10^6 dis/(sec g). D'Agostino and Kuehne[2] applied this technique to the measurement of metallic wear products in hydraulic systems. The samples in these studies consisted of filter patches through which hydraulic fluids had been passed to remove the particulate wear products. This excellent paper demonstrates the procedure for handling multicomponent systems in the determination of activation conditions and subsequent analysis of the gamma-ray

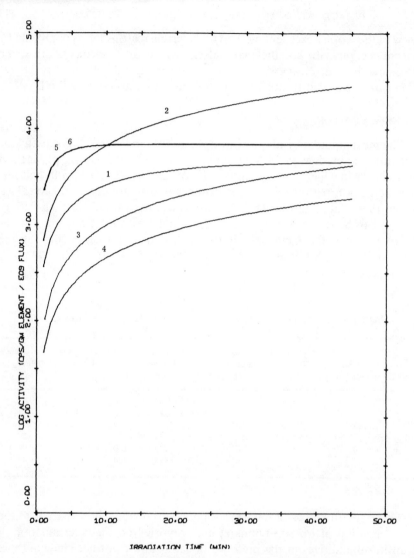

Fig. 7.91. Iron—activation curves (see Table 7.31).

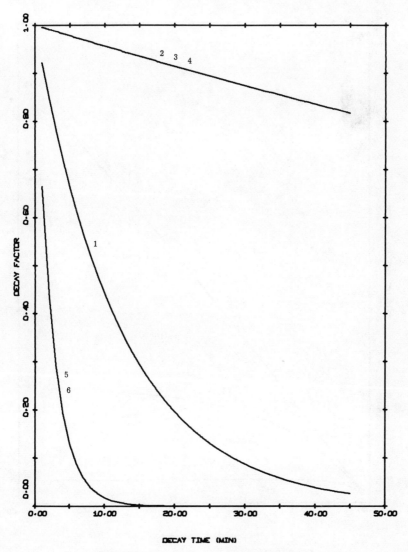

Fig. 7.92. Iron—decay curves (see Table 7.31).

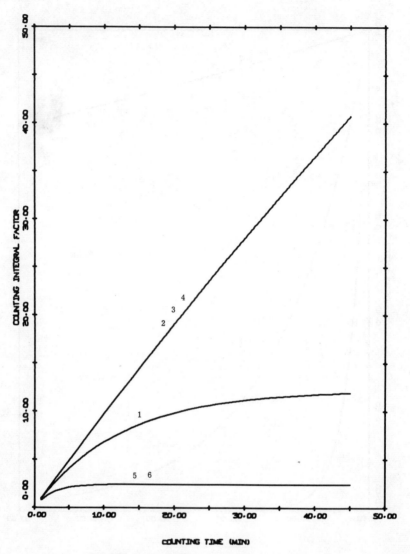

Fig. 7.93. Iron—integral counting curves (see Table 7.31).

spectra. In this work, the interference from aluminum, which also gives a 0.85-MeV gamma ray, was handled by solving a simultaneous equation set. The authors applied the method to the determination of iron over the range of 0.3 to 900 μg of iron per filter patch. The estimated minimum detectable amount was 7 μg. The attractiveness of this technique for rapid, automatic analysis has led to its application for the measurement of iron in minerals. Mathur and Oldham[58] reported a detection limit of 100 to 500 μg of iron in samples weighing 1.5 to 2 g with a neutron flux of 5×10^8 neutrons/(cm^2 sec). Irradiation times in these studies were 10 min. Standard deviations for the measurement of iron were rather large, ranging from 3 to 8% relative, depending on the matrix. The high background from other elements in the matrix contributed significantly to the poor precision. Kehler and Monaghan[17] reported on the use of this technique for the measurement of iron and other elements such as silicon, aluminum, magnesium, and sodium in ocean-bottom core samples. Sattarov et al.[59] measured iron in bituminous coal samples using this technique. Wood et al.[7] demonstrated the use of this method for the simultaneous determination of iron, aluminum, and magnesium. Chiba[29] measured iron oxide-barium oxide ratios in studying the magnetic characteristics of barium ferrite. In this work 100-mg samples were irradiated and identical samples were analyzed by wet chemical methods to establish the accuracy of the activation method. The precision of the measurements was found to be approximately 2%, whereas the comparison of the activation and chemical methods gave a correspondence of 95%; there is 5% deviation from the cross-calibration line.

Two elements may interfere with the determination of iron by this method; they are cobalt and manganese which, upon irradiation with fast neutrons, give the same nuclide as does iron. The extent of interference can be seen in the activation curves in Fig. 7.94. Figures 7.95 and 7.96 depict the decay and integral counting curves. The pertinent activation parameters for these interferences are given in Table 7.32. It can be seen that only samples containing extreme ratios of cobalt and manganese relative to the iron present would cause significant interference.

If the positron emitter ^{53}Fe is to be used for analysis, Tables 7.76 to 7.78 should be consulted for possible interfering matrix elements.

3-MeV AND THERMAL NEUTRONS

Neutron activation analysis with 14-MeV neutrons is far more sensitive than irradiation with neutrons of lower energy, either 3-MeV or thermalized neutrons. The sensitivity of detection with these lower energy neutrons is too low to be of analytical value. Waggoner and Knox[13] reported the determination of iron using a neutron inelastic scattering technique. The

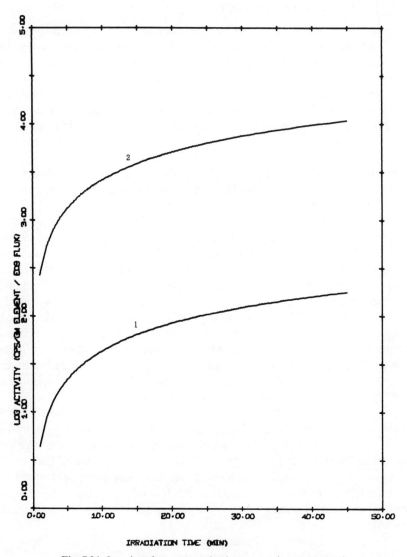

Fig. 7.94. Iron interferences—activation curves (see Table 7.32).

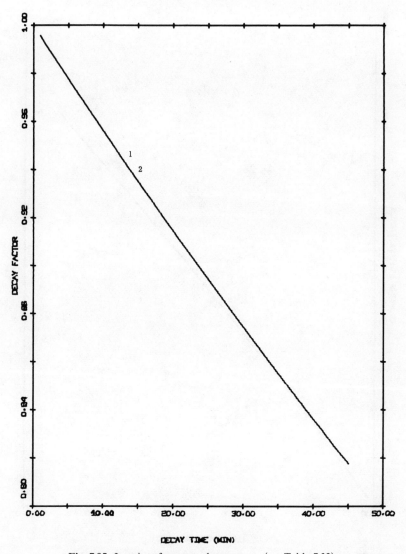

Fig. 7.95. Iron interferences—decay curves (see Table 7.32).

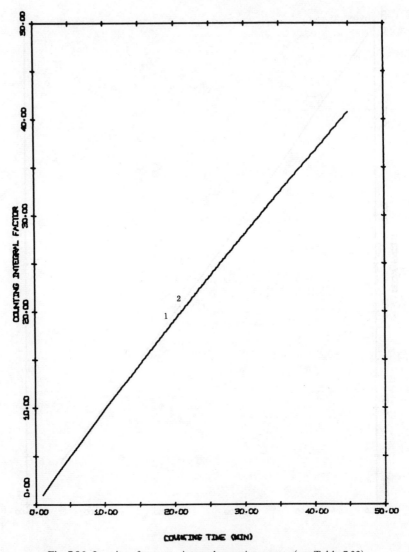

Fig. 7.96. Iron interferences—integral counting curves (see Table 7.32).

III. ACTIVATION ANALYSIS CONDITIONS FOR THE ELEMENTS 419

TABLE 7.32. Nuclear Data for Reactions with 14-MeV Neutrons—Iron Interferences

Nuclear Reaction	Target Isotope Abundance (%)	Cross Section (mb)	Half-life (min)	Gamma Energy (MeV)	Gamma Yield (%)	Detector Efficiency (%)	Peak/Total Ratio
$1 = {}^{55}\text{Mn}(n, \gamma){}^{56}\text{Mn}$	100.00	0.60	154.56	0.85	99	34	0.44
$2 = {}^{59}\text{Co}(n, \alpha){}^{56}\text{Mn}$	100.00	39.10	154.56	0.85	99	34	0.44

gamma ray measured was at 0.84 MeV. The authors imply that this technique can give higher sensitivity than the time-delayed activation method, and that the experimental difficulties can be resolved. This method is particularly well suited to remote analysis situations such as planetary surfaces using a pulsed source of neutrons.

LEAD

The four isotopes of lead which occur in nature have masses of 204, 206, 207, and 208. There are a wide variety of analytical methods that have been applied to the determination of lead. Chemical, electrochemical, as well as flame spectroscopic methods have been used to measure lead at the microgram level. X-Ray fluorescence methods have recently been widely used for nondestructively measuring lead in paint pigments. With the availability of many other analytical procedures for lead, the neutron activation analysis techniques for lead are not particularly attractive.

14-MeV NEUTRONS

The important parameters for the 14-MeV neutron activation analysis of lead are listed in Table 7.33. Figures 7.97 through 7.99 contain the activation, decay, and counting integral curves for the activation of lead with 14-MeV neutrons. The most sensitive reaction of lead with 14-MeV neutrons is the (n, 2n) reaction on ^{208}Pb yielding ^{207m}Pb. Cuypers and

TABLE 7.33. Nuclear Data for Reactions with 14-MeV Neutrons—Lead

Nuclear Reaction	Target Isotope Abundance (%)	Cross Section (mb)	Half-life (min)	Gamma Energy (MeV)	Gamma Yield (%)	Detector Efficiency (%)	Peak/Total Ratio
$1 = {}^{208}\text{Pb}(n, 2n){}^{207m}\text{Pb}$	52.30	1700	0.01	0.57	98	39	0.59
$2 = {}^{208}\text{Pb}(n, 2n){}^{207m}\text{Pb}$	52.30	1700	0.01	1.06	83	32	0.40
$3 = {}^{208}\text{Pb}(n, p){}^{208}\text{Tl}$	52.30	1	3.10	0.58	86	38	0.58
$4 = {}^{208}\text{Pb}(n, p){}^{208}\text{Tl}$	52.30	1	3.10	2.60	100	25	0.23

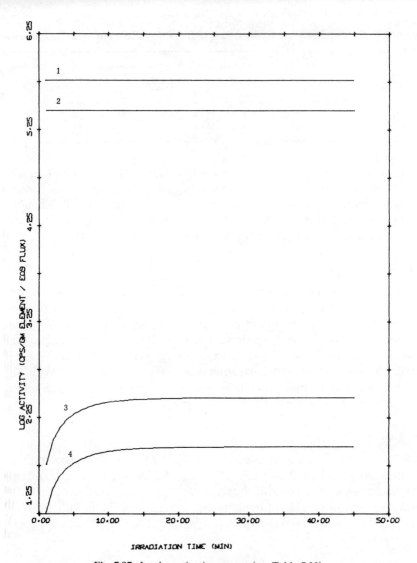

Fig. 7.97. Lead—activation curves (see Table 7.33).

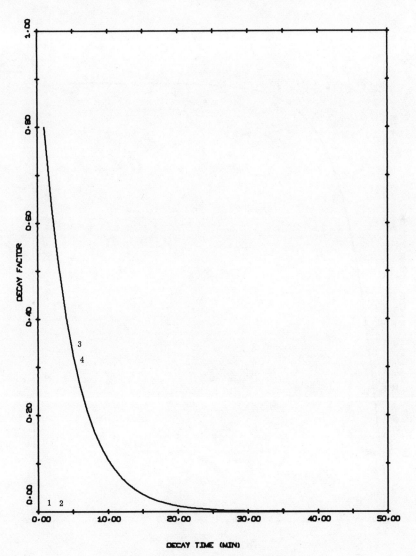

Fig. 7.98. Lead—decay curves (see Table 7.33).

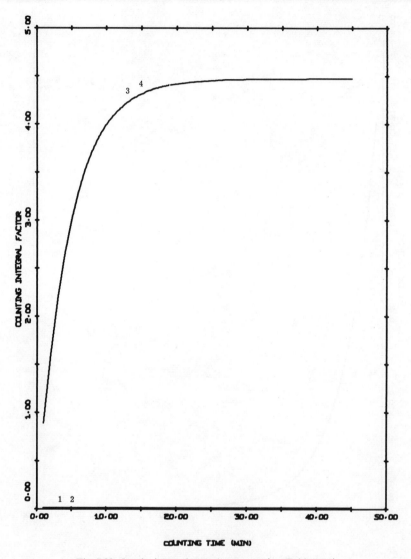

Fig. 7.99. Lead—integral counting curves (see Table 7.33).

Cuypers[1] have listed a sensitivity of approximately 5 mg of lead by this method using a cyclic irradiation technique of 2.4-sec duration, 0.8-sec decay, and a 2.4-sec counting time. Golanski[60] has utilized this same reaction in a cyclic irradiation technique utilizing a pulsed neutron generator and irradiation pulses of 5 msec.

In his technique Golanski uses a 5-msec irradiation, a decay time of 2 msec, a detection time of 40 msec, and a final decay time of 3 msec. Utilizing this cycle he has reported that in a 1-hr total experiment time, using 1000 cycles, he obtains a detection limit for lead of 3.3 g with a flux of 10^5 neutrons/(cm^2 sec). The standard deviation of these measurements was 3.5%. This detection limit could presumably be improved considerably by utilizing a generator with a higher flux output. The principal interference which Golanski found in his work was from standard aluminum in the neutron generator. In this particular case, he observed significant activity from ^{24m}Na($T_{1/2} = 19$ msec) which produces a γ-ray at 0.475 MeV. Since the detection system must be in close proximity to the sample and target area, materials of construction become exceedingly important using the pulse technique. The sensitivities of other reactions of lead with neutrons are considerably lower than those leading to the very short-lived species. For example, the long-lived ^{203}Pb, resulting from an (n, 2n) reaction on ^{204}Pb, results in a detection limit as reported by Cuypers and Cuypers[1] of 217 mg of Pb. In general, 14-MeV neutron activation analysis of lead does not provide an attractive alternate for the measurement of lead where a choice of other analytical methods is available.

3-MeV AND THERMAL NEUTRONS

Nargolwalla et al.[23] examined the suitability of determining lead by activation with 3-MeV neutrons. In general, the sensitivities reported by them were low and not particularly useful for the measurement of even milligram quantities of lead. Macroconstituent analyses for several grams of lead might be possible utilizing the inelastic scattering reaction on ^{204}Pb to produce ^{204m}Pb, which gave an activity of 110 counts/g for a flux of 10^6 neutrons/(cm^2 sec). Similar sensitivities were obtained in the production of shorter lived ^{207m}Pb by a capture reaction or by inelastic scattering reactions. Likewise, activation of lead with thermal neutrons does not yield useful levels of induced activity.

LUTETIUM

The two naturally occurring isotopes of lutetium are ^{175}Lu and ^{176}Lu. While both isotopes undergo a number of nuclear reactions, only one has sufficient sensitivity to be useful in activation analysis with fast neutrons.

Lutetium, not being a common element, has not been studied widely for its analytical reactions. Nondestructive methods such as neutron activation analysis recommend themselves for the measurement of rare elements when the sensitivity of the method is adequate.

14-MeV NEUTRONS

Pertinent activation parameters for the activation of lutetium with 14-MeV neutrons are given in Table 7.34, and the activation, decay, and integral counting curves are given in Figs. 7.100 to 7.102. The primary reaction observed in the activation of lutetium is the inelastic scattering reaction of ^{176m}Lu which yields a 0.088-MeV γ-emission. In addition, a lutetium x-ray at 0.055 MeV is also observed. Cuypers and Cuypers[1] report a detection limit of 7.5 mg for lutetium using the x-ray emission and 14.5 mg using the peak γ-emission at 0.088 MeV. There have been no reported applications of the 14-MeV technique to the determination of lutetium.

TABLE 7.34. Nuclear Data for Reactions with 14-MeV Neutrons—Lutetium

Nuclear Reaction	Target Isotope Abundance (%)	Cross Section (mb)	Half-life (min)	Gamma Energy (MeV)	Gamma Yield (%)	Detector Efficiency (%)	Peak/Total Ratio
1 = ^{175}Lu(n, α)^{174m}Lu	97.40	1600	201600	0.07	100	50	1.00
2 = ^{175}Lu(n, α)^{174m}Lu	97.40	1600	201600	0.18	100	50	0.94
3 = ^{175}Lu(n, α)^{174m}Lu	97.40	1600	201600	0.27	100	48	0.87
4 = ^{175}Lu(n, α)^{174m}Lu	97.40	1600	201600	0.99	100	32	0.42
5 = ^{175}Lu(n, γ)^{176m}Lu	97.40	2	222.0	0.09	10	50	1.00

3-MeV AND THERMAL NEUTRONS

Nargolwalla et al.[23] reported that the inelastic scattering reaction for lutetium which produced ^{176m}Lu was observed with 3-MeV neutrons and also the x-ray emission at 0.055 MeV. Using a 20-min irradiation, a 30-sec decay, and a 20-min counting time, they observed a sensitivity of 1.7×10^4 counts/g of lutetium using a flux of 10^6 neutrons/(cm² sec). A threefold decrease in sensitivity was observed with the 0.088-MeV peak. These sensitivities are slightly lower than those observed with 14-MeV neutrons when allowance is made for a 100-fold difference in available flux for the two energies.

The most sensitive reaction for the measurement of lutetium is activation with thermal neutrons to form ^{176m}Lu. Using the 0.055-MeV emission, Dibbs[24] reports a sensitivity of 15 μg [10^8-neutron/(cm² sec) flux].

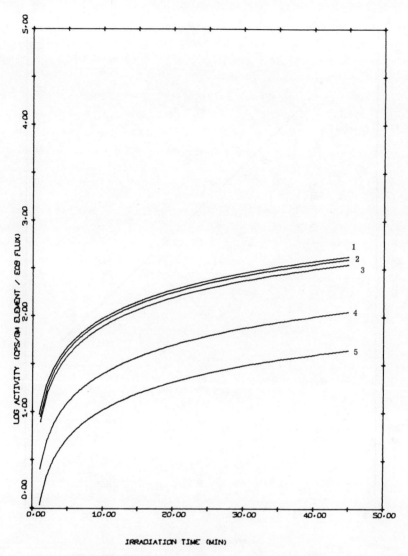

Fig. 7.100. Lutetium—activation curves (see Table 7.34).

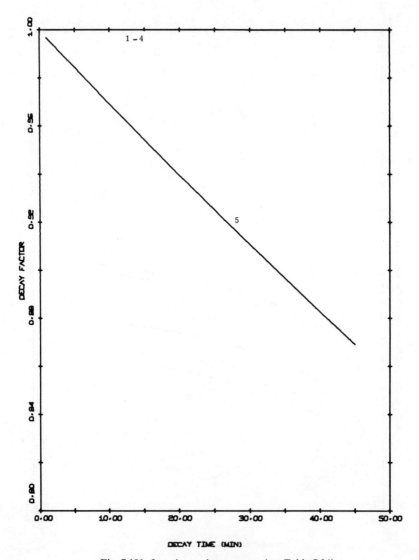

Fig. 7.101. Lutetium—decay curves (see Table 7.34).

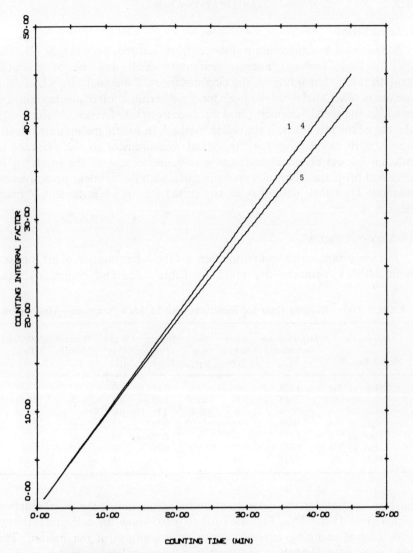

Fig. 7.102. Lutetium—integral counting curves (see Table 7.34).

MAGNESIUM

Magnesium has three naturally occurring isotopes of mass 24, 25, and 26. All three undergo nuclear reactions which may be of potential analytical use, depending on the circumstances of the analysis. While there are many other analytical methods for the determination of magnesium, few can offer the speed, simplicity, and freedom from interference obtained with the use of the fast neutron activation method. In nature magnesium usually occurs with calcium, and its analytical measurement in the presence of calcium by wet chemical methods is difficult because of the similarity in chemical properties. No such problem exists with the fast neutron activation method. Thus this method is ideally suited for the measurement of magnesium in ores.

14-MeV NEUTRONS

The important activation parameters for the determination of magnesium with 14-MeV neutrons are given in Table 7.35. The commonly used

TABLE 7.35. Nuclear Data for Reactions with 14-MeV Neutrons—Magnesium

Nuclear Reaction	Target Isotope Abundance (%)	Cross Section (mb)	Half-life (min)	Gamma Energy (MeV)	Gamma Yield (%)	Detector Efficiency (%)	Peak/Total Ratio
1 = $^{26}Mg(n, \alpha)^{23}Ne$	11.29	89	0.63	0.44	33	42	0.69
2 = $^{24}Mg(n, p)^{24}Na$	78.60	186	900.00	1.37	100	30	0.35
3 = $^{24}Mg(n, p)^{24}Na$	78.60	186	900.00	2.75	100	25	0.22
4 = $^{25}Mg(n, p)^{25}Na$	10.11	44.9	1.00	0.39	14	43	0.75
5 = $^{25}Mg(n, p)^{25}Na$	10.11	44.9	1.00	0.58	14	38	0.58
6 = $^{25}Mg(n, p)^{25}Na$	10.11	44.9	1.00	0.98	15	33	0.42
7 = $^{25}Mg(n, p)^{25}Na$	10.11	44.9	1.00	1.61	6	28	0.30

nuclear reaction for the determination of magnesium is the (n, p) reaction which gives 15-hr ^{24}Na. Figures 7.103 to 7.105 show the activation, decay, and integral counting curves for the determination of magnesium. The photopeak at 1.37 MeV is the most useful for analysis in matrices which contain oxygen. As little as 1 mg of magnesium can be detected. Activation parameters suggest that in the absence of oxygen, the activity from short-lived ^{23}Ne might be the most sensitive reaction for measuring sodium.

Two common interferences in the activation of magnesium are aluminum and large amounts of sodium. The activation parameters for these interferences are given in Table 7.36, and the working curves are shown in Figs. 7.106 to 7.108. Methods have been developed to minimize interference from these elements and are discussed later.

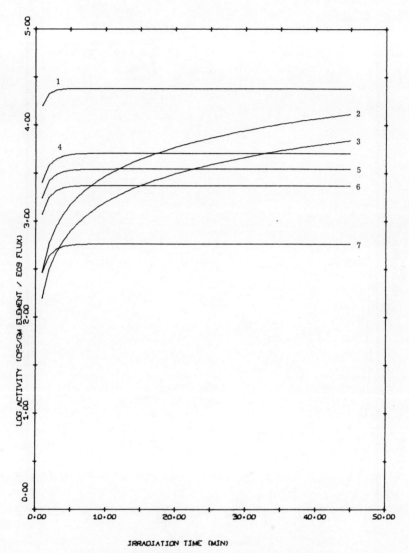

Fig. 7.103. Magnesium—activation curves (see Table 7.35).

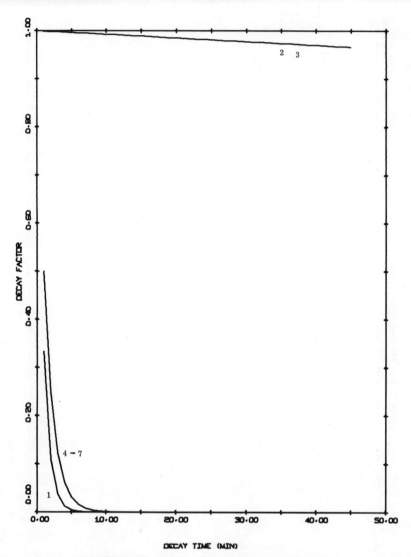

Fig. 7.104. Magnesium—decay curves (see Table 7.35).

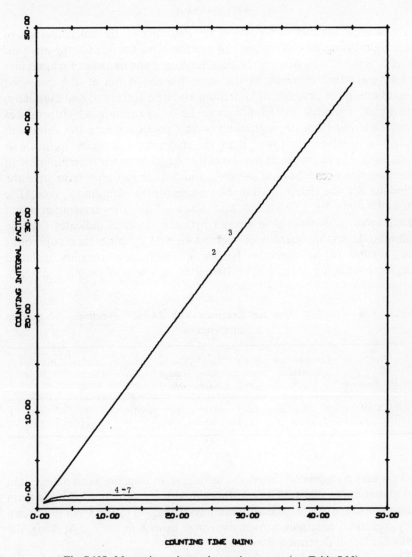

Fig. 7.105. Magnesium—integral counting curves (see Table 7.35).

There are a number of reports in the literature on the determination of magnesium using this technique. All use the ^{24}Na nuclide for the analysis. Most of these studies involve the simultaneous measurement of magnesium and several other elements in the same matrix. Wood et al.[7] measured magnesium in the presence of aluminum and iron and found that aluminum constituted a serious interference from the (n, α) reaction which produces ^{24}Na. Determination of magnesium is only possible where the aluminum content is significantly lower than the magnesium content. Kehler and Monaghan[17] have proposed the use of this reaction for the determination of magnesium in ocean-bottom samples, and Sattarov et al.[59] have used the technique for the determination of magnesium in bituminous coal. The extensive study by D'Agostino and Kuehne[2] on the determination of metallic wear components in aircraft hydraulic systems indicates that the minimum detectable quantity of magnesium is 11 μg under their conditions using a flux of 10^9 neutrons/(cm^2 sec). Their measurements included samples containing from 0.3 to 100 μg of magnesium on filter patches.

TABLE 7.36. Nuclear Data for Reactions with 14-MeV Neutrons—Magnesium Interferences

Nuclear Reaction	Target Isotope Abundance (%)	Cross Section (mb)	Half-life (min)	Gamma Energy (MeV)	Gamma Yield (%)	Detector Efficiency (%)	Peak/Total Ratio
1 = ^{27}Al(n, α)^{24}Na	100.00	120	900.00	1.37	100	30	0.35
2 = ^{23}Na(n, p)^{23}Ne	100.00	33.9	0.63	0.44	33	42	0.69

Breynat and co-workers[61] used the technique to measure both magnesium and aluminum in plants. These authors used the production of ^{27}Mg from aluminum to correct for the aluminum contribution to the ^{24}Na produced. They measured magnesium over the range from 4 to 75 mg. At 4 mg, the deviation for 10 determinations was 7%.

The only interference in the use of the 1.37-MeV photopeak for the measurement of magnesium is aluminum, which also produces ^{24}Na upon irradiation with neutrons. The contribution from this interference can be corrected on the basis of other nuclides which are also produced in the irradiation of aluminum. The 1.78-MeV peak of ^{28}Al can be used to correct for aluminum contribution at 0.84 MeV.

Waggoner and Knox[13] have proposed the use of neutron inelastic scattering to measure magnesium. They have applied this technique to the measurement of magnesium-to-silicon ratios in igneous rocks.

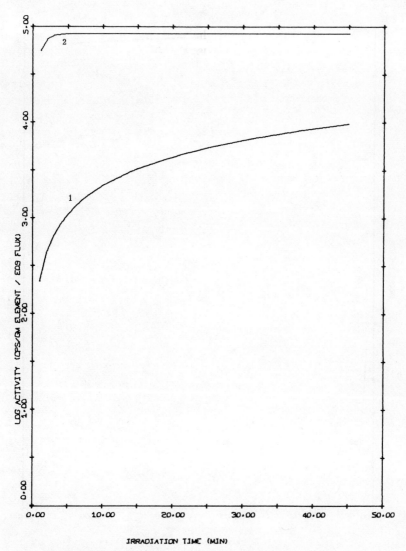

Fig. 7.106. Magnesium interferences—activation curves (see Table 7.36).

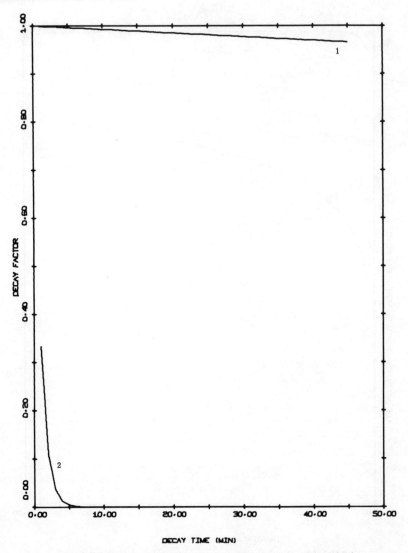

Fig. 7.107. Magnesium interferences—decay curves (see Table 7.36).

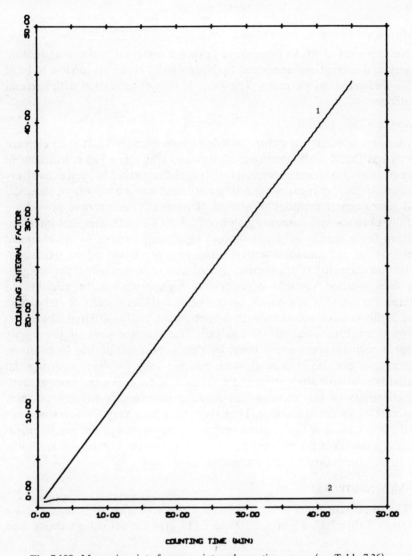

Fig. 7.108. Magnesium interferences—integral counting curves (see Table 7.36).

3-MeV AND THERMAL NEUTRONS

Neutrons of 3-MeV undergo a capture reaction with magnesium; however, the sensitivity measured by Nargolwalla et al.[23] is too low to be of value for analytical purposes. The same is true of activation with thermal neutrons.

MANGANESE

Manganese occurs in nature as one isotope of mass 55. It is an element which has found wide use in metallurgy and which is a key constituent in certain steels. For these reasons, its analytical chemistry has been extensively studied. Indeed, manganese can be determined using a variety of chemical and instrumental methods of analysis. Chemically, manganese provides a variety of oxidation states ranging from 2+ to 7+ with chemical interconversion being readily achieved. Thus a significant number of wet chemical methods for the measurement of manganese are based on its oxidation-reduction chemistry. For example, the reductimetric titration of permanganate is a method capable of extremely high precision. In addition to volumetric methods of analysis, several of the oxidation states of manganese give highly colored solutions with extinction values that are high enough to provide sensitive methods of analysis. The measurement of the highly colored permanganate anion provides a good example of this. In addition, manganese can be measured with relative ease by such instrumental methods as atomic absorption and x-ray fluorescence. The extensive analytical chemistry of this element has made it unnecessary to seek out new methods for its determination. Thus there have been very few investigations of the application of fast neutron activation techniques to the measurement of manganese. Yet the element has nuclear parameters which are favorable for its determination by this technique.

14-MeV NEUTRONS

The important nuclear parameters for the activation of manganese are given in Table 7.37. Figures 7.109 to 7.111 give the activation, decay, and

TABLE 7.37. Nuclear Data for Reactions with 14-MeV Neutrons—Manganese

Nuclear Reaction	Target Isotope Abundance (%)	Cross Section (mb)	Half-life (min)	Gamma Energy (MeV)	Gamma Yield (%)	Detector Efficiency (%)	Peak/Total Ratio
$1 = {}^{55}Mn(n, \alpha){}^{52}V$	100.00	52.5	3.76	1.43	100	29	0.34
$2 = {}^{55}Mn(n, \gamma){}^{56}Mn$	100.00	0.6	1545.60	0.85	99	34	0.44
$3 = {}^{55}Mn(n, \gamma){}^{56}Mn$	100.00	0.6	1545.60	1.81	29	28	0.28
$4 = {}^{55}Mn(n, \gamma){}^{56}Mn$	100.00	0.6	1545.60	2.11	15	26	0.26
$5 = {}^{55}Mn(n, 2n){}^{54}Mn$	100.00	825	436320	0.84	100	34	0.44

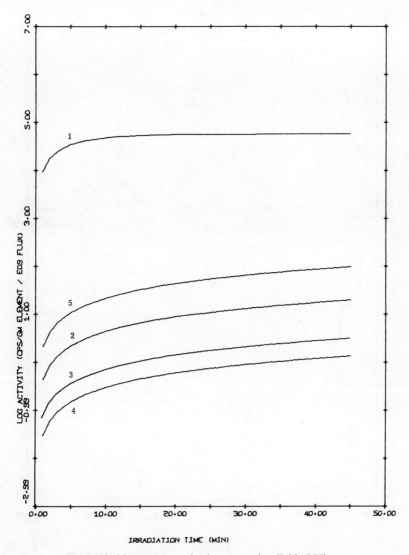

Fig. 7.109. Manganese—activation curves (see Table 7.37).

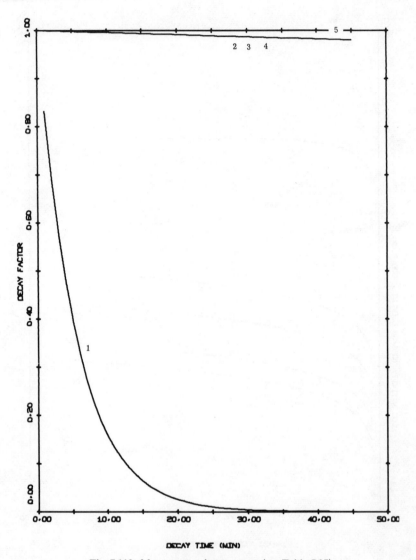

Fig. 7.110. Manganese—decay curves (see Table 7.37).

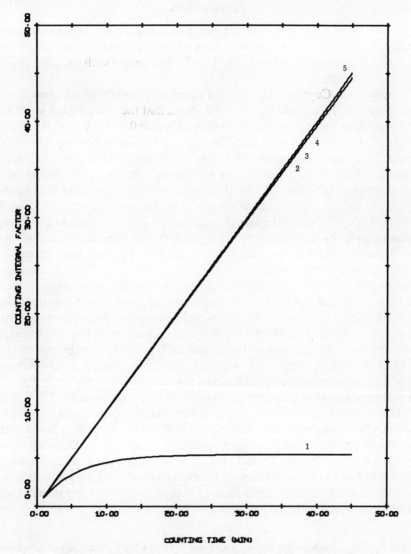

Fig. 7.111. Manganese—integral counting curves (see Table 7.37).

integral counting curves for the activation of manganese. The most sensitive reaction for the determination of manganese using 14-MeV neutrons is the (n, α) reaction which produces ^{52}V. This isotope decays with the emission of a 1.43-MeV gamma ray.

Cuypers and Cuypers[1] have listed a detection limit of 0.49 mg of manganese using this photopeak. Figure 7.109 shows that the second most sensitive photopeak for the measurement of manganese is at 0.84 MeV from the decay of the ^{54}Mn and ^{56}Mn isotopes; ^{54}Mn decays with an emission at 0.84 MeV while ^{56}Mn has a photopeak at 0.85 MeV. Cuypers and Cuypers[1] indicate a detection limit of 3.8 mg using this photopeak. Pierce et al.[149] utilized the photopeaks at 0.84 and 1.43 MeV to measure manganese as well as vanadium, chromium, and iron in mixtures. These authors demonstrated that it was possible to measure the four elements which yield photopeaks at both of these energies by irradiating two samples with neutrons of different energy such that the effective cross sections for the four elements were varied. In order to determine the four adjacent elements—vanadium, chromium, manganese, and iron—in the same sample, the samples were counted twice in each activation cycle, once after decay time of a half-minute to obtain the ^{52}V count, and again when the short-lived activity had decayed, to measure the ^{56}Mn activity. The manganese and iron concentrations were calculated first and then the manganese figure was used to apply a correction to the 1.43-MeV γ-yield, which provided for the measurement of chromium. The only other reported application of the neutron activation technique to the determination of manganese has been given by Boreisha et al.[62]

The common interferences in the determination of manganese are iron, cobalt, chromium, and vanadium, elements which are very often found with manganese either in manufactured products or in nature. Data for these interferences are listed in Table 7.38, and the working curves are displayed in Figs. 7.112 to 7.114. These elements interfere by producing the same nuclides as those produced in the activation of manganese. Thus their interference must be removed in the activation step. The technique described by Pierce et

TABLE 7.38. Nuclear Data for Reactions with 14-MeV Neutrons—Manganese Interferences

Nuclear Reaction	Target Isotope Abundance (%)	Cross Section (mb)	Half-life (min)	Gamma Energy (MeV)	Gamma Yield (%)	Detector Efficiency (%)	Peak/Total Ratio
1 = ^{56}Fe(n, p)^{56}Mn	91.68	103	154.56	0.85	99	34	0.44
2 = ^{59}Co(n, α)^{56}Mn	100.00	39.10	154.56	0.85	99	34	0.44
3 = ^{52}Cr(n, p)^{52}V	83.76	78	3.76	1.43	100	29	0.34
4 = ^{51}V(n, γ)^{52}V	99.76	0.35	3.76	1.43	100	29	0.34

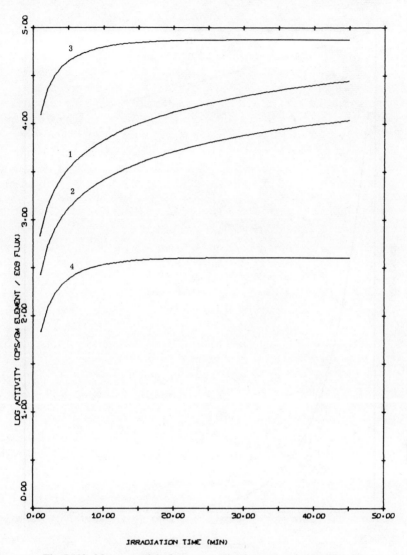

Fig. 7.112. Manganese interferences—activation curves (see Table 7.38).

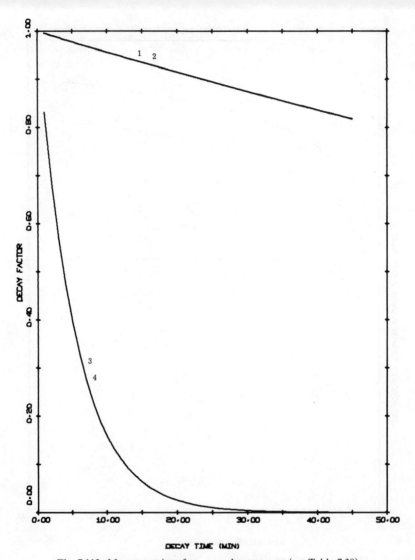

Fig. 7.113. Manganese interferences—decay curves (see Table 7.38).

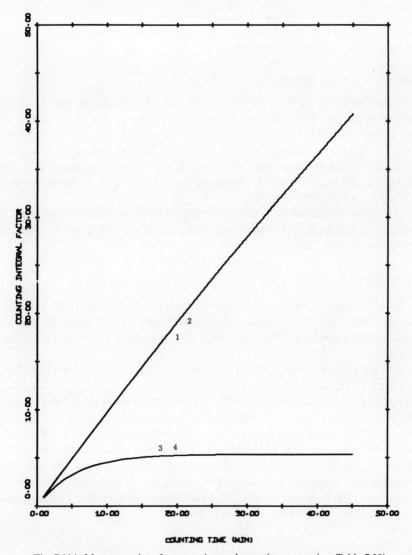

Fig. 7.114. Manganese interferences—integral counting curves (see Table 7.38).

al.[149] demonstrated that it does provide effective discrimination of the four elements when present in a mixture.

3-MeV AND THERMAL NEUTRONS

Nargolwalla et al.[23] reported on the activation of manganese using 3-MeV neutrons. These authors found a sensitivity of 9×10^3 counts/g of manganese utilizing a flux of 10^6 neutrons/(cm^2 sec). The irradiation time used in these experiments was 20 min. The 3-MeV technique is not as sensitive for the measurement of manganese as is the technique utilizing 14-MeV neutrons. No further reports of the activation of manganese using 3-MeV neutrons are available.

Generator-produced thermal neutrons give the most sensitive reaction for measuring manganese. Dibbs[24] reported a sensitivity of 10 μg at a flux of 10^8 neutrons/(cm^2 sec).

MERCURY

Mercury occurs in nature in seven isotopic forms of mass 196, 198 through 202, and 204. Only one of these isotopes, that of mass 200, undergoes reactions which may be useful for analytical purposes with fast neutrons.

Because of its industrial utility and human toxicity, a good deal of effort has been expended in analytical methods for mercury. Recently, much effort has gone into the measurement of mercury at the trace level in the environment and in food supplies. Neutron activation analysis utilizing the thermal flux of a reactor has been an important tool in these studies. Other methods such as flameless atomic absorption have also been used for these measurements. Activation analysis with fast neutrons does not have the required sensitivity for trace analysis of mercury. In fact, it is only useful for macroconstituent levels of mercury when other means such as wet chemical and x-ray methods are perhaps more suitable.

14-MeV NEUTRONS

The most useful reactions for the neutron activation of mercury result in the nuclide ^{199m}Hg, which has emissions at 0.16 and 0.38 MeV. Cuypers and Cuypers[1] list a detection limit of 0.46 mg for the 0.16-MeV photopeak and 3.8 mg using the one at higher energy, utilizing a flux of 5×10^8 neutrons/(cm^2 sec). Using somewhat longer irradiation and counting intervals, Dibbs[24] reported a sensitivity of 60 μg for mercury. Table 7.39 gives the reactions of importance for the activation of mercury, and Figs. 7.115 to 7.117 show the activation, decay, and counting integral curves. There have

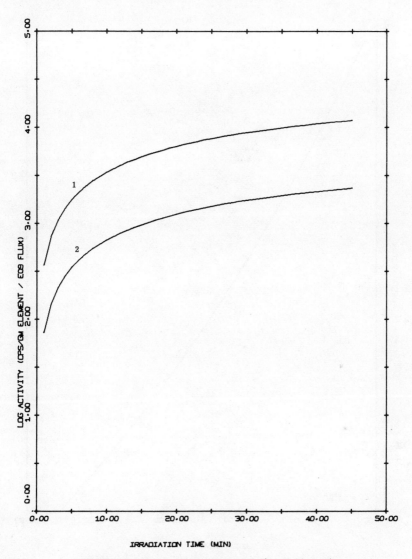

Fig. 7.115. Mercury—activation curves (see Table 7.39).

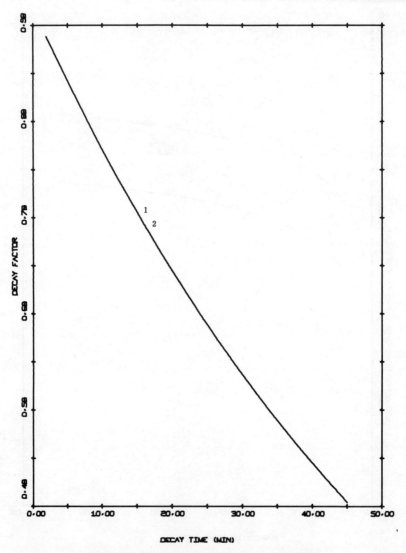

Fig. 7.116. Mercury—decay curves (see Table 7.39).

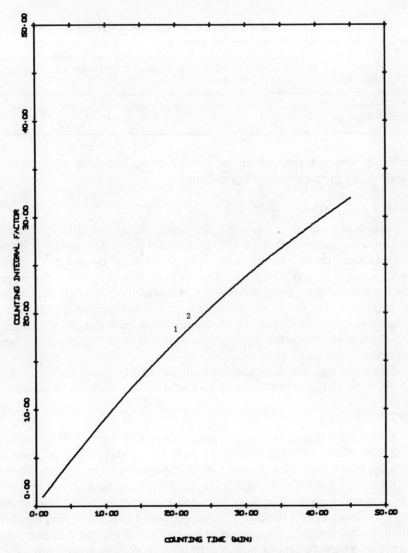

Fig. 7.117. Mercury—integral counting curves (see Table 7.39).

TABLE 7.39. Nuclear Data for Reactions with 14-MeV Neutrons—Mercury

Nuclear Reaction	Target Isotope Abundance (%)	Cross Section (mb)	Half-life (min)	Gamma Energy (MeV)	Gamma Yield (%)	Detector Efficiency (%)	Peak/Total Ratio
$1 = {}^{200}Hg(n, 2n){}^{199m}Hg$	23.13	130	43.00	0.16	53	50	0.96
$2 = {}^{200}Hg(n, 2n){}^{199m}Hg$	23.13	130	43.00	0.38	15	44	0.75

been no reported applications of the activation technique with 14-MeV neutrons to the measurement of mercury.

3-MeV AND THERMAL NEUTRONS

Nargolwalla et al.[23] reported an experimental sensitivity of 1.1×10^3 counts/g using a flux of 10^6 neutrons/(cm² sec) for mercury. These authors used the emissions from ^{199m}Hg produced by neutron capture reactions on ^{198}Hg. When corrected for sample attenuation effects the experimental sensitivity for the 0.16-MeV emission was approximately four orders of magnitude higher. Thus the 3-MeV technique may be useful for the measurement of semimicrolevels of mercury using small samples to minimize self-absorption. No other applications of the 3-MeV method have been reported.

Activation with the thermal neutron flux available from a generator does not provide sufficient sensitivity to be useful.

MOLYBDENUM

Molybdenum exists in nature in seven isotopes of mass 92, 94 through 98, and 100. Those of mass 92 and 97 produce 14-MeV neutron reactions of analytical significance. Because of its importance in ferrous metallurgy, molybdenum has been extensively studied and many methods are available for its determination. The more common methods are either colorimetric (molybdenum blue formation) or volumetric based on reaction with iron(III). Both methods require dissolution of a sample and are more cumbersome than instrumental methods such as neutron activation or x-ray fluorescence.

14-MeV NEUTRONS

The nuclear reactions of importance are listed in Table 7.40. Activation of ^{92}Mo yields two positron emitters, ^{91}Mo and ^{91m}Mo, with somewhat different half-lives. Cuypers and Cuypers[1] list the sensitivity of the determination of molybdenum at 1.0 mg using ^{91}Mo. Dibbs[24] has reported somewhat better sensitivity (0.13 mg) using this same reaction. From reaction parameters, it would appear that use could be made of the ^{91m}Mo positron

III. ACTIVATION ANALYSIS CONDITIONS FOR THE ELEMENTS

TABLE 7.40. Nuclear Data for Reactions with 14-MeV Neutrons—Molybdenum

Nuclear Reaction	Target Isotope Abundance (%)	Cross Section (mb)	Half-life (min)	Gamma Energy (MeV)	Gamma Yield (%)	Detector Efficiency (%)	Peak/Total Ratio
1 = ^{91}Mo(n, 2n)^{91}Mo	15.86	190	15.49	0.51	200	40	0.63
2 = ^{92}Mo(n, 2n)^{91m}Mo	15.86	190	1.10	0.51	276	40	0.63
3 = ^{97}Mo(n, p)^{97}Nb	9.45	108	72.00	0.67	98	37	0.53
4 = ^{100}Mo(n, p)^{100}Nb	96.20	2.20	11.00	0.54	60	39	0.61

emitter to differentiate molybdenum from other positron emitters (cf. Tables 7.76–7.78) which might be present in a matrix. ^{91m}Mo has the shortest half-life among the common positron emitters. Thus decay curve analysis could be used to determine molybdenum in the presence of such elements as phosphorus, praseodymium, and bromine. Activation, decay, and integral counting curves are given in Figs. 7.118 to 7.120.

Crambes[25] reported on a radiochemical separation procedure for separating molybdenum from complex matrices after irradiation with 14-MeV neutrons. This author used 8-quinolinol in a rapid radiochemical separation of the molybdenum and counted the annihilation radiation at 0.511 MeV. An experimental sensitivity of 228 counts/(min mg) of molybdenum was reported using an 8-min irradiation in a flux of 5.5×10^8 neutrons/(cm^2 sec).

3-MeV AND THERMAL NEUTRONS

Nargolwalla et al.[23] applied 3-MeV neutrons to the determination of molybdenum utilizing the (n, γ) capture reaction on ^{100}Mo. ^{101}Mo gives a spectrum of γ-ray emissions in the range from 0.19 to 2.08 MeV. It further decays to ^{101}Tc, which gives a photopeak at 0.307 MeV. Utilizing this for the analytical determination of molybdenum, Nargolwalla et al.[23] report an experimental sensitivity of 2.1×10^3 counts/g of molybdenum when using a flux of 10^6 neutrons/(cm^2 sec).

Dibbs[24] predicted a sensitivity of 830 μg using the (n, γ) reaction on ^{100}Mo.

NEODYMIUM

Neodymium occurs in nature in seven different isotopes, those of mass 142, 143, 144 to 146, 148, and 150. Neodymium usually occurs in nature with other rare earth elements; thus its determination in mixtures of rare earths is the usual requirement for an analytical method. For such applications the

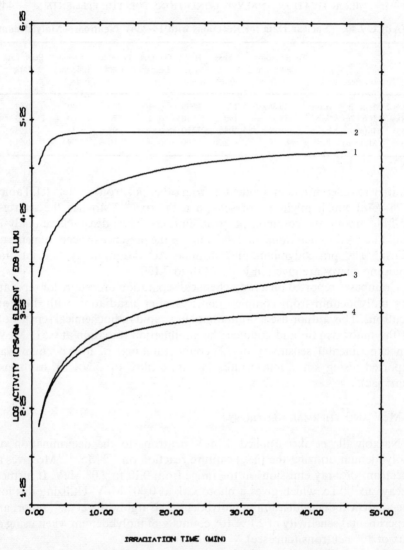

Fig. 7.118. Molybdenum—activation curves (see Table 7.40).

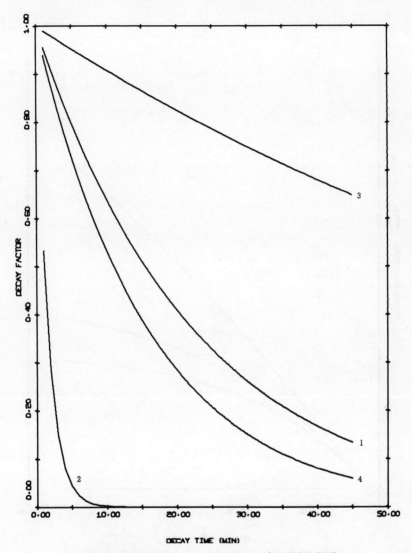

Fig. 7.119. Molybdenum—decay curves (see Table 7.40).

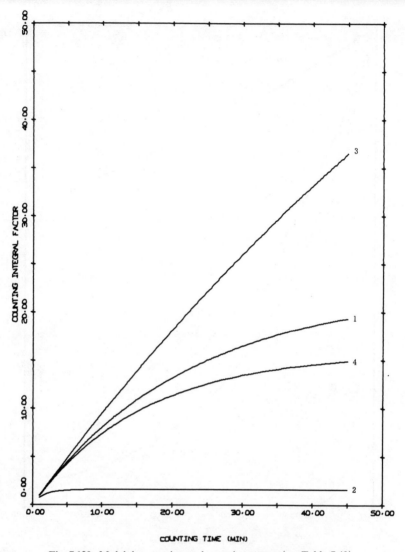

Fig. 7.120. Molybdenum—integral counting curves (see Table 7.40).

neutron activation procedure for neodymium is not highly selective since it is interfered with by cerium and samarium which give nuclides whose emissions are in the same general energy region. In view of the excellent alternate methods which exist for the chemical separation and determination of rare earths by chemical, x-ray fluorescence, and spectrophotometric techniques, the neutron activation method is not highly recommended. Only under circumstances in which one had a simple matrix would this method be useful.

14-MeV NEUTRONS

The most useful nuclear reaction for the determination of neodymium with fast neutrons is the (n, 2n) reaction on ^{142}Nd. This reaction produces 64-sec ^{141m}Nd which decays with an emission of 0.76-MeV gamma radiation. Other fast neutron reactions on neodymium produce at least an order of magnitude less activity than this reaction. The important activation parameters for neodymium are given in Table 7.41, and Figs. 7.121 to 7.123

TABLE 7.41. Nuclear Data for Reactions with 14-MeV Neutrons—Neodymium

Nuclear Reaction	Target Isotope Abundance (%)	Cross Section (mb)	Half-life (min)	Gamma Energy (MeV)	Gamma Yield (%)	Detector Efficiency (%)	Peak/Total Ratio
1 = ^{142}Nd(n, α)^{139m}Ce	27.13	10	0.92	0.75	93	35	0.50
2 = ^{142}Nd(n, 2n)^{141}Nd	27.13	2060	150.00	0.51	6	40	0.63
3 = ^{142}Nd(n, 2n)^{141m}Nd	27.13	545	1.07	0.76	100	35	0.49
4 = ^{150}Nd(n, 2n)^{149}Nd	5.60	2200	148.00	0.27	26	48	0.87

show the activation, decay, and integral counting curves for the determination of neodymium. Sensitivity for the determination of neodymium with 14-MeV neutrons is reported by Cuypers and Cuypers[1] as 320 μg at a flux of approximately 5×10^8 neutrons/(cm² sec). Cerium and samarium constitute an interference in the determination of neodymium. The nature of the interference is shown in Table 7.42 and the sensitivity of these reactions is illustrated in the activation, decay, and integral counting curves in Figs. 7.124 to 7.126.

Despite these limitations, two applications have been reported in the literature for the determination of neodymium. One by Plaksin et al.[67] reported a sensitivity for the determination of neodymium using a flux of 5×10^8 neutrons/(cm² sec) of 4 μg. This differs considerably from the value reported by Cuypers and Cuypers[1] and from that calculated based on activation parameters. The average error these authors reported for the

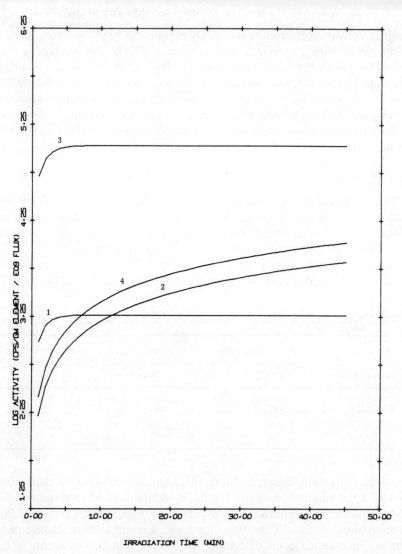

Fig. 7.121. Neodymium—activation curves (see Table 7.41).

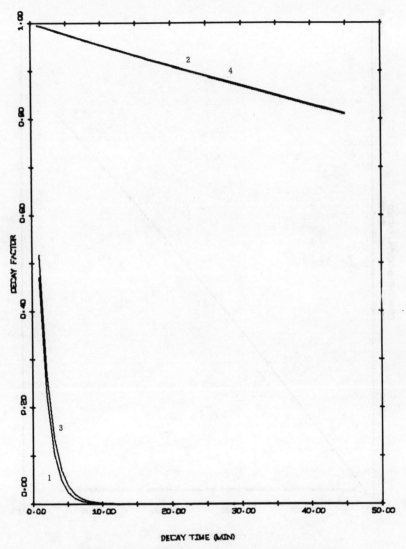

Fig. 7.122. Neodymium—decay curves (see Table 7.41).

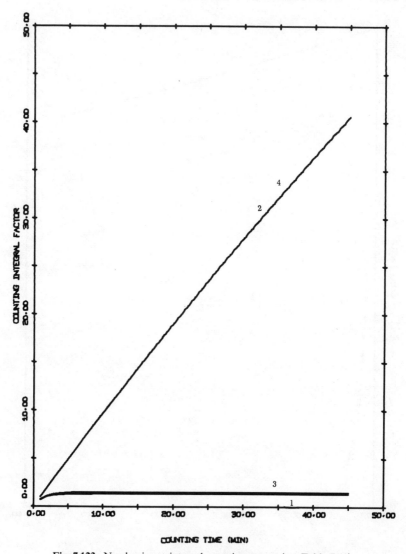

Fig. 7.123. Neodymium—integral counting curves (see Table 7.41).

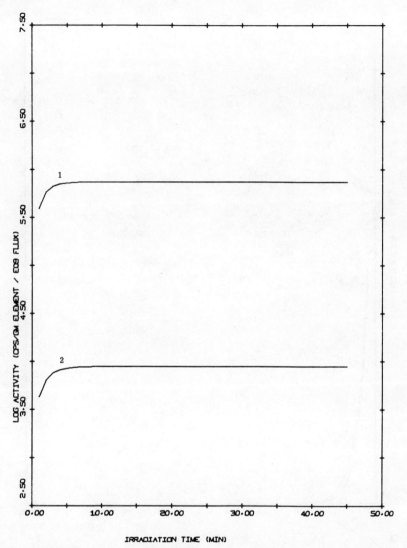

Fig. 7.124. Neodymium interferences—activation curves (see Table 7.42).

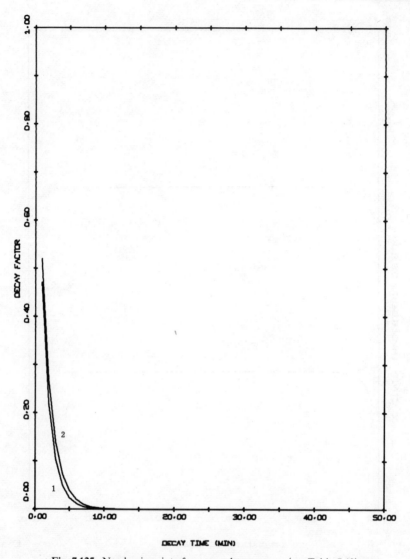

Fig. 7.125. Neodymium interferences—decay curves (see Table 7.42).

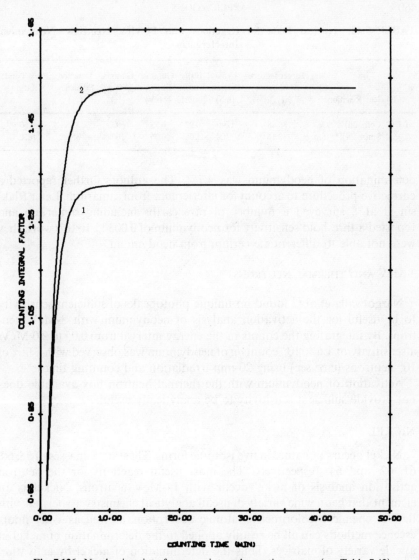

Fig. 7.126. Neodymium interferences—integral counting curves (see Table 7.42).

TABLE 7.42. Nuclear Data for Reactions with 14-MeV Neutrons—Neodymium Interferences

Nuclear Reaction	Target Isotope Abundance (%)	Cross Section (mb)	Half-life (min)	Gamma Energy (MeV)	Gamma Yield (%)	Detector Efficiency (%)	Peak/Total Ratio
$1 = {}^{140}Ce(n, 2n){}^{139m}Ce$	88.48	1200	0.92	0.75	93	35	0.50
$2 = {}^{144}Sm(n, 2n){}^{143m}Sm$	3.16	400	1.06	0.75	100	35	0.50

determination of neodymium was 4.7%. The authors further reported a corrective procedure to account for interference from samarium. Later Plaksin et al.[68] surveyed a number of rare earths including samarium and reported a threshold sensitivity for neodymium of 0.003%. In this work they were not able to differentiate cerium from neodymium.

3-MeV AND THERMAL NEUTRONS

Nargolwalla et al.[23] found no unique photopeaks of sufficient sensitivity to be useful for the activation analysis of neodymium with 3-MeV neutrons. By integrating the counts in the energy interval from 0.05 to 2.0 MeV, a sensitivity of 1.6×10^3 counts/g of neodymium was observed with a flux of 10^6 neutrons/(cm² sec) using 20-min irradiation and counting times.

Activation of neodymium with the thermal neutron flux available does not provide sufficient sensitivity to be analytically useful.

NICKEL

Nickel occurs in nature in five isotopic forms. These have masses of 58, 60, 61, 62, and 64, respectively. The most useful reactions for the neutron activation analysis of nickel occur with 14-MeV neutrons. Nickel is an element that has a wide variety of useful analytical chemistry associated with it. Wet chemical, colorimetric, atomic absorption, as well as x-ray fluorescence methods can all be readily applied to the determination of nickel in a wide variety of matrices. Thus the determination of nickel by activation analysis has not received extensive study.

14-MeV NEUTRONS

The important parameters for the activation of nickel with fast neutrons are given in Table 7.43. Figures 7.127 to 7.129 show the activation, decay, and integral counting curves for the determination of nickel. According to Cuypers and Cuypers,[1] the most sensitive reaction for the determination of nickel with 14-MeV neutrons is the (n, 2n) reaction on ^{58}Ni. They give a

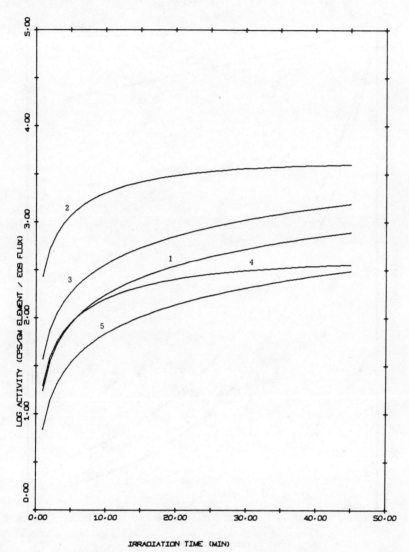

Fig. 7.127. Nickel—activation curves (see Table 7.43).

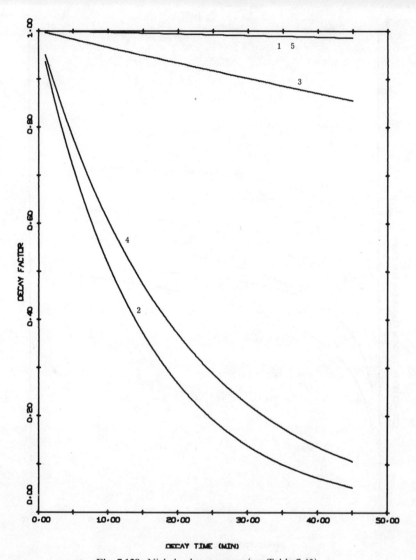

Fig. 7.128. Nickel—decay curves (see Table 7.43).

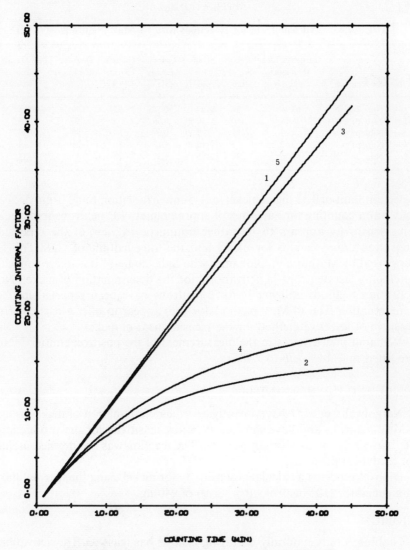

Fig. 7.129. Nickel—integral counting curves (see Table 7.43).

TABLE 7.43. Nuclear Data for Reactions with 14-MeV Neutrons—Nickel

Nuclear Reaction	Target Isotope Abundance (%)	Cross Section (mb)	Half-life (min)	Gamma Energy (MeV)	Gamma Yield (%)	Detector Efficiency (%)	Peak/Total Ratio
1 = ^{58}Ni(n, 2n)^{57}Ni	67.76	34	2160	0.51	92	40	0.63
2 = ^{60}Ni(n, p)^{60m}Co	26.16	150	10.50	0.06	2	50	1.00
3 = ^{61}Ni(n, p)^{61}Co	1.25	181	199.20	0.07	89	50	1.00
4 = ^{62}Ni(n, p)^{62}Co	3.62	5	13.90	1.17	180	31	0.38
5 = ^{58}Ni(n, 2n)^{57}Ni	67.76	34	2160	1.37	86	30	0.35

detection limit of 0.52 mg of nickel for a 5-min irradiation time, 1-min decay, and 5-min counting time at a flux of approximately 10^8 neutrons/(cm^2 sec). This sensitivity appears to be rather optimistic in view of the reported activation cross sections for nickel and the long half-life of ^{57}Ni. This is supported by Mathur and Oldham[58] who indicate that "it is uneconomical to employ fast neutrons in activation" for the determination of nickel. No activation methods utilizing 14-MeV neutrons have been reported for its determination. The 14-MeV method does not appear to offer a particularly sensitive or selective method for the measurement of nickel.

Potential interferences in the measurement of the positron emitter ^{57}Ni, are listed in Tables 7.76 to 7.78.

3-MeV AND THERMAL NEUTRONS

Nargolwalla et al.[23] have investigated the determination of nickel using 3-MeV neutrons and have shown a very low sensitivity for the production of ^{58}Co via an (n, p) reaction on ^{58}Ni. The reaction was not deemed useful for analytical purposes.

Dibbs[24] reported a calculated sensitivity for nickel using the thermal flux of a generator (10^8 neutrons/(cm^2-sec)) of 940 μg.

NIOBIUM

Niobium's only naturally occurring isotope has mass 93. The activation analysis technique is not suited to the determination of small amounts of niobium because of low cross sections. Since there are alternative and more sensitive colorimetric as well as instrumental methods for the determination of this element, very little work has been done on its determination with fast neutrons.

14-MeV NEUTRONS

The important activation parameters for the determination of niobium with 14-MeV neutrons are listed in Table 7.44. Figures 7.130 to 7.132 show

III. ACTIVATION ANALYSIS CONDITIONS FOR THE ELEMENTS

TABLE 7.44. Nuclear Data for Reactions with 14-MeV Neutrons—Niobium

Nuclear Reaction	Target Isotope Abundance (%)	Cross Section (mb)	Half-life (min)	Gamma Energy (MeV)	Gamma Yield (%)	Detector Efficiency (%)	Peak/Total Ratio
$1 = {}^{93}\text{Nb}(n, \alpha){}^{90m}\text{Y}$	100.00	5.9	186	0.20	97	50	0.93
$2 = {}^{93}\text{Nb}(n, \alpha){}^{90m}\text{Y}$	100.00	5.9	186	0.48	91	41	0.65
$3 = {}^{93}\text{Nb}(n, 2n){}^{92m}\text{Nb}$	100.00	5.0	609.60	0.93	99	33	0.43
$4 = {}^{93}\text{Nb}(n, n'\alpha){}^{89m}\text{Y}$	100.00	2.5	0.27	0.91	99	33	0.43

the activation, decay, and counting integral curves. The most sensitive useful reaction for the determination of niobium with 14-MeV neutrons is the (n, n'α) reaction which yields ^{89m}Y. However, because of its short half-life, the nuclide is difficult to measure without interference, particularly if oxygen is present in the sample. Cuypers and Cuypers[1] list a sensitivity of 18 mg as the detection limit for niobium using the photopeak at 0.20 MeV from ^{90m}Y. Using somewhat longer irradiation Dibbs[24] obtained a sensitivity of 1.8 mg using the same nuclide. There are no other reported methods for the determination of niobium using 14-MeV neutrons.

3-MeV AND THERMAL NEUTRONS

Nargolwalla et al.[23] report that the (n, γ) reaction with 3-MeV neutrons on niobium produces ^{94m}Nb, but with relatively low sensitivity. Reaction with the thermal flux available from a generator is likewise too insensitive to be useful.

NITROGEN

Nitrogen occurs in nature in two isotopes, those of mass 14 and 15. Because nitrogen occurs in a wide variety of naturally occurring materials, and is important in synthetic-organic and -inorganic chemistry, its precise analytical determination is of great importance. The most widely used methods for the determination of nitrogen are the classical Kjeldhal and Dumas procedures. Both are chemical methods which depend on the decomposition of the sample and measurement of ammonia and nitrogen, respectively. The importance of this determination has resulted in the development of automated equipment for the determination of nitrogen by these procedures. Today these analyses can be accomplished in approximately 15 min and yield a high degree of precision and accuracy. Fast neutron activation analysis methods may compare favorably with these procedures for the determination of macroconstituent nitrogen in certain types of matrices.

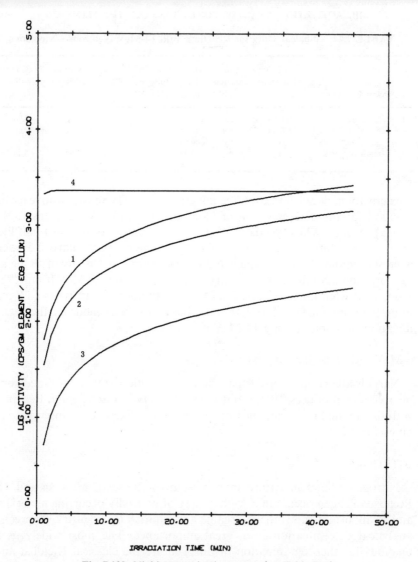

Fig. 7.130. Niobium—activation curves (see Table 7.44).

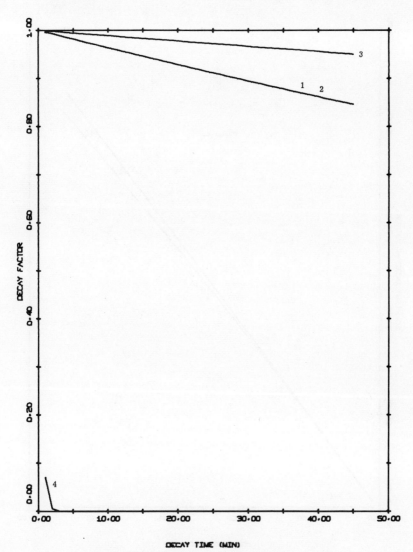

Fig. 7.131. Niobium—decay curves (see Table 7.44).

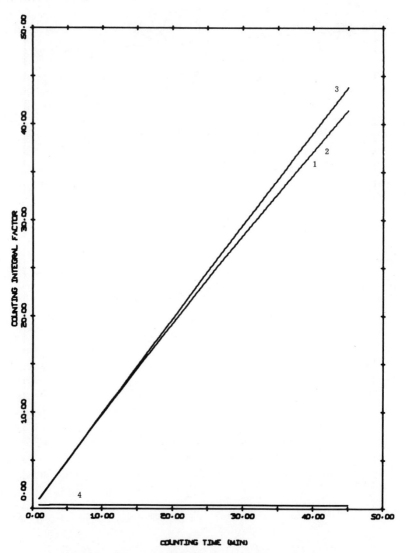

Fig. 7.132. Niobium—integral counting curves (see Table 7.44).

III. ACTIVATION ANALYSIS CONDITIONS FOR THE ELEMENTS

14-MeV NEUTRONS

The primary reaction of 14-MeV neutrons with ^{14}N is the (n, 2n) reaction which produces ^{13}N, a positron emitter ($t_{1/2}$ = 9.96 min). The activation, decay, and integral counting curves for this nuclide are shown in Figs. 7.133 to 7.135 and the nuclear parameters for the primary reaction are given in Table 7.45. Because it yields a positron emitter, there are a number of limitations in the determination of nitrogen in matrices which may yield other positron emitters with similar half-lives. Such elements are bromine, potassium, iron, samarium, and copper. Positron emitters with longer or shorter half-lives may be distinguished from nitrogen by decay curve analysis. The pertinent data for establishing conditions for differentiating positron emitters are given in Tables 7.76 to 7.78.

TABLE 7.45. Nuclear Data for Reactions with 14-MeV Neutrons—Nitrogen

Nuclear Reaction	Target Isotope Abundance (%)	Cross Section (mb)	Half-life (min)	Gamma Energy (MeV)	Gamma Yield (%)	Detector Efficiency (%)	Peak/Total Ratio
1 = ^{14}N(n, 2n)^{13}N	99.63	5.70	9.96	0.51	200	40	0.63

A second limitation in the determination of nitrogen occurs in hydrocarbon matrices where the 14-MeV neutrons can produce sufficient proton recoil to yield ^{13}N from the proton reaction on ^{13}C. Gilmore and Hull[65-67] have studied this reaction extensively and report that using high-purity butane containing no nitrogen, an activity equivalent to 460 ppm of nitrogen was obtained from the proton recoil reaction. Similar proton recoil can occur on oxygen, ^{16}O(p, α)^{13}N. Rison et al.[68] studied the production of ^{13}N via proton recoil reactions on carbon and oxygen and compared this with the yield obtained from the (n, 2n) reaction on ^{14}N. They found on a weight basis that the reaction with carbon was approximately 1/1000 as sensitive as that with nitrogen, whereas the reaction on oxygen was approximately 8/1000 as sensitive as that with nitrogen. The occurrence of proton recoil reactions on carbon and oxygen to yield ^{13}N precludes the use of the neutron activation method for the determination of small amounts of nitrogen in hydrocarbon, aqueous, or other similar matrices, unless the composition of the matrix is constant and nitrogen levels can be calibrated in the same matrix as the samples to be analyzed. In the determination of larger amounts of nitrogen, the error due to the proton recoil reaction becomes relatively small and can be tolerated. Rison et al.[68] utilized this procedure for the determination of nitrogen in explosives and propellants.

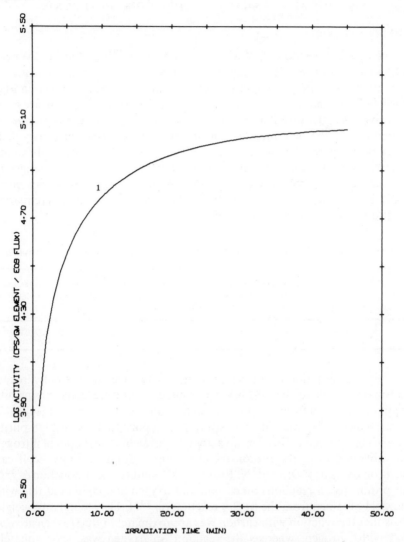

Fig. 7.133. Nitrogen—activation curves (see Table 7.45).

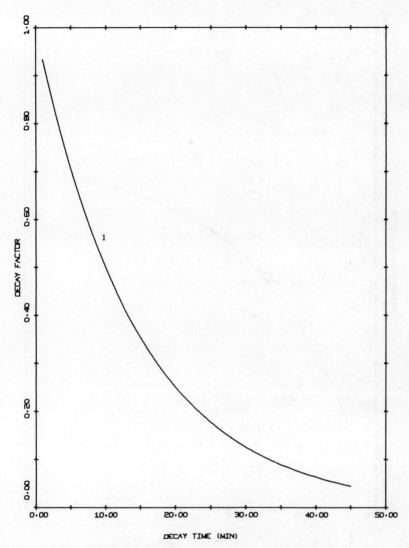

Fig. 7.134. Nitrogen—decay curves (see Table 7.45).

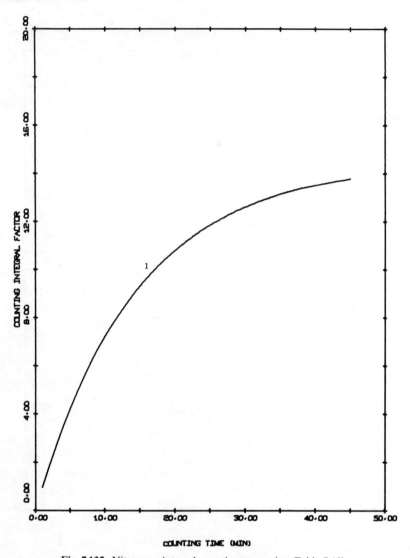

Fig. 7.135. Nitrogen—integral counting curves (see Table 7.45).

III. ACTIVATION ANALYSIS CONDITIONS FOR THE ELEMENTS 473

To obtain the 1% accuracy required, standards that were similar in composition to the sample to be analyzed were used. Tsuji[69] utilized this procedure for the analysis of nitrogen in organic compounds and measured as little as 0.2% nitrogen with a relative accuracy of 5%. Wood[70] carried out an extensive study on the application of this technique to the determination of nitrogen in grain products with a precision of 0.3 to 0.5%. He proposed this method as highly competitive with the Kjeldahl determination of nitrogen in grain products. Analysis times were as short as 4 min with an average precision of 2% of the nitrogen in the sample. However, this precision could be improved by using a higher output generator. Shamaev[71] reported the determination of nitrogen in organic compounds using strontium as an internal standard for flux normalization. This author also measured the extent of the proton recoil reaction and showed that irradiation of pure water induced ^{13}N activity equivalent to 1.6 mg nitrogen/g. Using hexane, he found an equivalent of 0.2 mg nitrogen/g hexane. Walker and Eggebraaten[72] used this procedure for the determination of nitrogen over the range of 3 to 5% in rubberized-sealant material. They demonstrated that considerable savings were achieved by replacing the Kjeldahl technique with this method. Since standards similar in composition to the samples were used, no correction was necessary for the proton recoil reaction. Crambes and his associates[73] reported the elemental analysis of proteins using the 14-MeV technique. Their results for nitrogen on two pure amino acids were between 1 and 3% lower than theory. No study was reported as to the cause of the lower results. The method has been proposed[19] for the determination of nitrogen in fertilizers, to replace the Kjeldahl method. In general, it would appear that the neutron activation method for nitrogen is best applied to routine analysis in a matrix whose composition is reasonably well defined. In these circumstances, contributions from the proton recoil reactions can be corrected by appropriate calibration. The method is, of course, useful for the determination of nitrogen in a wide variety of matrices not containing significant amounts of hydrogen. For low levels of nitrogen in unknown matrices, caution must be taken to avoid interferences both from proton recoil reactions and other positron emitters.

3-MeV AND THERMAL NEUTRONS

There are no useful nuclear reactions involving neutrons of these energies with nitrogen.

OXYGEN

Oxygen occurs in nature in the three isotopes of mass 16, 17, and 18. The only isotope of importance for activation analysis with neutron generators is that of mass 16. Oxygen is not an easy element to measure by classical chemical or instrumental methods of analysis. Such techniques as vacuum fusion or the Unterzaucher methods have a very narrow range of application and are not easy to use. Thus it is not surprising to find oxygen often being "determined" as the difference from 100% after all of the other elements in a matrix have been quantitatively measured. Obviously, a good method for oxygen was badly needed for some time. This was uniquely satisfied with the development of the fast neutron activation technique.

14-MeV NEUTRONS

Without question, the 14-MeV neutron activation technique is best suited for the measurement of oxygen. This is one of those coincidences of nature where all factors combine almost perfectly to provide a method of analysis which meets all the criteria for an ideal method. Although oxygen occurs in nature in the three isotopic forms of mass 16, 17, and 18, only the mass 16 isotope is present in sufficient abundance (99.76%) to be important for activation purposes. This isotope undergoes an (n, p) reaction ($\sigma = 42$ mb) with 14-MeV neutrons to produce ^{16}N which has a 7.2-sec half-life and decays with the emission of γ-rays predominantly of 6.1- and 7.1-MeV energy. This combination of parameters makes the method for oxygen almost unique. The short half-life provides for rapid saturation during irradiation, with 30 sec being sufficient. On the other hand, counting must commence as soon as possible after the termination of irradiation to maximize the sensitivity of the method. The nuclear data concerning the activation of oxygen with 14-MeV neutrons are given in Table 7.46. Figures 7.136 to 7.138 give the activation, decay, and integral counting curves for oxygen.

TABLE 7.46. Nuclear Data for Reactions with 14-MeV Neutrons—Oxygen

Nuclear Reaction	Target Isotope Abundance (%)	Cross Section (mb)	Half-life (min)	Gamma Energy (MeV)	Gamma Yield (%)	Detector Efficiency (%)	Peak/Total Ratio
$1 = {}^{16}O(n, p){}^{16}N$	99.76	42.00	0.12	6.13	69	23	0.14

The high-energy emissions from ^{16}N provide two distinct advantages. First, the method is remarkably free from interference by other matrix elements. The only common interference is from fluorine which produces the same isotope via an (n, α) reaction. Interference may occur from large amounts of boron or uranium present in the sample. In addition, the

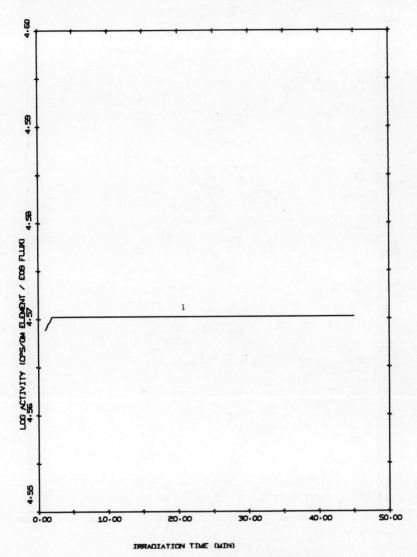

Fig. 7.136. Oxygen—activation curves (see Table 7.46).

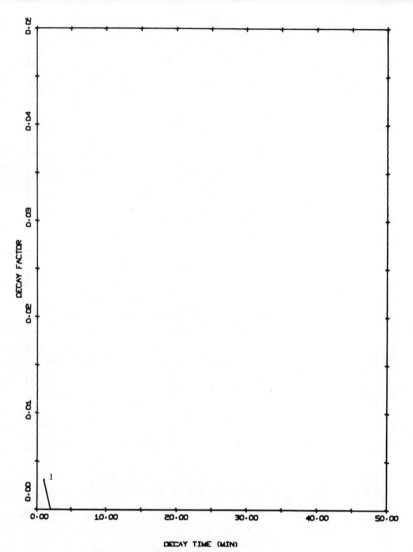

Fig. 7.137. Oxygen—decay curves (see Table 7.46).

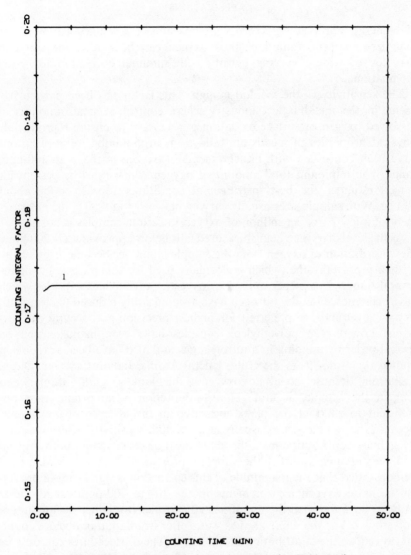

Fig. 7.138. Oxygen—integral counting curves (see Table 7.46).

high-energy γ-rays emitted by ^{16}N are less subject to self-absorption effects than less energetic radiation; thus oxygen can be readily measured in matrices of widely varying density with minimal correction for γ-ray attenuation.

The sensitivity of the method is high. This factor has been particularly useful in the metallurgical industry where control of small amounts of dissolved oxygen or surface oxidation is necessary to ensure reproducible physical properties of metals and alloys. A large number of applications have been concerned with measurements of oxygen in the 1- to 100-ppm range. The minimum total amount of oxygen which must be present for detection under the best instrumental conditions appears to be about 100 μg. With sample sizes usually varying between 1 and 50 g, the measurement of a few parts per million of oxygen in certain samples is possible.

As the level of oxygen being measured falls below approximately 200 ppm, the contribution of oxygen from the sample container becomes significant. Ordinary polyethylene (which is normally used for this purpose) contains several hundred parts per million of oxygen; polyethylene capsules which have been molded under nitrogen have a significantly reduced (factor of 10) oxygen level and are preferred for highest precision and accuracy at low levels. Low-oxygen polyethylene capsules have been manufactured on special order by molding in a nitrogen gas and also have been available in limited quantities from the Gulf General Atomic Laboratory, San Diego, California. Because the level of oxygen in the blank is significant relative to that in the sample, accurate blank subtraction is important. It is not sufficient to subtract the blank measured in the absence of the sample. Because the sample size is usually large in high-sensitivity determinations, the activity contribution from the blank must be corrected for both neutron and γ-ray attenuation when the sample is present. Nargolwalla et al.[74] have demonstrated that the magnitude of this correction is significant in providing accurate oxygen measurements in the 10- to 100-ppm range. Other workers have reported the use of partial capsules (to reduce the amount of polyethylene) or the removal of capsules after irradiation but before counting to reduce the contribution of the blank. These procedures can only be used for rod samples and require specialized transfer equipment. Some authors have used sample carriers fabricated from high-purity metals with low oxygen content. Guinn[75] has recommended the use of high-purity copper, and Priest and Burns[76] have made capsules from high-purity blister nickel. While the technique lends itself well to the determination of low levels of oxygen in a variety of materials, it also can be used successfully for the determination of macroconstituent oxygen.

The usefulness of the 14-MeV technique for the measurement of oxygen has received much attention. Wide application has been reported in the

III. ACTIVATION ANALYSIS CONDITIONS FOR THE ELEMENTS 479

literature in all types of matrices. A recent compendium[77] of literature references lists over 115 reports of the use of this technique for oxygen. Oxygen has been measured from macroconstituent levels down to trace amounts.

The neutron activation technique should be compared with several classical methods for the measurement of oxygen: the vacuum fusion method for oxygen in metals, the Unterzaucher method, and the combustion technique. The comparison of the 14-MeV technique with vacuum fusion has been made by a number of authors.[78-81] These comparisons have generally been favorable to the activation method due to increased speed of the analysis, usually with improved precision and accuracy. Results via neutron activation have been reported to provide slightly higher results on certain samples. This bias is viewed as the inability of the extraction method to decompose stable oxides completely and to losses due to "creeping" during vacuum fusion.[79] The National Bureau of Standards has utilized the neutron activation method for the certification of low oxygen levels in certain steel samples. Specific application of this method to steel samples has been reported by a number of authors.[79, 80, 82-86] The method has also been used for the measurement of low levels of oxygen in magnesium,[84] titanium,[84, 87] potassium,[88] aluminum,[89, 90] beryllium,[91-94] copper,[87, 95] iron,[87] nickel,[87] chromium,[92] zirconium,[92] tantalum,[81, 92] molybdenum,[96, 97] niobium,[81] and cesium.[98] While far from comprehensive, this diverse list of elements testifies to the broad utility of the method for measuring low levels of oxygen in virtually any metallic matrix.

In applying this method to macroconstituent oxygen analysis, the list of accomplishments is equally long. For these matrices the method should be contrasted with the more laborious Unterzaucher and combustion methods. Veal and Cook[99] noted that the 14-MeV technique provided a fivefold increase in speed over the conventional Unterzaucher method for oxygen in organic compounds. This method has been used in various mineral studies.[100-102] Volborth[102-105] has reported extensive studies with this technique for the measurement of the oxygen content of rocks. This technique combined with x-ray methods was reported to give a total nondestructive analysis of rocks. The method has also been applied to the measurement of macrolevels of oxygen in titanium and vanadium oxides,[106] silica, alumina, and calcium carbonate,[103] as well as organic compounds.[107] While this summary of applications for oxygen measurement may appear overly optimistic, the authors' own experience bears out the enthusiastic reports.

In the early use of the (n, p) reaction for measuring oxygen, well-type scintillation crystals without discriminator levels were used. This type of detector system does not provide sufficient discrimination from other γ-emissions, and nuclides which emit high-energy β-particles were found to interfere also. Anders and Briden[107] suggested the use of an energy window from 4.5 to 7.5 MeV using a 3 × 3-in. NaI crystal with a 1-cm-thick plastic β-shield. This basic detector system has come to be accepted for oxygen determinations and avoids interference from all elements except fluorine, boron, and uranium.

The fluorine interference can be handled by separately measuring the other nuclides that are produced by bombardment of fluorine with 14-MeV neutrons (see the section on fluorine) and correcting for its contribution to the ^{16}N activity produced. While this procedure is acceptable in the measurement of large amounts of oxygen in the presence of small amounts of fluorine, it is not useful for the inverse ratio of these two elements.

Oxygen is not activated with low-energy neutrons. The threshold value for the (n, p) reaction is 9.63 MeV. Thus, in the activation analysis of certain other elements which produce nuclides of short half-life where oxygen may interfere, the interference can be eliminated by using lower energy neutrons.

3-MeV AND THERMAL NEUTRONS

Oxygen is not activated by 3-MeV and thermal neutrons.

PALLADIUM

Palladium exists in nature in six isotopic forms, those of mass 102, 104 to 106, 108, and 110. Palladium can be readily measured by wet chemical and instrumental methods, although dissolving of palladium metal may pose a problem in preparing certain samples for chemical analysis. X-Ray fluorescence methods for palladium are satisfactory and of similar sensitivity to the fast neutron activation methods. Thus there has not been much work done in applying fast neutron activation procedures to the measurement of palladium.

14-MeV NEUTRONS

The important activation parameters for the determination of palladium with fast neutrons are given in Table 7.47. Figures 7.139 to 7.141 give the

TABLE 7.47. Nuclear Data for Reactions with 14-MeV Neutrons—Palladium

Nuclear Reaction	Target Isotope Abundance (%)	Cross Section (mb)	Half-life (min)	Gamma Energy (MeV)	Gamma Yield (%)	Detector Efficiency (%)	Peak/Total Ratio
1 = $^{105}Pd(n, p)^{105m}Rh$	22.60	800	0.75	0.13	100	50	0.98
2 = $^{106}Pd(n, p)^{106}Rh$	27.30	17	0.50	0.51	88	40	0.63
3 = $^{108}Pd(n, 2n)^{107m}Pd$	26.70	460	0.36	0.22	100	50	0.91
4 = $^{110}Pd(n, 2n)^{109m}Pd$	13.50	971	4.80	0.19	58	50	0.93

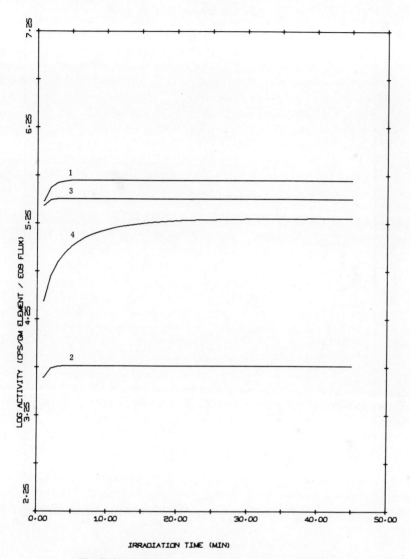

Fig. 7.139. Palladium—activation curves (see Table 7.47).

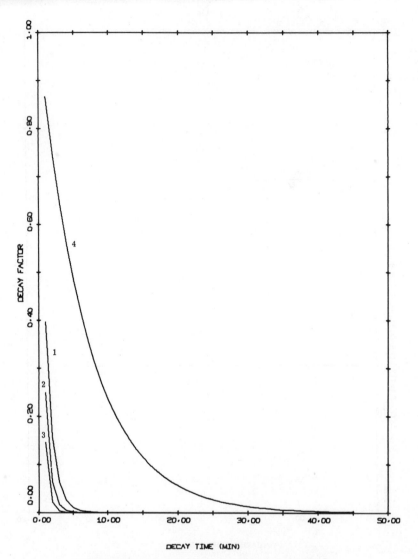

Fig. 7.140. Palladium—decay curves (see Table 7.47).

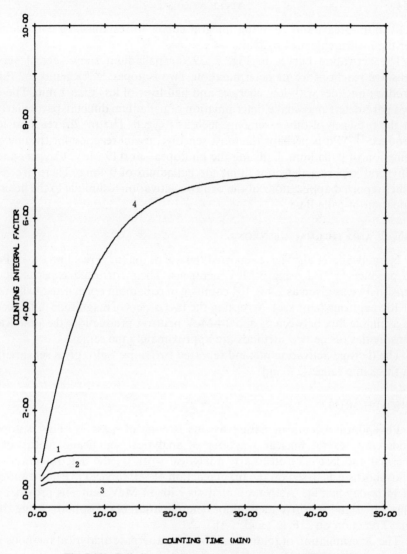

Fig. 7.141. Palladium—integral counting curves (see Table 7.47).

activation, decay, and counting integral curves for establishing conditions for the measurement of palladium.

The activation curves in Fig. 7.139 for palladium show several very sensitive reactions for its determination. Two isotopes, ^{105m}Rh and ^{107m}Pd, produce nuclides with low energies and half-lives of less than 1 min. These two parameters may make determination of palladium difficult, particularly in the presence of other elements such as oxygen. The (n, 2n) reaction to produce ^{109m}Pd is perhaps the most sensitive, useful reaction for the determination of palladium. Utilizing the photopeak at 0.19 MeV, Cuypers and Cuypers[1] report a detection limit for palladium of 0.4 mg. There are no other reported applications of the neutron activation technique to the determination of palladium.

3-MeV AND THERMAL NEUTRONS

Nargolwalla et al.[23] have reported the use of the (n, γ) reaction on ^{108}Pd to produce ^{109m}Pd using 3-MeV neutrons. Their corrected experimental sensitivity was given as 2.4×10^3 counts/g of palladium normalized to a flux of 10^6 neutrons/(cm^2 sec). Accepting the two orders of magnitude difference in available flux between 3- and 14-MeV neutron production, the practical sensitivities of the two methods are approximately the same.

The thermal activation method reported by Dibbs[24] also gives sensitivity in this same range (230 μg).

PHOSPHORUS

Phosphorus occurs in nature as one isotope of mass 31. This isotope undergoes several nuclear reactions of analytical significance with both thermal and fast neutrons. Early analytical work on the determination of phosphorus was based on the (n, γ) reaction of ^{31}P to yield ^{32}P, a β-emitter of 14.28-day half life. Activation analysis with 14-MeV neutrons provides a very rapid and sensitive method for determination of phosphorus using the (n, α) reaction on ^{31}P to yield ^{28}Al.

The determination of total phosphorus by alternate analytical methods is not altogether straightforward. In most wet chemical procedures, the phosphorus is first converted to phosphate by oxidation procedures from which it is then either precipitated as ammonium phosphomolybdate or magnesium ammonium phosphate, then dried or ignited to constant weight. Titrimetric methods for phosphorus are usually indirect methods based on the backtitration of unreacted metal ions that can be used to precipitate phosphorus. Some of the more sensitive photometric methods for determination of total phosphorus are based on the molybdenum blue determination which results from the initial formation of the phosphomolybdate acids.

Lanthanum chloranilate methods for the determination of orthophosphate have also been reported which are considerably simpler although not as selective as the phosphomolybdate acid methods. Emission spectrographic procedures for the determination of phosphorus use the 2535.65-Å line and are especially valuable for the determination of phosphorus in petroleum products. It is reported that a single determination of phosphorus using the emission method requires approximately 30 min. Flame emission methods based on the determination of the emission from the phosphorus-oxygen species have been reported to give accuracies of the order of 1% provided other interfering elements are absent.

14-MeV NEUTRONS

In comparing the 14-MeV neutron activation method for phosphorus with the foregoing techniques, it can clearly be said to be competitive. The method is reasonably selective and highly sensitive in addition to providing a method that is quite rapid. The important parameters for activation analysis of phosphorus with 14-MeV neutrons are given in Table 7.48.

TABLE 7.48. Nuclear Data for Reactions with 14-MeV Neutrons—Phosphorus

Nuclear Reaction	Target Isotope Abundance (%)	Cross Section (mb)	Half-life (min)	Gamma Energy (MeV)	Gamma Yield (%)	Detector Efficiency (%)	Peak/Total Ratio
$1 = {}^{31}P(n, \alpha){}^{28}Al$	100.00	150	2.31	1.78	100	28	0.29
$2 = {}^{31}P(n, p){}^{31}Si$	100.00	83	157.20	1.26	1	30	0.37
$3 = {}^{31}P(n, 2n){}^{30}P$	100.00	11	2.50	0.51	100	40	0.63

Figures 7.142 to 7.144 show the activation, decay, and integral counting curves. Two nuclides are important in the measurement of phosphorus: ^{28}Al, formed by the (n, α) reaction, is the most sensitive with a reported detection limit of 40 μg and ^{30}P, a positron emitter that gives a sensitivity of approximately 900 μg of phosphorus. The choice between these two nuclides for the measurement of phosphorus will, of course, be determined by the composition of the matrix in which the phosphorus is measured and the needed sensitivity of the measurement. Using the 1.78-MeV photopeak from ^{28}Al, only silicon and aluminum are known interferences and corrections for their presence in a sample can be made. If aluminum is known to be present in the sample, wrapping the sample with 0.02-in. cadmium foil reduces the aluminum interference to an insignificant level even where there is 1000 times as much aluminum as phosphorus in the sample. Conversely, the sensitivity of phosphorus is not appreciably reduced by the presence of

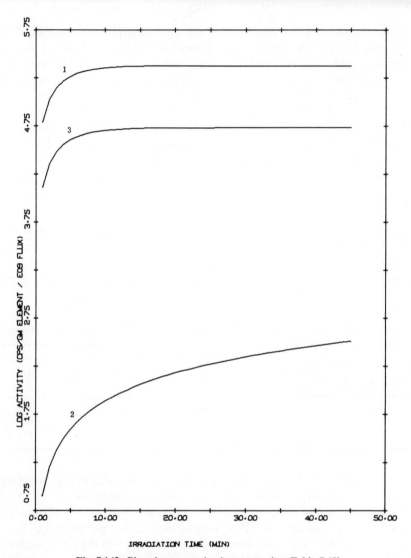

Fig. 7.142. Phosphorus—activation curves (see Table 7.48).

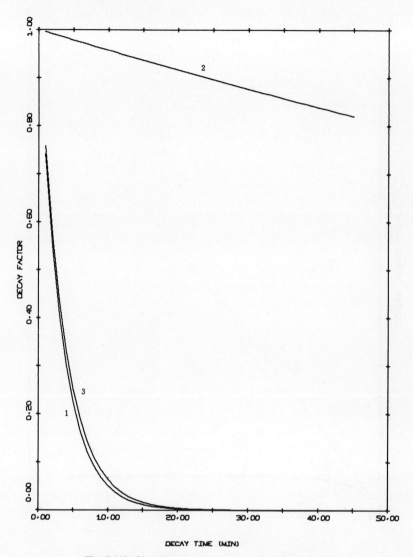

Fig. 7.143. Phosphorus—decay curves (see Table 7.48).

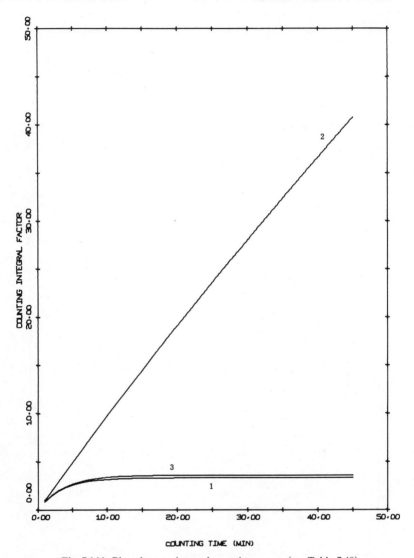

Fig. 7.144. Phosphorus—integral counting curves (see Table 7.48).

the cadmium. The interference from silicon is more severe since the cross section for silicon activation is approximately twice that for the (n, α) reaction on phosphorus. This interference can be corrected for by use of the ^{28}Al photopeak at 1.28 MeV which is produced from the ^{29}Si reaction, as a correction.

The major interfering reactions of silicon and aluminum in the determination of phosphorus are given in Table 7.49. Activation, decay, and integral counting curves for these interferences are given in Figs. 7.145 to 7.147. For possible interferences with the measurement of positron-emitting ^{30}P see Tables 7.76 to 7.78.

TABLE 7.49. Nuclear Data for Reactions with 14-MeV Neutrons—Phosphorus Interferences

Nuclear Reaction	Target Isotope Abundance (%)	Cross Section (mb)	Half-life (min)	Gamma Energy (MeV)	Gamma Yield (%)	Detector Efficiency (%)	Peak/Total Ratio
1 = ^{28}Si(n, p)^{28}Al	92.27	235	2.31	1.78	100	28	0.29
2 = ^{27}Al(n, γ)^{28}Al	100.00	0.50	2.31	1.78	100	28	0.29

A large number of methods have been reported in the literature for the determination of phosphorus using 14-MeV neutrons. Selected references are surveyed here to indicate the kinds of applications that have been made and the corrective procedures that have been used to compensate for matrix interferences. To-on and co-workers[108] first reported the determination of phosphorus by fast neutron activation analysis using the (n, α) reaction. They calculated a limit of detection of 48 ppm of phosphorus based on an activity of 40 dis/sec as being the minimum detection limit from a flux of 10^8 neutrons/(cm^2 sec) and a sample size of 1 g. These authors studied the interference of aluminum and silicon in the reaction and found that the contribution of aluminum to the 1.78-MeV photopeak was negligible for a weight ratio of aluminum to phosphorus of 5. For larger quantities of aluminum, the interference was reduced to a negligible level by surrounding the sample with a 0.020-in. cadmium sheet. Interference from silicon was found to contribute a 1% error in the analysis of phosphorus when the silicon-to-phosphorus ratio was 0.65. These authors found it advisable to volatilize the silicon in samples containing silicon oxide with hydrofluoric acid prior to activation. An extensive study of the determination of phosphorus in a variety of matrices including pure samples of ammonium dihydrogen phosphate, lubricating oils, soluble phosphate materials, biological materials, fertilizers, and phosphate rock samples is given by

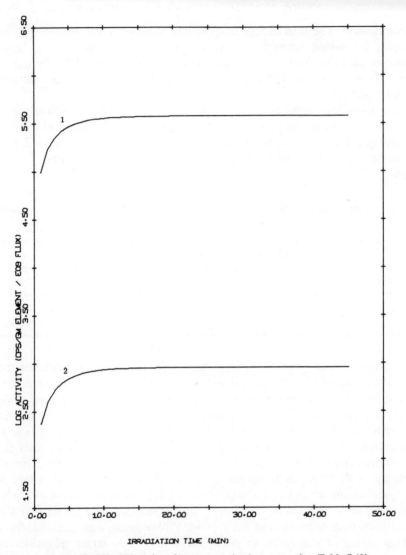

Fig. 7.145. Phosphorus interferences—activation curves (see Table 7.49).

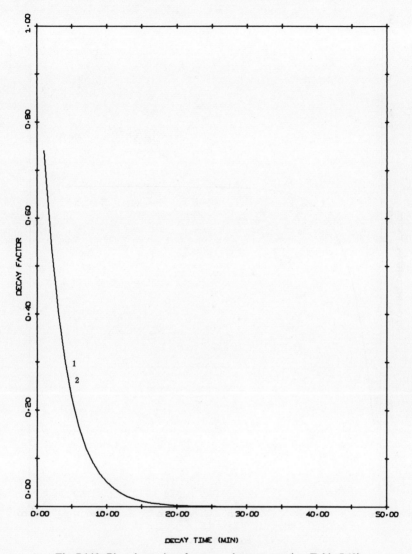

Fig. 7.146. Phosphorus interferences—decay curves (see Table 7.49).

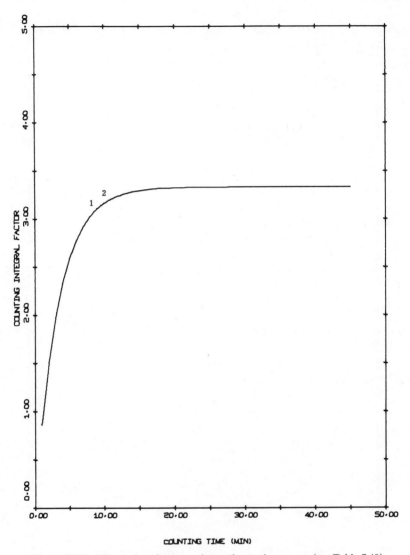

Fig. 7.147. Phosphorus interferences—integral counting curves (see Table 7.49).

To-on.[109,110] The technique of volatilizing silicon was used in the determination of phosphorus in fertilizer and phosphate rock samples. Results for phosphorus were reported to be low by as much as 4.7% under these conditions. Prapuolenis and Bakes[111] determined phosphorus and nitrogen in herbage flour by 14-MeV neutron activation analysis. These authors used ^{29}Al at 1.28 MeV as a measure of silicon contribution to the phosphorus measurement at 1.78 MeV. Unfortunately, the authors did not show a quantitative comparison of this correction method for the contribution from silicon. In principle there is no reason why the presence of modest amounts of silicon relative to the phosphorus cannot be properly taken into account using this approach.

Several authors have proposed the use of 14-MeV neutron activation analysis to determine the elemental composition of pure organic compounds. Ricci[112] illustrates the use of this technique for the determination of phosphorus in a typical phosphine oxide derivative, while Crambes et al.[73] proposed this technique for the determination of phosphorus in certain proteins. Indications are that the method has the required precision and accuracy to compete favorably with classical methods for phosphorus determination which are used to establish empirical formulas of organic compounds. Schramel[38] has reported the determination of phosphorus in a complex biological matrix containing silicon, chlorine, potassium, calcium, and aluminum. Decay curve analysis was used for the determination of phosphorus and a correction for the presence of silicon based on the counting of ^{29}Al was used. Palmer et al.[113] discussed the feasibility of determining phosphorus *in vivo* using the ^{28}Al nuclide. Indications are that sufficient activity can be obtained using an exposure of 16 mrad of 14-7-MeV neutrons. A whole-body counter was used to measure the γ-rays produced. These calculations were based on uniform whole-body radiation.

Broadhead et al.[12] reported on the determination of phosphorus in molten salts. These authors report a sensitivity of 10 ppm for the determination of phosphorus in a 10-g sample. Correction for silicon interference was made based on emission spectrographic analysis. Chomel[114] reported on the determination of phosphorus in the presence of sulfur using this technique. A reported sensitivity of 10 μg of phosphorus was given. Hull and Gilmore[28] illustrated the determination of a complex mixture of elements using 14-MeV activation followed by γ-ray spectral analysis using computer techniques. Among the elements determined was phosphorus. The authors used a scheme of decay curve analysis to separate interferences. The matrix used in these studies was lubricating oil containing calcium, barium, zinc, nitrogen, phosphorus, oxygen, and chlorine. The general procedures are outlined in detail in their paper; however, only limited statistical data are given to indicate the precision of the method. Gibbons et al.[45] determined phosphorus in lubricating oils in the range of 0.01 to 1%. These authors studied the simultaneous determination of barium and phosphorus and found that decay curve analysis gave higher results than analysis by direct γ-ray spectrometry. Their results for the determination of phosphorus show relative deviations ranging from 2.7 to 6.7% for phosphorus. While Gibbons and his associates recognize the interference of silicon, they made no correction for its contribution. Tsuji[69] reported the measurement of phosphorus in a variety of materials including bone powder, synthetic fertilizer, fish powders, calcium phosphate, triphenyl phosphate and benzoylthioamino-o-monophosphate. He has also shown comparative data using a chemical analysis procedure for the determination of phosphorus. In the analysis of calcium phosphate, the activation method gave results that were approximately 2% high, which may be attributed to interference from silicon impurities. Comparative analyses on other materials showed remarkably good agreement between the methods except in the analysis of synthetic fertilizer where presumably silicon was present. Babikova et al.[115] reported on the determination of impurities (including phosphorus) in steel.

VanGrieken et al.[116] have reported on a simultaneous determination of silicon and phosphorus in cast iron. These authors use a coincidence technique to measure the activity at 0.51 MeV. Copper was found to interfere in this procedure although the samples studied by these authors were low in copper so that corrections for its presence did not have to be made. VanGrieken and co-workers present a method of simultaneous equations for correcting for the presence of copper if it is present in the material. Other interferences commonly found in the steel matrix were studied and it was determined that 1% of chromium introduces an error of 0.002% phosphorus, and 1% molybdenum causes an error of 0.05%. The effect of the presence of aluminum, carbon, cobalt, lead, magnesium, manganese, nickel, nitrogen, oxygen, sulfur, tin, titanium, tungsten, vanadium, or zirconium appeared to be negligible for the concentrations normally occurring in iron and steel. The maximum interference of any of these elements at the 1% level was 0.003% in phosphorus. Excellent agreement was found between chemical analyses and the neutron activation results, with agreement being generally within 1%. The experimental standard deviations in the phosphorus determination were in good agreement with those calculated from counting statistics. However, because of the low level of phosphorus, the standard deviation on a single determination was found to be quite high. Whereas the total analysis time of 10 min is quite short in comparison to chemical methods for the determination of phosphorus in steel, the technique described by these authors gave only sufficient precision when more than 0.8% phosphorus was present in the sample. In addition, the precision was unfavorably influenced by the presence of copper.

III. ACTIVATION ANALYSIS CONDITIONS FOR THE ELEMENTS

3-MeV AND THERMAL NEUTRONS

The sensitivity of the 3-MeV and thermal neutron reactions with phosphorus is too low to be of analytical value.[23, 24]

POTASSIUM

The two isotopes of potassium which exist in nature are ^{39}K and ^{41}K. Potassium is easily determined in solution by flame emission and atomic absorption techniques. Alternate solution methods are not available because potassium does not form complexes or insoluble salts very readily. The ease with which potassium can be determined by flame spectroscopy has not created a need for alternate methods of analysis. On the other hand, for remote location analyses such as extraterrestrial explorations and bore-hole logging, the neutron activation technique may provide a useful approach.

14-MeV NEUTRONS

The pertinent nuclear parameters for the activation of potassium with fast neutrons are given in Table 7.50, and Figs. 7.148 to 7.150 give the activation, decay, and counting integral curves for the determination of potassium.

The most sensitive reaction for the determination of potassium by activation with 14-MeV neutrons is the (n, 2n) reaction on ^{39}K which yields ^{38}K ($T_{1/2} = 7.71$ min). Cuypers and Cuypers[1] list a detection limit of 3.4 mg of potassium as measured by this reaction. Dibbs[24] reported significantly higher sensitivity of 0.19 mg, using somewhat longer irradiation and decay times. Garrec[121] has used this reaction for the nondestructive analysis of plant materials for their potassium content. Possible interferences with this reaction are listed in Tables 7.76 to 7.78. Potassium activates to form nuclides which have useful emissions at 1.29 MeV from ^{41}Ar and a second ^{38}K emission at 2.20 MeV. The sensitivity of these peaks for the determination of potassium, however, is about fivefold less than that of the annihilation radiation.

TABLE 7.50. Nuclear Data for Reactions with 14-MeV Neutrons—Potassium

Nuclear Reaction	Target Isotope Abundance (%)	Cross Section (mb)	Half-life (min)	Gamma Energy (MeV)	Gamma Yield (%)	Detector Efficiency (%)	Peak/Total Ratio
1 = ^{41}K(n, α)^{38}Cl	6.91	31.4	37.30	0.66	100	37	0.54
2 = ^{39}K(n, 2n)^{38}K	93.09	4	7.71	0.51	200	40	0.63
3 = ^{39}K(n, 2n)^{38}K	93.09	4	7.71	2.17	100	26	0.25
4 = ^{41}K(n, p)^{41}Ar	6.91	81.2	109.80	1.29	99	30	0.36

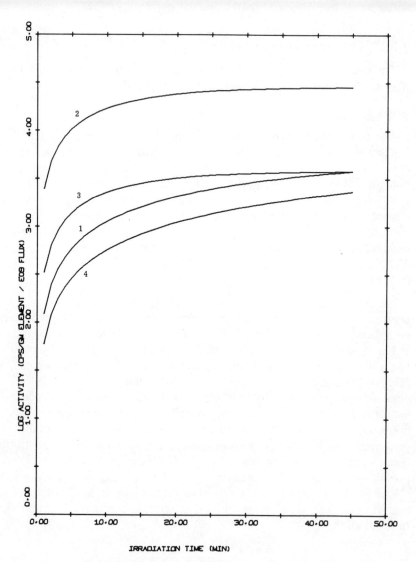

Fig. 7.148. Potassium—activation curves (see Table 7.50).

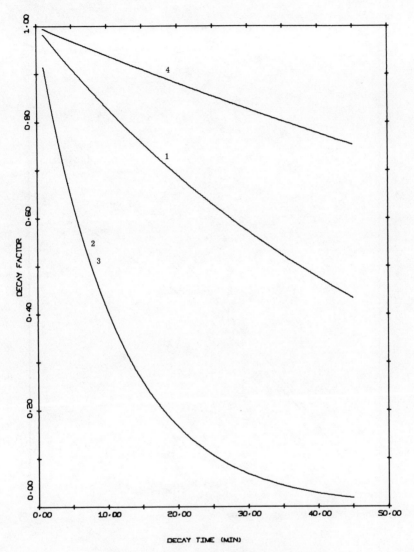

Fig. 7.149. Potassium—decay curves (see Table 7.50).

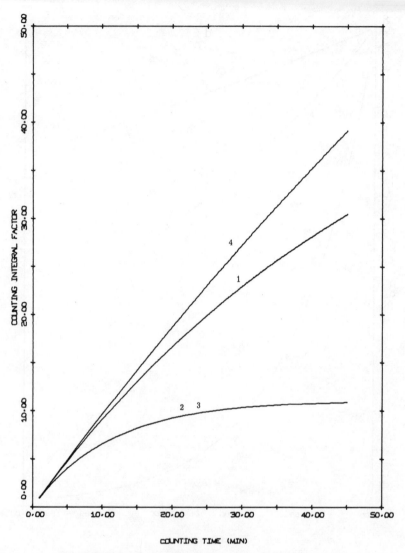

Fig. 7.150. Potassium—integral counting curves (see Table 7.50).

III. ACTIVATION ANALYSIS CONDITIONS FOR THE ELEMENTS

Waggoner and Knox[13] have used activation with 14-MeV neutrons to measure potassium using inelastic scattering of neutrons. For potassium, neutron inelastic scattering yields a γ-emission at 2.81 MeV. These authors built a prototype device to measure potassium as well as other elements in remote locations such as the lunar surface. The sensitivity for potassium measurement was such that levels above 4% were detectable in soil samples.

3-MeV AND THERMAL NEUTRONS

Sensitivity for the activation analysis of potassium with 3-MeV or thermal neutrons appears to be too low to be of value in activation analysis.[23, 24]

PRASEODYMIUM

Praseodymium occurs in nature in one isotope of mass 141. This isotope has a good reaction cross section for 14-MeV neutrons. Neutron activation methods for the determination of praseodymium are attractive because it can be selectively determined in the presence of other rare earths with which it commonly occurs in nature.

Because of the similarity of solution chemistry of many of the rare earths, a direct determination of praseodymium in the presence of the other rare earths is difficult by most chemical methods. Perhaps the most popular method for carrying out such analyses is by ion-exchange separation of the rare earths and then spectral or gravimetric determination. On the other hand, spectroscopic methods such as atomic absorption and emission spectroscopy do not provide a high degree of sensitivity for the determination of praseodymium, with the atomic absorption detection limit placed at about 10 ppm of praseodymium solution.

With these limitations, the neutron activation method takes on some attraction as an alternate method despite the fact that the nuclide measured from the activation of praseodymium is a positron emitter of 3.39-min half-life. The most common interferences for praseodymium will be those elements such as phosphorus, bromine, and potassium which produce positron emitters whose half-lives are similar to that of ^{140}Pr (cf. Tables 7.76–7.78). The interference from phosphorus is perhaps the most serious.

14-MeV NEUTRONS

Pertinent data for 14-MeV neutron activation are listed in Table 7.51. The activation, decay, and integral counting curves are shown in Figs. 7.151 to 7.153. The primary applications reported for the 14-MeV neutron activation of praseodymium have been made on ores and minerals. Cuypers and Menon[39] demonstrated the utility of this technique for the determination of cerium and praseodymium in ores and minerals. Their results showed a sensitivity of at least 150 μg for praseodymium using the annihilation radiation. Both praseodymium and cerium were determined simultaneously after a single irradiation. These authors reported some interference from the

TABLE 7.51. Nuclear Data for Reactions with 14-MeV Neutrons—Praseodymium

Nuclear Reaction	Target Isotope Abundance (%)	Cross Section (mb)	Half-life (min)	Gamma Energy (MeV)	Gamma Yield (%)	Detector Efficiency (%)	Peak/Total Ratio
1 = ^{141}Pr(n, 2n)^{140}Pr	100.00	2080	3.39	0.51	100	40	0.63
2 = ^{141}Pr(n, 2n)^{140}Pr	100.00	2080	3.39	1.60	1	29	0.31

^{31}P(n, 2n)^{30}P reaction due to the presence of phosphate in certain monazite ores. Calculation showed that the contribution to the praseodymium measurement was less than 10%. A more detailed survey of various ores examined by this technique was reported by the same authors.[40] Plaksin and his associates[64] also demonstrated the utility of this technique for the determination of a variety of rare earth elements including praseodymium. These authors made use of an anticoincident spectrometer to reduce contributions to low-energy peaks from the 1.78-MeV peak due to silicon, aluminum, and phosphorus, usually found in such samples. The authors reported significant improvement in the determination of the very low energy peaks due to terbium, erbium, and praseodymium. They report an error (standard deviation) of 1.2% in the determination of samarium by this technique. Using 10-g samples, the detection sensitivity for praseodymium was 4×10^{-4}% praseodymium in the sample of oxide. In another report on the determination of praseodymium in rare earth ores,[63] a sensitivity of 2 μg of praseodymium was reported using a flux of 5×10^8 neutrons/(cm^2 sec). In this application, correction was made for the contribution of samarium, but lanthanum, europium, and gadolinium were found not to interfere. Fujii et al.[118, 119] reported a rapid method for the determination of praseodymium in rare earth ores over the concentration range of 0 to 20% with an accuracy of 3%. These authors noted that the method is comparable in accuracy and sensitivity with the conventional photometric and gravimetric methods. They utilized a coincidence method to minimize interference from other elements. Broadhead and co-workers[12] noted the application of this method for metallurgical studies of vanadium and titanium alloys. In such matrices the sensitivity for praseodymium was reported to be 300 ppm. No details of the analysis technique used are given in this publication.

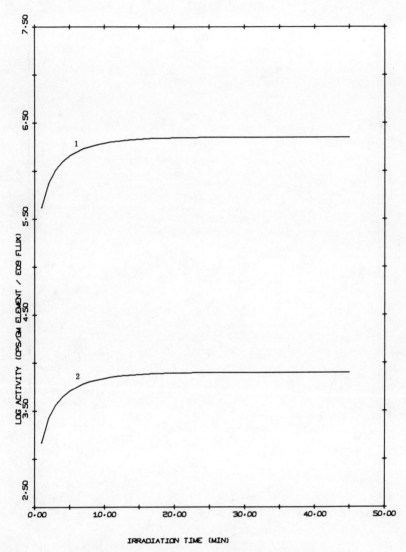

Fig. 7.151. Praseodymium—activation curves (see Table 7.51).

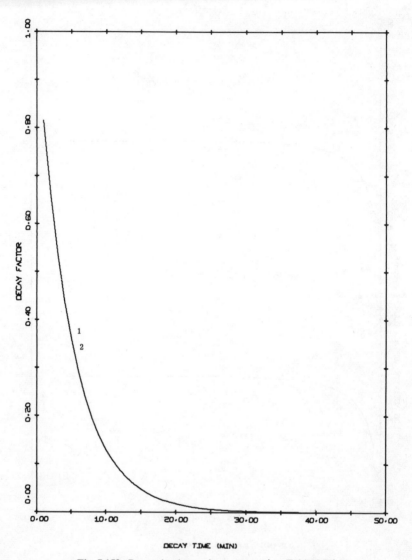

Fig. 7.152. Praseodymium—decay curves (see Table 7.51).

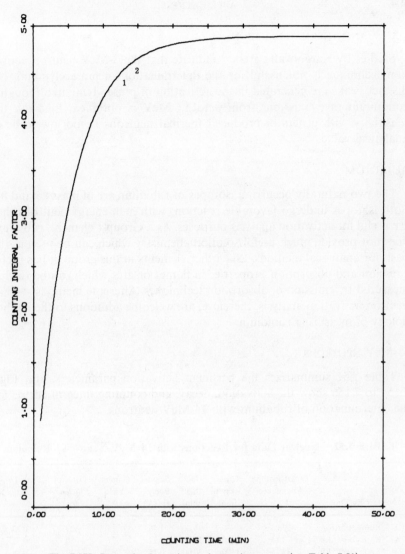

Fig. 7.153. Praseodymium—integral counting curves (see Table 7.51).

3-MeV AND THERMAL NEUTRONS

Studies by Nargolwalla et al.[23] indicate that the 3-MeV neutron activation technique is not useful for the determination of praseodymium. No distinct peaks are generated in the activation of praseodymium although a continuum of γ-emissions from 0 to 1.6 MeV is observed. Likewise the sensitivity with generator-produced thermal neutrons is too low to be of analytical value.[24]

RUBIDIUM

The two naturally occurring isotopes of rubidium are of mass 85 and 87. Both isotopes undergo favorable reactions with high-energy neutrons that are useful for activation analysis purposes. As a Group I element, rubidium does not provide much useful solution chemistry which can be used as the basis for analytical methods. Like other elements in this group, it has useful emission and absorption properties in flames or arcs which can be readily measured by emission or absorption techniques. Alternate methods such as neutron activation analysis, therefore, are welcome additions to the methodology of measuring rubidium.

14-MeV NEUTRONS

Table 7.52 summarizes the pertinent activation parameters, and Figs. 7.154 to 7.156 show the activation, decay, and counting integral curves for the determination of rubidium with 14-MeV neutrons.

TABLE 7.52. Nuclear Data for Reactions with 14-MeV Neutrons—Rubidium

Nuclear Reaction	Target Isotope Abundance (%)	Cross Section (mb)	Half-life (min)	Gamma Energy (MeV)	Gamma Yield (%)	Detector Efficiency (%)	Peak/Total Ratio
1 = ^{87}Rb(n, α)^{84}Rb	27.85	38.9	6.00	1.46	75	29	0.33
2 = ^{85}Rb(n, 2n)^{84m}Rb	72.15	1300	20.00	0.22	37	50	0.91
3 = ^{85}Rb(n, 2n)^{84m}Rb	72.15	1300	20.00	0.25	65	49	0.88
4 = ^{85}Rb(n, 2n)^{84m}Rb	72.15	1300	20.00	0.46	32	42	0.67
5 = ^{87}Rb(n, 2n)^{86m}Rb	27.85	950	1.04	0.56	100	39	0.59

The most sensitive reaction to use for the determination of rubidium is the (n, 2n) reaction on ^{85}Rb producing ^{84m}Rb. Utilizing both photopeaks at 0.22 and 0.25 MeV as the basis for analysis, Cuypers and Cuypers[1] reported a detection limit of 0.12 mg of rubidium. Dibbs[24] reported a tenfold better

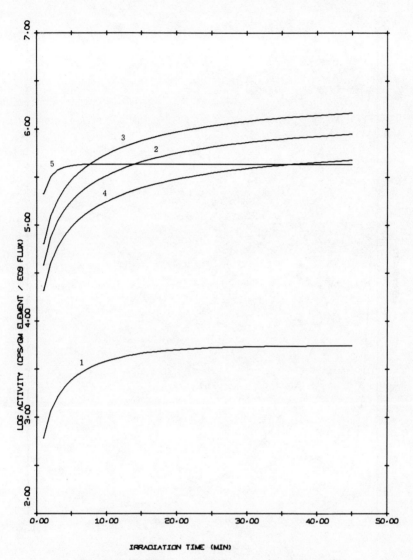

Fig. 7.154. Rubidium—activation curves (see Table 7.52).

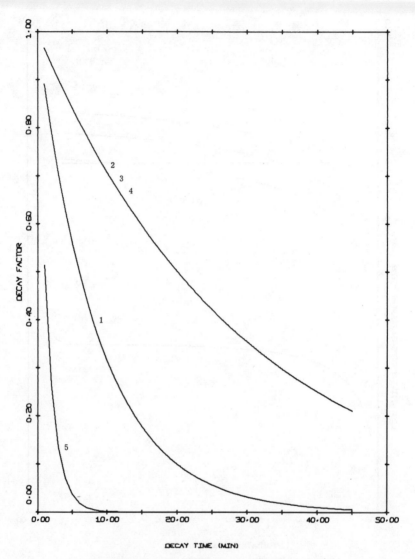

Fig. 7.155. Rubidium—decay curves (see Table 7.52).

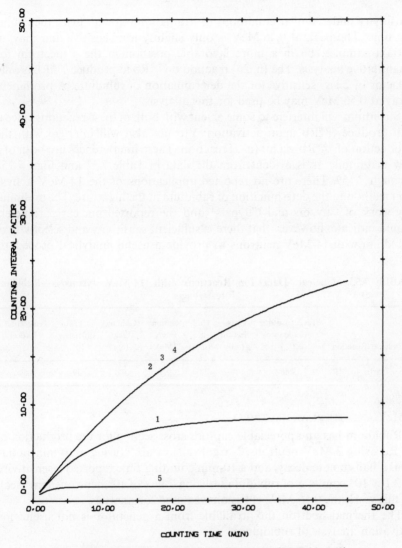

Fig. 7.156. Rubidium—integral counting curves (see Table 7.52).

sensitivity under similar conditions of irradiation but with twice the counting time. The peak at 0.46 MeV is only slightly less sensitive and may, in certain samples, be in a more favorable position in the γ-spectrum for quantitative analysis. The (n, 2n) reaction on ^{87}Rb to produce ^{86m}Rb is only a factor of 2 less sensitive for the determination of rubidium. A prominent γ-ray at 0.56 MeV may be used for this analysis.

Strontium will interfere to some extent with both of these reactions since it will produce ^{86m}Rb upon activation. Yttrium also will interfere with the production of ^{86m}Rb via an (n, α) reaction. These interferences are both of a low magnitude as is evident from the data in Table 7.53 and Figs. 7.157 through 7.159. There are no reported applications of the 14-MeV activation method to the determination of rubidium in the literature. The preliminary work of Cuypers and Cuypers[1] and the reported nuclear parameters would indicate, however, that there is sufficient sensitivity and selectivity of rubidium with 14-MeV neutrons to provide a useful analytical procedure.

TABLE 7.53. Nuclear Data for Reactions with 14-MeV Neutrons—Rubidium Interferences

Nuclear Reaction	Target Isotope Abundance (%)	Cross Section (mb)	Half-life (min)	Gamma Energy (MeV)	Gamma Yield (%)	Detector Efficiency (%)	Peak/Total Ratio
$1 = {}^{86}Sr(n, p){}^{86m}Rb$	9.86	39	1.04	0.56	100	39	0.59

3-MeV AND THERMAL NEUTRONS

Rubidium has an appreciable capture cross section for the production of ^{86m}Rb using 3-MeV neutrons. Nargolwalla et al.,[23] using a 10-min radiation, a half-minute decay, and a 10-min counting time reported a sensitivity of 2.1×10^3 counts/g of rubidium utilizing a flux of 10^6 neutrons/(cm^2 sec). In this work the 0.56-MeV emission was used for analysis.

The thermal neutron flux available from a generator is not useful for activation analysis of rubidium.[24]

RUTHENIUM

Ruthenium exists in nature in seven isotopes of mass 96, 98 through 102, and 104.

14-MeV NEUTRONS

The important activation parameters for ruthenium are listed in Table 7.54. The activation, decay, and integral counting curves are given in Figs.

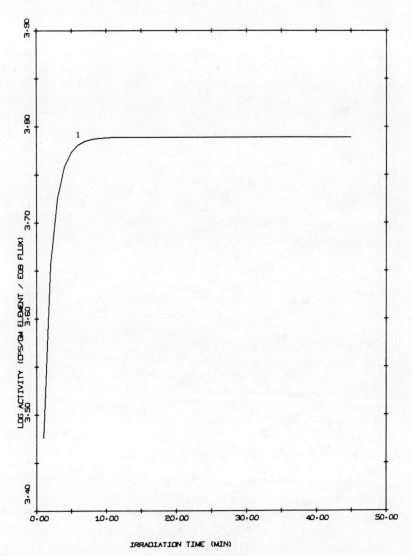

Fig. 7.157. Rubidium interferences—activation curves (see Table 7.53).

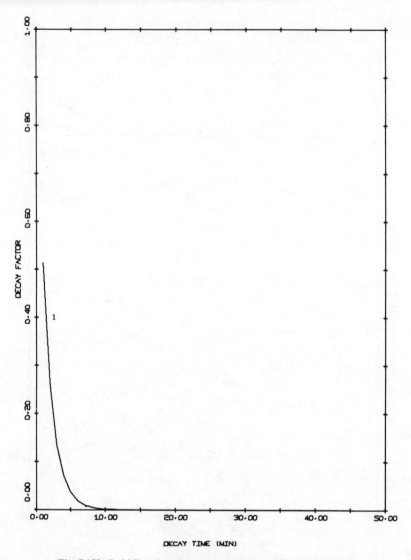

Fig. 7.158. Rubidium interferences—decay curves (see Table 7.53).

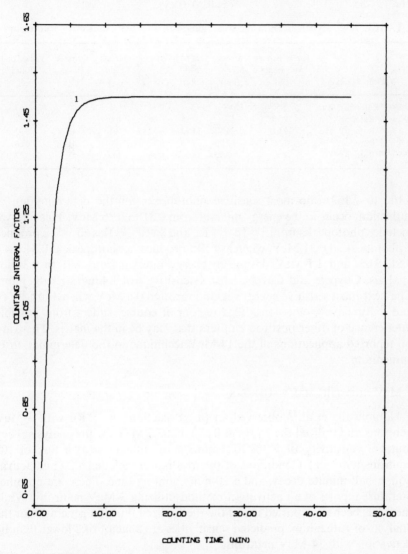

Fig. 7.159. Rubidium interferences—integral counting curves (see Table 7.53).

TABLE 7.54. Nuclear Data for Reactions with 14-MeV Neutrons—Ruthenium

Nuclear Reaction	Target Isotope Abundance (%)	Cross Section (mb)	Half-life (min)	Gamma Energy (MeV)	Gamma Yield (%)	Detector Efficiency (%)	Peak/Total Ratio
1 = ^{96}Ru(n, 2n)^{95}Ru	5.50	1200	102	0.51	30	40	0.63
2 = ^{100}Ru(n, p)^{100}Tc	12.70	11	0.28	0.60	100	38	0.57
3 = ^{101}Ru(n, p)^{101}Tc	17.01	36	14.00	0.31	91	46	0.83
4 = ^{102}Ru(n, p)^{102}Tc	31.50	2.70	4.50	0.47	100	41	0.66
5 = ^{104}Ru(n, p)^{104}Tc	18.50	7.20	18.00	0.36	100	45	0.77

7.160 to 7.162. The most sensitive photopeaks for the determination of ruthenium occur in the energy interval from 0.31 to 0.36 MeV. This interval includes photopeaks from ^{101}Tc, ^{104}Tc, and ^{95}Ru. ^{101}Tc and ^{104}Tc produce a photopeak at 0.31 MeV, whereas ^{95}Ru produces a photopeak at 0.34, 0.47, 0.51, 0.63, and 1.1 MeV. Using an energy interval from 0.31 to 0.34 for analysis, Cuypers and Cuypers[1] list a sensitivity of 1.4 mg for ruthenium. The next most sensitive energy region is around 0.5 MeV where both ^{102}Tc and ^{95}Ru have γ-emissions. This region, of course, suffers from possible interference of other positron emitters that may be in the matrix. There are no reported applications of the 14-MeV technique to the determination of ruthenium.

3-MeV AND THERMAL NEUTRONS

Nargolwalla et al.[23] observed an (n, γ) reaction on ^{104}Ru with 3-MeV neutrons and utilized the γ-rays at 0.67 and 0.73 MeV for the analysis. They found a sensitivity of 7.3×10^2 counts/g of ruthenium at a flux of 10^6 neutrons/(cm^2 sec). Conditions of the irradiation included a 20-min activation, a half-minute decay, and a 20-min counting time. There are no other literature reports of an activation method utilizing 3-MeV neutrons. While there has been no reported use of the thermalized flux of a generator for the analysis of ruthenium, predicted sensitivities are a factor of 2 lower than by activation with 14-MeV neutrons.[24]

SAMARIUM

The seven isotopes of samarium which exist in nature have masses of 144, 147 through 150, 152, and 154, respectively. Samarium belongs to the group of rare earth elements whose properties have sufficient similarity that they are difficult to differentiate by wet chemical and certain instrumental methods of analysis. Thus for certain matrices, neutron activation analysis may provide a good means of differentiation.

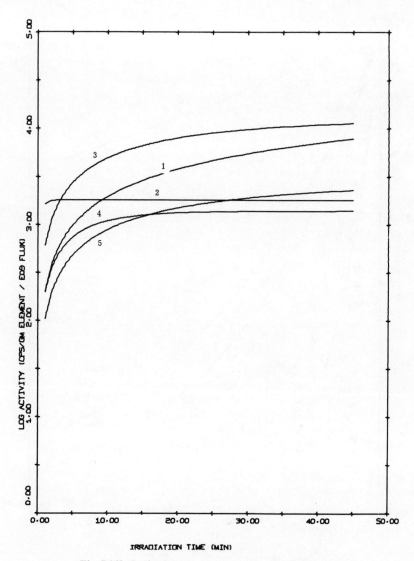

Fig. 7.160. Ruthenium—activation curves (see Table 7.54).

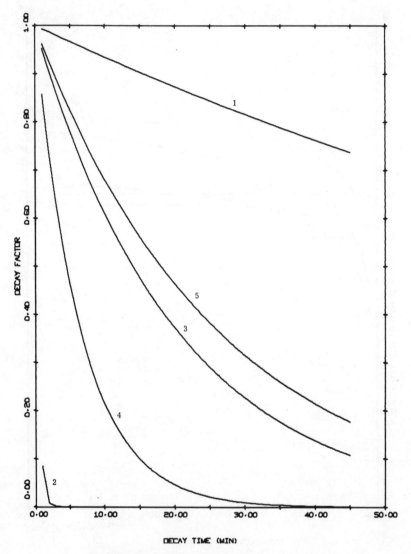

Fig. 7.161. Ruthenium—decay curves (see Table 7.54).

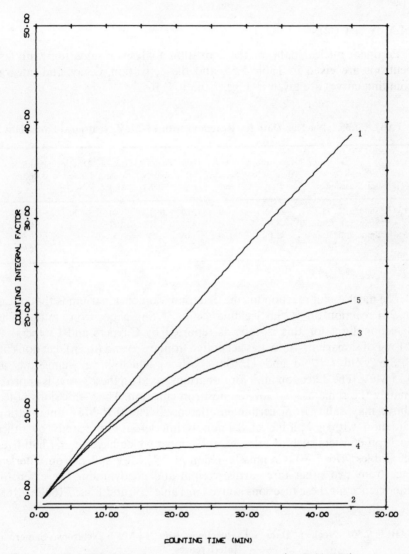

Fig. 7.162. Ruthenium—integral counting curves (see Table 7.54).

14-MeV NEUTRONS

Pertinent nuclear data for the activation analysis of samarium with fast neutrons are given in Table 7.55, and the activation, decay, and integral counting curves are given in Figs. 7.163 to 7.165.

TABLE 7.55. Nuclear Data for Reactions with 14-MeV Neutrons—Samarium

Nuclear Reaction	Target Isotope Abundance (%)	Cross Section (mb)	Half-life (min)	Gamma Energy (MeV)	Gamma Yield (%)	Detector Efficiency (%)	Peak/Total Ratio
1 = ^{154}Sm(n, α)^{151}Nd	22.53	9	12.00	0.11	40	50	1.00
2 = ^{144}Sm(n, 2n)^{143}Sm	3.16	1200	8.90	0.51	100	40	0.63
3 = ^{144}Sm(n, 2n)^{143m}Sm	3.16	400	1.06	0.75	100	35	0.50
4 = ^{152}Sm(n, p)^{152}Pm	26.63	3.7	6.00	0.12	100	50	0.99

The most useful reaction for the determination of samarium is that of the (n, 2n) reaction on ^{144}Sm yielding 9-min ^{143}Sm, a positron emitter. The detection limit for this isotope as reported by Cuypers and Cuypers[1] is 2.1 mg of samarium. Activities generated from (n, α) and (n, 2n) reactions on ^{144}Sm yield ^{141m}Nd and ^{143m}Sm, both of which have γ-emissions at 0.75 MeV. The detection limit for samarium based on these γ-rays is approximately the same as that for the positron emission. The γ-emissions noted above may suffer from certain interferences. The 0.51-MeV annihilation radiation with a half-life of 8.9 min is interfered with readily by other positron emitters such as nitrogen and copper which have similar half-lives (cf. Tables 7.76–7.78). Gamma emission at 0.75 MeV suffers from interference from two other rare earths, cerium and neodymium. The extent of interference for these reactions is given in Table 7.56 and Figs. 7.166 to 7.168.

TABLE 7.56. Nuclear Data for Reactions with 14-MeV Neutrons—Samarium Interferences

Nuclear Reaction	Target Isotope Abundance (%)	Cross Section (mb)	Half-life (min)	Gamma Energy (MeV)	Gamma Yield (%)	Detector Efficiency (%)	Peak/Total Ratio
1 = ^{140}Ce(n, 2n)^{130m}Ce	88.48	1200	0.92	0.75	93	35	0.50
2 = ^{142}Nd(n, 2n)^{141m}Nd	27.13	545	1.05	0.76	100	35	0.49
3 = ^{142}Nd(n, α)^{139m}Ce	27.13	10	0.92	0.75	93	35	0.50
4 = ^{63}Cu(n, 2n)^{62}Cu	69.10	550	9.80	0.51	195	40	0.63
5 = ^{14}N(n, 2n)^{13}N	99.63	5.7	9.96	0.51	200	40	0.63

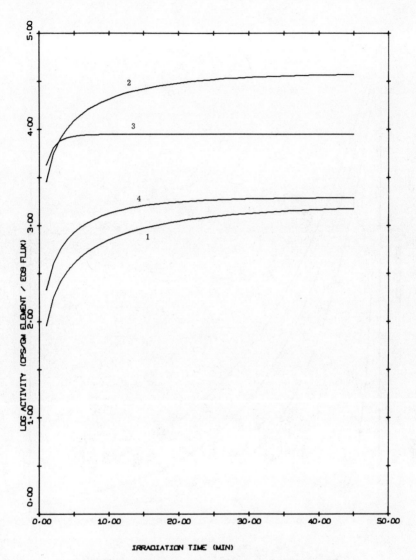

Fig. 7.163. Samarium—activation curves (see Table 7.55).

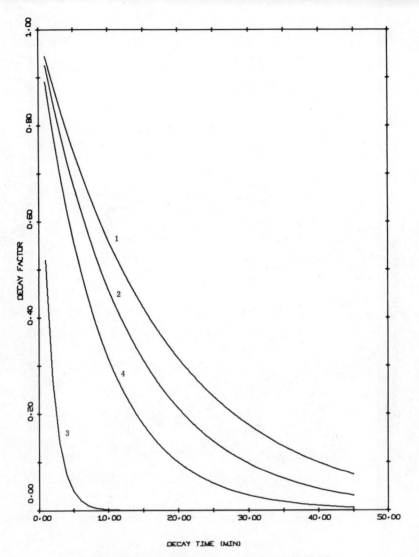

Fig. 7.164. Samarium—decay curves (see Table 7.55).

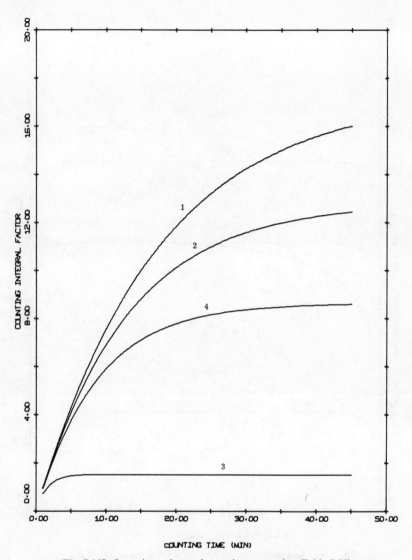

Fig. 7.165. Samarium—integral counting curves (see Table 7.55).

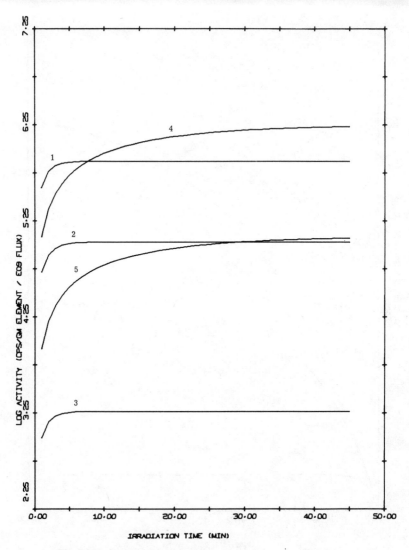

Fig. 7.166. Samarium interferences—activation curves (see Table 7.56).

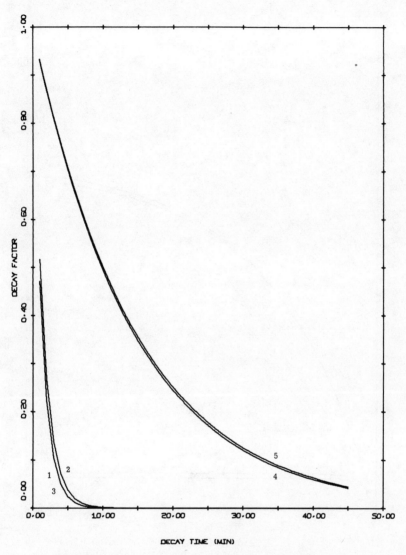

Fig. 7.167. Samarium interferences—decay curves (see Table 7.56).

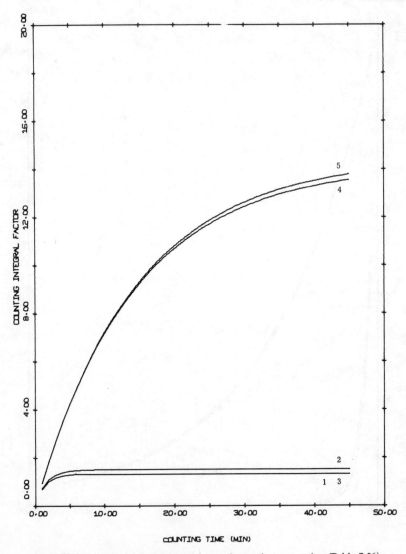

Fig. 7.168. Samarium interferences—integral counting curves (see Table 7.56).

Plaksin and co-workers[68] made use of the annihilation radiation to measure samarium. They report an overall standard deviation of 2.4% for the determination. In addition, they report a threshold sensitivity of 0.03% samarium for a 10-g total sample. This sensitivity is identical with that reported by Cuypers and Cuypers.[1]

3-MeV AND THERMAL NEUTRONS

Nargolwalla et al.[23] reported reasonably good sensitivity for the determination of samarium via activation with 3-MeV neutrons. They reported the (n, γ) reaction on ^{154}Sm producing 23-min ^{155}Sm to be reasonably useful. The γ-emission at 0.104 MeV gave a sensitivity of 6.3×10^3 counts/g of samarium at a flux of 10^6 neutrons/(cm^2 sec). It is likely that some interference may occur from other rare earth elements which produce low-energy emissions on being activated with 3-MeV neutrons. Typical of these would be dysprosium, erbium, terbium, and ytterbium.

The activation analysis of samarium with thermal neutrons produced by a neutron generator is more sensitive than with 14-MeV neutrons. Dibbs[24] reported a sensitivity of 60 μg via ^{155}Sm($T_{1/2} = 23$ min).

SELENIUM

Selenium exists in nature in six isotopes of mass 74, 76, 77, 78, 80, and 82. While the analytical chemistry of selenium offers a number of options for its determination at macro- and microlevels, the neutron activation analysis method may compete successfully with these methods in certain types of matrices. Selenium occurs often in nature in the form of mixed selenides which are refractory in nature and present problems in sample dissolution if wet chemical methods of analysis are to be used. For such samples in which selenium levels are at least a few percent, this technique should be very useful.

14-MeV NEUTRONS

The important parameters for the activation of selenium with fast neutrons are given in Table 7.57. Figures 7.169 to 7.171 show the activation, decay, and integral counting curves for the determination of selenium.

The most sensitive reaction for the determination of selenium is the reaction producing ^{79m}Se and ^{81m}Se. Both of these nuclides have emissions at 0.10 MeV which are useful for the quantitative measurement. The photopeak at 0.16 MeV resulting from the decay of ^{77m}Se is somewhat lower than sensitivity, although it may be useful in those matrices where short

TABLE 7.57. Nuclear Data for Reactions with 14-MeV Neutrons—Selenium

Nuclear Reaction	Target Isotope Abundance (%)	Cross Section (mb)	Half-life (min)	Gamma Energy (MeV)	Gamma Yield (%)	Detector Efficiency (%)	Peak/Total Ratio
1 = ^{74}Se(n, 2n)^{73}Se	0.87	383	426.0	0.51	130	40	0.63
2 = ^{78}Se(n, p)^{78}As	23.52	38	91.0	0.61	60	38	0.56
3 = ^{80}Se(n, α)^{77m}Ge	49.82	37	0.90	0.22	21	50	0.91
4 = ^{80}Se(n, 2n)^{79m}Se	49.82	300	3.90	0.10	9	50	1.00
5 = ^{82}Se(n, 2n)^{81m}Se	9.19	1050	0.95	0.10	8	50	1.00

irradiation times are preferred because of the activation of other matrix elements. Cuypers and Cuypers[1] have listed a detection sensitivity of 0.24 mg of selenium using the 0.10-MeV peak from ^{79}Se and ^{81m}Se and a detection sensitivity of 0.44 mg for the 0.16-MeV photopeaks of ^{71m}Se. Tustanovskii and Orifkhodzhaer[120] have applied the 14-MeV activation technique for the measurement of selenium in ores. In their application they avoided the activation of matrix elements by utilizing short irradiation times. For this ore they recommend a 30-sec irradiation in which case the interference from nickel can be neglected. They note a factor of 10^4 lower activation of the nickel and the selenium in such materials. The activation of copper and tellurium introduces an error of 2 to 5% in the determination of selenium.

3-MeV AND THERMAL NEUTRONS

Activation of selenium with 3-MeV neutrons provides an alternate way of measuring macroconstituent levels of selenium. Nargolwalla et al.[23] have reported a detection sensitivity of 1×10^4 counts/g of selenium at a flux of 10^6 neutrons/(cm^2 sec). This was obtained using the 0.16-MeV photopeak emission of ^{77m}Se ($T_{1/2} = 18$ sec).

Dibbs[24] predicts that the sensitivity for selenium using the thermal neutrons from a generator should be 27 μg.

SILICON

Silicon occurs in nature as three isotopes of mass 28, 29, and 30. By far the most important isotope from the standpoint of neutron activation analysis is ^{28}Si which undergoes an (n, p) reaction with 14-MeV neutrons to produce ^{28}Al. Under certain circumstances the (n, p) reaction on the less abundant ^{29}Si isotope may also be used.

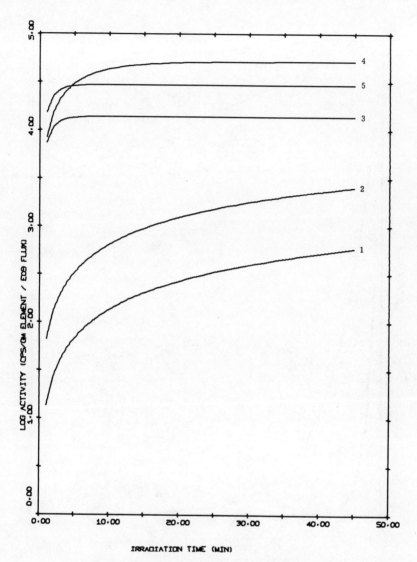

Fig. 7.169. Selenium—activation curves (see Table 7.57).

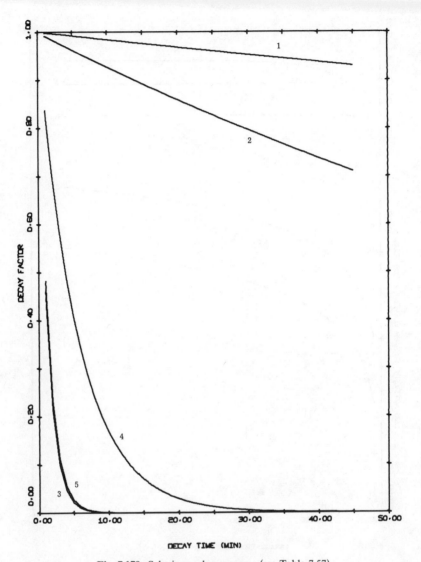

Fig. 7.170. Selenium—decay curves (see Table 7.57).

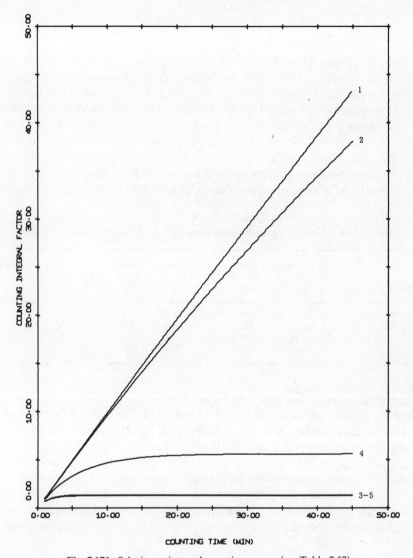

Fig. 7.171. Selenium—integral counting curves (see Table 7.57).

14-MeV NEUTRONS

Parameters important for the 14-MeV neutron activation analysis of silicon are listed in Table 7.58. Activation, decay, and integral counting curves for silicon are listed in Figs. 7.172 to 7.174. Silicon has some activation parameters ideally suited for its determination with 14-MeV neutrons. It has a large cross section for fast neutrons and yields product nuclei of relatively short half-life which decay by γ-emissions. Unfortunately, the analysis scheme utilizing the (n, p) reaction on ^{28}Si suffers from two direct-interferences, one from phosphorus and the second from aluminum, both of which produce the same isotope via an (n, α) and (n, γ) reaction, respectively.

TABLE 7.58. Nuclear Data for Reactions with 14-MeV Neutrons—Silicon

Nuclear Reaction	Target Isotope Abundance (%)	Cross Section (mb)	Half-life (min)	Gamma Energy (MeV)	Gamma Yield (%)	Detector Efficiency (%)	Peak/Total Ratio
1 = ^{30}Si(n, α)^{27}Mg	3.05	45.9	9.50	0.84	70	34	0.44
2 = ^{28}Si(n, p)^{28}Al	92.27	235	2.31	1.78	100	28	0.29
3 = ^{29}Si(n, p)^{29}Al	4.68	100	6.60	1.28	94	30	0.36
4 = ^{30}Si(n, p)^{30}Al	3.12	60	0.05	2.23	61	26	0.25

These interferences are given in Table 7.59. The activation, decay, and integral counting curves are in Figs. 7.175 to 7.177. Interference from phosphorus is quite serious, with the sensitivity of the (n, α) reaction on ^{31}P being approximately half of that for the ^{28}Si reaction. The interference from aluminum is much less significant, with its activation being about two orders of magnitude less sensitive than for the equivalent amount of phosphorus. In both cases of interference, γ-rays from other nuclides produced in the activation of the phosphorus and aluminum may be used to correct for the contribution of the silicon activation. The utility of this approach will

TABLE 7.59. Nuclear Data for Reactions with 14-MeV Neutrons—Silicon Interferences

Nuclear Reaction	Target Isotope Abundance (%)	Cross Section (mb)	Half-life (min)	Gamma Energy (MeV)	Gamma Yield (%)	Detector Efficiency (%)	Peak/Total Ratio
1 = ^{27}Al(n, γ)^{28}Al	100.00	0.50	2.31	1.78	100	28	0.29
2 = ^{31}P(n, α)^{28}Al	100.00	150	2.31	1.78	100	28	0.29

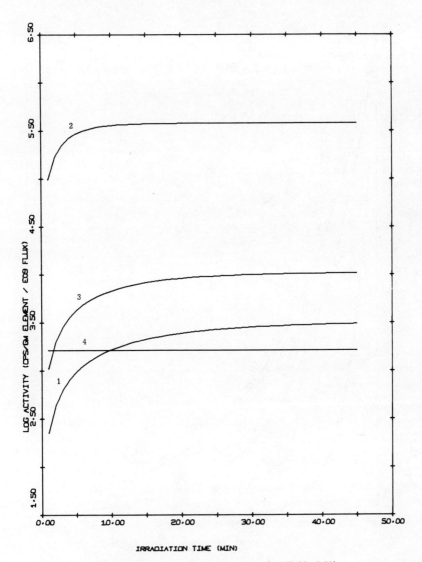

Fig. 7.172. Silicon—activation curves (see Table 7.58).

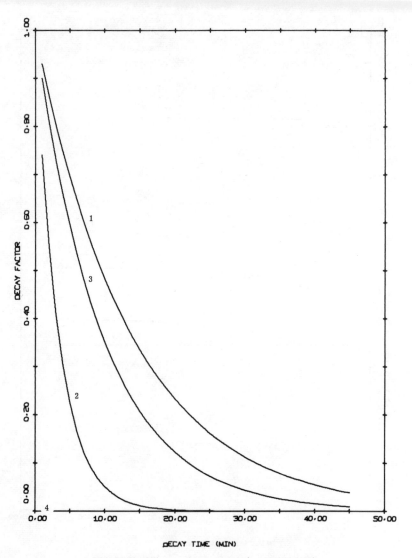

Fig. 7.173. Silicon—decay curves (see Table 7.58).

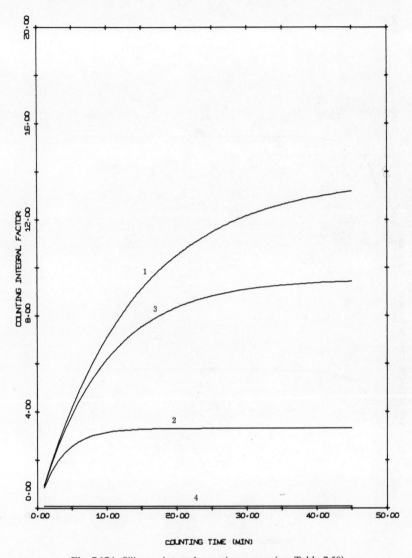

Fig. 7.174. Silicon—integral counting curves (see Table 7.58).

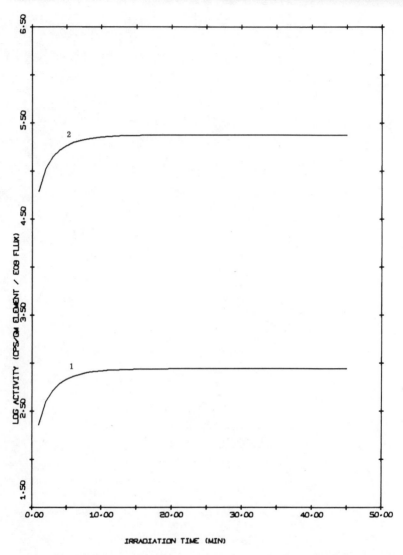

Fig. 7.175. Silicon interferences—activation curves (see Table 7.59).

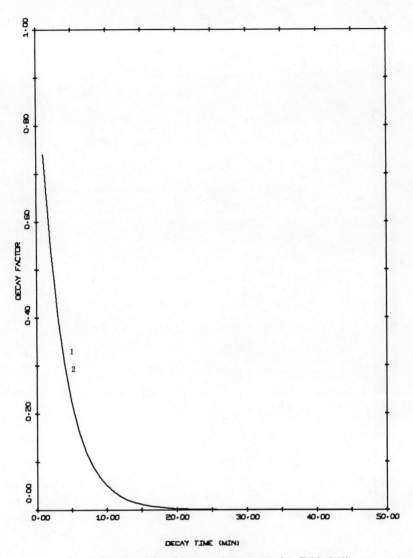

Fig. 7.176. Silicon interferences—decay curves (see Table 7.59).

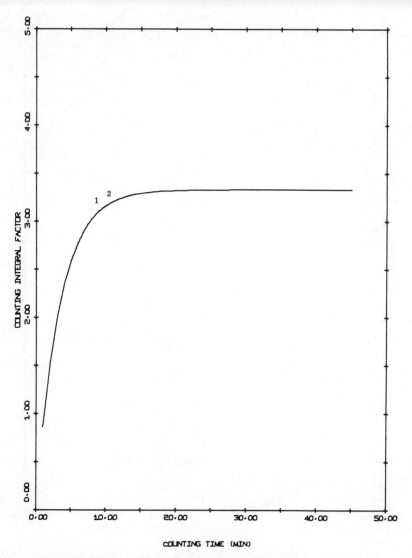

Fig. 7.177. Silicon interferences—integral counting curves (see Table 7.59).

III. ACTIVATION ANALYSIS CONDITIONS FOR THE ELEMENTS

depend on the relative amounts of the interferences present compared with the amount of silicon.

The neutron activation analysis method with 14-MeV neutrons compares favorably with alternate methods for the determination of silicon. The method is rapid, sensitive, and capable of high precision provided the above-mentioned interferences are absent. Most of the methods that are presently in use for the determination of silica are based on classical wet chemical procedures. The most common methods are the gravimetric determination as the silica-12-molybdate and titrimetric methods involving the hydrolysis of the hexafluorosilicate ion. Silicate analysis has remained unchanged for the last 60 years and it has been only recently that techniques such as the emission spectrograph and x-ray fluorescence have found acceptance for the determination of silica. Because of the great need for more rapid and precise methods for silicate analysis, the neutron activation method has found popularity for the determination of this element.

Turner[11] first pointed out the use of 14-MeV neutron activation analysis for the determination of silicon and aluminum as applied to silica samples and shale deposits. In 1956 he pointed out, "Although its accuracy is less than conventional wet chemical means, the considerable saving in time and effort justifies its [neutron activation analysis] use." At that time he reported the method to be accurate to within 5% for samples containing 0.4 g of silica. Advances in the technology of neutron activation analysis which provide more accuracy and precision have caused a rapid multiplication of reported applications.

Vogt and Ehmann[121] determined silicon in 107 chondrites and 11 achondrites by fast neutron activation analysis. These authors reported the precision of the silicon determinations based on four separate analyses of each meteorite to be better than $\pm 3\%$, and the absolute accuracy of the silicon values to be better than 5% for most of the determinations. The authors used the (n, p) reaction for the determination of silicon and found that the levels of aluminum and phosphorus normally found in meteorites would contribute less than 0.2% of the activity of the 1.78-MeV γ-ray. They concluded that an appreciable interference from alumina or phosphorus was not believed to occur in these types of materials. Vincent and Volborth[122,123] determined silicon in geological samples using 2- to 3-g samples and a flux of 3×10^7 neutrons/(cm² sec). These authors used four cycles, each consisting of a 25-sec irradiation, a 90-sec decay, and a 4-min count to constitute a single determination. Each determination took 40 min to complete. They reported that values obtained by this method for standard samples agreed well with those obtained by classical gravimetric analysis. Mean values obtained for silicon in six U.S. Geological Survey standard rocks showed an average standard deviation of only 0.12% silicon. Interference due to iron by its activation to yield a 1.8-MeV γ-ray from ^{56}Mn was negligible in the samples studied and phosphorus interference was corrected for by noting that 1% phosphorus is equal to 0.21% silicon. Aluminum up to 20% by weight of aluminum oxide in the sample was found to give negligible interference in the silicon determination.

Ehmann and co-workers[124,125] extended the study of meteorites and standard rocks by activation analysis and showed excellent agreement for a series of eight standard rocks with analyses obtained by classical methods. Schmitt et al.[126] reported on silicon abundances in meteoric chondrules utilizing the neutron activation method. They have determined silicon in some 275 chondrules using sample sizes from 0.1 to 70 mg. In the very smallest samples examined in this study, the authors found it advantageous to remove the samples from the polyethylene containers in which they were irradiated in order to avoid contributions of aluminum or silicon contaminants in the containers. Significant interference from aluminum or phosphorus was not found in this work. Santos and Wainerdi[127] determined silicon in rocks using an internal standard of barium acetate. These authors found that, by incorporating barium into the samples and relating it to a prepared calibration curve, they were able to avoid an external neutron monitor or sample rotation system. This approach, of course, is limited to those samples in which the internal flux monitor can be dispersed in the samples. The 0.62-MeV peak of ^{137m}Ba was used to monitor the integrated neutron flux. This was put into a ratio with the 1.78-MeV peak for ^{28}Al. The authors report a standard deviation of 2.9 to 5.4% for a series of rock samples. In addition, a detection limit of 0.07 mg of silicon was observed for a neutron flux of 2×10^8 neutrons/(cm^2 sec). Santos and Wainerdi note that the observed precision was considered acceptable for silicon analyses.

Volborth et al.[128] reported on a total nondestructive analysis of caas syenite, using a combination of x-ray emission and fast neutron activation methods. These authors have compared the x-ray and fast neutron activation method for silicon in this standard rock and found that the neutron activation method had equivalent precision to the x-ray method but was more accurate. The authors suggest that a combination of neutron activation and x-ray emission techniques provide a total nondestructive analytical scheme for the determination of 35 trace and 8 major elements in granite rock. Glasson et al.[129] have used the fast neutron activation method for the analysis of sedimentary formations on the Siberian platform. These authors report relative standard deviations ranging from 1.6 to 15%, depending on the nature of the sample. Garrec et al.[130] determined silicon in dry homogeneous soil with an error of 10% for the silicon determination. In addition, aluminum and iron were determined simultaneously. Oldham and Mathur[100] have summarized the use of 14-MeV neutrons in the activation analysis of minerals. These authors reported an absolute error of 0.19% in the determination of silicon in a silicon-zinc ore with a relative standard deviation of 6.5%. On a sample of ore containing oxygen, silicon, potassium, aluminum, and iron, the error was up to 1.94%. Morgan and Ehmann[131] reported improvements in the determination of oxygen and silicon by 14-MeV neutron activation. A single transfer system was employed to attain a precision approaching that of biaxial rotator systems.

The determination of silicon and iron in steels has also been extensively studied. Kusaka and Tsuji[132] have determined small amounts of silicon and iron in steel. Since, in this type of sample, iron causes some interferences due to the overlap of the 1.81-MeV peak of ^{56}Mn with the 1.78-MeV peak of ^{28}Al, some corrective procedures must be applied. These authors subtracted contributions from the long-lived component from the activity at 1.78 MeV by an instrumental complement subtraction method. They reported a detection limit of 5 mg of silicon in a 10-g sample. Kusaka and Tsuji found that if phosphorus was present in amounts less than 10% of the silicon, there was no need to correct for its presence. VanGrieken et al.[133] reported an internal standard method for the determination of silicon in steel. These authors utilize the iron content of the steel which is precisely known to provide an internal standard. Since the 1.81-MeV photopeak of ^{56}Mn cannot be distinguished from the 1.78-MeV photopeak of ^{28}Al, the authors take advantage of the difference in decay times of these two elements and apply two successive countings of the 1.81-MeV activity to resolve the iron and silicon activities. In addition, they use either the 0.84-MeV or the 2.11-MeV photopeaks of ^{56}Mn for internal standardization purposes. Where possible, due to the absence of interfering γ-rays, the authors prefer the 0.84-MeV peaks since it offers better counting statistics. VanGrieken and his associates found, using these procedures, that a coefficient of variation of 1.18% was observed for a single determination, whereas for three determinations the coefficient of variation was 0.68%. The authors have taken many precautions to minimize systematic errors in irradiation and counting of the samples. When 3% silicon is present in the steel, phosphorus, if present at a level of 0.1%, was found to cause a relative error of +1.9%, and aluminum was found to cause an error of −1.8% at the same level. Chromium, copper, and manganese were found to contribute less than 1% relative error when present at these same levels. Other trace elements normally found in steel gave relative errors an order of magnitude lower than these. An excellent discussion of the interferences found for the determination of silicon in steel is given by VanGrieken et al. in another publication.[134] They report a sensitivity of 0.02 to 0.05% of silicon in steel samples. It is interesting to note that these authors have studied both the internal standard and direct calibration methods for the determination of silicon in steel and at the same laboratory have reported a precision of 2 to 3% for steels containing about 1% silicon by the direct calibration method. The internal standardization procedure, on the other hand, provides significantly higher precision.

Eichelberger et al.[135] reported the determination of silicon in standard steel samples issued by the National Bureau of Standards, Washington, D.C., and show excellent agreement between the certified values and values obtained by the neutron activation method. In this procedure the interference due to ^{56}Mn was removed by the complementary subtraction method. Even higher accuracy was expected by these authors based on improvements in existing counting equipment. VanGrieken et al.[116] reported on the simultaneous determination of silicon and phosphorus in cast iron. These authors measured the silicon via the 1.78-MeV peak and corrected it for ^{56}Mn activity by counting the 1.8-MeV peak at two time intervals after the decay of the ^{28}Al was complete. Since ^{28}Al is also produced by phosphorus, the annihilation radiation at 0.51 MeV from ^{30}P is counted by two opposed sodium iodide detectors in coincidence. Again two successive coincidence measurements are carried out in order to account for the ^{53}Fe activity from the (n, 2n) reaction on iron. The ^{28}Al measurement is corrected for the computed phosphorus content. Precision for the silicon determination was found to be 5% relative for 0.2% silicon and 2.5% for silicon contents above 1% in the steel samples. These authors show a cross-comparison of an x-ray fluorescence and a gravimetric method for silicon in a series of steel and cast iron samples compared with a neutron activation method. The neutron activation method in both cases correlates extremely well. Wood and Roper[136] devised a fast neutron activation method for silicon and iron in the range of 0.5 to 10% by weight of silicon. These authors were the first to use the iron activation as an internal flux monitor and have proposed a design for an automated system for carrying out such analyses. Analysis time was short and a 2σ precision of 1.25% of the silicon content was reported.

A number of applications of this method to the determination of sea-floor compositions have been reported. Santos et al.[137] have demonstrated the determination of silicon marine sediments. They propose that the analysis could be carried out in situ from a submersible vessel or from the surface by using a portable pulse height analyzer. Control system and readout equipment could be aboard the research vessel and the neutron source and detection equipment could be in a remotely controlled pressure chamber. Shigematsu et al.[138] also carried out the analysis of marine sediment for silica using this technique. They compared the values with colorimetric values on identical samples and showed an error of about 5%. Kehler and Monaghan[17] developed a system for performing neutron activation analysis of ocean-bottom core samples. A series of simultaneous equations were used to resolve spectra containing photopeaks from silicon, aluminum, magnesium, sodium, and iron. Since the samples vary in density and water content, corrections are necessary because of moderation of the neutron flux and attenuation of the gamma emissions.

Silicon has also been measured in oil samples from aircraft hydraulic systems using this technique. D'Agostino and Kuehne[2] developed a routine procedure for the determination of metallic wear products filtered out of aircraft hydraulic fluids. In this case, the material was filtered onto a patch which was then irradiated and its pulse height spectrum analyzed. This technique was useful in detecting as little as 1 µg of silicon in hydraulic fluids that contained from 1 to 600 µg/50 cc of hydraulic fluid. This paper presents an excellent discussion of techniques used to correct for interference by decay curve analysis. This work was also reported earlier at the Second Accelerator Symposium.[2] Eden[139] used this technique to identify rapidly the presence of traces of silicon in some vacuum pump oils. Many commercial vacuum pump oils are based on silicon-containing compounds and the detection of the presence of silicon is a suitable indicator of pump valve leakage. The samples were collected as smears on tissue paper and no preliminary treatment was given before irradiation. In these circumstances only a qualitative pulse height spectrum was needed for confirmation of the presence of silicon. Contaminants normally present in used vacuum oils, such as chromium, iron, aluminum, nickel, and manganese, were not observed. Interference from aluminum was ruled out since no ^{27}Mg was observed, and likewise there was no 0.84-MeV peak for ^{56}Mn, which would have been observed if the 1.78-peak was enhanced by the presence of iron. Thus Eden concluded that the detection of ^{28}Al revealed the presence of silicon in significant trace amounts. Daniel et al.[140] reported on the quantitative determination of silicon in fuels utilizing the same reaction. These workers noted a precision of $\pm 5\%$ with an error of approximately 1%.

The 14-MeV neutron technique has been used to measure silicon in a number of matrices. Martin et al.[18] have used the method to measure a number of trace elements including silicon in coal. Silicon and aluminum were determined simultaneously using the 0.51-MeV annihilation irradiation for aluminum and the 1.78-MeV peak for the combination of silicon and aluminum. Martin et al.[141] devised an on-line process control approach for the measurement of silicon in cement raw mix, bauxite, coal tar sands, copper ore, and taconite. These authors note that the use of activation analysis for on-line determination of oxygen, aluminum, and silicon appears extremely promising and far superior to x-ray fluorescence in terms of simplicity of sample presentation. Martin and co-workers seem to favor the presentation of a "slurry sample" on a conveyor belt for irradiation and counting.

Broadhead et al.[12] noted the application of this technique to electrometallurgical samples where the presence of silicon needs to be measured. Sattarov et al.[142] have determined silicon in carborundum. Wood et al.[7] demonstrated the use of the sealed-tube neutron generator for the determination of major mineral constituents such as oxygen, silicon, and aluminum. Thus far, silicon has been determined with an accuracy of 2% in $\frac{1}{2}$-g samples, utilizing this technique. ^{28}Al was used for the silicon determination in standard rock samples G1, W1, and R117, and no interference from other elements was observed from these matrices. Buczkó et al.[143] developed a method for determining silicon in aluminum. These authors used a Ge(Li) detector to separate the 1.37- and 2.75-MeV lines from the product of the (n, α) reaction on aluminum yielding ^{24}Na and the 1.78-MeV line from the (n, p) reaction on silicon. In the case of aluminum-silicon alloys, 100 μg of silicon can be determined with an error of 10% in an aluminum sample of 1 g. The phosphorus content of the aluminum sample was checked by the 0.51-MeV annihilation peak and was not found within the sensitivity of this measurement (0.04%). Tsuji and Kusaka[144] measured silicon and aluminum in silicates. The analytical errors by this method were 2% for the silicon with a neutron flux of 1×10^7 neutrons/(cm^2 sec).

Silicon in various biological samples has also been determined using the fast neutron technique. Schramel[38] has measured silicon in the presence of chlorine, potassium, phosphorus, calcium, and aluminum in biological samples. Decay curve analysis was used to analyze the spectra, the 0.51-MeV peak was utilized to correct for the interferences from phosphorus, and the 0.83-MeV peak was used for the correction for the presence of aluminum. In another biological application Sardi and Tomcsanyi[145] reported the measurements of silicon in sputum by this technique. Here the technique was used to correlate silicon content of sputum with patients suffering from silicosis. In order to cope with the viscosity of the sample a trypsin enzymic digest was carried out on the sputum followed by trichloroacetic acid digestion to cause precipitation of the silicon. Phosphorus, which was left in the supernatant liquid as phosphoric acid, was removed and the precipitate was washed several times prior to activation. A positive correlation was found between silicon levels and the incidence of silicosis.

3-MeV AND THERMAL NEUTRONS

Silicon cannot be usefully determined with 3-MeV or thermal neutrons. The reaction cross sections are too low to give useful results with the fluxes available from neutron generators.[23, 24]

III. ACTIVATION ANALYSIS CONDITIONS FOR THE ELEMENTS

SILVER

Silver exists in nature in two isotopes of mass 107 and 109. Both isotopes undergo useful reactions for neutron activation analysis with 14- and 3-MeV neutrons.

Because of the large number of other methods for the determination of silver, including volumetric, colorimetric, and instrumental, the neutron activation analysis procedure for silver has not been highly developed. Silver has some rather ideal solution chemistry which can be used for its precise quantitative determination. Since silver can be readily dissolved into solution from its ores and salts, wet chemical methods are widely used. Instrumental methods based on x-ray spectrometry also provide a useful precise method for the estimation of silver in a variety of matrices.

14-MeV NEUTRONS

The primary reaction for the determination of silver with high sensitivity is the (n, 2n) reaction on ^{107}Ag to produce ^{106}Ag, a positron emitter with a 24-min half-life. Cuypers and Cuypers[1] report a sensitivity of 0.08 mg of silver for this reaction. The nuclear parameters for the activation of silver are given in Table 7.60, and Figs. 7.178 to 7.180 give the activation, decay, and integral counting curves for establishing analytical conditions for the measurement of silver. Although silver forms a number of other nuclides upon activation with 14-MeV neutrons, they yield γ-rays of low energy which are more difficult to measure. The most serious limitation of the neutron activation analysis method for silver is that it provides a nuclide that is a positron emitter and which can experience interference from other positron emitters in the same sample. The relatively long half-life of ^{106}Ag does, however, provide for differentiation from other positron emitters that may be in the sample (cf. Tables 7.76-7.78). Senftle et al.[150] proposed the use of the

TABLE 7.60. Nuclear Data for Reactions with 14-MeV Neutrons—Silver

Nuclear Reaction	Target Isotope Abundance (%)	Cross Section (mb)	Half-life (min)	Gamma Energy (MeV)	Gamma Yield (%)	Detector Efficiency (%)	Peak/Total Ratio
1 = ^{107}Ag(n, n'γ)^{107m}Ag	51.35	140	0.74	0.09	5	50	1.00
2 = ^{107}Ag(n, 2n)^{106}Ag	51.35	600	24.00	0.51	140	40	0.63
3 = ^{109}Ag(n, 2n)^{108}Ag	48.65	700	2.42	0.43	1	42	0.71
4 = ^{109}Ag(n, 2n)^{108}Ag	48.65	700	2.42	0.51	1	40	0.63
5 = ^{109}Ag(n, p)^{109m}Pd	48.65	13.70	4.70	0.19	58	50	0.93

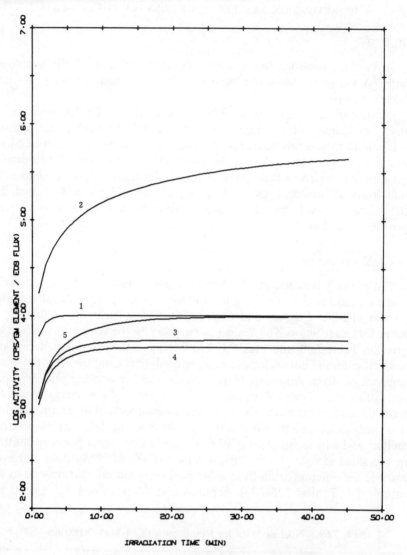

Fig. 7.178. Silver—activation curves (see Table 7.60).

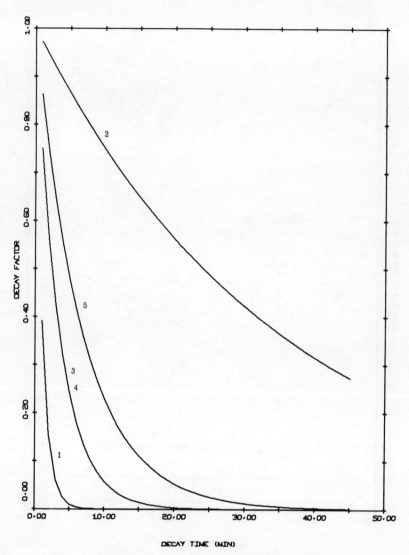

Fig. 7.179. Silver—decay curves (see Table 7.60).

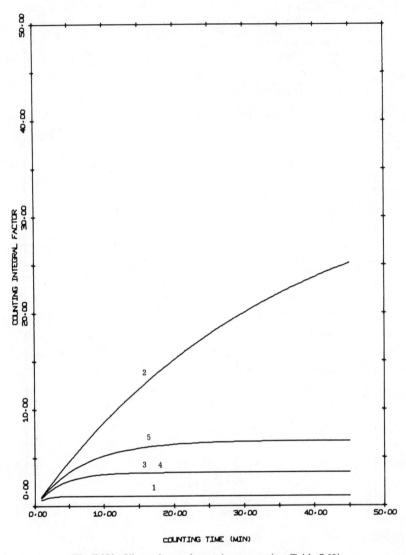

Fig. 7.180. Silver—integral counting curves (see Table 7.60).

neutron generator system for prospecting for silver deposits. These authors found it preferable to thermalize the 14- or 3-MeV neutron flux to promote the (n, γ) reaction on ^{109}Ag to produce ^{110m}Ag activity. Because of the freedom from activation of many matrix elements which are activated with 14-MeV neutrons, Senftle et al. found it preferable to use the 2.8-MeV neutron source for silver exploration. They noted that the fluxes available from 3-MeV sources [5×10^7 neutrons/(cm^2 sec)] yielding a thermal flux of approximately 10^6 neutrons/(cm^2 sec) were a little marginal for locating low-content silver deposits. Since the background from the 3-MeV neutron activation is low, however, this appeared to be the preferred source. These authors reported a lower limit of detection of 1 oz/ton. Crambes[25] has reported on the activation of silver using thermalized 14-MeV neutrons. In his studies he found it necessary to carry out radiochemical separations of the silver from other matrix elements such as antimony, copper, gallium, molybdenum, and zinc.

3-MeV AND THERMAL NEUTRONS

Nargolwalla et al.[23] reported excellent sensitivity for the determination of silver using 3-MeV neutrons. Inelastic scattering reactions on both ^{107}Ag and ^{109}Ag produce short-lived isotopes which have γ-emissions at 0.094 and 0.088 MeV. Sensitivity for this reaction is 1.8×10^6 counts/g at a flux of 10^6 neutrons/(cm^2 sec). The (n, γ) reaction to produce ^{110m}Ag has a sensitivity of 5.3×10^2 counts/g of element. The reactions forming ^{110m}Ag or ^{108}Ag with either 3-MeV or thermal neutrons are not as sensitive as the 14-MeV technique using the annihilation radiation at 0.51 MeV.

SODIUM

Sodium occurs naturally as one isotope of mass 23. This isotope has excellent cross sections for reactions with both thermal and fast neutrons and can be readily measured by neutron activation techniques.

Activation analysis of sodium has some advantages over alternate methods for sodium determination. While sodium has good spectroscopic characteristics for its determination by flame photometry or atomic absorption, macrolevels of sodium are difficult to determine directly using other instrumental methods of analysis or wet chemical techniques. The solution chemistry of sodium is very limited. Sodium does not undergo good gravimetric or compleximetric reactions which can be employed for its determination. On the other hand, atomic absorption provides an excellent approach to the determination of microamounts of sodium in aqueous or nonaqueous media. Macroconstituent analysis for sodium by neutron activation analysis has been used in the standardization of certain reference materials at the National Bureau of Standards. It has been shown that this technique is capable of better than 1% precision.

14-MeV NEUTRONS

The pertinent nuclear data for 14-MeV neutron activation are given in Table 7.61, and the activation, decay, and integral counting curves are shown in Figs. 7.181 to 7.183. The most sensitive reaction for the determination of sodium is the (n, α) reaction, which produces ^{20}F. Cuypers and Cuypers[1] report a detection limit of 0.50 mg using this reaction. Iddings and Wade[151] made use of this reaction in determining sodium in organic materials. These authors showed that as little as 1 mg of sodium could be measured with good precision using this technique. Precision of 10% was obtained in the determination of as little as 0.3% sodium in polymers. They report an improvement in precision as the level of sodium exceeds 1%. ^{28}Al from the presence of silicon was noted to be the worst interference (1.78-MeV photopeak). Iddings and Wade indicated successful use of the spectrum-stripping technique to correct for the presence of silicon. They also note that if the sample is wrapped in cadmium foil before irradiation, aluminum does not produce any interference. Fluorine was also noted not to interfere. The irradiation and counting conditions for this analysis are such that the Compton continuum from oxygen in the sample falls under the 1.63-MeV peak. The authors used a simple base-line technique to correct for any contributions from high-energy peaks.

TABLE 7.61. Nuclear Data for Reactions with 14-MeV Neutrons—Sodium

Nuclear Reaction	Target Isotope Abundance (%)	Cross Section (mb)	Half-life (min)	Gamma Energy (MeV)	Gamma Yield (%)	Detector Efficiency (%)	Peak/Total Ratio
1 = ^{23}Na(n, α)^{20}F	100.00	222	0.19	1.63	100	28	0.30
2 = ^{23}Na(n, p)^{23}Ne	100.00	33.9	0.63	0.44	33	42	0.69
3 = ^{23}Na(n, p)^{23}Ne	100.00	33.9	0.63	1.64	1	28	0.30

Kehler and Monaghan[17] reported the use of a sealed-tube neutron generator for the activation analysis of ocean-bottom cores. In this analysis they standardized irradiation and counting times and utilized the photopeak at 0.44 MeV from ^{23}Ne for the analysis. This analysis was carried out as part of a simultaneous determination of silicon, aluminum, magnesium, sodium,

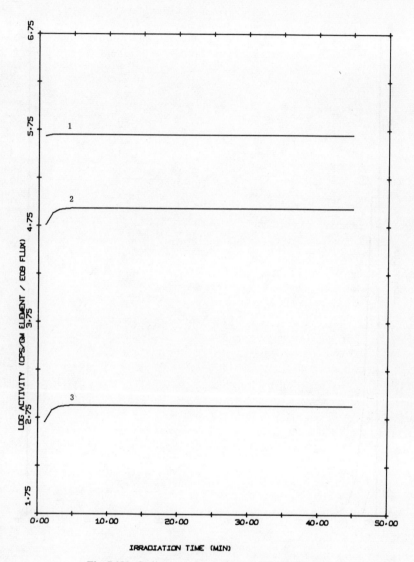

Fig. 7.181. Sodium—activation curves (see Table 7.61).

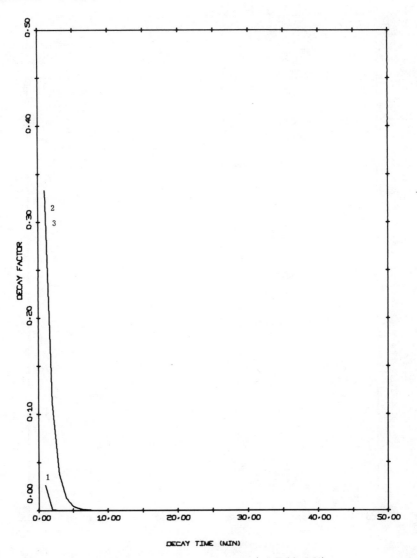

Fig. 7.182. Sodium—decay curves (see Table 7.61).

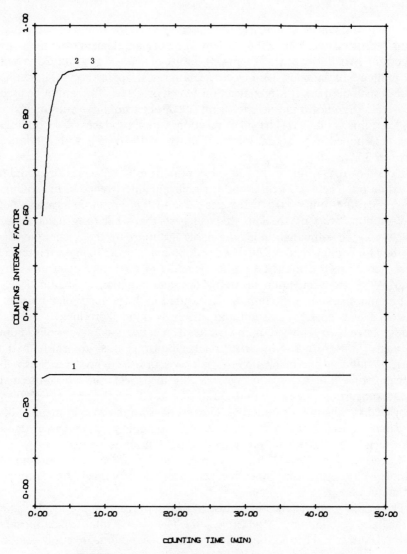

Fig. 7.183. Sodium—integral counting curves (see Table 7.61).

and iron. Mathematical matrices were used to solve the simultaneous equations necessary to calculate the concentration of each element from the γ-ray spectrum. Broadhead et al.[12] discussed the use of neutron activation analysis on molten salt samples from electrometallurgical applications. They note that in such matrices the (n, α) reaction, which produces ^{20}F, cannot be used because of significant interference from silicon. Instead, these authors chose to determine sodium via the (n, p) reaction, which produces ^{23}Ne. Sodium was determined over a concentration range of 18 to 35% with a relative error of 3.7%, using 2-g samples.

Gorski et al.[152] utilized the 0.44-MeV peak from ^{23}Ne to measure sodium in crude oil. These workers found a significant interference in the sodium determination from the 0.51-MeV peak due to ^{13}N produced as a result of the reaction ^{13}C(p, n)^{13}N. This study suggests that while oxygen and sulfur levels may be estimated with reasonable accuracy by this technique, the method for sodium requires further development in order to make it interference free in such a matrix. Certain elements in coal were determined by 14-MeV neutron activation including the determination of sodium levels in bituminous samples. Sodium was measured in the presence of aluminum, silicon, iron, copper, magnesium, and barium as an (n, γ) product.[153] Several studies have been reported on in vivo neutron activation analysis of sodium in man.[37, 43, 113] A 14-MeV neutron generator is used to irradiate the patient; sufficient thermalization of the flux occurs due to body water as well as from some added polyethylene shielding to provide the usable thermal flux. Exposures in the range of 0.05 rad were used and the patient was then subjected to whole-body counting. The results from these experiments indicate that the technique provides whole-body sodium figures that are in good agreement with those from previous isotope dilution techniques.

Waggoner and Knox[13] have proposed the determination of sodium in remote locations by using a neutron generator and prompt elastic scattering techniques.

The primary interference in the use of the (n, p) reaction producing ^{23}Ne is the (n, α) reaction on ^{26}Mg (Table 7.62). The ratio of interference appears to be about 3 : 1. That is, about three parts of magnesium are equivalent to

TABLE 7.62. Nuclear Data for Reactions with 14-MeV Neutrons—Sodium Interferences

Nuclear Reaction	Target Isotope Abundance (%)	Cross Section (mb)	Half-life (min)	Gamma Energy (MeV)	Gamma Yield (%)	Detector Efficiency (%)	Peak/Total Ratio
1 = ^{26}Mg(n, α)^{23}Ne	11.29	89	0.63	0.44	33	42	0.69

III. ACTIVATION ANALYSIS CONDITIONS FOR THE ELEMENTS

about one part of sodium if both are present in the same sample. This interfering reaction is listed in Table 7.62, and its activation, decay, and integral counting curves are shown in Figs. 7.184 to 7.186.

3-MeV AND THERMAL NEUTRONS

Activation analysis of sodium using 3-MeV neutrons can be performed.[23] However, the sensitivity for its determination with neutrons of this energy is very low. An experimental sensitivity of approximately 150 counts/g of sodium using a 10^6-neutron/(cm^2 sec) flux was reported.

The thermal neutron flux from a generator can be used to activate sodium with good sensitivity. Dibbs[24] has reported a sensitivity of 0.24 mg using this technique.

STRONTIUM

The four naturally occurring isotopes of strontium are of mass 84, 86, 87, and 88. As one of the alkaline earth elements, strontium has a variety of analytical chemistry which is useful for its determination. Strontium can be determined by flame spectroscopic methods, x-ray fluorescence methods, as well as by wet chemical procedures. Wet chemical procedures are not selective enough to discriminate strontium from other alkaline elements. The availability of these alternative analytical methods for strontium probably accounts for the lack of reported applications for the neutron activation analysis procedure.

14-MeV NEUTRONS

The important parameters for the activation analysis of strontium with 14-MeV neutrons are given in Table 7.63. Figures 7.187 to 7.189 give the activation, decay, and integral counting curves to establish conditions for

TABLE 7.63. Nuclear Data for Reactions with 14-MeV Neutrons—Strontium

Nuclear Reaction	Target Isotope Abundance (%)	Cross Section (mb)	Half-life (min)	Gamma Energy (MeV)	Gamma Yield (%)	Detector Efficiency (%)	Peak/Total Ratio
1 = ^{86}Sr(n, 2n)^{85m}Sr	9.86	21	70.00	0.23	85	50	0.90
2 = ^{86}Sr(n, p)^{86m}Rb	9.86	39	1.04	0.56	100	39	0.59
3 = ^{88}Sr(n, α)^{85m}Kr	82.56	64	264.00	0.15	74	50	0.97
4 = ^{88}Sr(n, α)^{85m}Kr	82.56	64	264.00	0.31	13	47	0.83
5 = ^{88}Sr(n, 2n)^{87m}Sr	82.56	225	169.80	0.39	80	44	0.75
6 = ^{88}Sr(n, p)^{88}Rb	82.56	17	17.80	0.90	13	33	0.43
7 = ^{88}Sr(n, p)^{88}Rb	82.56	17	17.80	1.86	21	27	0.28

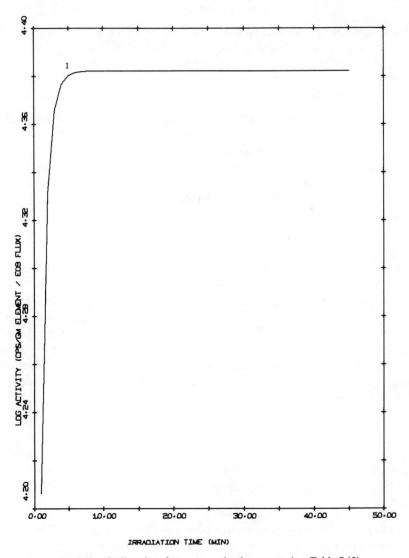

Fig. 7.184. Sodium interferences—activation curves (see Table 7.62).

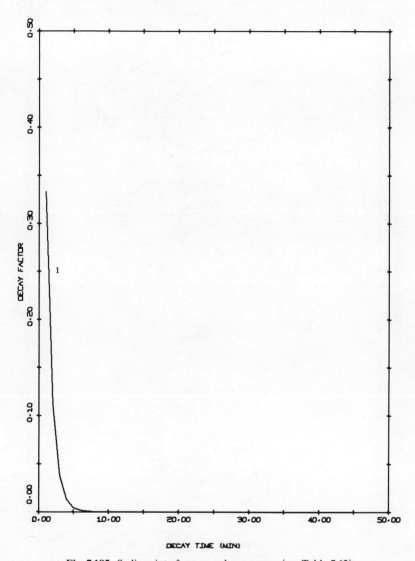

Fig. 7.185. Sodium interferences—decay curves (see Table 7.62).

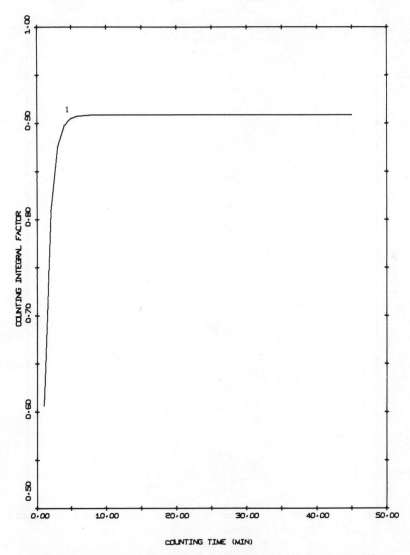

Fig. 7.186. Sodium interferences—integral counting curves (see Table 7.62).

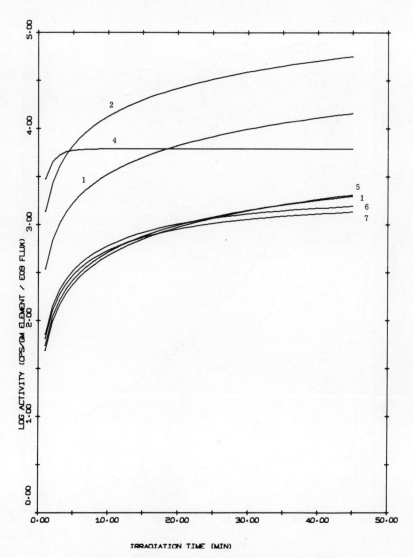

Fig. 7.187. Strontium—activation curves (see Table 7.63).

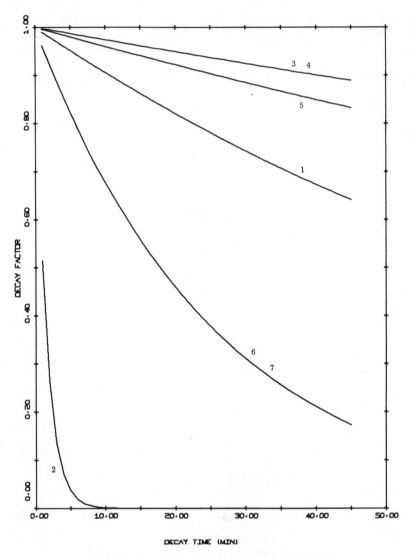

Fig. 7.188. Strontium—decay curves (see Table 7.63).

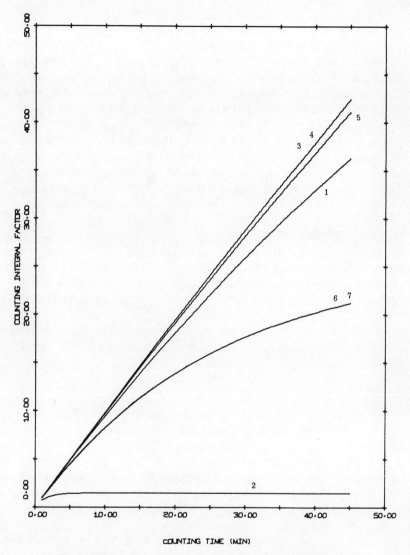

Fig. 7.189. Strontium—integral counting curves (see Table 7.63).

the determination of strontium. The most sensitive reaction for the determination of strontium is the (n, p) reaction on ^{86}Sr to give ^{86m}Rb. Two other reactions, one which yields ^{85m}Sr from ^{86}Sr (n, 2n), and the (n, α) reaction on ^{88}Sr to give ^{85m}Kr, have sensitivities which make the neutron activation method attractive for the determination of fairly small amounts of strontium. Cuypers and Cuypers[1] list a detection limit of 1.8 mg for strontium using the gamma ray at 0.28 MeV from ^{85m}Sr. The most sensitive reaction listed in their tabulation is that utilizing the 0.39-MeV photopeak from ^{87m}Sr. Nuclear parameters from other sources do not predict this to be the most sensitive reaction by which to measure strontium. (See Fig. 7.187.) There are no reported applications of this technique to the determination of strontium.

3-MeV AND THERMAL NEUTRONS

Nargolwalla et al.[23] found that the production of ^{87m}Sr with 3-MeV neutrons provided an attractive method for the determination of macroamounts of strontium. Utilizing the photopeak at 0.39 MeV they reported a sensitivity of 1×10^4 counts/g of strontium for a normalized flux of 10^6 neutrons/(cm^2 sec).

Activation of strontium with the thermal neutron flux from a generator provides the best sensitivity for determining strontium. Dibbs[24] reported a sensitivity of 0.21 mg for strontium by this technique.

SULFUR

Sulfur exists in nature as three isotopes, those of mass 32, 34, and 36. Sulfur occurs in nature in a variety of oxidation states ranging from 2− to 6+.

The analytical chemistry of sulfur and the ease with which it can be determined are highly dependent on the state and combination in which it is found. A large number of analytical methods exist for the determination of sulfur, many of which depend on the reduction of sulfur to the 2− state followed by a colorimetric determination or, conversely, oxidation of the sulfur to a 6+ state (sulfate) and its determination with titrimetric or colorimetric procedures. Nondestructive elemental analysis of sulfur is not readily achieved. In the x-ray fluorescence techniques, sulfur has emissions which are soft enough to require vacuum-path spectrometers, and the radiation is easily attenuated by other matrix elements. Flame emission methods for sulfur exist based on the S–O band emission; however, these are in the near-ultraviolet and are not always accessible, depending on the matrix composition. A nondestructive technique such as neutron activation analysis, therefore, would have a high degree of applicability, particularly for nondestructive analysis.

III. ACTIVATION ANALYSIS CONDITIONS FOR THE ELEMENTS

14-MeV NEUTRONS

The important nuclear parameters for the activation analysis of sulfur with 14-MeV neutrons are given in Table 7.64. Figures 7.190 to 7.192 give the activation, decay, and integral counting curves for sulfur.

TABLE 7.64. Nuclear Data for Reactions with 14-MeV Neutrons—Sulfur

Nuclear Reaction	Target Isotope Abundance (%)	Cross Section (mb)	Half-life (min)	Gamma Energy (MeV)	Gamma Yield (%)	Detector Efficiency (%)	Peak/Total Ratio
$1 = {}^{34}S(n, \alpha){}^{31}Si$	4.20	138	157.20	1.26	1	30	0.37
$2 = {}^{34}S(n, p){}^{34}P$	4.20	85.2	0.21	2.13	25	28	0.26
Sulfur Interferences							
$3 = {}^{37}Cl(n, \alpha){}^{34}P$	24.60	52.4	0.21	2.13	25	28	0.26

There are two useful reactions for the determination of sulfur with 14-MeV neutrons. The first is the (n, p) reaction on ^{34}S to yield 12.4-sec ^{34}P. This has a γ-emission at 2.1 MeV and also copious β-emission. This reaction, however, suffers interference by the (n, α) reaction on ^{37}Cl which produces the same product isotope. A reaction of considerably lower sensitivity is the (n, α) reaction on ^{34}S to yield ^{31}Si. This reaction is interfered with by phosphorus which, through an (n, p) reaction, gives the same isotope. There have been several applications of the neutron activation method for the determination of sulfur. Crambes et al.[73] proposed the determination of sulfur in proteins by utilizing the (n, p) reaction and counting the bremsstrahlung from the β-emissions. It was reported that in order to obtain 1% standard deviation (1σ) a minimum of 13 mg of sulfur had to be present in the sample. Gorski et al.[148] studied the feasibility of determining sulfur in crude oil utilizing this technique. There was a good deal of scatter in the calibration curves from synthetic mixtures of oil. Chomel[114] found the sensitivity limit for the determination of sulfur at 1 mg utilizing a flux of 10^8 neutrons/(cm^2 sec).

3-MeV AND THERMAL NEUTRONS

Nargolwalla et al.[23] showed that there was no activation of sulfur using 3-MeV neutrons. Likewise sulfur is not activated by the thermal neutron flux available from a generator.

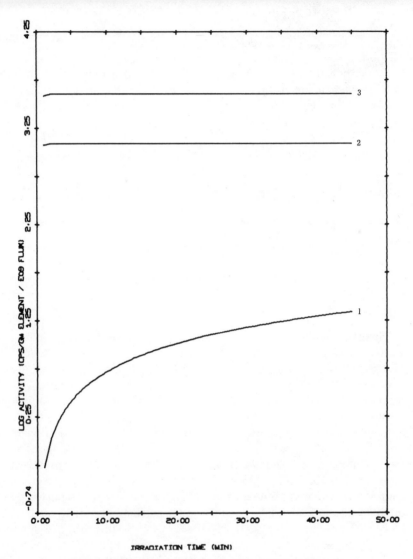

Fig. 7.190. Sulfur—activation curves (see Table 7.64).

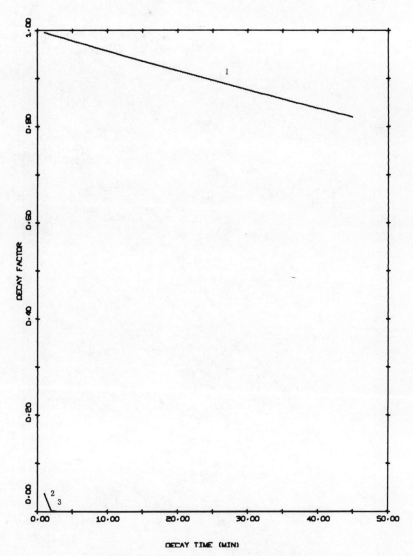

Fig. 7.191. Sulfur—decay curves (see Table 7.64).

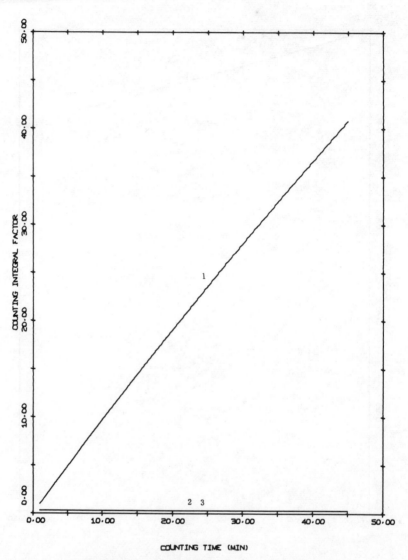

Fig. 7.192. Sulfur—integral counting curves (see Table 7.64).

TANTALUM

Tantalum occurs in nature in one isotope of mass 180. While tantalum is a difficult metal to determine by wet chemical methods, instrumental methods, particularly x-ray fluorescence, are well suited to its determination. Because the element is difficult to dissolve from some of its naturally occurring ores, chemical methods of analysis are sometimes quite laborious. Unfortunately, the determination of tantalum with fast neutrons does not offer a highly attractive alternative.

14-MeV NEUTRONS

The most sensitive activation reaction for tantalum is listed in Table 7.65, and its activation, decay, and counting characteristics are shown in Figs. 7.193 to 7.195. The most useful emission for the determination of tantalum is x-ray at 0.06 MeV. Cuypers and Cuypers[1] list a detection limit of 0.48 mg.

TABLE 7.65. Nuclear Data for Reactions with 14-MeV Neutrons—Tantalum

Nuclear Reaction	Target Isotope Abundance (%)	Cross Section (mb)	Half-life (min)	Gamma Energy (MeV)	Gamma Yield (%)	Detector Efficiency (%)	Peak/Total Ratio
$1 = {}^{181}\text{Ta}(n, 2n){}^{180m}\text{Ta}$	100.00	870	486.0	0.09	4	50	1.00

Unfortunately, other significant activation reactions for tantalum with 14-MeV neutrons yield ^{180m}Ta with relatively low yield due to the 8.1-hr half-life. The activation methods for tantalum are not competitive with alternate instrumental methods for determining this element. There have been no reported applications of the 14-MeV neutron technique to the determination of tantalum.

3-MeV AND THERMAL NEUTRONS

The (n, γ) reaction on ^{181}Ta occurs with sufficient sensitivity to provide a possible means of determining tantalum with 3-MeV neutrons. Nargolwalla et al.[23] utilized a single-channel analyzer including photopeaks at 0.147, 0.172, and 0.184 MeV to measure tantalum, and found an experimental sensitivity of 5.7×10^3 counts/g of tantalum using a normalized flux of 10^6 neutrons/(cm^2 sec). This method might be useful for the macroconstituent analysis of tantalum.

Dibbs[24] has shown that the thermalized neutron flux of a generator can be used to activate tantalum with the same sensitivity as is obtained with the 14-MeV technique.

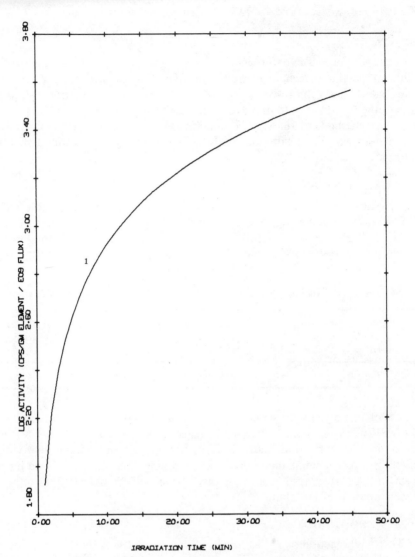

Fig. 7.193. Tantalum—activation curves (see Table 7.65).

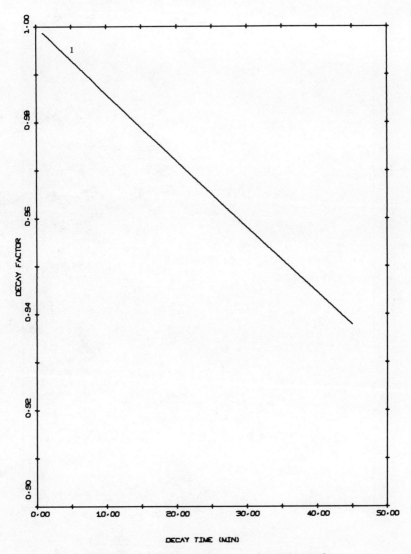

Fig. 7.194. Tantalum—decay curves (see Table 7.65).

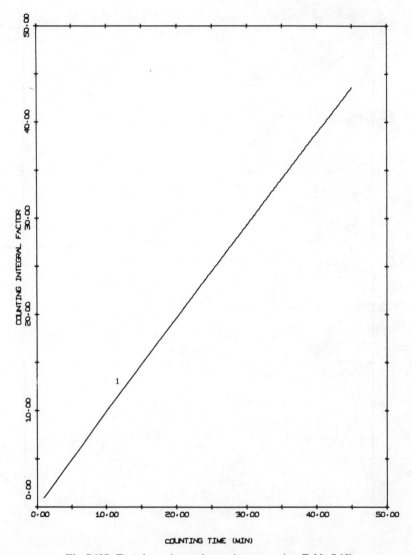

Fig. 7.195. Tantalum—integral counting curves (see Table 7.65).

III. ACTIVATION ANALYSIS CONDITIONS FOR THE ELEMENTS 567

TERBIUM

Terbium exists in nature in one isotopic form of mass number 159. It is one of the rarer elements in terms of its natural abundance, and its analytical chemistry has not been highly developed. The activation parameters for the determination of terbium with fast neutrons, however, are not favorable enough to make this method competitive with other methods.

14-MeV NEUTRONS

The activation parameters important for the determination of terbium are listed in Table 7.66. Activation, decay, and integral counting curves for this element are given in Figs. 7.196 to 7.198. The most sensitive reaction for the determination of terbium by activation is the (n, 2n) reaction resulting in ^{158m}Tb. In view of the short half-life of this isotope, the photopeak at 0.11 MeV is in an unfavorable region. X-ray emission at 0.044 MeV is listed as the useful reaction for the determination of terbium by Cuypers and Cuypers.[1] Their listed detection limit is 0.93 mg of terbium utilizing this reaction. Again the low energy of this x-ray makes the activation method for terbium of dubious value.

TABLE 7.66. Nuclear Data for Reactions with 14-MeV Neutrons—Terbium

Nuclear Reaction	Target Isotope Abundance (%)	Cross Section (mb)	Half-life (min)	Gamma Energy (MeV)	Gamma Yield (%)	Detector Efficiency (%)	Peak/Total Ratio
1 = ^{159}Tb(n, 2n)^{158m}Tb	100.00	160	0.20	0.11	1	50	1.00
2 = ^{159}Tb(n, p)^{159}Gd	100.00	2.2	1080	0.06	3	50	1.00
3 = ^{159}Tb(n, p)^{159}Gd	100.00	2.2	1080	0.36	9	45	0.77

3-MeV AND THERMAL NEUTRONS

Nargolwalla et al.[23] investigated the utility of 3-MeV neutrons for the determination of terbium. They found only a general increase in activity in the range of 0.012 to 0.5 MeV and used a gross-counting technique. This procedure is not deemed of sufficient analytical selectivity to provide a useful method for the measurement of terbium.

Activation with the thermal neutron flux available from a generator does not produce useful sensitivity for measuring terbium.

TIN

The 10 isotopes of tin which occur in nature are of mass 112, 114 through 120, 122, and 124. Tin is an element that is determined with ease using a

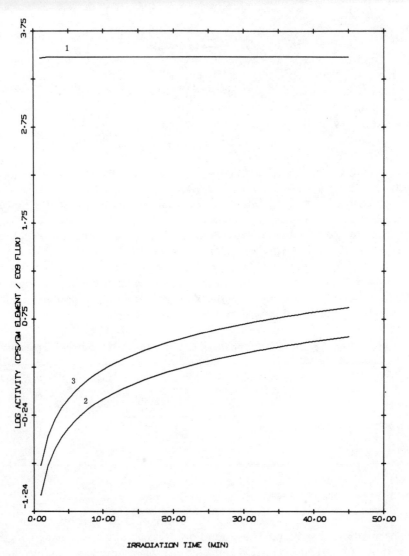

Fig. 7.196. Terbium—activation curves (see Table 7.66).

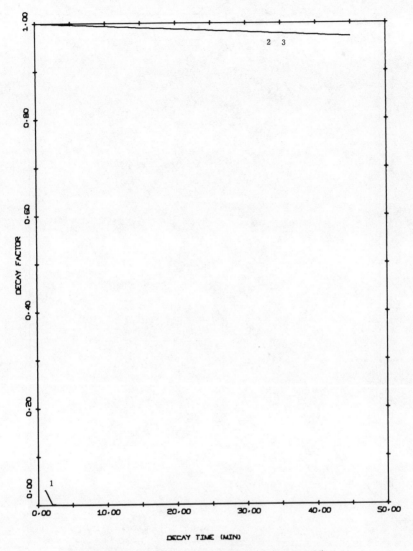

Fig. 7.197. Terbium—decay curves (see Table 7.66).

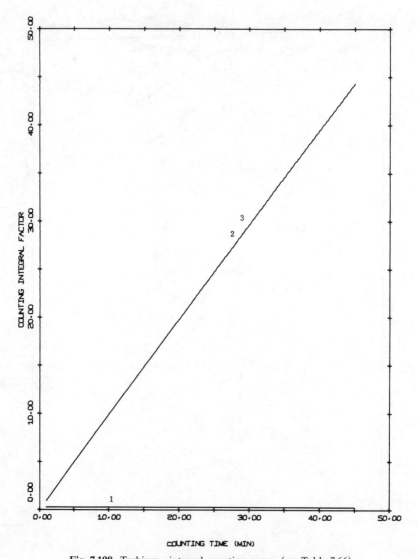

Fig. 7.198. Terbium—integral counting curves (see Table 7.66).

III. ACTIVATION ANALYSIS CONDITIONS FOR THE ELEMENTS   571

variety of wet chemical and instrumental methods of analysis. Probably for this reason, the neutron activation analysis for this element has not been widely explored.

14-MeV NEUTRONS

Table 7.67 gives the important parameters for the activation analysis of tin using fast neutrons. The activation, decay, and integral counting curves are shown in Figs. 7.199 to 7.201. Nuclear parameters for the activation of tin with 14-MeV neutrons indicate that the primary reactions of interest are the (n, p) reactions which occur with the mass 118, 119, and 120 isotopes.

TABLE 7.67. Nuclear Data for Reactions with 14-MeV Neutrons—Tin

Nuclear Reaction	Target Isotope Abundance (%)	Cross Section (mb)	Half-life (min)	Gamma Energy (MeV)	Gamma Yield (%)	Detector Efficiency (%)	Peak/Total Ratio
1 = ^{112}Sn(n, 2n)^{111}Sn	0.95	1500	35.00	0.51	54	40	0.63
2 = ^{118}Sn(n, p)^{118}In	24.01	28	0.10	1.23	97	31	0.37
3 = ^{118}Sn(n, p)^{118}In	24.01	28	0.10	1.05	80	32	0.41
4 = ^{119}Sn(n, p)^{119m}In	8.58	14	18.00	0.30	100	47	0.84
5 = ^{120}Sn(n, p)^{120m}In	32.97	1.00	0.05	1.17	15	31	0.38
6 = ^{120}Sn(n, p)^{120m}In	32.97	3.80	0.76	1.17	100	31	0.38

The most useful reaction indicated here (Table 7.67) is with the isotope of mass 120 which yields ^{120}In which has a photopeak of 1.17 MeV and a half-life of 0.76 min. Cuypers and Cuypers,[1] on the other hand, report that the most sensitive reaction for the measurement of tin using 14-MeV neutrons results from the (n, γ) reaction of ^{122}Sn, giving a γ-ray at 0.153 MeV. The detection limit quoted by Cuypers and Cuypers is 0.69 mg. A reaction of somewhat lesser sensitivity reported by these workers is the (n, 2n) reaction on ^{112}Sn giving the positron emitter ^{111}Sn. There are no further reports on the measurement of tin using 14-MeV neutrons.

3-MeV AND THERMAL NEUTRONS

Nargolwalla et al.[23] reported (n, γ) reactions occurring with sufficient sensitivity on the mass 116, 117, and 122 isotopes of tin to produce γ-emissions in the range of 0.15 to 0.16 MeV. Utilizing a 10-min irradiation time and a 10-min counting time, they reported an experimental sensitivity of 2.7×10^3 counts/g utilizing a flux of 10^6 neutrons/(cm² sec). The (n, γ) reaction on ^{124}Sn also occurs with sufficient sensitivity to give a photopeak at 0.325 MeV with an experimental sensitivity of 8.3×10^2 counts/g of tin

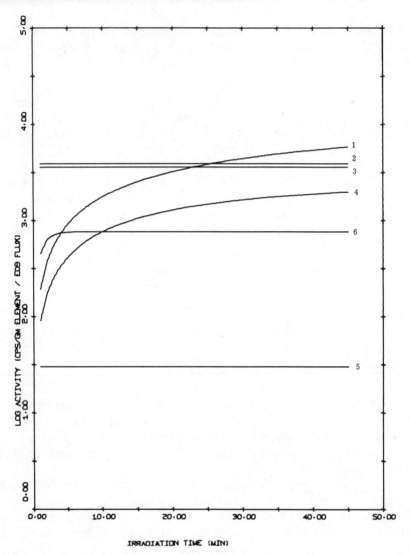

Fig. 7.199. Tin—activation curves (see Table 7.67).

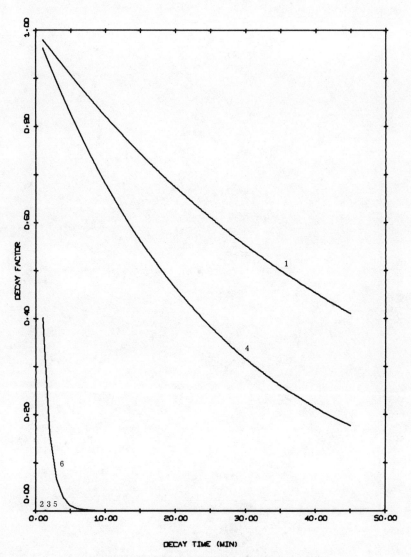

Fig. 7.200. Tin—decay curves (see Table 7.67).

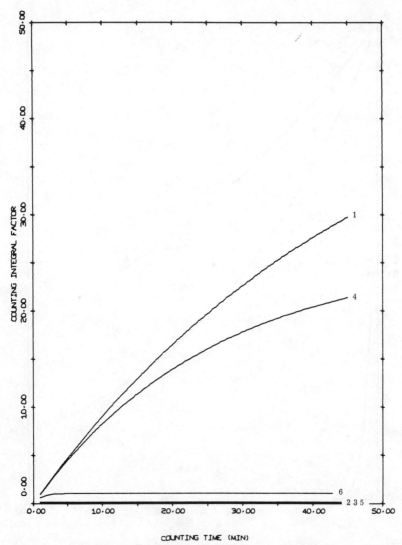

Fig. 7.201. Tin—integral counting curves (see Table 7.67).

III. ACTIVATION ANALYSIS CONDITIONS FOR THE ELEMENTS

for the same normalized flux. There are no reported applications of the 3-MeV technique to the determination of tin.

The sensitivity of measuring tin with the thermalized neutron flux of a generator is too low to be of value.[24]

TITANIUM

Titanium occurs in nature in five isotopes, ranging in mass from 46 to 50. There are a number of chemical and instrumental methods for the determination of titanium. It has been determined by polarographic, colorimetric, emission spectrographic, and x-ray spectrographic methods. The primary difficulty in applying wet chemical methods to the determination of titanium is the need for sample dissolution. Certain oxides of titanium can only be dissolved in hot hydrofluoric acid, and occasionally fusion techniques must be applied to ores, minerals, and various titanates. In these circumstances a nondestructive technique such as neutron activation or x-ray spectrography is preferred. The x-ray fluorescence method suffers from the soft x-rays given by titanium. For most applications a helium path should be used to improve sensitivity. Matrix elements can cause serious interference in the x-ray fluorescence analysis of titanium.

14-MeV NEUTRONS

The nuclear parameters important in the activation analysis of titanium with fast neutrons are given in Table 7.68. Figures 7.202 to 7.204 show the activation, decay, and counting integral curves for establishing activation conditions. By far the most sensitive reaction for the determination of titanium is that producing ^{46m}Sc which gives a prominent γ-ray at 0.14 MeV. Being in the low-energy portion of the spectrum, it is easily interfered with by Compton and bremsstrahlung contributions from other matrix elements.

TABLE 7.68. Nuclear Data for Reactions with 14-MeV Neutrons—Titanium

Nuclear Reaction	Target Isotope Abundance (%)	Cross Section (mb)	Half-life (min)	Gamma Energy (MeV)	Gamma Yield (%)	Detector Efficiency (%)	Peak/Total Ratio
1 = ^{46}Ti(n, p)^{46m}Sc	7.95	530	0.33	0.14	100	50	0.97
2 = ^{48}Ti(n, p)^{48}Sc	73.45	92.7	2635	0.98	100	32	0.42
3 = ^{48}Ti(n, p)^{48}Sc	73.45	92.7	2635	1.04	100	32	0.41
4 = ^{48}Ti(n, p)^{48}Sc	73.45	92.7	2635	1.31	100	30	0.36
5 = ^{49}Ti(n, p)^{49m}Sc	5.51	97	57.50	1.76	1	28	0.29
6 = ^{50}Ti(n, p)^{50}Sc	5.25	48	1.72	0.52	100	40	0.62
7 = ^{50}Ti(n, p)^{50}Sc	5.25	48	1.72	1.12	100	31	0.39
8 = ^{50}Ti(n, p)^{50}Sc	5.25	48	1.72	1.55	100	28	0.32

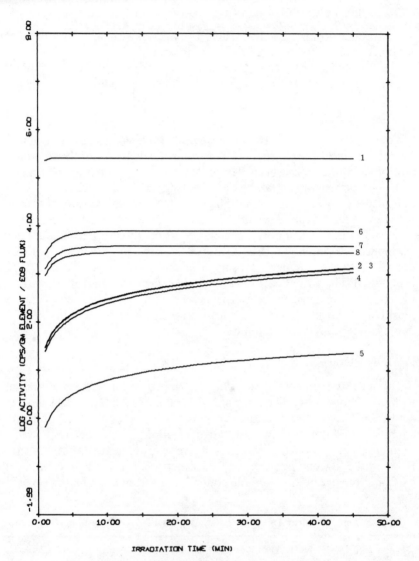

Fig. 7.202. Titanium—activation curves (see Table 7.68).

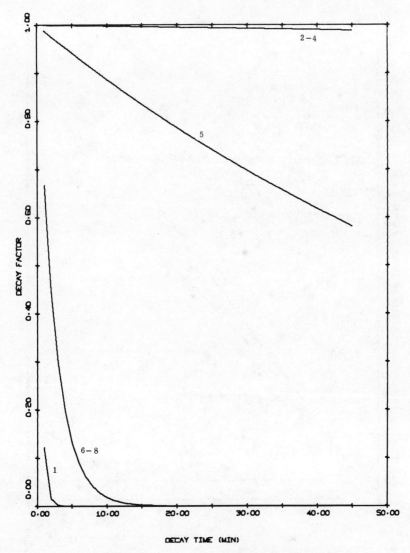

Fig. 7.203. Titanium—decay curves (see Table 7.68).

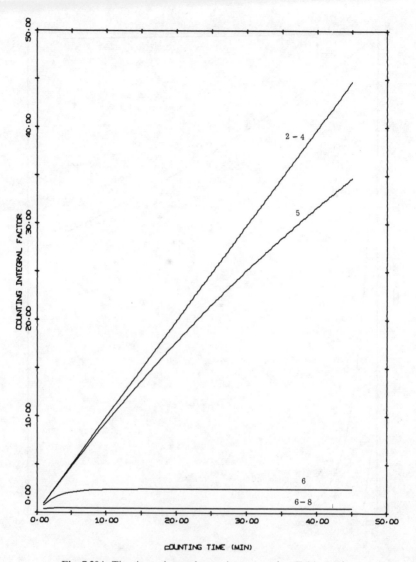

Fig. 7.204. Titanium—integral counting curves (see Table 7.68).

III. ACTIVATION ANALYSIS CONDITIONS FOR THE ELEMENTS

Cuypers and Cuypers[1] report a detection limit of 2.5 mg of titanium using the 0.14-MeV peak and an irradiation time of 1 min. Utilizing a longer radiation time of 5 min, the 1.76-MeV peak from ^{49m}Sc gave a detection limit of 7.1 mg. While titanium does not have ideal parameters for determination by neutron activation analysis, it can be determined in certain matrices with good precision and accuracy.

Presumably because of the availability of alternative methods for the determination of titanium, there have been relatively few reported applications of neutron activation for this measurement. Hull[16] surveyed the application of fast neutron activation analysis to petroleum samples and reported the determination of titanium in such materials. More recently, Persiani and Cosgrove[106] applied this method to the determination of the stoichiometry of titanium and vanadium oxides. Although these authors do not give the details of the analytical procedure used, the irradiation and counting times imply that they used the 0.14-MeV photopeak from ^{46m}Sc to measure the titanium. Under these conditions a 3 to 5% relative error at the 95% confidence limit was found for the determination of titanium. Comparison of neutron activation and chemical methods for the determination of titanium showed excellent agreement, with less than 0.2% deviation on titanium oxide and less than 0.9% deviation on mixed titanium-vanadium oxides in the measurement of titanium.

3-MeV AND THERMAL NEUTRONS

Nargolwalla et al.[23] reported on the determination of titanium using 3-MeV neutrons and note a sensitivity of 250 counts/g at a flux of 10^6 neutrons/(cm^2 sec) utilizing the (n, γ) reaction on ^{50}Ti. A slightly lower sensitivity was observed for the ^{47}Ti(n, p)^{47}Sc reaction. This nuclide has an emission at 0.16 MeV. The 3-MeV neutron activation technique could be used only for macroconstituent analysis of titanium in high-percentage titanium ores.

Activation with the thermal flux available from a generator does not produce useful sensitivity for the measurement of titanium.

TUNGSTEN

Tungsten occurs in nature in five isotopes of mass 180, 182, 183, 184, and 186. Tungsten is an element that is not readily determined by wet chemical methods, with perhaps the phosphotungstic acid precipitation or colorimetric reaction being the most popular. Some of the classical methods for the determination of tungsten involve precipitation of tungstic acid and colorimetric determination, although as applied to steel analysis these have been largely supplanted by the x-ray fluorescence or atomic absorption methods of analysis.

14-MeV NEUTRONS

The pertinent nuclear parameters for the activation of tungsten are given in Table 7.69, and Figs. 7.205 to 7.207 give the activation, decay, and integral counting curves for the determination of tungsten. The activation of tungsten with 14-MeV neutrons has favorable cross sections; however, the nuclides produced have relatively short half-lives and produce photopeaks of low energies. The latter two characteristics are not ideally suited for its measurement because of the potential interference from nuclides which emit higher energy gamma rays. Cuypers and Cuypers[1] report that the best sensitivity (1.3 mg) for the determination of tungsten results from the emission at 0.059 MeV, which is a tungsten x-ray. The somewhat longer lived ^{185m}W, which yields a 0.17-MeV photopeak, has almost equivalent sensitivity and is perhaps the more useful gamma-ray peak because of its higher energy. There are no reported applications for the 14-MeV technique to the determination of tungsten in practical samples.

TABLE 7.69. Nuclear Data for Reactions with 14-MeV Neutrons—Tungsten

Nuclear Reaction	Target Isotope Abundance (%)	Cross Section (mb)	Half-life (min)	Gamma Energy (MeV)	Gamma Yield (%)	Detector Efficiency (%)	Peak/Total Ratio
$1 = {}^{183}$W(n, n)^{183m}W	14.40	170	0.09	0.05	11	50	1.00
$2 = {}^{184}$W(n, 2n)^{183m}W	30.60	750	0.09	0.05	11	50	1.00
$3 = {}^{184}$W(n, 2n)^{185m}W	30.60	750	1.60	0.10	32	50	1.00
$4 = {}^{184}$W(n, 2n)^{185m}W	30.60	750	1.60	0.16	6	50	0.96
$5 = {}^{186}$W(n, 2n)^{185m}W	28.40	500	1.60	0.13	38	50	0.98
$6 = {}^{186}$W(n, 2n)^{185m}W	28.40	500	1.60	0.17	55	50	0.95

3-MeV AND THERMAL NEUTRONS

Nargolwalla et al.[23] reported a useful (n, γ) reaction on ^{186}W which produces emissions at 0.48 and 0.69 MeV. Using a 20-min irradiation time and 20-min counting time, a sensitivity of 4×10^2 counts/g using a flux of 10^6 neutrons/(cm^2 sec) was reported. There have been no other reported applications of the 3-MeV technique to the determination of tungsten.

Dibbs[24] reported a sensitivity of 0.65 mg for the measurement of tungsten using a thermal flux of 10^8 neutrons/(cm^2 sec). He utilized the 0.48-MeV emission of ^{187}W.

VANADIUM

The two naturally occurring isotopes of vanadium are ^{50}V and ^{51}V. Because of its widespread use in steel, vanadium has been an element of

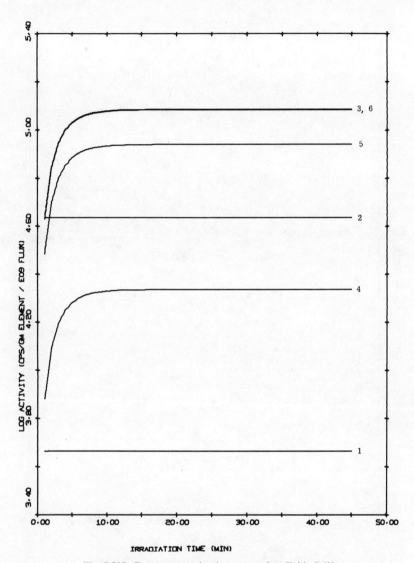

Fig. 7.205. Tungsten—activation curves (see Table 7.69).

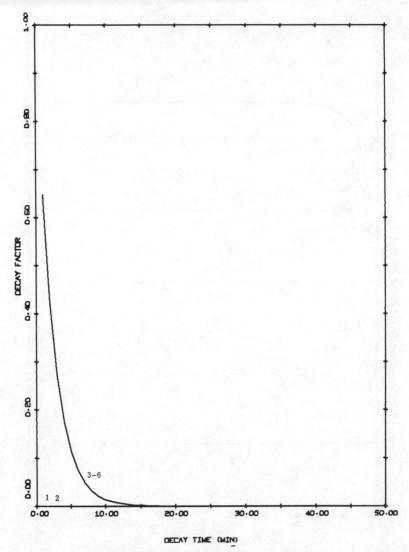

Fig. 7.206. Tungsten—decay curves (see Table 7.69).

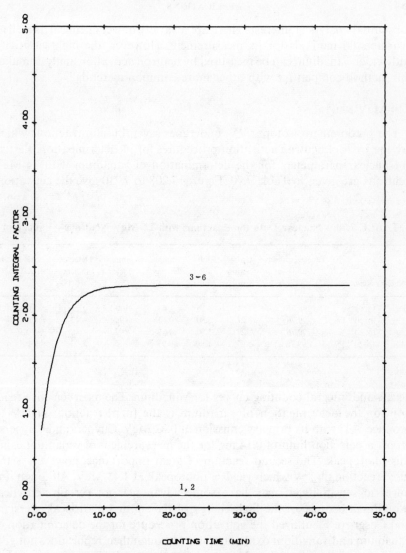

Fig. 7.207. Tungsten—integral counting curves (see Table 7.69).

substantial analytical interest. There are a variety of wet chemical as well as instrumental methods for its measurement. However, the high sensitivity with which vanadium can be measured by neutron activation analysis makes this method competitive with other more common methods.

14-MeV NEUTRONS

The predominant isotope, ^{51}V, undergoes several useful reactions which are the basis of neutron activation procedures for its determination. Pertinent nuclear parameters for the determination of vanadium with 14-MeV neutrons are given in Table 7.70. Figures 7.208 to 7.210 give the activation,

TABLE 7.70. Nuclear Data for Reactions with 14-MeV Neutrons—Vanadium

Nuclear Reaction	Target Isotope Abundance (%)	Cross Section (mb)	Half-life (min)	Gamma Energy (MeV)	Gamma Yield (%)	Detector Efficiency (%)	Peak/Total Ratio
1 = ^{51}V(n, γ)^{52}V	99.76	1	3.76	1.43	100	29	0.34
2 = ^{51}V(n, p)^{51}Ti	99.76	27	5.80	0.32	95	42	0.81
3 = ^{51}V(n, p)^{51}Ti	99.76	27	5.80	0.61	2	38	0.56
4 = ^{51}V(n, p)^{51}Ti	99.76	27	5.80	0.93	5	33	0.43

decay, and integral counting curves for vanadium. The most sensitive reaction for the determination of vanadium is the (n, p) reaction on ^{51}V to produce ^{51}Ti with its primary emission at 0.32 MeV. Cuypers and Cuypers[1] report a detection limit of 0.14 mg for the measurement of vanadium using this photopeak. The second reaction of great importance, however, is the (n, γ) reaction on ^{51}V which yields a photopeak at 1.43 MeV. Although it is lower in sensitivity, it may have some advantages for certain analytical procedures because of the high energy of the emitted photopeak. Persiani and Cosgrove[106] utilized the activation procedure for the determination of vanadium and vanadium oxide samples. Although their report does not give details of the analytical procedures, it appears that the photopeak at 0.32 MeV from ^{51}Ti was the reaction used in their measurement. Their data show excellent agreement between the amounts of vanadium added and that found by activation methods. These studies were carried out with a mixed vanadium-titanium oxide sample as well as with pure vanadium oxide. Persiani and Cosgrove also proposed a rapid way of determining the oxygen-to-vanadium ratio simultaneously using the 14-MeV technique. Pierce et al.[149] used neutron moderation to differentiate between vanadium and other matrix elements where there were direct interferences in the

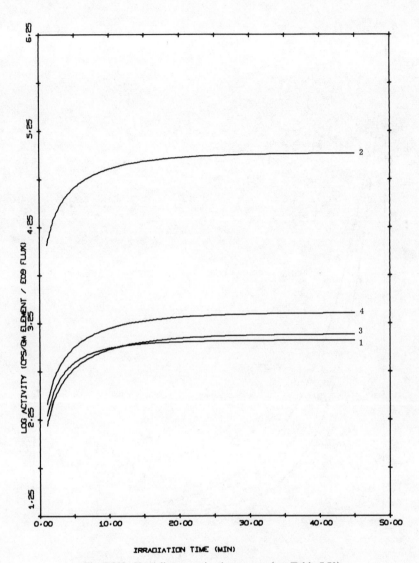

Fig. 7.208. Vanadium—activation curves (see Table 7.70).

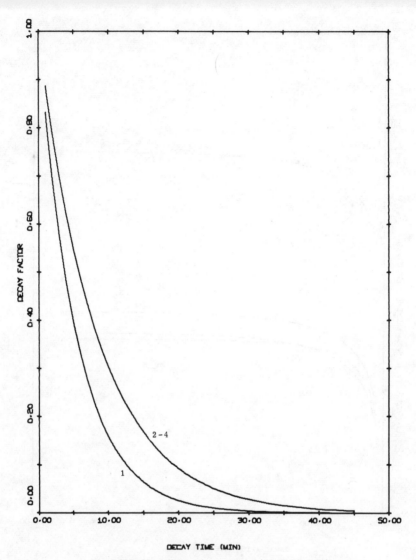

Fig. 7.209. Vanadium—decay curves (see Table 7.70).

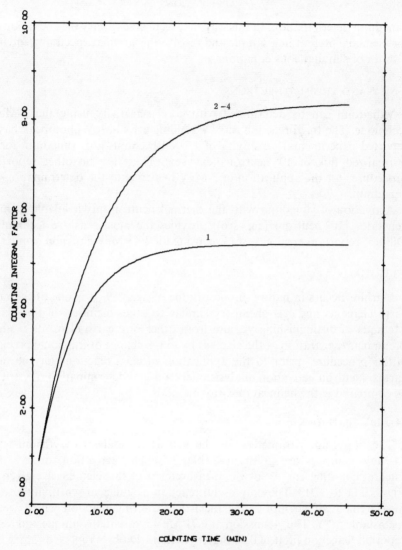

Fig. 7.210. Vanadium—integral counting curves (see Table 7.70).

vanadium determination. In this way they were able to vary the activation of the elements present in a sample and resolve the complex spectra by solving a series of simultaneous equations.

3-MeV AND THERMAL NEUTRONS

Vanadium can be determined with good sensitivity using the 3-MeV technique. The (n, γ) reaction on ^{51}V yielding a 1.43-MeV photopeak has a reported experimental sensitivity of 4.7×10^3 counts/g of vanadium for a normalized flux of 10^6 neutrons/(cm^2 sec). There are no other reported procedures for the application of 3-MeV neutrons to the determination of vanadium.

Activation of vanadium with the thermal neutron flux available from a generator [10^8 neutrons/(cm^2 sec)] provides the most sensitive technique. Dibbs[24] reports a sensitivity of 13 μg using the 1.43-MeV emission from ^{52}V.

YTTRIUM

Yttrium occurs in nature in one isotope of mass 89. It is one of the rare earth elements and has chemistry similar to other members of this class. Methods of distinguishing yttrium from other rare earth elements require preliminary separation of the element by ion-exchange or fractional precipitation procedures prior to the application of class reactions for the rare earths. Neutron activation analysis offers a useful determination of yttrium as compared with chemical or x-ray methods.

14-MeV NEUTRONS

The important parameters for the activation analysis of yttrium with 14-MeV neutrons are given in Table 7.71. The activation, decay, and integral counting curves for the measurement of this element are given in Figs. 7.211 to 7.213. The only useful reaction of sufficient sensitivity for the measurement of yttrium is the inelastic scattering reaction which yields metastable ^{89m}Y. The γ-emission at 0.91 MeV is essentially unique and has a reported detection limit of 0.58 mg of yttrium.[1] Dibbs[24] reported somewhat

TABLE 7.71. Nuclear Data for Reactions with 14-MeV Neutrons—Yttrium

Nuclear Reaction	Target Isotope Abundance (%)	Cross Section (mb)	Half-life (min)	Gamma Energy (MeV)	Gamma Yield (%)	Detector Efficiency (%)	Peak/Total Ratio
1 = ^{89}Y(n, 2n)^{88}Y	100.00	600	151200.03	0.90	91	33	0.43
2 = ^{89}Y(n, 2n)^{88}Y	100.00	600	151200.03	1.84	100	28	0.28
3 = ^{89}Y(n, n'γ)^{89m}Y	100.00	400	0.27	0.91	99	33	0.43

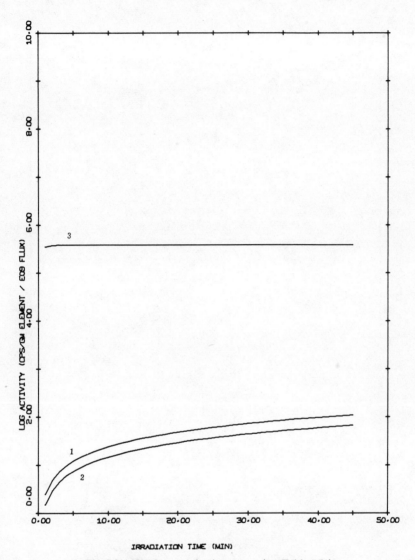

Fig. 7.211. Yttrium—activation curves (see Table 7.71).

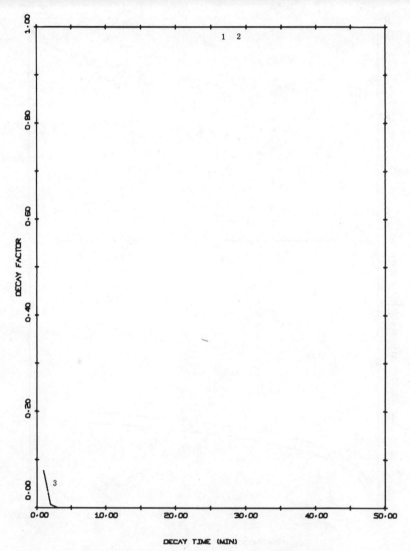

Fig. 7.212. Yttrium—decay curves (see Table 7.71).

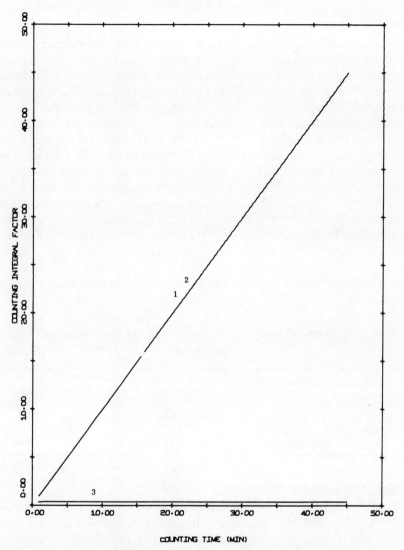

Fig. 7.213. Yttrium—integral counting curves (see Table 7.71).

higher sensitivity for the same reaction (0.1 mg). Because of its short half-life, measurement of this γ-ray may be interfered with by the presence of large amounts of oxygen in the sample due to a high Compton background under the photopeak. There is, however, sufficient difference in half-life between ^{16}N and ^{89m}Y to permit resolution of this background contribution. Broadhead et al.[12] noted in their survey of fast neutron reactions that yttrium could be determined with excellent sensitivity utilizing this reaction in molten salt media. Plaksin et al.[63, 64] utilized this reaction for the determination of yttrium in rare earth mixtures. These have been the only reported applications of this method thus far. In comparison with other destructive methods for the determination of yttrium in rare earth ores or in metallurgical products, it appears that the neutron activation analysis method can compete successfully. There are two potential interferences from zirconium and niobium in the determination of yttrium by this method. Both elements yield metastable ^{89m}Y product by different reactions. The nuclear data for the interfering reactions are given in Table 7.72 and the activation, decay, and integral counting curves for zirconium and niobium are shown in Figs. 7.214 to 7.216.

TABLE 7.72. Nuclear Data for Reactions with 14-MeV Neutrons—Yttrium Interferences

Nuclear Reaction	Target Isotope Abundance (%)	Cross Section (mb)	Half-life (min)	Gamma Energy (MeV)	Gamma Yield (%)	Detector Efficiency (%)	Peak/Total Ratio
1 = $^{90}Zr(n, np)^{89m}Zr$	51.46	770.00	4.18	0.91	99	33	0.43
2 = $^{93}Nb(n, n\alpha)^{89m}Y$	100.00	2.50	0.27	0.91	99	33	0.43

3-MeV AND THERMAL NEUTRONS

Nargolwalla et al.[23] have reported excellent sensitivity for the production of ^{89m}Y utilizing 3-MeV neutrons. Their work indicates that upon short irradiation (2 min) they achieved a sensitivity of 1.2×10^4 counts/g of yttrium with a flux of 10^6 neutrons/(cm^2 sec). For macroconstituent determinations of yttrium in ores this reaction could be useful since other matrix elements might not be activated as readily with the low-energy neutrons. These authors also noted the generation of some ^{90m}Y resulting from the capture reaction with emissions at 0.20 and 0.48 MeV. However, the sensitivity for this reaction was a factor of 50 lower than the production of ^{89m}Y. There have been no other applications reported of the use of 3-MeV neutrons for the determination of yttrium.

Thermalized neutrons are not useful for the activation analysis of yttrium.

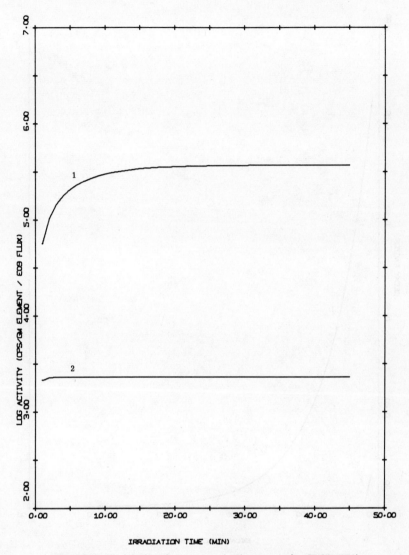

Fig. 7.214. Yttrium interferences—activation curves (see Table 7.72).

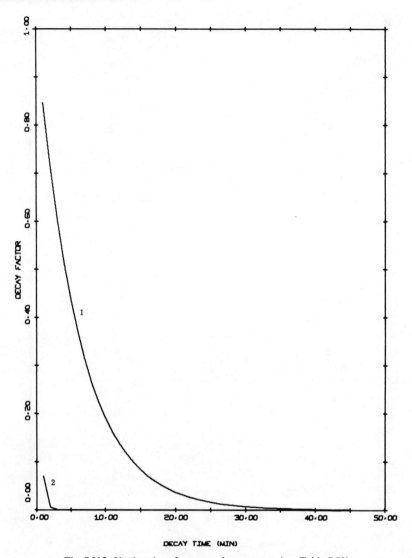

Fig. 7.215. Yttrium interferences—decay curves (see Table 7.72).

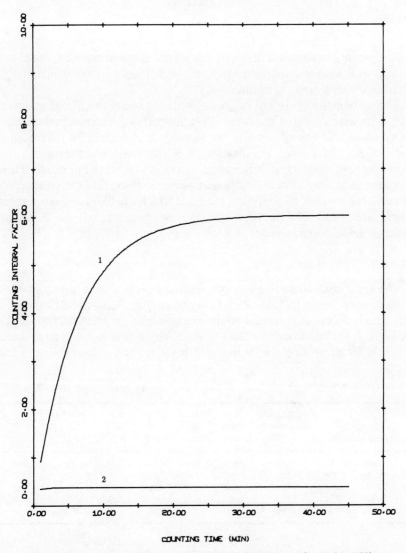

Fig. 7.216. Yttrium interferences—integral counting curves (see Table 7.72).

ZINC

Zinc occurs in nature in five isotopic forms, those of mass 64, 66, 67, 68, and 70. It is an element which has found wide industrial use in metallurgy, pharmaceuticals, and pigments.

There are many analytical methods for the measurement of zinc in ores as well as in manufactured products. Wet chemical and instrumental methods based on atomic absorption or x-ray methods are excellent for the measurement of zinc. Moreover, the dissolution of zinc and the separation of zinc from matrices in which it may appear are relatively straightforward. There has been little need to develop alternative methods for zinc analysis because of the large number of analytical methods which exist for its measurement. There are also useful reactions for the measurement of zinc using fast neutron activation analysis.

14-MeV NEUTRONS

The most useful reactions for the measurement of zinc using 14-MeV neutrons are shown in Table 7.73. In addition, Figs. 7.217 to 7.219 show the activation, decay, and integral counting curves for the measurement of this element. It is clear from an examination of these data that the most useful reaction for measuring zinc is that which yields the positron emitter ^{63}Zn.

TABLE 7.73. Nuclear Data for Reactions with 14-MeV Neutrons—Zinc

Nuclear Reaction	Target Isotope Abundance (%)	Cross Section (mb)	Half-life (min)	Gamma Energy (MeV)	Gamma Yield (%)	Detector Efficiency (%)	Peak/Total Ratio
1 = ^{64}Zn(n, 2n)^{63}Zn	48.89	167	38.40	0.51	186	40	0.63
2 = ^{64}Zn(n, 2n)^{63}Zn	48.89	167	38.40	0.67	8	37	0.53
3 = ^{64}Zn(n, p)^{64}Cu	48.89	386	768.00	0.51	38	40	0.63
4 = ^{66}Zn(n, p)^{66}Cu	27.81	101	5.10	1.04	9	32	0.41
5 = ^{68}Zn(n, p)^{68}Cu	18.56	25	0.50	1.08	95	31	0.40

All the other reactions are of considerably lower sensitivity and probably of little interest. The reaction which yields ^{63}Zn may be interfered with by a number of other positron emitters of similar half-lives. Most notable interferences would be chlorine, chromium, and gallium. Readers are referred to Tables 7.76 to 7.78 for consideration of possible positron-emitting interferences for the determination of zinc. Gibbons et al.[45] have utilized the fast neutron activation technique to measure zinc in lubricating oils in the presence of other trace elements. Kuehne and D'Agostino[2] have used a simplified technique for the analysis of metallic wear products and have used

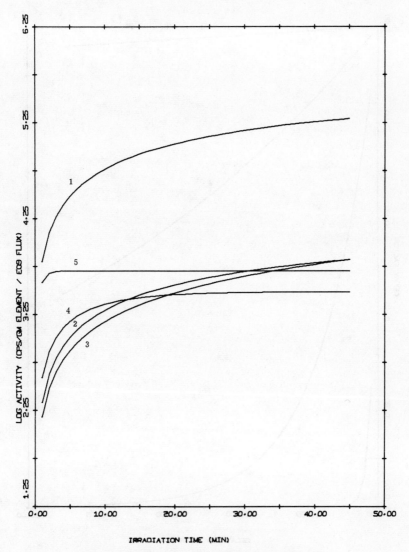

Fig. 7.217. Zinc—activation curves (see Table 7.73).

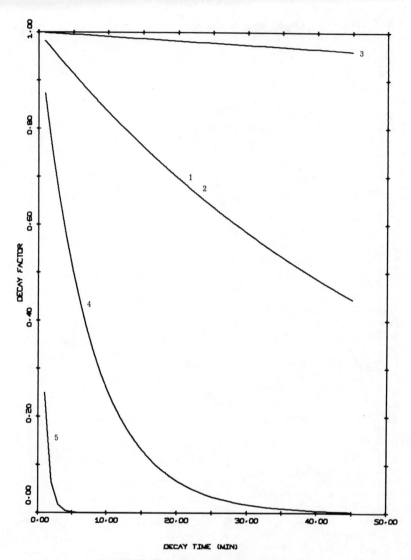

Fig. 7.218. Zinc—decay curves (see Table 7.73).

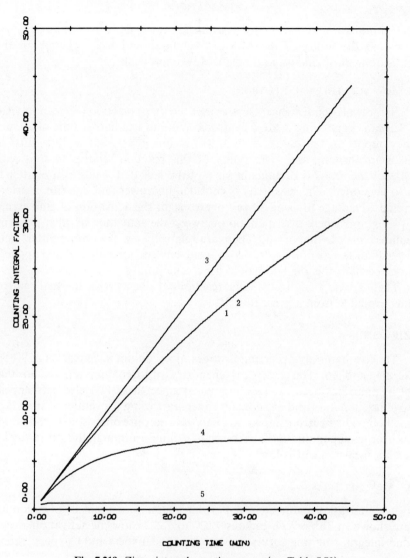

Fig. 7.219. Zinc—integral counting curves (see Table 7.73).

the annihilation radiation to measure zinc in such oils. Crambes[25] has surveyed the utility of the 14-MeV technique and has reported on the measurement of zinc using the 0.51-MeV photopeak.

3-MeV AND THERMAL NEUTRONS

Nargolwalla et al.[23] have shown that the (n, p) reaction on ^{64}Zn to yield ^{64}Cu can occur using 3-MeV neutrons. With an irradiation time of 20 min they report a sensitivity of 6×10^2 counts/g of zinc with a flux of 10^6 neutrons/(cm^2 sec). The utility of this reaction relative to that with 14-MeV neutrons is dubious in view of the fact that a positron emitter is also generated. The sensitivity is considerably lower and one can, a priori, see no advantage to using 3-MeV neutrons for the activation of zinc except in situations where discrimination between the activation of other positron emitters may be important. There are relatively few positron emitters generated in the activation with 3-MeV neutrons and for such isolated applications perhaps this method would have some utility.

There are no analytically useful reactions of zinc with the thermal neutron flux available from a generator.[24]

ZIRCONIUM

The five naturally occurring isotopes of zirconium have masses of 90, 91, 92, 94, and 96. The analytical chemistry of zirconium relative to that of hafnium has been extensively studied since these two elements appear together in nature and are difficult to separate using conventional analytical methods. The neutron activation analysis procedure offers some promise for measuring zirconium in the presence of hafnium, although the sensitivity for such a measurement is low.

14-MeV NEUTRONS

The activation parameters important for the measurement of zirconium are shown in Table 7.74. Figures 7.220 to 7.222 show the activation, decay, and integral counting curves for this element. Cuypers and Cuypers[1] report

TABLE 7.74. Nuclear Data for Reactions with 14-MeV Neutrons—Zirconium

Nuclear Reaction	Target Isotope Abundance (%)	Cross Section (mb)	Half-life (min)	Gamma Energy (MeV)	Gamma Yield (%)	Detector Efficiency (%)	Peak/Total Ratio
1 = ^{90}Zr(n, α)^{87m}Sb	51.46	194	169.80	0.39	80	44	0.75
2 = ^{90}Zr(n, 2n)^{89m}Zr	51.46	770	4.18	0.51	44	40	0.63
3 = ^{90}Zr(n, 2n)^{89m}Zr	51.46	770	4.18	0.91	99	33	0.43

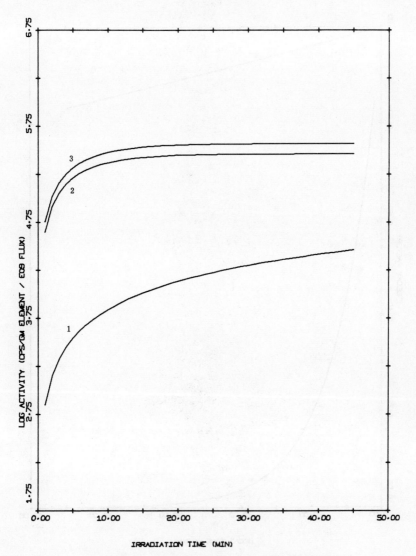

Fig. 7.220. Zirconium—activation curves (see Table 7.74).

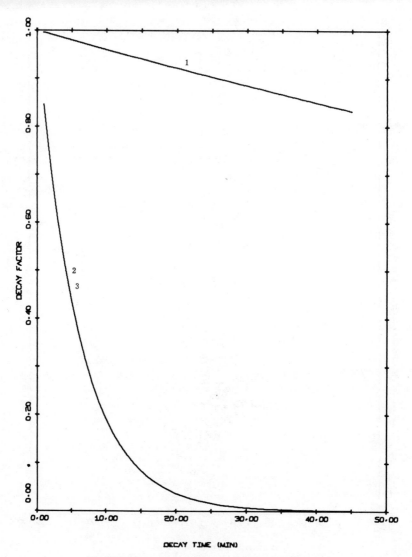

Fig. 7.221. Zirconium—decay curves (see Table 7.74).

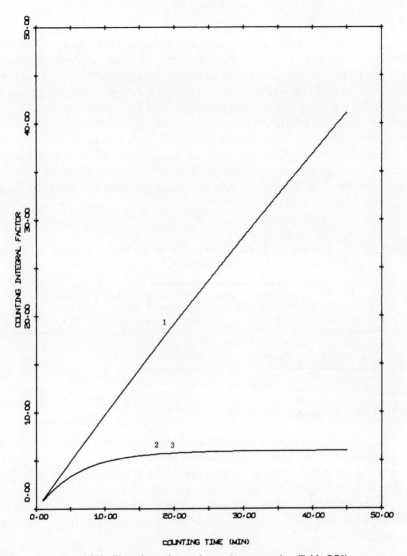

Fig. 7.222. Zirconium—integral counting curves (see Table 7.74).

the most sensitive reaction for the determination of zirconium to be that yielding ^{89m}Zr which has a γ-emission at 0.59 MeV. They report a detection limit of 0.25 mg of zirconium utilizing a flux of approximately 5×10^8 neutrons/(cm^2 sec). Other reactions for zirconium are of lesser importance for its determination. There have been no reported applications of the 14-MeV technique for the determination of zirconium. Yttrium constitutes a significant interference in the measurement of zirconium (Table 7.75) although if present its contribution to the 0.91-MeV activity (from ^{89m}Y) can be minimized by appropriate choice of decay time. Niobium interferes if present in large amounts also by producing ^{89m}Y. Data for these interferences are listed in Table 7.75, and the working curves are shown in Figs. 7.223 to 7.225.

TABLE 7.75. Nuclear Data for Reactions with 14-MeV Neutrons—Zirconium Interferences

Nuclear Reaction	Target Isotope Abundance (%)	Cross Section (mb)	Half-life (min)	Gamma Energy (MeV)	Gamma Yield (%)	Detector Efficiency (%)	Peak/Total Ratio
1 = ^{89}Y(n, n')^{89m}Y	100.00	400	0.27	0.91	99	33	0.43
2 = ^{93}Nb(n, n'α)^{89m}Y	100.00	2.5	0.27	0.91	99	33	0.43

3-MeV AND THERMAL NEUTRONS

Nargolwalla et al.[23] have reported a very low activation sensitivity for the measurement of zirconium using 3-MeV neutrons. Through an inelastic scattering reaction ^{90m}Zr can be generated, but of sensitivity too low to be useful for most analytical applications. There have been no other reported applications of the 3-MeV technique for the measurement of zirconium. There are no useful reactions with the thermal flux available from a neutron generator.

POSITRON EMITTERS

Listed in Tables 7.76 to 7.78 are the positron-emitting nuclides that are generated from the activation of a variety of elements using 14-MeV neutrons. These elements are tabulated here as well as in the section on the particular element in order to provide a convenient way for examining other possible interfering reactions. The accompanying working curves for activation, decay, and integral counting (Figs. 7.226–7.234) should permit the

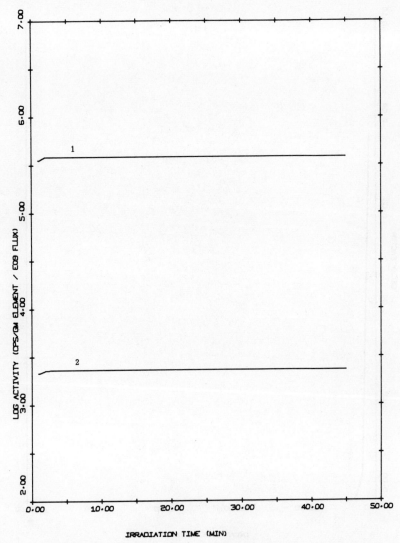

Fig. 7.223. Zirconium interferences—activation curves (see Table 7.75).

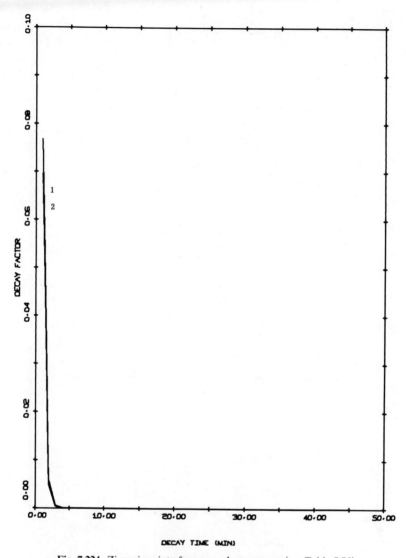

Fig. 7.224. Zirconium interferences—decay curves (see Table 7.75).

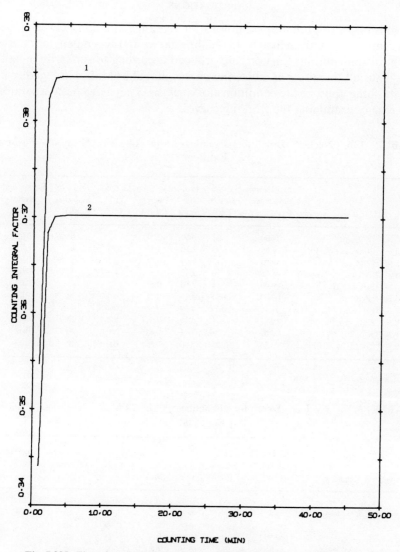

Fig. 7.225. Zirconium interferences—integral counting curves (see Table 7.75).

establishment of optimum conditions for the differentiation between positron emitters in a given matrix. In the three tables and accompanying figures the positron-emitting reactions are grouped according to similarity in half-life. Thus all three sets of tables and curves should be looked at together in establishing appropriate conditions for analysis. They have been separated for ease of examining the working curves.

TABLE 7.76. Nuclear Data for Reactions with 14-MeV Neutrons—Positron Emitters[a]

Nuclear Reaction	Target Isotope Abundance (%)	Cross Section (mb)	Half-life (min)	Gamma Energy (MeV)	Gamma Yield (%)	Detector Efficiency (%)	Peak/Total Ratio
1 = ^{106}Pd(n, p)^{106}Rh	27.30	17	0.50	0.51	88	40	0.63
2 = ^{92}Mo(n, 2n)^{91m}Mo	15.86	190	1.10	0.51	276	40	0.63
3 = ^{71}Ga(n, p)^{71}Zn	39.50	10	2.40	0.51	13	40	0.63
4 = ^{109}Ag(n, 2n)^{108}Ag	48.65	700	2.42	0.51	1	40	0.63
5 = ^{31}P(n, 2n)^{30}P	100.00	11	2.50	0.51	100	40	0.63
6 = ^{141}Pr(n, 2n)^{140}Pr	100.00	2080	3.39	0.51	100	40	0.63
7 = ^{90}Zr(n, 2n)^{89m}Zr	51.46	770	4.18	0.51	44	40	0.63
8 = ^{79}Br(n, 2n)^{78}Br	50.52	1141	6.50	0.51	184	40	0.63
9 = ^{39}K(n, 2n)^{38}K	93.09	4	7.71	0.51	200	40	0.63
10 = ^{144}Sm(n, 2n)^{143}Sm	3.16	1200	8.90	0.51	100	40	0.63

[a] Cf. Figs. 7.226, 7.227, and 7.228.

TABLE 7.77. Nuclear Data for Reactions with 14-MeV Neutrons—Positron Emitters[a]

Nuclear Reaction	Target Isotope Abundance (%)	Cross Section (mb)	Half-life (min)	Gamma Energy (MeV)	Gamma Yield (%)	Detector Efficiency (%)	Peak/Total Ratio
1 = ^{54}Fe(n, 2n)^{53}Fe	5.84	15	8.50	0.51	196	40	0.63
2 = ^{63}Cu(n, 2n)^{62}Cu	69.10	550	9.80	0.51	195	40	0.63
3 = ^{14}N(n, 2n)^{13}N	99.63	5	10.00	0.51	200	40	0.63
4 = ^{113}In(n, 2n)^{112}In	4.23	1500	14.00	0.51	44	40	0.63
5 = ^{91}Mo(n, 2n)^{91}Mo	15.86	190	15.49	0.51	200	40	0.63
6 = ^{121}Sb(n, 2n)^{120}Sb	57.25	750	15.90	0.51	87	40	0.63
7 = ^{81}Br(n, 2n)^{80}Br	49.48	440	17.60	0.51	5	40	0.63
8 = ^{107}Ag(n, 2n)^{106}Ag	51.35	600	24.00	0.51	140	40	0.63
9 = ^{35}Cl(n, 2n)^{34m}Cl	75.40	3.50	32.00	0.51	100	40	0.63
10 = ^{64}Zn(n, 2n)^{63}Zn	48.89	167	38.40	0.51	186	40	0.63

[a] Cf. Figs. 7.229, 7.230, and 7.231.

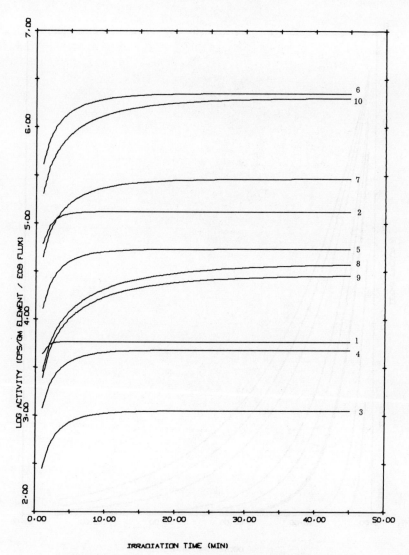

Fig. 7.226. Positron emitters—activation curves (see Table 7.76).

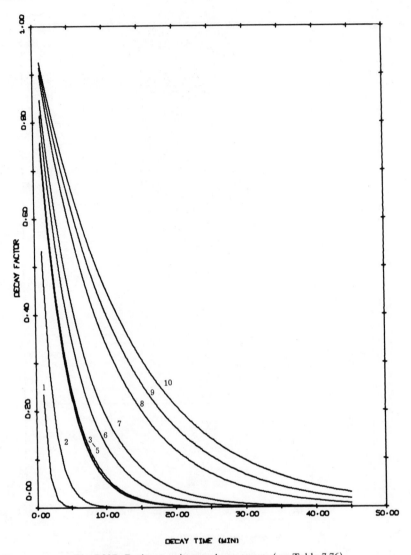

Fig. 7.227. Positron emitters—decay curves (see Table 7.76).

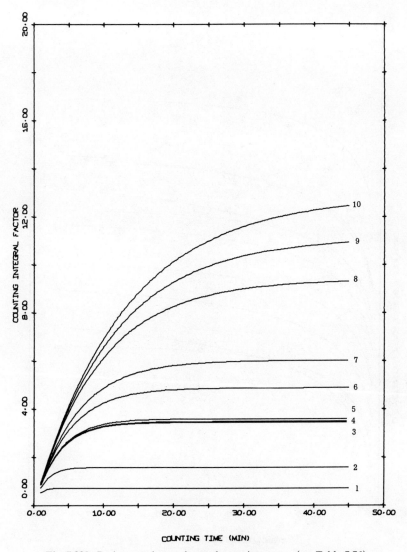

Fig. 7.228. Positron emitters—integral counting curves (see Table 7.76).

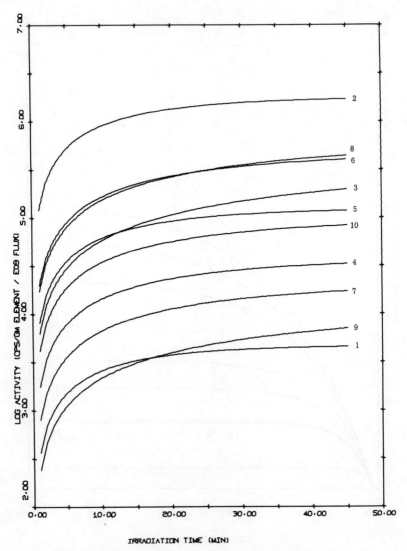

Fig. 7.229. Positron emitters—activation curves (see Table 7.77).

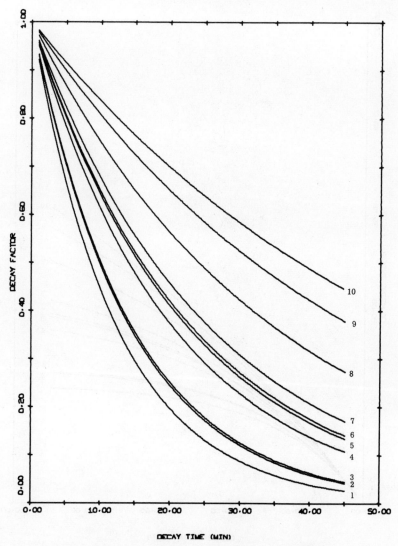

Fig. 7.230. Positron emitters—decay curves (see Table 7.77).

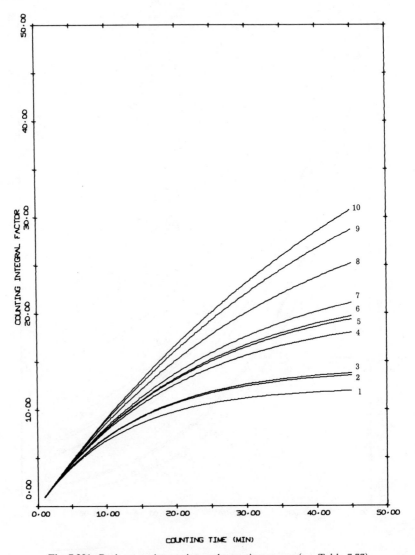

Fig. 7.231. Positron emitters— integral counting curves (see Table 7.77).

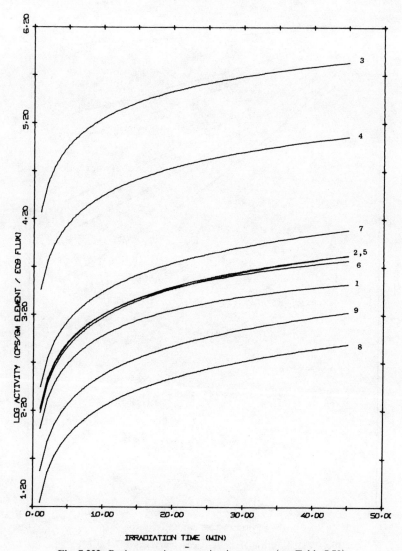

Fig. 7.232. Positron emitters—activation curves (see Table 7.78).

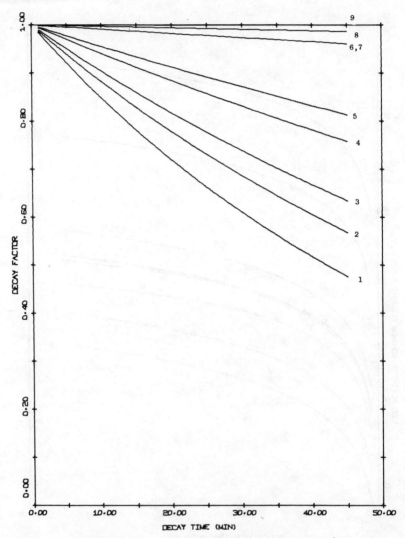

Fig. 7.233. Positron emitters—decay curves (see Table 7.78).

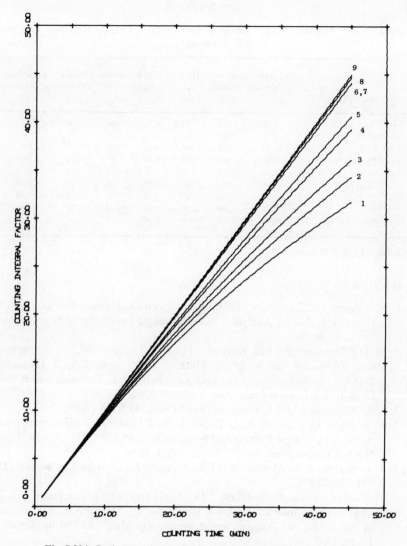

Fig. 7.234. Positron emitters—integral counting curves (see Table 7.78).

617

TABLE 7.78. Nuclear Data for Reactions with 14-MeV Neutrons—Positron Emitters[a]

Nuclear Reaction	Target Isotope Abundance (%)	Cross Section (mb)	Half-life (min)	Gamma Energy (MeV)	Gamma Yield (%)	Detector Efficiency (%)	Peak/Total Ratio
1 = ^{50}Cr(n, 2n)^{49}Cr	4.31	27	41.9	0.51	186	40	0.63
2 = ^{106}Cd(n, 2n)^{105}Cd	1.21	827	55.0	0.51	100	40	0.63
3 = ^{69}Ga(n, 2n)^{68}Ga	60.50	800	68.3	0.51	176	40	0.63
4 = ^{19}F(n, 2n)^{18}F	100.00	60.6	112	0.51	97	40	0.63
5 = ^{142}Nd(n, 2n)^{141}Nd	27.13	2060	150	0.51	6	40	0.63
6 = ^{64}Zn(n, p)^{64}Cu	48.89	386	768	0.51	38	40	0.63
7 = ^{65}Cu(n, 2n)^{64}Cu	30.90	1100	768	0.51	38	40	0.63
8 = ^{58}Ni(n, 2n)^{57}Ni	67.76	34	2160	0.51	92	40	0.63
9 = ^{75}As(n, 2n)^{74}As	100.00	1110	24480	0.51	59	40	0.63

[a] Cf. Figs. 7.232, 7.233, and 7.234.

REFERENCES

1. M. Cuypers and J. Cuypers, "Gamma Ray Spectra and Sensitivities for 14-MeV Neutron Activation Analysis," Texas A and M University, College Station, Texas, April, 1966.
2. M. D. D'Agostino and F. J. Kuehne, "The Use of a Simplified Neutron Activation Technique for Analyzing Metallic Wear from Aircraft Hydraulic Systems," in *Radioisotopes for Aerospace*, Part I, J. C. Dempsey and P. Polishuk (Eds.), Plenum Press, New York, 1966, pp. 346–379.
3. A. E. Richardson and A. Harrison, *Anal. Chem.*, **41**, 1396 (1969).
4. K. R. Blake, C. V. Parker, L. O. England, and I. L. Morgan, "Elemental Trace Analysis by Charge Particle and Neutron Activation," ORD-2980-14, Texas Nuclear Corporation, Austin, Texas, July 1, 1966.
5. T. C. Martin, S. C. Mathur, and I. L. Morgan, *Int. J. Appl. Radiat. Isot.*, **15**, 331–338 (1964).
6. The Texas Nuclear Corporation, "The Application of Nuclear Techniques in Coal Analysis, Internal Technical Report No. 2," Second National Meeting of the Society for Applied Spectroscopy, San Diego, California, October 14–18, 1963.
7. J. D. L. H. Wood, D. W. Downton, and J. M. Bakes, "A Fast Neutron Activation Analysis System with Industrial Applications," Proceedings, 1965 International Conference on Modern Trends in Activation Analysis, College Station, Texas, April 19–22, 1965, pp. 175–181.
8. E. Gorin and B. M. Yavorsky, "Critical Evaluation of the Status of the Neutron Activation Method for Coal Analysis," Report No. 1, TID-22474, June 16, 1964.
9. T. C. Martin, I. L. Morgan, and J. D. Hall, "Nuclear Analysis System for Coal," Proceedings, 1965 International Conference on Modern Trends in Activation Analysis, College Station, Texas, April 19–22, 1965, pp. 71–75.

10. D. E. Fisher and R. L. Currie, "Instrumental Activation Analysis of Meteorites Involving both Thermal and Fast Neutrons," in *Radiochemical Methods of Analysis*, Vol. 1, IAEA, Vienna, 1965, pp. 217–228.
11. S. E. Turner, *Anal. Chem.*, **28**, 1457 (1956).
12. K. G. Broadhead, D. E. Shanks, and H. H. Heady, "Fast-Neutron Activation Analysis in Molten Salt Electrometallurgical Research," Proceedings, 1965 International Conference on Modern Trends in Activation Analysis, College Station, Texas, April 19–22, 1965, pp. 39–43.
13. J. A. Waggoner and R. J. Knox, "Elemental Analysis Using Neutrons in Elastic Scattering," in *Radioisotopes for Aerospace*, Part II, J. C. Dempsey and P. Polishuk (Eds.), Plenum Press, New York, 1966, pp. 270–291.
14. J. S. Hislop and R. E. Wainerdi, "Extraterrestrial Neutron Activation Analysis," *Anal. Chem.*, **39**, 28A–39A (February 1967).
15. L. E. Fite, E. L. Steele, R. E. Wainerdi, E. Ibert, P. Jimenez, C. Samson, and M. To-on, "An Investigation of Remote Lunar Analysis by Nuclear Activation Analysis," TID-19999, Texas A and M University, College Station, Texas April 30, 1963.
16. R. L. Hull, *Proc. Am. Pet. Inst.*, **44** (III), 264–270 (1964).
17. P. Kehler and R. Monaghan, "Activation Analysis of Ocean Bottom Cores," TID-18125, Dresser Research, Tulsa, Oklahoma November 1962.
18. T. C. Martin, G. T. Prud'homme, and I. L. Morgan, "Activation Analysis in Process Control Applications," in *Developments in Applied Spectroscopy*, Vol. 5, Plenum Press, New York, 1966, pp. 485–494.
19. D. Gibbons and G. Olive, "Applications of a Neutron Generator in Radioactivation Analysis," AERE-R-4576, Wantage Research Laboratory, Wantage, England November 15, 1963.
20. Y. F. Babikova, V. M. Mineev, and V. T. Samosadnyi, *Zavod. Lab.*, **32**, 47–49 (1966).
21. F. B. Gray, "Use of a Portable Neutron Generator in Industrial Analyses," Kaman Nuclear Corporation Technical Report, Kaman Nuclear Corporation, Colorado Springs, Colorado, 1960, pp. 9–18.
22. A. Golanski, *J. Radioanal. Chem.*, **3**, 161–173 (1969).
23. S. S. Nargolwalla, J. Niewodniczanski, and J. E. Suddueth, *J. Radioanal. Chem.*, **5**, 403 (1970).
24. H. P. Dibbs, "Activation Analysis with a Neutron Generator," Department of Mines Research Report R155, Ottawa, Canada, February 1965.
25. M. Crambes, "Determination of Sb, Ag, Cu, Ga, Mo, Zn by 14-MeV Neutron Activation," CEA Report R2965, Center for Nuclear Studies, Grenoble, France, April 1966.
26. D. E. Fisher and R. L. Currie, "Instrumental Activation Analysis of Meteorites Involving Both Thermal and Fast Neutrons," in *Radiochemical Methods of Analysis*, Vol. 1, IAEA, Vienna, 1965, pp. 217–228.
27. R. L. Hull, *Proc. Am. Pet. Inst.*, **44** (III), 264–279 (1964).
28. D. Hull and G. T. Gilmore, "Practical Computer Routines for Neutron Activation Analysis," Society for Applied Spectroscopy, Second National Meeting, San Diego, California, October 14–18, 1963.

29. M. Chiba, *J. Radioanal. Chem.*, **2**, 415–423 (1969).
30. The Mound Laboratory Quarterly Progress Report, July–September 1966, MLM-1388 Mound Laboratory, Mound Laboratory, Miamisburg, Ohio, September 30, 1966.
31. N. C. Rasmussen and T. J. Thompson, "Neutron and Gamma-Ray Spectroscopy in Activation Analysis," Final Report, AD-633252, Massachusetts Institute of Technology, Cambridge, Massachusetts, January 1966.
32. P. D. LaFleur, Ed., National Bureau of Standards (U.S.), Technical Note 548, 1970.
33. S. Nagatsuka, H. Suzuki, K. Nakajima, and M. Kobayashi, *Radioisotopes (Tokyo)*, **16**, 504–508 (1967).
34. G. Oldham and G. G. Darrall, *Anal. Chim. Acta*, **40**, No. 2, 330–333 (1968).
35. N. Ishibashi and S. Kamata, *Radioisotopes (Tokyo)*, **13**, 7–12 (1964).
36. T. G. Broadhead and D. E. Shanks, *J. Appl. Radiat. Isot.*, **18**, No. 5, 279–283 (1967).
37. J. Anderson, S. B. Osborn, R. W. S. Tomlinson, D. Newton, J. Rundo, L. Salmon, and J. W. Smith, *Lancet*, **2**, 1201–1205 (1964).
38. P. Schramel, *J. Radioanal. Chem.*, **3**, 29–36 (1969).
39. M. Y. Cuypers and M. P. Menon, *Trans. Am. Nucl. Soc.*, **7**, 330 (1964).
40. M. P. Menon and M. Y. Cuypers, *Anal. Chem.*, **37**, 1057 (1965).
41. D. E. Hull and J. T. Gilmore, *Anal. Chem.*, **36**, 2072 (1964).
42. E. P. Przybylowicz, G. W. Smith, J. E. Suddueth, and S. S. Nargolwalla, *Anal. Chem.*, **41**, 819 (1969).
43. C. K. Battye, R. W. S. Tomlinson, J. Anderson, and S. B. Osborn, "Experiments Relating to Whole Body Activation Analysis in Man *In Vivo* Using 14-MeV Incident Neutrons," Proceedings of a Symposium on Nuclear Activation Techniques in the Life Sciences, Amsterdam, May 8–12, 1967.
44. E. L. Steele and W. W. Meinke, "Fast Neutron Activation Analysis," Proceedings, International Conference on Modern Trends in Activation Analysis, College Station, Texas, December 1961, pp. 161–165.
45. D. Gibbons, W. J. McCabe, and G. Olive, *Radiochemical Methods of Analysis*, Vol. 1, IAEA, Vienna, 1965, pp. 297–322.
46. L. Gorski, W. Kusch, and J. Wojtkowska, "A Statement of the Applicability of Fast Neutron Activation Analysis for the Determination of Copper in Copper Deposits," PAN-487/IA, Polish Academy of Sciences, Institute of Nuclear Research, Warsaw, Poland, December 1963.
47. L. Gorski, W. Kusch, and J. Wojtkowska, *Talanta*, **11**, 1135 (1964).
48. P. J. Daly, K. J. Hofstetter, and F. Schmidt-Bleek, *J. Chem. Educ.*, **44**, 412–413 (1967).
49. L. C. Nelson, Jr., and H. Bussell, New Brunswick Laboratory, AEC Annual Progress Report for the Period of July 1964 through June 1965, NBL-230, December 1965.
50. J. K. Perry, Colorado School of Mines Research Foundation, Inc., "Feasibility Experiments Related to Trace Analysis of Fluorine in Zinc Electro-Refining Solutions by Fast Neutron Activation and Coincidence Counting Techniques," Interim Report No. 3 TID 22662, February 23, 1966.

51. S. Nargolwalla and R. E. Jervis, *Trans. Am. Nucl. Soc.*, **8**, 86–87 (May 1965).
52. S. S. Nargolwalla, "A Study of 14 MeV Neutrons Induced Reactions for the Analysis of Trace Fluorides," Ph.D. Thesis, University of Toronto, Toronto, Canada, 1965.
53. R. Blackburn, "Determination of Fluorine in Organic Compounds by Fast Neutron Activation Analysis," *Anal. Chem.*, **36**, 669–671 (1964).
54. E. A. M. England, G. B. Hornsby, W. T. Jones, and D. R. Terrey, *Anal. Chem. Acta*, **40**, 365–371 (1968).
55. R. Debiard, A. Fourcy, and J. P. Carrec, *C. R.*, Ser. D, **264**, No. 23, 2668 (1967).
56. P. Bussiere, *J. Radiochim. Anal.*, **1965**, 31–38.
57. E. Ricci and T. H. Handley, *Anal. Chem.*, **42**, 378 (1970).
58. S. C. Mathur and G. Oldham, *Nucl. Energy*, September–October 1967, pp. 136–141.
59. M. Sattarov, Y. N. Talanin, and T. Khalikov, *Izv. Akad. Nauk Uzb. SSR, Ser. Fiz.-Mat. Nauk*, **12**, 74–77 (1968).
60. A. Golanski, *J. Radioanal. Chem.*, **3**, 161 (1969).
61. J. Breynat, A. Fourcy, and J. P. Garrec, Proceedings of a Symposium on Nuclear Activation Techniques in the Life Sciences, Amsterdam, May 1967, IAEA, Vienna, Austria, 1967, pp. 81–98.
62. E. G. Boreisha, V. V. Kravtsov, and L. A. Sokolov, *Fiz. Khim.*, **7**, 57–61 (1967).
63. I. N. Plaksin, L. P. Starchik, and V. T. Tustanovskii, *Dokl. Akad. Nauk SSR*, **165**, 1095 (1965).
64. I. N. Plaksin, L. P. Starchik, and V. T. Tustanovskii, *Ind. Lab.*, **33**, 1297 (1967).
65. J. T. Gilmore and D. E. Hull, "Nitrogen-13 in Hydrocarbons Irradiated with Fast Neutrons," Proceedings, International Conference on Modern Trends in Activation Analysis, College Station, Texas, December 1961.
66. J. T. Gilmore and D. E. Hull, *Anal. Chem.*, **34**, 187 (1962).
67. D. E. Hull and J. P. Gilmore, "Computer Routines for Neutron Activation Analysis of Lubricating Oils," *Anal. Chem.*, **36**, 2072 (1964).
68. M. H. Rison, W. H. Barber, and P. E. Wilkniss, *Radiochim. Acta*, **7**, 196 (1967).
69. H. Tsuji, *Bunseki Kagaku*, **15**, 263 (1966).
70. D. E. Wood, "Fast Neutron Activation Analysis for Nitrogen in Grain Products," KN-65-186, Kaman Nuclear Corporation, Colorado Springs, Colorado, August 1965.
71. V. I. Shamaev, "Determination of Nitrogen in Organic Compounds by Fast Neutron Activation," AEC-PR-6639, 32-35, Proceedings of the First All Union Coordinating Conference, Tashkent, October 1962.
72. E. J. Walker and V. L. Eggebraaten, "Correlation of Physical Properties of Rubber with Nitrogen Content by Means of Neutron Activation Analysis," Proceedings, 1965 International Conference on Modern Trends in Activation Analysis, College Station, Texas, April 19–22, 1965.
73. M. Crambes, S. S. Nargolwalla, and L. May, *Trans. Am. Nucl. Soc.*, **10**, 63 (1967).
74. S. S. Nargolwalla, E. P. Przybylowicz, J. E. Suddueth, and S. L. Birkhead, *Anal. Chem.*, **41**, 168 (1969).
75. V. Guinn, private communication.

76. H. L. Priest and F. C. Burns, private communication.
77. National Bureau of Standards Bibliography on Activation Analysis, NBS Technical Note 467, May 1971.
78. R. F. Colemen, *The Analyst*, **87**, 590 (1962).
79. L. C. Pasztor and D. E. Wood, *Talanta*, **13**, 389 (1966).
80. D. E. Wood and L. Z. Pasztor, Proceedings, 1965 International Conference on Modern Trends in Activation Analysis, College Station, Texas, April 19–22, 1965, pp. 259–264.
81. A. V. Andreev, I. Y. Barit, and I. M. Pronman, *Ind. Lab.*, **33**, 1306 (1967).
82. A. Metcalf, *Steel Times*, **188**, 90 (1964).
83. B. L. Twitty and K. M. Fritz, "Internal Standard Techniques in 14-MeV Activation Analysis," NLCO-979, National Lead Company of Ohio, Cincinnati, Ohio, June 1, 1966.
84. B. L. Twitty and K. M. Fritz, *Anal. Chem.*, **39**, 527 (1967).
85. P. Schramel, *Oesterr. Akad. Wiss., Math.-Naturwiss. Kl., Sitzungsber. Abt. II*, **174**, 535 (1965).
86. K. Miyagawa, I. Ichijima, A. Asai, E. Nomura, and I. Mishima, The 9th Japan Conference on Radioisotopes, Nippon, Toshi Center, Kozimachi Kaikan, May 13–15, 1969.
87. I. Fujii, H. Muto, and K. Miyoshi, *Bunseki Kagaku*, **13**, 249 (1964).
88. R. F. Gahn and L. Rosenblum, *Anal. Chem.*, **38**, 1014 (1966).
89. D. Brune and K. Jirlow, *J. Radioanal. Chem.*, **2**, 49 (1969).
90. D. Gibbons, G. Oliver, P. Sevier, and J. E. Deutschman, *J. Inst. Met.*, **95**, 280 (1967).
91. R. F. Coleman, *UKAEA Prog. Rep.*, **171**, 73 (1960).
92. M. Del Milebro Perez, *Energ. Space Nucl. (Madrid)*, **11**, 537 (1967).
93. V. N. Karev, G. P. Dolya, N. V. Sivokon, A. I. Tutubanlin, N. F. Khalin, and A. S. Zabornyi, *Ind. Lab.*, **34**, 1724 (1968).
94. U. Eisner and H. B. Mark, Jr., *Talanta*, **16**, 27 (1969).
95. J. Janczyszyn, L. Loska, and S. Taczanowski, *Chem. Anal.*, **14**, 391 (1969).
96. A. V. Andreev, I. Y. Barat, R. M. Musaelyan, and I. M. Pronman, *J. Anal. Chem. USSR*, **21**, 1292 (1966).
97. S. R. Abeurakahmanova, V. A. Kirev, L. V. Navalikhin, and Y. N. Talanin, *J. Anal. Chem. USSR*, **23**, 1043 (1968).
98. O. U. Anders and D. W. Briden, *Anal. Chem.*, **37**, 530 (1965).
99. D. J. Veal and C. F. Cook, *Anal. Chem.*, **34**, 178 (1962).
100. G. Oldham and S. C. Mathur, "The Use of 14-MeV Neutrons in Activation Analysis of Minerals," *Nucl. Energy*, March–April 1969.
101. J. R. Vogt and W. D. Ehmann, *Radiochim. Acta*, **4**, 24 (1965).
102. A. Volborth, Golden Gate Metals Conference, February 13–15, 1964, American Society of Metals, Berkeley, California, 1964, Vol. II, p. 117.
103. A. Volborth, *Fortschr. Mineral.*, **43**, 10 (1966).
104. A. Volborth, Nevada Bureau of Mines, Report No. 6, B-1-B-13, 1963.
105. A. Volborth and H. E. Banta, *Anal. Chem.*, **35**, 2203 (1963).
106. C. Persiani and J. F. Cosgrove, *Anal. Chem.*, **40**, 1350 (1968).
107. O. U. Anders and D. W. Briden, *Anal. Chem.*, **36**, 287 (1964).

108. M. To-on, F. Sicilio, and R. E. Wainerdi, *Food Technol.*, **17**, 17–22 (1963).
109. M. To-on, "The Determination of Phosphorous by Fast Neutron Activation Analysis," TID 20472 Texas A and M University, College Station, Texas, 1963.
110. M. To-on, "The Determination of Phosphorous by Fast Neutron Activation Analysis," TEES 2671-4, v-1-v-37 Texas A and M University, College Station, Texas, January 1, 1965.
111. A. A. Prapuolenis and J. M. Bakes, *Radiochem. Radioanal.*, **1**, 19–23 (1969).
112. E. Ricci, "Determination of Oxygen, Phosphorous and Nitrogen in a Typical Phosphene Oxide Derivative," in *Guide to Activation Analysis*, W. S. Lyon, Jr. (Ed.), Van Nostrand, Princeton, New Jersey, 1964, pp. 133–142.
113. H. E. Palmer, W. B. Nelp, R. Murano, and C. Rich, *Phys. Med. Biol.*, **13**, 269 (1968).
114. N. Chomel, "The Determination of Sulfur and Phosphorus by Radioactivation," Thesis, Lyon University, France, 1964.
115. J. F. Babikova, V. M. Minaev, and V. T. Samosadnyi, *Zavodsk. Lab.*, **32**, 47 (1966).
116. R. VanGrieken, A. Specke, and J. Hoste, *J. Radioanal. Chem.*, **6**, 385 (1970).
117. J. P. Garrec, CEA-R-3636 Report, Commisariat, A. L., Energie Atomique, Centre D'Etudes Space Nucleair, Grenoble, France, November 1968.
118. K. Tada, I. Fujii, T. Adachi, and K. Ogawa, AEC-TR-5637, April 8, 1963.
119. I. Fujii, A. Tani, H. Muto, K. Ogawa, and M. Sato, *J. At. Energy Soc. Japan*, **5**, 218 (1963).
120. V. T. Tustanovskii and U. Orifkhodzhaev, *Sov. At. Energy*, **26**, 437 (1969).
121. J. R. Vogt and W. D. Ehmann, *Geochim. Cosmochim. Acta*, **29**, 373–383 (1965).
122. H. A. Vincent and A. Volborth, *Trans. Am. Nucl. Soc.*, **10**, 26 (June 1967).
123. H. A. Vincent and A. Volborth, *Nucl. Appl.*, **3**, 753–757 (December 1967).
124. J. R. Vogt and W. D. Ehmann, Proceedings, 1965 International Conference on Modern Trends in Activation Analysis, College Station, Texas, April 19–22, 1965, pp. 82–85.
125. W. D. Ehmann and D. R. Durbin, *Geochim. Cosmochim. Acta*, **32**, No. 4, 461–464 (1968).
126. R. A. Schmitt, R. H. Smith, W. D. Ehmann, and D. McKown, *Geochim. Cosmochim. Acta*, **31**, 1975–1985 (October 1967).
127. G. G. Santos and R. E. Wainerdi, *J. Radioanal. Chem.*, **1**, 509–514 (November 1968).
128. A. Volborth, B. P. Fabbi, and H. A. Vincent, *Advances in X-Ray Analysis*, Vol. 11, Plenum Press, New York, 1968, pp. 158–163.
129. V. V. Glasson, M. M. Mandelbaum, and K. S. Turitsyn, *Geol. Geofiz.*, No. 5, 127–132 (1969).
130. J. P. Garrec, A. Fer, and A. Fourcy, *C. R. Acad. Sci.*, Paris, Ser. D, **268**, No. 25, 3021–3024 (1969).
131. J. W. Morgan and W. D. Ehmann, *Anal. Chim. Acta*, **49**, 287–299 (1970).
132. Y. Kusaka and H. Tsuji, *Nippon Kagaku Zasshi*, **86**, 733–736 (July 1965).
133. R. VanGrieken, R. Gijbels, A. Speecke, and J. Hoste, *Anal. Chim. Acta*, **43**, No. 3, 381–395 (1968).

134. R. VanGrieken, R. Gijbels, A. Speecke, and J. Hoste, *Anal. Chim. Acta*, **43**, No. 2, 199–209 (1968).
135. J. F. Eichelberger, G. R. Grove, and L. V. Jones, Mound Laboratory Progress Report for November 1964, MLM-1227, 37P., Mound Laboratory, Miamisburg, Ohio, November 30, 1964.
136. D. E. Wood and N. J. Roper, Fast Neutron Activation Analysis for Silicon in Iron, KN-65-140(R), 21P., Kaman Nuclear Corporation, Colorado Springs, Colorado, April 15, 1965.
137. G. G. Santos, L. E. Fite, W. E. Kuykendall, R. E. Wainerdi, A. H. Bouma, and W. R. Bryant, *Nuclear Techniques and Mineral Resources*, IAEA, Vienna, 1969, pp. 463–487.
138. T. Shigematsu, O. Fujino, and T. Honjo, *Bull. Inst. Chem. Res., Kyoto Univ.*, **45**, No. 4–5, 299–306 (1967).
139. Y. Eden, *Nucl. Instrum. Methods*, **49**, No. 2. 352–354 (1967).
140. R. Daniel, W. Maerdi, and D. Monnier, *Chimia*, **21**, No. 11, 544–546 (1967).
141. T. C. Martin, J. R. Rhodes, and J. B. Waters, Continuous On-Line Nuclear Analysis Measurements in Process Control Applications, ORO-2980-16, 88P., Texas Nuclear Corporation, Austin, Texas, July 31, 1967.
142. M. Sattarov, Y. N. Talanin, and M. Yunusov, *Izv. Akad. Nauk Uzb. SSR, Ser. Fiz.-Mat. Nauk*, **11**, No. 6, 50–53 (1967).
143. M. Buczkó, J. Csikai, and G. Varga, *J. Radioanal. Chem.*, in press.
144. H. Tsuji and Y. Kusaka, *Bunseki Kagaku*, **17**, No. 7, 864–870 (1968).
145. A. Sardi and A. Tomcsanyi, *Analyst (London)*, **92**, No. 1097, 529–531 (1967).
146. F. E. Senftle, P. Sarigianis, and P. W. Philbin, *Geol. Surv. Can. Econ. Geol. Rep.*, **26**, 462–469 (1967).
147. F. A. Iddings and J. P. Wade, Proceedings, 1965 International Conference on Modern Trends in Activation Analysis, College Station, Texas, April 19–22, 1965, pp. 149–151.
148. L. Gorski, J. Janczyszyn, L. Loska, *Radiochem. Radioanal. Lett.*, **1**, 99–109 (1969).
149. T. B. Pierce, J. W. Edwards, and K. Haines, *Talanta*, **15**, 1153 (1968).

APPENDICES

INTRODUCTION

In attempting to fulfill the objectives of this treatise, it would be appropriate to include in the Appendices all aspects of neutron generator development and application. However, selection of useful information from the voluminous data reported in literature presents a considerable problem due to the diversity and sometimes contradictory nature of published studies. It was therefore necessary to evaluate appropriate data critically and select only those considered both comprehensive and useful to the user from an analytical standpoint.

As a short prefex to each Appendix, an effort is made to include pertinent details to guide proper use of data. The reader is reminded that several companion references to those included in the Appendices have already been mentioned in Chapters 2, 3, and 7. Moreover, the working curves given for each element in Chapter 7 include the nuclear data given in these Appendices. It is recommended that these be carefully studied to obtain additional information for the sake of completeness and meaningful comparisons. It should be realized, however, that data describing analytical sensitivities in particular, be they calculated or experimentally determined, should be used with some reservation since, in many instances, important factors influencing the sensitivity, such as irradiation and counting conditions, flux-monitoring methods, and adequate consideration of systematic errors, are either omitted or inadequately described in the principal report.

In our judgment complete reliance on any single, presently available tabulation as a guide for selecting conditions for activation analysis is risky. The authors have strived to compile in Chapter 7 a composite of data from a number of sources. Several existing tabulations are useful in assessing qualitatively possible sources of interferences and thus provide a reasonable basis for determining the feasibility of a particular analytical procedure. The latter approach can be further improved by supplemental information concerning the analytical device at the disposal of the analyst. A thorough understanding of the analyst's own irradiation and counting facilities provides a firm foundation for investigation and subsequent development of analytical procedures.

APPENDIX

I

ACTIVATION CROSS SECTIONS FOR D–T NEUTRONS[1]

The activation cross sections for the reactions (n, 2n), (n, p), and (n, α) versus the target proton number, Z, for incident neutrons of energy $\simeq 14.7$ MeV are plotted in Figs. A.1a to e, A.2a to e, and A.3a to d, respectively. Data given represent a

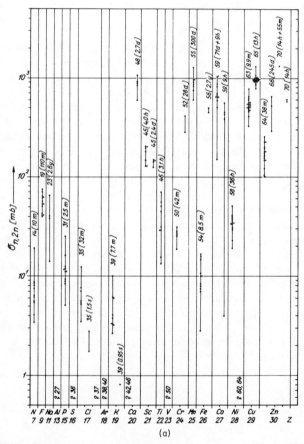

Fig. A.1. (a)–(e) Activation cross sections for (n, 2n) reactions for $\simeq 14.7$-MeV neutrons. The points for each nucleous show the experimental values obtained by various laboratories.

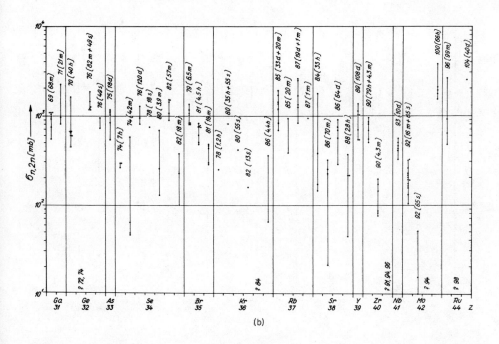

(b)

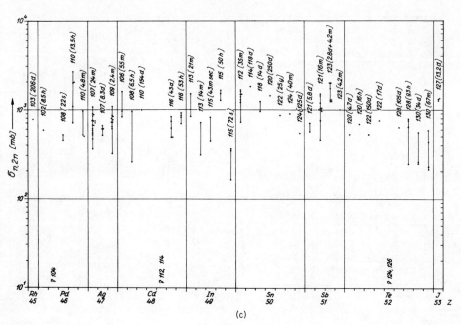

(c)

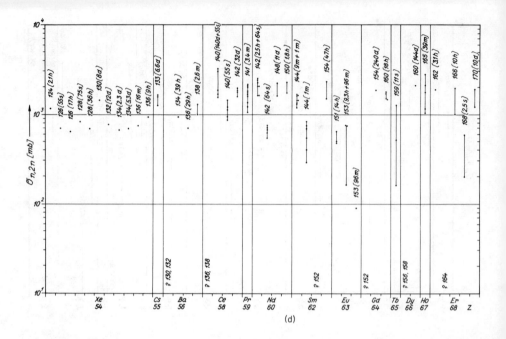

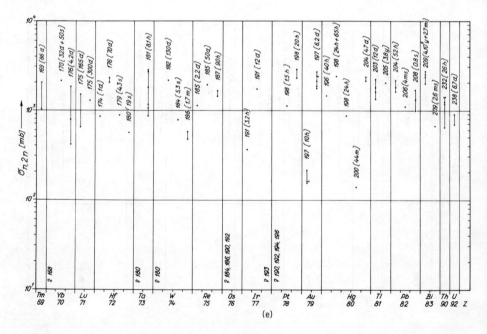

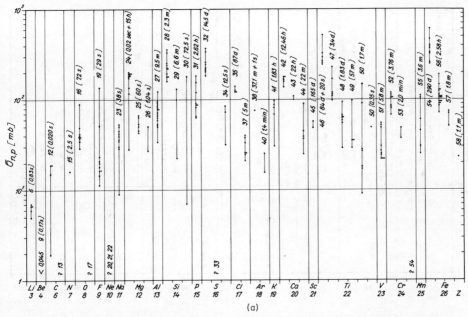

Fig. A-2. (a)–(e) Activation cross sections for (n, p) reactions for $\simeq 14.7$ MeV neutrons. The points for each nucleus show the experimental values obtained by various laboratories.

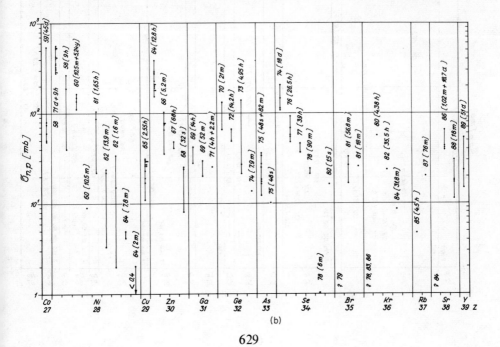

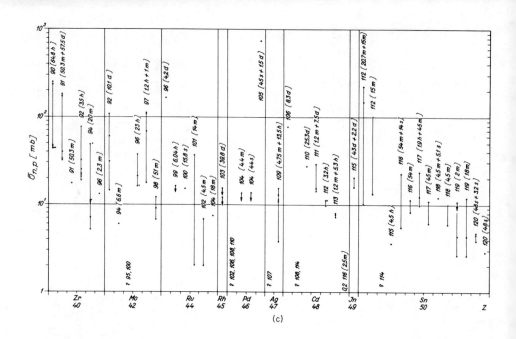

(c)

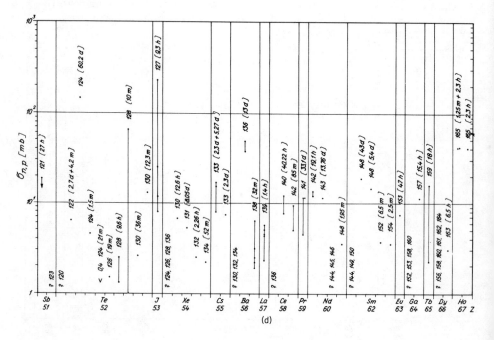

(d)

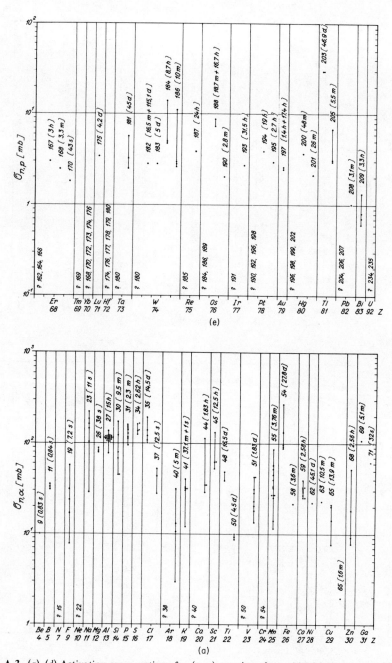

Fig. A.3. (a)–(d) Activation cross sections for (n, α) reactions for $\simeq 14.7$ MeV neutrons. The points for each nucleus show the experimental values obtained by various laboratories.

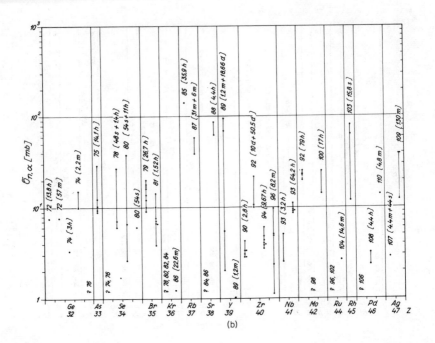

(b)

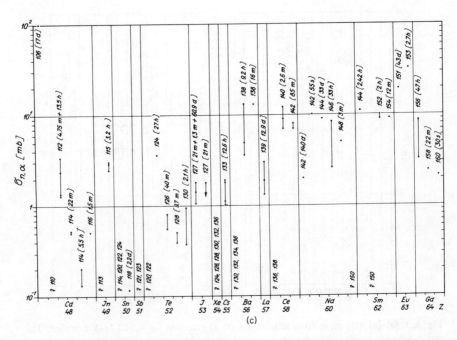

(c)

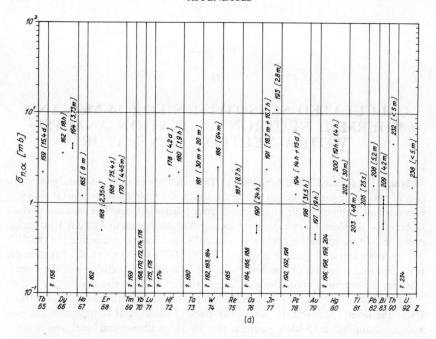

(d)

compilation[1] concluded about June 1969. In these figures the target mass numbers and half-lives (in parentheses) of residual nuclei are shown above the experimental points. Mass numbers indicated above question marks (?) denote those for which data were unavailable. A limitation on the half-life of less than one year has been placed on the selection of data given in Figs. A.1 to A.3. In the case of isomeric pairs, two data are presented. Since in many cases the differences between data from several laboratories for a given nuclide are considerably larger than the experimental error of individual points, error limits have been excluded from the graphs.

REFERENCE

1. J. Csikai, M. Buczkó, Z. Bödy, and A. Demény, "Nuclear Data for Neutron Activation Analysis," *Atomic Energy Review*, Vol. VII, No. 4, IAEA, Vienna, 1969.

APPENDIX

II

CALCULATED SENSITIVITIES FOR 14-MeV AND THERMAL NEUTRON ACTIVATION ANALYSIS WITH A NEUTRON GENERATOR

Although the selected[1] data tabulated in Table A.1 are limited, since only the reaction of maximum yield for each element is listed, the table does give useful information relevant to the possible interferences from other reactions. The reader is advised to examine supplementary reports[2,3] to augment data given in Table A.1. An interesting two-dimensional mapping of short-lived nuclides according to half-life (0.1 msec–60 sec) and gamma energy (0.01–10 MeV) by Nagy et al.[4] has recently been reported. The aim of this survey was to establish the usefulness of these reactions for activation analysis using different types of neutron sources.

The following explanatory statements need to be made relative to this tabulation. Using the listed cross-section values, half-lives, Q values, and isotopic abundances, and assuming an $\simeq 15$-MeV neutron flux of 10^9 neutrons/(cm^2 sec) and a thermalized flux of 10^7 neutrons/(cm^2 sec), the specific activity in dis/(sec g), for each element, for the reaction giving the maximum yield, is calculated.

In the listing of interfering reactions, (n, n'α) processes have been omitted because of the lack of data. The flux levels chosen above are considered representative of those available from the average neutron generator. The calculated specific activity is that obtainable at termination of a 10-hr irradiation. It should be recognized that 10-hr irradiations with neutron generators are not economical. However, the listed specific activities can easily be adjusted for any irradiation time considered suitable for comparison.

TABLE A.1. List of the Reactions of Maximum Yield for Each Element and Interfering Reactions[a]

Element	Reaction	Half-life	σ (mb)	dps/g	Interfering Reactions
$_3$Li	$^7_3$Li(n, d)$^6_2$He	0.83 sec	14	1.1×10^6	$^6_3$Li(n, p); $^9_4$Be(n, α)
$_4$Be	$^9_4$Be(n, α)$^6_2$He	0.83 sec	10	6.7×10^5	$^6_3$Li(n, p); $^7_3$Li(n, d)
$_5$B	$^{11}_5$B(n, α)$^8_3$Li	0.84 sec	33	1.5×10^6	$^7_3$Li(n, γ)
$_6$C	$^{12}_6$C(n, p)$^{12}_5$B	0.02 sec	17	8.5×10^5	$^{11}_5$B(n, γ); $^{15}_7$N(n, α)
$_7$N	$^{14}_7$N(n, 2n)$^{13}_7$N	10 min	8.5	3.6×10^5	
$_8$O	$^{16}_8$O(n, p)$^{16}_7$N	7.2 sec	39	1.4×10^5	$^{15}_7$N(n, γ); $^{17}_8$O(n, d); $^{18}_8$O(n, ^{3}H); $^{19}_9$F(n, α)
$_9$F	$^{19}_9$F(n, 2n)$^{18}_9$F	110 min	53	1.7×10^6	
$_{10}$Ne	$^{20}_{10}$Ne(n, p)$^{20}_9$F	11 sec	104 Calc.	2.8×10^6	$^{19}_9$F(n, γ); $^{21}_{10}$Ne(n, d); $^{23}_{11}$Na(n, α)
$_{11}$Na	$^{23}_{11}$Na(n, α)$^{20}_9$F	11 sec	160	4.2×10^6	$^{19}_9$F(n, γ); $^{20}_{10}$Ne(n, p); $^{21}_{10}$Ne(n, d)
$_{12}$Mg	$^{24}_{12}$Mg(n, p)$^{24}_{11}$Na	15 hr	190	1.4×10^6	$^{23}_{11}$Na(n, γ); $^{25}_{12}$Mg(n, d); $^{27}_{13}$Al(n, α)
$_{13}$Al	$^{27}_{13}$Al(n, p)$^{27}_{12}$Mg	9.5 min	80	1.7×10^6	$^{26}_{12}$Mg(n, γ); $^{28}_{14}$Si(n, ^{3}He); $^{30}_{14}$Si(n, α)
$_{14}$Si	$^{28}_{14}$Si(n, p)$^{28}_{13}$Al	2.3 min	220	4.2×10^6	$^{27}_{13}$Al(n, γ); $^{29}_{14}$Si(n, d); $^{30}_{14}$Si(n, ^{3}H); $^{31}_{15}$P(n, α)
$_{15}$P	$^{31}_{15}$P(n, α)$^{28}_{13}$Al	2.3 min	150	2.9×10^6	$^{27}_{13}$Al(n, γ); $^{28}_{14}$Si(n, p); $^{29}_{14}$Si(n, d); $^{30}_{14}$Si(n, ^{3}H)
$_{16}$S	$^{32}_{16}$S(n, α)$^{29}_{14}$Si	2.6 hr	150	1×10^5	$^{30}_{14}$Si(n, γ); $^{31}_{15}$P(n, p); $^{32}_{16}$S(n, ^{3}He)
$_{17}$Cl	$^{37}_{17}$Cl(n, α)$^{34}_{15}$P	12.5 sec	40	1.6×10^5	$^{34}_{16}$S(n, p); $^{36}_{16}$S(n, ^{3}H)
$_{18}$Ar	$^{40}_{18}$Ar(n, p)$^{40}_{17}$Cl	1.4 min	20	3×10^5	
$_{19}$K	$^{41}_{19}$K(n, p)$^{41}_{18}$Ar	1.85 hr	80	7.5×10^5	$^{40}_{18}$Ar(n, γ); $^{44}_{20}$Ca(n, α); $^{43}_{20}$Ca(n, ^{3}He)
$_{20}$Ca	$^{44}_{20}$Ca(n, p)$^{44}_{19}$K	22 min	40	1.2×10^4	$^{46}_{20}$Ca(n, ^{3}H)
$_{21}$Sc	$^{45}_{21}$Sc(n, 2n)$^{44}_{21}$Sc	4 hr	160	1.7×10^5	$^{45}_{22}$Ti(n, ^{3}H)
$_{22}$Ti	$^{48}_{22}$Ti(n, p)$^{48}_{21}$Sc	1.8 days	62	8×10^4	$^{47}_{21}$Sc(n, γ); $^{47}_{22}$Ti(n, d); $^{49}_{22}$Ti(n, ^{3}H); $^{50}_{23}$V(n, nα)
$_{23}$V	$^{51}_{23}$V(n, p)$^{51}_{22}$Ti	5.8 min	33	3.6×10^5	$^{50}_{22}$Ti(n, γ); $^{52}_{24}$Cr(n, ^{3}He); $^{54}_{24}$Cl(n, α)
$_{24}$Cr	$^{52}_{24}$Cr(n, p)$^{52}_{23}$V	3.77 min	100	9×10^3	$^{51}_{23}$V(n, γ); $^{52}_{24}$Cr(n, d); $^{54}_{24}$Cr(n, ^{3}H); $^{55}_{25}$Mn(n, α)
$_{25}$Mn	$^{55}_{25}$Mn(n, γ)$^{56}_{25}$Mn	2.58 hr	12,900	1.3×10^6	$^{56}_{26}$Fe(n, p); $^{57}_{26}$Fe(n, d); $^{58}_{26}$Fe(n, ^{3}H); $^{59}_{27}$Co(n, α)
$_{26}$Fe	$^{56}_{26}$Fe(n, p)$^{56}_{25}$Mn	2.58 hr	105	9.6×10^5	$^{55}_{25}$Mn(n, γ); $^{57}_{26}$Fe(n, d); $^{58}_{26}$Fe(n, ^{3}H); $^{59}_{27}$Co(n, α)
$_{27}$Co	$^{59}_{27}$Co(n, γ)$^{60}_{27}$Co	10.5 min	20,000	2.0×10^6	$^{60}_{28}$Ni(n, p); $^{61}_{28}$Ni(n, d); $^{62}_{28}$Ni(n, ^{3}H); $^{63}_{29}$Cu(n, α)
$_{28}$Ni	$^{58}_{28}$Ni(n, p)$^{58}_{27}$Co	9 hr	170	5.9×10^5	$^{59}_{27}$Co(n, 2n); $^{60}_{28}$Ni(n, ^{3}H)
$_{29}$Cu	$^{63}_{29}$Cu(n, 2n)$^{62}_{29}$Cu	9.9 min	540	3.5×10^6	$^{64}_{30}$Zn(n, ^{3}H)
$_{30}$Zn	$^{64}_{30}$Zn(n, 2n)$^{63}_{30}$Zn	38 min	170	7.5×10^5	—
$_{31}$Ga	$^{69}_{31}$Ga(n, 2n)$^{68}_{31}$Ga	68 min	1000	4.7×10^6	$^{72}_{32}$Ge(n, ^{3}H)
$_{32}$Ge	$^{70}_{32}$Ge(n, 2n)$^{75}_{32}$Ge	49 sec	820	5.25×10^5	$^{74}_{32}$Ge(n, γ); $^{73}_{33}$As(n, p); $^{77}_{34}$Se(n, ^{3}He); $^{78}_{34}$Se(n, α)
		82 min	380	2.4×10^5	
$_{33}$As	$^{75}_{33}$As(n, 2n)$^{74}_{33}$As	18 days	1100	1.2×10^6	$^{74}_{34}$Se(n, p); $^{76}_{34}$Se(n, ^{3}H)
$_{34}$Se	$^{82}_{34}$Se(n, 2n)$^{81}_{34}$Se	57 min	1400	9.6×10^5	$^{80}_{34}$Se(n, γ); $^{81}_{35}$Br(n, p); $^{83}_{36}$Kr(n, ^{3}He); $^{84}_{36}$Kr(n, α)
		18 min	220	1.3×10^5	
$_{35}$Br	$^{81}_{35}$Br(n, 2n)$^{80}_{35}$Br	4.5 hr	750	2.1×10^6	$^{79}_{35}$Br(n, γ); $^{80}_{36}$Kr(n, p); $^{82}_{36}$Kr(n, ^{3}H)
		18 min	400	1.4×10^6	
$_{36}$Kr	$^{82}_{36}$Kr(n, γ)$^{83}_{36}$Kr	1.86 hr	3000	2.5×10^4	$^{84}_{36}$Kr(n, 2n); $^{85}_{37}$Rb(n, ^{3}H); $^{86}_{38}$Sr(n, α); $^{87}_{38}$Sr(n, nα)
$_{37}$Rb	$^{85}_{37}$Rb(n, 2n)$^{84}_{37}$Rb	20 min	600	3×10^5	$^{84}_{38}$Sr(n, p); $^{86}_{38}$Sr(n, ^{3}H)
		33 days	700	3.2×10^4	
$_{38}$Sr	$^{88}_{38}$Sr(n, α)$^{85}_{36}$Kr	4.4 hr	86	3.9×10^5	$^{84}_{36}$Kr(n, γ); $^{86}_{36}$Kr(n, 2n); $^{85}_{37}$Rb(n, p); $^{87}_{37}$Rb(n, ^{3}H); $^{87}_{38}$Sr(n, ^{3}He)
$_{39}$Y	$^{89}_{39}$Y(n, 2n)$^{88}_{39}$Y	108 days	1000	2×10^4	$^{90}_{40}$Zr(n, ^{3}H)
$_{40}$Zr	$^{90}_{40}$Zr(n, 2n)$^{89}_{40}$Zr	4.3 min	125	4.2×10^5	$^{92}_{42}$Mo(n, α)
		79 hr	580	3.2×10^5	
$_{41}$Nb	$^{93}_{41}$Nb(n, 2n)$^{92}_{41}$Nb	10 days	460	8.2×10^4	$^{92}_{42}$Mo(n, p); $^{94}_{42}$Mo(n, ^{3}H)
$_{42}$Mo	$^{100}_{42}$Mo(n, 2n)$^{99}_{42}$Mo	66 hr	2000	1.2×10^5	$^{92}_{42}$Mo(n, γ); $^{101}_{44}$Ru(n, ^{3}He); $^{102}_{44}$Ru(n, α)
$_{44}$Ru	$^{96}_{44}$Ru(n, 2n)$^{95}_{44}$Ru	99 min	1000	3.3×10^5	
$_{45}$Rh	$^{103}_{45}$Rh(n, γ)$^{104}_{45}$Rh	4.4 min	12,000	7×10^5	$^{104}_{46}$Pd(n, p); $^{105}_{46}$Pd(n, d); $^{106}_{46}$Pd(n, ^{3}H); $^{107}_{47}$Ag(n, α)
		42 sec	140,000	8×10^6	
$_{46}$Pd	$^{110}_{46}$Pd(n, 2n)$^{109}_{46}$Pd	4.8 min	600	4×10^5	$^{108}_{46}$Pd(n, γ); $^{109}_{47}$Ag(n, p); $^{111}_{48}$Cd(n, ^{3}He); $^{112}_{48}$Cd(n, α); $^{113}_{48}$Cd(n, nα)
		13.5 hr	1800	2.4×10^5	
$_{47}$Ag	$^{109}_{47}$Ag(n, 2n)$^{108}_{47}$Ag	2.4 min	780	2×10^6	$^{107}_{47}$Ag(n, γ); $^{108}_{48}$Cd(n, p); $^{110}_{48}$Cd(n, ^{3}H)

(*Continued*)

[a] (n, γ) relates to thermal energy, the others to D + T neutrons.

(TABLE A.1 cont.)

Element	Reaction	Half-life	σ (mb)	dps/g	Interfering Reactions
$_{48}$Cd	$^{106}_{48}$Cd(n, 2n)$^{105}_{48}$Cd	55 min	1000	6.3×10^4	
$_{49}$In	$^{115}_{49}$In(n, γ)$^{116}_{49}$In	2.5 sec	80,000	4×10^6	$^{116}_{50}$Sn(n, p); $^{117}_{50}$Sn(n, d); $^{118}_{50}$Sn(n, ^{3}H)
		54 min	148,000	7×10^6	
		14 sec	47,000	2×10^6	
$_{50}$Sn	$^{124}_{50}$Sn(n, 2n)$^{123}_{50}$Sn	40 min	900	2.7×10^5	$^{123}_{50}$Sn(n, γ); $^{123}_{51}$Sb(n, p); $^{123}_{52}$Te(n, ^{3}He); $^{126}_{52}$Te(n, α)
$_{51}$Sb	$^{121}_{51}$Sb(n, 2n)$^{120}_{51}$Sb	16 min	1000	2.6×10^6	$^{120}_{52}$Te(n, p); $^{122}_{52}$Te(n, ^{3}H)
$_{52}$Te	$^{132}_{52}$Te(n, 2n)$^{131}_{52}$Te	67 min	410	6.6×10^5	$^{132}_{53}$Te(n, γ); $^{131}_{54}$Xe(n, ^{3}He); $^{134}_{54}$Xe(n, α)
$_{53}$I	$^{127}_{53}$I(n, γ)$^{128}_{53}$I	25 min	6000	2.8×10^5	$^{128}_{54}$Xe(n, p); $^{129}_{54}$Xe(n, d); $^{130}_{54}$Xe(n, ^{3}H)
$_{54}$Xe	$^{136}_{54}$Xe(n, 2n)$^{135}_{54}$Xe	15 min	750	3.1×10^5	$^{134}_{54}$Xe(n, γ)
		9.2 hr	950	1.9×10^5	
$_{55}$Cs	$^{133}_{55}$Cs(n, 2n)$^{132}_{55}$Cs	6.6 days	1500	4.5×10^6	$^{132}_{56}$Ba(n, p); $^{134}_{56}$Ba(n, ^{3}H)
$_{56}$Ba	$^{138}_{56}$Ba(n, 2n)$^{137}_{56}$Ba	2.6 min	1150	3.6×10^6	$^{136}_{56}$Ba(n, γ); $^{138}_{57}$La(n, d); $^{139}_{57}$La(n, ^{3}H); $^{140}_{58}$Ce(n, α)
$_{57}$La	$^{139}_{57}$La(n, γ)$^{140}_{57}$La	40.2 hr	8400	5.8×10^6	$^{140}_{58}$Ce(n, p); $^{142}_{58}$Ce(n, ^{3}H)
$_{58}$Ce	$^{140}_{58}$Ce(n, 2n)$^{139}_{58}$Ce	55 sec	1100	4.1×10^6	$^{138}_{58}$Ce(n, γ); $^{141}_{59}$Pr(n, ^{3}H); $^{142}_{60}$Nd(n, α); $^{143}_{60}$Nd(n, nα)
$_{59}$Pr	$^{141}_{59}$Pr(n, 2n)$^{140}_{59}$Pr	3.4 min	1800	7.6×10^6	$^{142}_{60}$Nd(n, ^{3}H); $^{143}_{60}$Nd(n, nt)
$_{60}$Nd	$^{150}_{60}$Nd(n, 2n)$^{149}_{60}$Nd	1.8 hr	2000	4.5×10^5	$^{148}_{60}$Nd(n, γ); $^{152}_{62}$Sm(n, α)
$_{62}$Sm	$^{144}_{62}$Sm(n, 2n)$^{143}_{62}$Sm	9 min	1000	1.2×10^5	
		1 min	550	6.6×10^4	
$_{63}$Eu	$^{153}_{63}$Eu(n, 2n)$^{152}_{63}$Eu	9.3 hr	650	6.2×10^5	$^{151}_{63}$Eu(n, γ); $^{152}_{64}$Gd(n, p); $^{154}_{64}$Gd(n, ^{3}H); $^{155}_{64}$Gd(n, nt)
		96 min	90	1.8×10^5	
$_{64}$Gd	$^{160}_{64}$Gd(n, 2n)$^{159}_{64}$Gd	18 hr	1600	4×10^5	$^{158}_{64}$Gd(n, γ); $^{159}_{65}$Tb(n, p); $^{161}_{66}$Dy(n, ^{3}He); $^{162}_{66}$Dy(n, α); $^{163}_{66}$Dy(n, nα)
$_{65}$Tb	$^{159}_{65}$Tb(n, 2n)$^{158}_{65}$Tb	11 sec	700	2.6×10^6	$^{158}_{66}$Dy(n, p); $^{160}_{66}$Dy(n, ^{3}H); $^{161}_{66}$Dy(n, np)
$_{66}$Dy	$^{164}_{66}$Dy(n, γ)$^{165}_{66}$Dy	1.3 min	2,000,000	2×10^7	$^{165}_{67}$Ho(n, p); $^{166}_{68}$Er(n, ^{3}He); $^{168}_{68}$Er(n, α)
$_{67}$Ho	$^{165}_{67}$Ho(n, 2n)$^{164}_{67}$Ho	39 min	1800	6.5×10^6	$^{164}_{68}$Er(n, p); $^{166}_{68}$Er(n, ^{3}H); $^{167}_{68}$Er(n, nt)
$_{68}$Er	$^{168}_{68}$Er(n, 2n)$^{167}_{68}$Er	2.5 sec	400	3.9×10^5	$^{166}_{68}$Er(n, γ); $^{169}_{69}$Tm(n, ^{3}H); $^{170}_{70}$Yb(n, α); $^{171}_{70}$Yb(n, nα)
$_{69}$Tm	$^{169}_{69}$Tm(n, 2n)$^{168}_{69}$Tm	86 days	1500	1.8×10^4	$^{168}_{70}$Yb(n, p); $^{170}_{70}$Yb(n, ^{3}H); $^{171}_{70}$Yb(n, nt)
$_{70}$Yb	$^{176}_{70}$Yb(n, γ)$^{177}_{70}$Yb	4.2 days	55,000	5.9×10^4	$^{176}_{70}$Yb(n, 2n); $^{175}_{71}$Lu(n, p); $^{177}_{71}$Lu(n, d); $^{177}_{72}$Hf(n, ^{3}He); $^{178}_{72}$Hf(n, α); $^{179}_{72}$Hf(n, nα)
		67 msec	46,000	5×10^5	
$_{71}$Lu	$^{175}_{71}$Lu(n, γ)$^{176}_{71}$Lu	3.7 hr	37,000	9.9×10^5	$^{176}_{72}$Hf(n, p); $^{177}_{72}$Hf(n, d); $^{180}_{73}$Ta(n, nα); $^{178}_{72}$Hf(n, ^{3}H); $^{179}_{72}$Hf(n, nt)
$_{72}$Hf	$^{178}_{72}$Hf(n, 2n)$^{179}_{72}$Hf	19 sec	50,000	4.5×10^5	$^{180}_{72}$Hf(n, 2n); $^{180}_{73}$Ta(n, d); $^{181}_{73}$Ta(n, ^{3}H); $^{182}_{74}$W(n, α); $^{183}_{74}$W(n, nα)
$_{73}$Ta	$^{181}_{73}$Ta(n, 2n)$^{180}_{73}$Ta	8.1 hr	1800	3.3×10^6	$^{180}_{74}$W(n, p); $^{182}_{74}$W(n, ^{3}H); $^{183}_{74}$W(n, nt)
$_{74}$W	$^{182}_{74}$W(n, γ)$^{183}_{74}$W	5.5 sec	20,000	1.7×10^5	$^{184}_{74}$W(n, 2n); $^{184}_{75}$Re(n, ^{3}H); $^{186}_{76}$Os(n, α); $^{187}_{76}$Os(n, nα)
$_{75}$Re	$^{187}_{75}$Re(n, 2n)$^{186}_{75}$Re	90 hr	1550	2.8×10^5	$^{185}_{75}$Re(n, γ); $^{186}_{76}$Os(n, p); $^{187}_{76}$Os(n, d); $^{188}_{76}$Os(n, ^{3}H); $^{189}_{76}$Os(n, nt)
$_{76}$Os	$^{192}_{76}$Os(n, γ)$^{193}_{76}$Os	31.5 hr	1600	4.1×10^3	$^{193}_{77}$Ir(n, p); $^{195}_{78}$Pt(n, ^{3}He); $^{196}_{78}$Pt(n, α)
$_{77}$Ir	$^{191}_{77}$Ir(n, γ)$^{192}_{77}$Ir	1.4 min	610,000	2.3×10^6	$^{193}_{77}$Ir(n, 2n); $^{192}_{78}$Pt(n, p); $^{194}_{78}$Pt(n, ^{3}H); $^{195}_{78}$Pt(n, nt)
$_{78}$Pt	$^{198}_{78}$Pt(n, 2n)$^{197}_{78}$Pt	1.5 hr	1130	2.5×10^5	$^{198}_{78}$Pt(n, γ); $^{197}_{79}$Au(n, p); $^{198}_{80}$Hg(n, ^{3}He); $^{200}_{80}$Hg(n, α)
		20 hr	2500	1.7×10^5	
$_{79}$Au	$^{197}_{79}$Au(n, 2n)$^{196}_{79}$Au	10 hr	150	1.7×10^5	$^{196}_{80}$Hg(n, p); $^{198}_{80}$Hg(n, ^{3}H); $^{199}_{80}$Hg(n, nt)
$_{80}$Hg	$^{198}_{80}$Hg(n, 2n)$^{197}_{80}$Hg	25 hr	880	1×10^5	$^{196}_{80}$Hg(n, γ)
		65 hr	1460	0.7×10^5	
$_{81}$Tl	$^{203}_{81}$Tl(n, 2n)$^{202}_{81}$Tl	12 days	1700	3.5×10^4	$^{204}_{82}$Pb(n, ^{3}H)
$_{82}$Pb	$^{208}_{82}$Pb(n, 2n)$^{207}_{82}$Pb	0.85 sec	1300	2×10^6	$^{206}_{82}$Pb(n, γ); $^{209}_{83}$Bi(n, ^{3}H)
$_{83}$Bi	$^{209}_{83}$Bi(n, 2n)$^{208}_{83}$Bi	2.6 msec	660	1.9×10^6	
$_{90}$Th	$^{232}_{90}$Th(n, 2n)$^{231}_{90}$Th	26 hr	1300	8.1×10^5	$^{234}_{92}$U(n, α); $^{235}_{92}$U(n, nα); $^{232}_{90}$Th(n, f)
$_{92}$U	$^{238}_{92}$U(n, 2n)$^{237}_{92}$U	6.75 days	800	8.2×10^4	$^{238}_{92}$U(n, f)

REFERENCES

1. J. Csikai, M. Buczkó, Z. Bödy, and A. Demény, "Nuclear Data for Neutron Activation Analysis," *Atomic Energy Review*, Vol. VII, No. 4, IAEA, Vienna, 1969.
2. M. Cuypers and J. Cuypers, "Gamma Ray Spectra and Sensitivities for 14-MeV Neutron Activation Analysis," Texas A and M, Unnumbered report, Activation Analysis Research Laboratory, College Station, Texas, April 12, 1966.
3. B. T. Kenna and F. J. Conrad, "Tabulation of Cross Sections, Q-Values, and Sensitivities for Nuclear Reactions of Nuclides with 14-MeV Neutrons," SC-RR-66-229, Sandia Laboratory, Albuquerque, New Mexico, June 1966.
4. Á. Z. Nagy, A. Csöke, and E. Szabó, "Two-Dimensional (T-1/2, E_γ) Mapping of Short-Lived Nuclides," Hungarian Academy of Sciences, Central Research Institute for Physics, Budapest, Hungary, Report KFKI-70-18 NAC, July 28, 1970.

APPENDIX

III

EXPERIMENTAL SENSITIVITIES FOR 14-MeV AND THERMAL ACTIVATION ANALYSIS WITH A NEUTRON GENERATOR[1]

Empirical sensitivity data have been obtained for 66 elements, under standard irradiation and counting conditions, by irradiating a known weight of either the element or a pure compound of the element. The weight of sample used for these tests was dependent on the neutron activation cross section of the particular element and was normally a few hundred milligrams. The sample for irradiation was contained in a standard, 5-ml capacity, polythene irradiation vial. In order for this relatively small weight of sample to occupy a volume representative of the whole capsule, the sample was placed in a polythene-tube insert in the vial. The internal diameter of the insert was about 0.1 in. and the outside diameter such that it was a close fit in the irradiation capsule.

All of the irradiations were performed using the pneumatic transfer facility. For an isotope with a short half-life, the sample was counted in the transfer tube, following the irradiation, with a decay time of about 1 sec. For an isotope with a reasonably long half-life, the sample was counted on top of the 3 in. × 3 in. NaI(Tl) scintillation detector, using a 1200-mg/cm^2 beta absorber. The decay time before counting, in this case, was usually 1 min.

The identities of the isotopes produced were confirmed, if necessary, by half-life as well as by gamma-ray energy measurements. The photopeak counts were then taken for the most intense gamma rays emitted by the isotope and normalized to a fast flux of 5×10^8 neutrons/(cm^2 sec) or a thermal flux of 1×10^8 neutrons/(cm^2 sec), depending on the irradiation position that was used. The nuclear reaction and the gamma-ray photopeak that offer the highest analytical sensitivity for a given element are listed in Table A.2. Where two photopeaks from the same isotope offer comparable sensitivities, both photopeaks are listed. Similarly, where two different reactions with the same element give comparable sensitivities, both reactions are listed. Other reactions with a given element which offer lesser sensitivities have not been included in Table A.2. The sensitivity data are usually the average of at least two determinations.

The irradiation and measurement times used in these empirical sensitivity tests do not represent the maximum times that could have been used. However, the various times that were used may be extended to other values by simple calculation.

The sensitivity data given in Table A.2 are relevant to the experimental arrangements used for these determinations. The use of a second scintillation detector, for example, would result in an increase in sensitivity. However, the relative sensitivities from this table should be approximately the same for other detector assemblies.

TABLE A.2. Empirical Sensitivity Data

Z	Target Element	F or T	Irradiation	Time Decay	Counting	Product Isotope	Half-life	Photopeak Measured (MeV)	Number of μg To Give 100 Photopeak Counts
5	B	F	40 sec	1.5 sec	40 sec	^{11}Be	13.6 sec	2.12	2.5×10^4
7	N	F	5 min	1 min	10 min	^{13}N	10 min	β^+	35
8	O	F	40 sec	1.5 sec	40 sec	^{16}N	7.14 sec	4.5–6.5	7×10^2
9	F	F	2 min	1 sec	2 min	^{19}O	29.1 sec	0.2	1.35×10^2
	F	F	10 min	4 min	15 min	^{18}F	109.7 min	β^+	31
11	Na	F	2 min	1 sec	2 min	^{23}Ne	38 sec	0.44	2.3×10^2
	Na	T	15 min	4 min	30 min	^{24}Na	15.05 hr	1.37	2.4×10^2
12	Mg	F	2 min	1 sec	2 min	^{23}Ne, ^{25}Na	38 sec, 60 sec	0.44, 0.4	1.4×10^3
	Mg	F	15 min	4 min	30 min	^{24}Na	15.05 hr	1.37	1.5×10^2
13	Al	F	5 min	1 min	10 min	^{27}Mg	9.5 min	0.84	50
								1.01	1.7×10^2
14	Si	F	2.3 min	1 min	4.6 min	^{28}Al	2.23 min	1.78	45
15	P	F	2.3 min	1 min	4.6 min	^{28}Al	2.23 min	1.78	70
	P	F	2.3 min	1 min	4.6 min	^{30}P	2.53 min	β^+	1.6×10^2
16	S	F	1 min	1 sec	1 min	^{34}P	12.4 min	2.1	8.7×10^4
17	Cl	F	10 min	1 min	20 min	^{34m}Cl	32.4 min	β^+	1.9×10^2
19	K	F	10 min	1 min	15 min	^{38}K	7.7 min	β^+	1.9×10^2
20	Ca	F	10 min	1 min	15 min	^{44}K	22 min	1.13	9×10^3
21	Sc	F	10 min	1 min	20 min	^{44}Sc	3.92 hr	β^+	26
	Sc	T	1 min	1 sec	1 min	^{46m}Sc	20 sec	0.14	27
22	Ti	F	10 min	1 min	20 min	^{50}Sc, ^{45}Ti	1.8 min, 3.08 hr	0.51, β^+	3.2×10^2
23	V	F	5 min	1 min	10 min	^{51}Ti	5.8 min	0.32	70
	V	T	5 min	1 min	10 min	^{52}V	3.76 min	1.43	13
24	Cr	F	3 min	1 min	3 min	^{52}V	3.76 min	1.43	3×10^2
25	Mn	T	10 min	1 min	20 min	^{56}Mn	2.58 hr	0.845	10
								1.81	75
26	Fe	F	10 min	1 min	20 min	^{56}Mn	2.58 hr	0.845	1.4×10^2
27	Co	T	5 min	1 min	10 min	^{60m}Co	10.35 min	0.059	40
28	Ni	F	10 min	1 min	20 min	^{60m}Co, ^{61}Co	10.35 min, 1.65 hr	0.059, 0.072	9.4×10^2
29	Cu	F	5 min	1 min	10 min	^{62}Cu	9.8 min	β^+	6

(*continued*)

(TABLE A.2 cont)

Z	Target Element	F or T	Time Irradiation	Time Decay	Time Counting	Product Isotope	Half-life	Photopeak Measured (MeV)	Number of μg To Give 100 Photopeak Counts
30	Zn	F	5 min	1 min	10 min	^{63}Zn	38.1 min	β^+	70
31	Ga	F	10 min	1 min	20 min	^{68}Ga	68 min	β^+	8
32	Ge	F	1 min	1 sec	2 min	^{75}Ge	49 sec	0.139	3.3×10^2
33	As	T	15 min	1 min	30 min	^{76}As	26.5 hr	0.56	2×10^2
34	Se	F	30 sec	1 sec	60 sec	^{77m}Se	17.7 sec	0.16	1.9×10^2
35	Br	F	20 sec	1 sec	20 sec	^{79m}Br	5.1 sec	0.21	4.6×10^2
	Br	F	3.2 min	1 min	6.4 min	^{78}Br	6.4 min	β^+	8
37	Rb	F	1 min	1 sec	2 min	^{86m}Rb	61 sec	0.56	1×10^2
	Rb	F	5 min	1 min	10 min	^{84m}Rb	21 min	0.22, 0.25	12
38	Sr	T	10 min	1 min	20 min	^{87m}Sr	2.8 hr	0.39	2.1×10^2
39	Y	F	1 min	1 sec	1 min	^{89m}Y	16 sec	0.91	1×10^2
40	Zr	F	5 min	1 min	10 min	^{89m}Zr	4.18 min	0.59	1.5×10^2
41	Nb	F	10 min	1 min	20 min	^{90m}Y	3.14 hr	0.2	1.8×10^3
42	Mo	F	5 min	1 min	10 min	^{91}Mo	15.6 min	β^+	1.3×10^2
45	Rh	T	1 min	1 sec	1 min	^{104}Rh	42.8 sec	0.56	80
46	Pd	T	4 min	1 min	8 min	^{109m}Pd	4.8 min	0.18	2.3×10^2
47	Ag	F	5 min	1 min	10 min	^{106}Ag	24 min	β^+	20
48	Cd	T	10 min	1 min	20 min	^{111m}Cd	49 min	0.25	1×10^2
49	In	T	10 min	1 min	20 min	^{116m}In	54 min	0.41	1.4
								1.09	1.5
								1.27	1.3
50	Sn	F	5 min	1 min	10 min	^{125m}Sn	9.7 min	0.33	2.2×10^3
	Sn	F	5 min	1 min	10 min	^{123}Sn	41 min	0.15	1.4×10^3
51	Sb	F	5 min	1 min	10 min	^{120}Sb	15.7 min	β^+	18

Z	El					Isotope			
52	Te	T	5 min	1 min	10 min	^{131}Te	25 min	0.15	1.8×10^3
53	I	T	10 min	1 min	20 min	^{128}I	25 min	0.44	32
55	Cs	T	15 min	1 min	30 min	^{134m}Cs	2.91 hr	0.127	1.5×10^2
56	Ba	F	2.6 min	1 min	5.2 min	^{137m}Ba	2.6 min	0.662	27
57	La	T	20 min	1 min	40 min	^{140}La	40.2 hr	0.49	2.3×10^2
58	Ce	F	1 min	1 sec	2 min	^{139m}Ce	1 min	0.75	70
59	Pr	F	3 min	1 min	6 min	^{140}Pr	3.4 min	β^+	8
60	Nd	F	2 min	1 min	4 min	^{141m}Nd	64 sec	0.76	3×10^2
62	Sm	T	5 min	1 min	10 min	^{155}Sm	22 min	0.104	60
63	Eu	T	15 min	1 min	30 min	^{152m}Eu	9.3 hr	0.12	3.5
64	Gd	T	3 min	1 min	6 min	^{161}Gd	3.7 min	0.32, 0.36	1.7×10^3
66	Dy	T	1 min	1 sec	2 min	^{165m}Dy	1.3 min	0.11	20
67	Ho	T	15 min	1 min	30 min	^{166}Ho	27 hr	0.08	48
68	Er	T	15 min	1 min	30 min	^{171}Er	7.5 hr	0.3	2.6×10^2
70	Yb	T	10 min	1 min	20 min	^{177}Yb	1.9 hr	0.15	1.3×10^3
71	Lu	T	15 min	1 min	30 min	^{176m}Lu	3.7 hr	0.055	15
72	Hf	T	1 min	1 sec	1 min	^{179m}Hf	18.6 sec	0.22	25
73	Ta	T	10 min	1 min	20 min	^{180m}Ta, ^{182m}Ta	8.1 hr, 16 min	0.057	4.5×10^2
74	W	T	15 min	1 min	20 min	^{187}W	24 hr	0.48	6.5×10^2
								0.68	7.5×10^2
75	Re	T	5 min	1 min	10 min	^{188m}Re	20 min	0.064	95
77	Ir	T	15 min	1 min	30 min	^{194}Ir	19 hr	0.33	25
78	Pt	T	10 min	1 min	20 min	^{199}Pt	30 min	0.48, 0.54	2.3×10^2
79	Au	F	40 sec	1.5 sec	40 sec	^{197m}Au	7.3 sec	0.28	4×10^2
	Au	T	20 min	1 min	75 min	^{198}Au	2.7d	0.41	8
80	Hg	F	10 min	1 min	20 min	^{199m}Hg	44 min	0.16	60
82	Pb	F	15 min	1 min	30 min	^{204m}Pb, ^{208}Tl	67 min, 3.1 min	0.9, 0.86	7×10^3
						^{204m}Pb	67 min	0.38	1.1×10^4

A number of fast neutron reactions were found to offer good analytical sensitivity but are not included in Table A.2 through lack of published cross-section data. These reactions are listed separately in Table A.3.

TABLE A.3. Empirically Determined Fast Neutron Reactions That Offer High Analytical Sensitivities

Z	Element	Reaction	Half-life	Gamma Energy (MeV)
28	Ni	^{60}Ni(n, p)^{60}Co	10.35 min	0.059 (99.7%), 1.33 (0.3%)
32	Ge	^{76}Ge(n, 2n)^{75m}Ge	49 sec	0.139 (100%)
35	Br	^{79}Br(n, n')^{79m}Br	5.1 sec	0.21
37	Rb	^{85}Rb(n, 2n)^{84m}Rb	21 min	0.216 (32)^x, 0.25 (62)^x, 0.46 (32)^x
		^{87}Rb(n, 2n)^{86m}Rb	61 sec	0.56
39	Y	^{89}Y(n, n')^{89m}Y	16 sec	0.91
41	Nb	^{93}Nb(n, α)^{90m}Y	3.14 hr	0.2, 0.48
50	Sn	^{124}Sn(n, 2n)^{123}Sn	40 min	0.15 (88%)
56	Ba	^{138}Ba(n, 2n)^{137m}Ba	2.6 min	0.662 (100%)
58	Ce	^{140}Ce(n, 2n)^{139m}Ce	1 min	0.75
60	Nd	^{142}Nd(n, 2n)^{141m}Nd	64 sec	0.76
79	Au	^{197}Au(n, n')^{197m}Au	7.3 sec	0.13 (99%), 0.28 (99%)
80	Hg	^{199}Hg(n, n')^{199m}Hg	44 min	0.16, 0.37
82	Pb	^{204}Pb(n, n')^{204m}Pb	67 min	0.38, 0.9

^{x} Relative gamma abundance.

REFERENCE

1. H. P. Dibbs, "Activation Analysis with a Neutron Generator," Department of Mines and Technical Surveys, Research Report R155, Ottawa, Canada, February 1965.

APPENDIX

IV

EXPERIMENTAL SENSITIVITIES FOR 3-MeV NEUTRON ACTIVATION ANALYSIS

During the last decade, the neutron generator has contributed significantly to analytical programs in many research, academic, and industrial laboratories. In its capacity as a tool for activation analysis, the principal emphasis has been on the use of 14-MeV neutrons. The preferential use of 14-MeV neutrons is dictated primarily by the high neutron flux obtained from the $^2H(^3H, n)^4He$ reaction. Depending on the acceleration voltage, target thickness, and the experimental configuration of the sample with respect to the neutron source, the 14-MeV neutron flux is approximately two orders of magnitude greater than 3-MeV fluxes from the $^2H(^2H, n)^3He$ reaction.

To explore the utility of 3-MeV neutron activation analysis fully, γ-ray spectra of most of the elements irradiated with 3-MeV neutrons were systematically investigated. The results of this study[1] have since been successfully applied to compositional analysis and for the prediction of possible interferences from the buildup of 3-MeV neutrons in 14-MeV neutron activation analysis. This buildup, described in Chapter 3, is due to neutron production from interaction of the primary deuteron beam with deuterons embedded in a tritium target from previous irradiations.

Data given in this Appendix have been abstracted from this study. In addition to the experimental sensitivities, detailed information regarding irradiation and counting geometries, sample attenuation corrections, and detector efficiencies have been included to facilitate expected sensitivity calculations for other systems on a semi-quantitative basis. For those elements where it was not possible to compute γ-ray peak areas because of low counting rates, integral counting data over selected energy intervals are given so that upper limits for the evaluation of interferences can be calculated. The complete spectral catalog including explanatory data sheets and digital readouts are available from Activation Analysis Section, Analytical Chemistry Division, Institute for Materials Research, National Bureau of Standards, U.S. Department of Commerce, Washington, D. C. 20234.

REFERENCE

1. S. S. Nargolwalla, J. Niewodniczanski, and J. E. Suddueth, *J. Radioanal. Chem.*, **5**, 403 (1970).

EXPERIMENTAL SENSITIVITIES FOR 3-MeV NEUTRON ACTIVATION ANALYSIS

The experimental sensitivity for 72 different elements using 3-MeV neutron activation has been investigated. Using a 200 kV Cockcroft-Walton neutron generator with a 3-MeV neutron flux of about $1.5 \cdot 10^6$ n $\cdot$ cm^{-2} $\cdot$ sec^{-1}, γ-ray spectra of 51 elements were obtained with a sufficient number of photopeak counts for sensitivity calculations using a photopeak integration method. A useful table summarizing the sensitivity results is given. That 3-MeV neutron activation analysis is practical, is demonstrated by the experimental sensitivities obtained.

EXPERIMENTAL FACILITIES

NEUTRON SOURCE

A neutron generator (Fig. A.4), capable of accelerating atomic deuterium ions up to 200 keV, was used. The maximum beam current obtainable is 2.5 mA. The accelerator is used exclusively for production of fast neutrons from the nuclear reactions

$$^{2}H + {}^{3}H = {}^{4}He + {}^{1}n + 17.6 \text{ MeV}$$

Fig. A.4. 2.5 mA neutron generator.

and

$$^2H + {}^2H = {}^3He + {}^1n + 3.25 \text{ MeV}$$

At 200 keV deuteron energy, the energies of fast neutrons emitted in the forward direction in the two cases are approximately 15 MeV and 3 MeV, respectively. With a new 5 Ci/in^2 tritium target, the maximum 15 MeV neutron flux at the sample position is approximately $5 \cdot 10^8$ n $\cdot$ cm^{-2} $\cdot$ sec^{-1}. When a 6 cm^3 per square inch deuterium target is used, the maximum 3 MeV neutron flux at the same location approximates $2 \cdot 10^6$ n $\cdot$ cm^{-2} $\cdot$ sec^{-1}. Although the 15 MeV neutron flux is over two orders of magnitude greater than the 3 MeV neutron flux, the deuterium target life is over two orders of magnitude longer than that for the tritium target because of continuous regeneration of the deuterium target by the bombarding deuterons. In general, irradiations were performed using beam currents between 1.5 and 2.5 mA. The thermal neutron flux from scattering processes in the shield and cooling-water jacket is estimated to be less than $5 \cdot 10^{-4}$ of all neutrons above the epithermal region. This estimate is based on experiments with bare and cadmium-covered indium foils. The location of the generator inside the biological shield is shown in Fig. A.5.

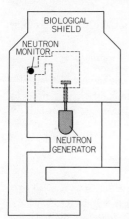

Fig. A.5. Plan view of biological shield showing location of neutron generator and neutron monitor.

SAMPLE IRRADIATION ASSEMBLY

This assembly has been described in detail elsewhere.[1] Only one position of the dual sample-biaxial rotating assembly was used for any one given irradiation. An earlier report[1] described the pneumatic transfer system used in conjunction with the sample irradiation unit. The average distance from the deuterium target to the center axis of a cylindrical sample is approximately 4 cm. For all irradiations, except for those in which rapid cyclic activation of samples was performed, the samples were rotated at 1000 rpm. The automatic mode of the sequence programmer[2] was used in special cases where the half-life was short enough to introduce errors from manual

Fig. A.6. Twin detector assembly inside counting shield. Sample to be counted is centrally placed in a Lucite β-absorber located between the detectors.

timing operations. For radioisotopes with half-lives less than one second, the samples were not rotated. For longer lived species (>10 min), a stop-watch was used to measure the irradiation and decay times, and the generator beam turned off manually.

COUNTING EQUIPMENT

Two $4'' \times 3''$ NaI(Tl) scintillation detectors, each mounted on a $5''$ diameter photomultiplier tube, formed the basis of the counting system. These detectors were closely matched for gain, gain shift with count rate, and resolution to within 0.25% of one another. The entire detector assembly (Fig. A.6) is placed inside a graded shield of lead, cadmium, paraffin and high density concrete to shield the detectors from the neutron source and reduce background. In this configuration no residual activity was measured immediately after beam shut-down. During irradiation, however, the detectors were sensitive to high-energy prompt γ-emission from the biological shield surrounding the neutron generator.

The detectors were connected to a 400-channel pulse height analyzer-readout system. The resolution of the spectrometer for the 0.662 MeV ^{137}Cs photopeak was 7.25%. To improve sensitivity, the output from each detector was summed via a summing preamplifier. The spectra were plotted using a fast point-plotter and peak analysis performed with the aid of a computer. Further information regarding the neutron generator facility can be obtained from a previous report.[2]

NEUTRON MONITOR

A spherical neutron dosimeter connected to a scaler was used to monitor the neutron flux for each irradiation. This detector consists of a 4 mm × 8 mm ^{6}LiI(Eu) scintillator surrounded by an 8" diameter sphere of polyethylene. The location of this detector inside the biological shield is shown in Fig. A.5.

EXPERIMENTAL PROCEDURE

SAMPLE PREPARATION

As far as possible, pure elements were irradiated. In cases where this was not possible, chemical compounds containing the element of interest were selected on the basis of matrix insensitivity to 3 MeV neutron activation. In general, samples were packed uniformly in our 8 cm^3 polyethylene vials using a hand-operated plunger device. Samples prepared in this manner have been assigned Code A. Where it was not possible to obtain large samples, smaller polyethylene vials of 1.6 cm^3 capacity were employed. After filling, these vials were placed centrally in the large 8 cm^3 vials for transport in the pneumatic tube facility. These samples are classified under Code B. In a few cases where the cost of samples was prohibitive, small amounts (<1 g) of the samples were loosely encapsulated in the 1.6 cm^3 volume vials. Such partially filled vials have been designated Code C. In this study the sample weights ranged from 100 mg to 60 g.

IRRADIATION OF SAMPLES

For the activation process, the irradiation time selected was governed by the half-life of the induced activity. For radioisotopes with half-lives greater than one hour, 20 min irradiations were made. Except for a few cases in which short lived radioisotopes were produced, the rest of the samples were irradiated for ten minutes each. For extremely short lived activities (<1 sec), cyclic activation was used; the activity being saturated within the irradiation time for each cycle. In the latter case the sample was not rotated. The decay time for most samples was 0.5 min, the exceptions being those where variable decay times were necessary for peak analysis and where the half-life of the induced radioactivity was very short. The environmental background was subtracted from all spectra and the samples generally were counted for a time period equal to their irradiation time.

3 MeV NEUTRON FLUX DETERMINATION

The relative neutron flux was measured by the neutron monitor which was checked daily against a sample of ammonium iodide which was irradiated and the ^{128}I activity determined. The day-to-day performance of the neutron detector system over a period of one month was found to be reproducible enough for the purpose of this study, and within the statistical precision dictated by counts accumulated. To avoid a measurement of the absolute neutron flux during each irradiation, the neutron detector counts were related to absolute measurements of the flux. For the absolute flux measurement, the foil activation technique was used, except that a chemical compound was used in place of a foil to closely simulate the geometry for sample counting. The 3 MeV neutron flux at the sample position was calculated from the equation,

$$\Phi = \frac{AM}{\sigma \cdot 10^{-24} WNPafE(e^{-\mu x}e^{-\Sigma y})(1 - e^{-\lambda t_i})(e^{-\lambda t_d})} \quad (A.1)$$

where

Φ—3 MeV neutron flux, n·cm^{-2}·sec^{-1}
A—measured count rate, counts·sec^{-1}
M—atomic weight of the target element
σ—3 MeV neutron activation cross-section, barns
W—weight of sample, g
P—weight fraction of target element in sample
a—isotopic abundance fraction of target isotope
f—γ-ray abundance from decay scheme
E—detection efficiency for γ-ray of interest
x—effective sample thickness for γ-ray attenuation in sample, cm
μ—total linear γ-ray absorption coefficient, cm^{-1}
y—effective sample thickness for 3 MeV neutron attenuation in sample, cm
N—Avogadro's number, $6.0023 \cdot 10^{23}$ atoms/g atom
Σ—total macroscopic removal cross-section for 3 MeV neutrons, cm^{-1}
t_i—irradiation time, min
t_d—decay time, min
λ—decay constant, min^{-1}

A sample of Al_2O_3 encapsulated in an 8 cm^3 polyethylene vial was used for absolute flux determinations. The 0.84 MeV photopeak from the decay of the 9.46 min ^{27}Mg activity produced by the ^{27}Al(n, p)^{27}Mg reaction was counted. The values of Σ and μ were obtained by interpolation of existing data given in references[3 and 4], respectively. The values of x and y were 1.04 and 0.97 cm, respectively (analogous values for these thicknesses for the 1.6 cm^3 vial are 0.68 and 0.5 cm, respectively). These were determined experimentally by a method previously described.[5] The 10.67 g sample of Al_2O_3 was irradiated for 20 min and counted for 20 min with a one minute delay in between. The detection efficiency E for our counting geometry (Fig. A.7) was measured using a standard source of ^{54}Mn ($E_\gamma = 0.835$ MeV) and was found to be 7.0%. An absolute flux of $1.3 \cdot 10^6$ n·cm^{-2}·sec^{-1} was calculated in this particular experiment using an activation cross-section of 1.5 mb.[6] During this irradiation the neutron count was taken and a flux normalization factor calculated. The average flux for all subsequent irradiations could be estimated by using the flux normalization factor R, where R = neutron flux/monitor cpm in the relationship

$$\Phi = O_c \cdot R \quad (A.2)$$

where

Φ—3 MeV neutron flux, n·cm^{-2}·sec^{-1}
O_c—observed monitor cpm
R—flux normalization factor, n·cm^{-2}·sec^{-1}/monitor cpm.

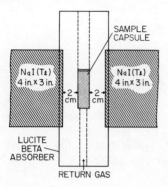

Fig. A.7. Schematic of counting geometry.

The value for R for the source-detector irradiation geometry used was approximately 60. The absolute 3 MeV flux was also determined using ^{89}Y(n, n'γ)^{89m}Y reaction ($\sigma = 150$ mb), and was found to be about 5% lower than that calculated from the reaction ^{27}Al(n, p)^{27}Mg. The precision of both these measurements estimated from Poisson counting statistics was about $\pm 5\%$ at the 95% confidence level for triplicate determinations.

CALCULATION OF EXPERIMENTAL SENSITIVITY

The spectra from the analyzer were read out on punched-paper tape and a simple computer program was used to integrate pertinent γ-ray peaks.[7] No attempt was made to smooth the spectra. The peak of interest was integrated and its area normalized to a constant element weight of one gram for a 3 MeV neutron flux of 10^6 n · cm^{-2} · sec^{-1}. To permit comparisons between sensitivities obtained under other laboratory conditions and those obtained in this work, the calculated sensitivities were corrected for both neutron and γ-ray attenuation for Code A and B samples, using the relationship

$$L = \frac{S}{e^{-\Sigma y} \cdot e^{-\mu x}} \quad (A.3)$$

where

L—corrected sensitivity,
S—uncorrected sensitivity,

and the exponential absorption terms for appropriate sample diameters have already been defined in Eq. (A.1). For Code C samples it has been assumed that the absorption terms in Eq. (A.3) are unity. In view of many approximations made in this study, and bearing in mind the fundamental purpose of this work as outlined in the Introduction, no attempt was made to evaluate the accuracy of the calculated sensitivity. An estimate of $\pm 10\%$ would, however, not be unreasonable.

PRESENTATION OF RESULTS

GAMMA-RAY SPECTRA

All spectra were plotted on semi-logarithmic co-ordinates. Four different energy scales (2.5, 5, 10 and 20 keV per channel) were used. On each spectrum, the γ-ray energies (MeV) were identified. It has been necessary at times to calculate the sensitivity for an element based on several closely located γ-ray lines as a whole. A typical example is illustrated in Fig. A.8 and the associated nuclear and counting data are given in Tables A.4 and A.5. Information regarding sample identity and size, and relevant activation and nuclear data[8] have also been included.

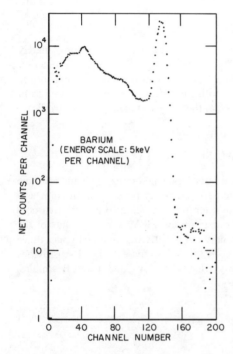

Fig. A.8. γ-Ray spectrum from 3-MeV neutron irradiation of barium.

TABLE A.4. Sample Data Sheet: Element Barium

Sample Data		
Element	Barium	Z = 56
Sample	BaCO$_3$	Energy scale: 5 keV per channel
Sample weight	16.0904 g	
Element weight	11.1976 g	
Sample code	A	

Activation Data

T$_{activation}$	T$_{decay}$	T$_{count}$	Neutron Flux (n · cm^{-2} · sec^{-1})
10 min	0.5 min	10 min	1.4 · 10^6

Experimental Sensitivity

Nuclear Reaction	Gamma Energy, MeV	Half-life	Net Peak Counts	Experimental Sensitivity (Net Peak counts/gram of element · 10^6 n · cm^{-2} · sec^{-1})	
				Uncorrected for Sample Attenuation	Corrected for Sample Attenuation
^{136}Ba(n, γ)^{137m}Ba ^{137}Ba(n, n'γ)^{137m}Ba	0.662	2.554 min	289260	1.8 · 10^4	2.3 · 10^4

TABLE A.5. Sample of Digital Output: Element Barium

```
Barium
00DATA00000,  00001,  00009,  00000,  00004,  00350,  03050,  04748,  04177,  03676
01DATA03629,  03713,  04012,  03286,  05369,  04937,  05270,  05571,  05745,  05949
02DATA06108,  06216,  06683,  06618,  06847,  07270,  07318,  07674,  07621,  07524
03DATA07365,  07431,  07451,  07622,  07476,  07360,  07777,  07727,  07897,  08245
04DATA08732,  09094,  09145,  09375,  09040,  08585,  08356,  08021,  07722,  07088
05DATA06946,  06667,  06245,  06210,  05931,  05823,  05502,  05436,  05202,  05057
06DATA04828,  04815,  04539,  04442,  04419,  04359,  04158,  04036,  04131,  03994
07DATA03916,  03907,  03829,  03750,  03778,  03549,  03605,  03494,  03554,  03549
08DATA03468,  03352,  03435,  03415,  03278,  03299,  03285,  03396,  03300,  03178
09DATA03056,  03081,  03022,  02917,  02857,  02721,  02607,  02429,  02276,  02283
10DATA02148,  02024,  01973,  01879,  01826,  01794,  01692,  01694,  01763,  01640
11DATA01720,  01633,  01594,  01600,  01583,  01625,  01613,  01630,  01595,  01632
12DATA01664,  01764,  01918,  02022,  02467,  03095,  04074,  05362,  07036,  09225
13DATA11662,  14965,  17594,  19931,  22095,  22847,  22697,  21288,  18932,  16146
14DATA13108,  10257,  07453,  05219,  03350,  02132,  01342,  00787,  00432,  00247
15DATA00121,  00056,  00046,  00037,  00032,  00036,  00016,  00023,  00023,  00024
16DATA00021,  00013,  00024,  00018,  00023,  00017,  00016,  00016,  00016,  00016
17DATA00019,  00019,  00021,  00022,  00019,  00009,  00015,  00019,  00026,  00025
18DATA00021,  00008,  00020,  00018,  00021,  00014,  00035,  00020,  00011,  00003
19DATA99998,  00006,  00009,  00009,  00004,  00005,  00016,  00006,  00010,  00007
```

EXPERIMENTAL SENSITIVITY

The results of sensitivity calculations for specific γ-ray energies have been tabulated in Table A.6. It may be observed that in cases where X-ray or low energy γ-ray peaks were counted for heavy samples of high atomic number elements, no corrected sensitivity has been given because of limited penetration of these radiations. In cases where insufficient counts for good statistical consideration of peak integration or identification were obtained, integral γ-counting data have been tabulated (Table A.7). In Table A.8 a list of elements which did not appear to be activated sufficiently to permit sensitivity calculations of any kind is found. Since interpolation of removal cross-section data for the elements is a long and tedious task, a tabulation is provided in Table A.9 which can be of general use to experimenters engaged in sensitivity determinations and 3 MeV neutron activation analysis of dense samples.

TABLE A.6. Experimental Sensitivities for 3-MeV Neutron Activation Analysis

Atomic Number, Z	Element	Nuclear Reaction	γ-Energy (MeV)	T_{act}	T_{dec}	T_{cnt}	Uncorrected	Corrected	Note
				Experimental Conditions			Experimental Sensitivity (counts/g · 10^6 n · cm^{-2} · sec^{-1}) Sample Attenuation Correction		
4	Be	Gross counts	0.1–4.0	6 s	2 s	6 s	$1.3 \cdot 10^3$		a
9	F	^{19}F(n, γ)^{20}F	1.63	2 m	3 s	2 m	120	130	b
11	Na	^{23}Na(n, γ)^{24}Na	1.369	20 m	0.5 m	20 m	130	150	
			2.754				70	80	
12	Mg	^{26}Mg(n, γ)^{27}Mg	0.84	10 m	0.5 m	10 m	40	60	
13	Al	^{27}Al(n, γ)^{28}Al	1.78	10 m	0.5 m	10 m	570	630	
		^{27}Al(n, p)^{27}Mg	0.84	10 m	0.5 m	10 m	400	470	
14	Si	Gross counts	0.05–2.0	20 m	0.5 m	20 m	100		
15	P	Gross counts	0.05–2.0	20 m	0.5 m	20 m	$3 \cdot 10^3$		
16	S	Gross counts	0.1–4.0	10 m	0.5 m	10 m	Background		
17	Cl	^{37}Cl(n, γ)^{38}Cl	1.60	10 m	0.5 m	10 m	35	40	
			2.17				27	30	
19	K	Gross counts	0.05–2.0	20 m	0.5 m	20 m	70		
20	Ca	Gross counts	0.05–4.0	20 m	0.5 m	20 m	50		
21	Sc	^{45}Sc(n, γ)^{46m}Sc	0.142	2 m	3 s	2 m	$7.7 \cdot 10^3$		c
22	Ti	^{50}Ti(n, γ)^{51}Ti	0.320	20 m	0.5 m	20 m	210	250	
		^{47}Ti(n, p)^{47}Sc	0.160	20 m	0.5 m	20 m	130	170	
23	V	^{51}V(n, γ)^{52}V	1.434	10 m	5 s	10 m	$0.4 \cdot 10^3$	$4.7 \cdot 10^3$	
24	Cr	Gross counts	0.05–1.0	20 m	0.5 m	20 m	100		
25	Mn	^{55}Mn(n, γ)^{56}Mn	0.847	20 m	0.5 m	20 m	$6.7 \cdot 10^3$	$9.0 \cdot 10^3$	
			1.811				730	900	
			2.110				330	430	
26	Fe	Gross counts	0.05–2.0	20 m	0.5 m	20 m	30		
27	Co	^{59}Co(n, γ)^{60m}Co	1.33	10 m	0.5 m	10 m	17	23	
28	Ni	^{58}Ni(n, p)^{58}Co	0.810	20 m	0.5 m	20 m	50	80	
29	Cu	^{63}Cu(n, γ)^{64}Cu	0.51	10 m	0.5 m	10 m	90	180	
			1.02						
		^{65}Cu(n, γ)^{66}Cu	1.039	10 m	0.5 m	10 m	240	400	
30	Zn	^{64}Zn(n, p)^{64}Cu	0.51	20 m	0.5 m	20 m	470	600	
			1.02				210	260	

(TABLE A.6. *continued*)

Z	Element	Reaction	Energy (MeV)	t₁	t₂	t₃	Sens 1	Sens 2
31	Ga	^{71}Ga(n, γ)^{72}Ga	0.601 ⎫ 0.630 ⎭	20 m	0.5 m	20 m	670	700
			0.835				$1.3 \cdot 10^3$	$1.4 \cdot 10^3$
			0.894					
32	Ge	^{74}Ge(n, γ)^{75m}Ge ^{76}Ge(n, γ)^{77m}Ge	0.139 ⎫ 0.159 ⎭	10 m	4 s	10 m	730	900
33	As	^{75}As(n, γ)^{76}As	0.559	20 m	0.5 m	20 m	$1.0 \cdot 10^5$	$1.3 \cdot 10^5$
			1.216 ⎫ 1.22 ⎭				90	110
34	Se	^{76}Se(n, γ)^{77m}Se						
		^{77}Se(n, n'γ)^{77m}Se	0.161	2 m	3 s	2 m	$5.3 \cdot 10^3$	$1.0 \cdot 10^4$
		^{78}Se(n, γ)^{79m}Se ^{80}Se(n, γ)^{81}Se	0.096 ⎫ 0.103 ⎭	10 m	3 m	10 m	370	
35	Br	^{79}Br(n, γ)^{80}Br	0.51	10 m	0.5 m	10 m	630	800
			0.618 ⎫ 0.666 ⎭				$2.2 \cdot 10^3$	$2.7 \cdot 10^3$
			1.02				370	430
37	Rb	^{85}Rb(n, γ)^{86}Rb	0.56	10 m	0.5 m	10 m	$1.9 \cdot 10^3$	$2.1 \cdot 10^3$
38	Sr	^{86}Sr(n, γ)^{87m}Sr ^{87}Sr(n, n'γ)^{87m}Sr	0.388	20 m	0.5 m	20 m	$8.3 \cdot 10^3$	$1.0 \cdot 10^4$
39	Y	^{87}Y(n, n'γ)^{89}Y	0.91	2 m	3 s	2 m	$1.0 \cdot 10^4$	$1.2 \cdot 10^4$
		^{89}Y(n, γ)^{90m}Y	0.202	20 m	3 m	20 m	210	310
			0.482				97	120
40	Zr	^{90}Zr(n, n'γ)^{90m}Zr	2.18	10 s	1.5 s	12 s	100	110 [d]
			2.32					
41	Nb	^{93}Nb(n, γ)^{94m}Nb	0.871	10 m	0.5 m	10 m	60	70
42	Mo	^{100}Mo(n, γ)^{101}Mo	0.191				630	$2.0 \cdot 10^3$
			0.51 0.545(^{101}Tc) ⎭				330	500
			0.59					
			0.70	10 m	0.5 m	10 m	100	150
			1.02				110	150
			2.08				50	70
		↓ ^{101}Tc	0.307				$1 \cdot 10^3$	$1.8 \cdot 10^3$
		^{100}Mo(n, γ)^{101}Mo	0.191				370	$1.2 \cdot 10^3$
			0.51 0.545(^{101}Tc) ⎭				190	280
			0.59					
			0.70	10 m	15 m	10 m	50	70
			1.02				60	80
			2.08				27	33
		↓ ^{101}Tc	0.307				$1.2 \cdot 10^3$	$2.1 \cdot 10^3$
44	Ru	^{104}Ru(n, γ)^{105}Ru	0.67	20 m	0.5 m	20 m	700	730
			0.726					
45	Rh	^{103}Rh(n, γ)^{104}Rh	0.56	10 m	5 s	10 m	$1.0 \cdot 10^3$	
46	Pd	^{108}Pd(n, γ)^{109m}Pd	0.188	20 m	0.5 m	20 m	$2.1 \cdot 10^3$	$2.4 \cdot 10^3$
47	Ag	^{107}Ag(n, n'γ)^{107m}Ag	0.094	2 m	3.5 s	2 m	$3.7 \cdot 10^3$	$1.8 \cdot 10^6$
		^{109}Ag(n, n'γ)^{109m}Ag	0.088					
		^{107}Ag(n, γ)^{108}Ag	0.615					
			0.632	2 m	3.5 s	2 m	400	530
		^{109}Ag(n, γ)^{110}Ag	0.658					
		^{107}Ag(n, γ)^{108}Ag	0.434	10 m	4 m	10 m	57	83
			0.51				30	43
			0.615				220	300
			0.632					
48	Cd	^{110}Cd(n, γ)^{111m}Cd	0.15	10 m	0.5 m	10 m	$2.4 \cdot 10^3$	$6.3 \cdot 10^4$
		^{111}Cd(n, n'γ)^{111m}Cd	0.247				$0.9 \cdot 10^4$	$3.7 \cdot 10^4$

(*continued*)

(TABLE A.6 continued)

Atomic Number, Z	Element	Nuclear Reaction	γ-Energy (MeV)	Experimental Conditions			Experimental Sensitivity (counts/g · 10^6 n · cm^{-2} · sec^{-1})		Note
							Sample Attenuation Correction		
				T_{act}	T_{dec}	T_{cnt}	Uncorrected	Corrected	
49	In	^{115}In(n, n'γ)^{115m}In	0.335	20 m	0.5 m	20 m	$5.7 \cdot 10^4$	$6 \cdot 10^4$	
		^{113}In(n, n'γ)^{113m}In	0.399				$0.5 \cdot 10^5$	$5.3 \cdot 10^4$	
		^{115}In(n, γ)^{116m1}In	0.417						
			0.819				$1.5 \cdot 10^4$	$1.6 \cdot 10^4$	
			1.09				$0.47 \cdot 10^5$	$4.7 \cdot 10^4$	
			1.293				$0.57 \cdot 10^5$	$5.7 \cdot 10^4$	
			2.111				$7.7 \cdot 10^3$	$7.7 \cdot 10^3$	
50	Sn	^{116}Sn(n, γ)^{117m}Sn							
		^{117}Sn(n, n'γ)^{117m}Sn	0.158	10 m	0.5 m	10 m	180	$2.7 \cdot 10^3$	
		^{122}Sn(n, γ)^{123m}Sn	0.160						
		^{124}Sn(n, γ)^{125m}Sn	0.325				400	830	
51	Sb	^{123}Sb(n, γ)^{124m1}Sb	0.505						
			0.603	10 m	0.5 m	10 m	210	310	
			0.644						
52	Te	^{130}Te(n, γ)^{131}Te	0.150	10 m	0.5 m	10 m	$0.5 \cdot 10^3$	$5.7 \cdot 10^3$	
			0.453				210	310	
			0.493						
			0.603				40	60	
			0.95				40	50	
			1.00						
			1.147				30	35	
53	I	^{127}I(n, γ)^{128}I	0.441	10 m	0.5 m	10 m	$7.7 \cdot 10^3$	$9.7 \cdot 10^3$	
			0.528				260	320	
			0.969				190	220	
55	Cs	^{133}Cs(n, γ)^{134m}Cs	0.128	20 m	0.5 m	20 m	370	$1.5 \cdot 10^4$	
56	Ba	^{136}Ba(n, γ)^{137m}Ba	0.662	10 m	0.5 m	10 m	$1.8 \cdot 10^4$	$2.3 \cdot 10^4$	
		^{137}Ba(n, n'γ)^{137m}Ba							
57	La	^{139}La(n, γ)^{140}La	0.487	20 m	0.5 m	20 m	170	240	
			1.596				120	150	
58	Ce	Gross counts	0.05–2.0	20 m	0.5 m	20 m	$0.4 \cdot 10^3$		
59	Pr	Gross counts	0.05–2.0	20 m	0.5 m	20 m	$1.3 \cdot 10^3$		
60	Nd	Gross counts	0.05–2.0	20 m	0.5 m	20 m	$1.6 \cdot 10^3$		
62	Sm	^{154}Sm(n, γ)^{155}Sm	X Eu	10 m	0.5 m	10 m	$2.7 \cdot 10^3$		
			0.104				$6.3 \cdot 10^3$		
63	Eu	Gross counts	0.025–1.65	20 m	0.5 m	20 m	$0.5 \cdot 10^5$		
64	Gd	^{158}Gd(n, γ)^{159}Gd	0.363	10 m	0.5 m	10 m	$2.2 \cdot 10^3$	$2.4 \cdot 10^3$	
		^{160}Gd(n, γ)^{161}Gd	0.361						
65	Tb	Gross counts	0.012–0.5	20 m	0.5 m	20 m	$2.7 \cdot 10^3$		
66	Dy	^{164}Dy(n, γ)^{165m1}Dy	X Dy	10 m	4 s	10 m	$6.3 \cdot 10^3$		
			0.108				$1.6 \cdot 10^3$		
67	Ho	Gross counts	0.05–2.0	20 m	0.5 m	20 m	$1.1 \cdot 10^3$		
68	Er	^{166}Er(n, γ)^{167m}Er }	0.208	24 s	3 s	24 s	$5.7 \cdot 10^3$		e
		^{167}Er(n, n'γ)^{167m}Er }							
		^{170}Er(n, γ)^{171}Er	0.296 } 0.308 }	20 m	0.5 m	20 m	$1.1 \cdot 10^3$		
69	Tm	Gross counts	0.012–0.5	20 m	0.5 m	20 m	$4.3 \cdot 10^3$		
70	Yb	Gross counts	0.025–0.25	2 m	3 s	2 m	150		
71	Lu	^{175}Lu(n, γ)^{176m}Lu }	X Hf	20 m	0.5 m	20 m	$1.7 \cdot 10^4$		
		^{176}Lu(n, n'γ)^{176m}Lu }	0.088				$0.5 \cdot 10^4$		

(continued)

(TABLE A.6 *continued*)

72	Hf	^{178}Hf(n, γ)^{179m}Hf ^{179}Hf(n, n'γ)^{179m}Hf	0.217	3 m	3 s	3 m	475	695
73	Ta	^{181}Ta(n, γ)^{182m}Ta	X Ta	20 m	0.5 m	20 m	$0.6 \cdot 10^3$	
			0.147					
			0.172				770	$5.7 \cdot 10^3$
			0.184					
			0.356				110	170
74	W	^{182}W(n, γ)^{183m}W	0.160	1 m	4 s	1 m	5	ca. $3 \cdot 10^5$
		^{186}W(n, γ)^{187}W	0.479	20 m	1 m	20 m	90	400
			0.686				65	140
75	Re	^{185}Re(n, γ)^{186}Re	X W, X Os					
		^{187}Re(n, γ)^{188}Re	X Os	20 m	0.5 m	20 m	$5.7 \cdot 10^3$	
		^{187}Re(n, γ)^{188}Re	X Re					
77	Ir	^{191}Ir(n, n'γ)^{191m}Ir	X Ir	1 m	3 s	1 m	$1.5 \cdot 10^3$	$0.5 \cdot 10^4$
			0.129				$0.5 \cdot 10^3$	$1.4 \cdot 10^3$
78	Pt	^{196}Pt(n, γ)^{197}Pt	X Au					
			0.077					
		^{198}Pt(n, γ)^{199}Pt	X Au	20 m	0.5 m	20 m	470	
			0.075					
		^{196}Pt(n, γ)^{197}Pt	0.191					
		^{198}Pt(n, γ)^{199}Pt	0.197				100	$0.4 \cdot 10^4$
			0.32				110	$0.4 \cdot 10^3$
			0.475				240	430
			0.540					
			0.715				40	60
			0.790					
			0.960				7	10
79	Au	^{197}Au(n, n'γ)^{197m}Au	X Au	1 m	3 s	1 m	$1.3 \cdot 10^3$	ca. $3 \cdot 10^7$
			0.279				$0.5 \cdot 10^4$	$2.4 \cdot 10^4$
		^{197}Au(n, γ)^{198}Au	0.412	20 m	2 m	20 m	$0.93 \cdot 10^3$	$1.8 \cdot 10^3$
80	Hg	^{198}Hg(n, γ)^{199m}Hg	0.158	10 m	0.5 m	10 m	$1.1 \cdot 10^3$	ca. $3 \cdot 10^7$
		^{199}Hg(n, n'γ)^{199m}Hg	0.375				$1.0 \cdot 10^3$	$0.6 \cdot 10^4$
81	Tl	Gross counts	0.05–2.0	10 m	0.5 m	10 m	670	
82	Pb	^{204}Pb(n, n'γ)^{204m}Pb	0.375	20 m	0.5 m	20 m	25	110
			0.90				45	70
		^{206}Pb(n, γ)^{207}Pb	0.570	10 s	2 s	12 s	55	120
		^{207}Pb(n, n'γ)^{207m}Pb	1.064				30	45 ^{f}
83	Bi	Gross counts	0.025–0.5	20 m	0.5 m	20 m	Background	

Note:

^{a} Cyclic activation; 10 cycles, 10 sec decay between cycles.

^{b} Cyclic activation; 5 cycles, 3 min decay between cycles.

^{c} Cyclic activation; 5 cycles, 6 min decay between cycles.

^{d} Cyclic activation; 10 cycles, 3 min decay between cycles.

^{e} Cyclic activation; 10 cycles, 2 min decay between cycles.

^{f} Cyclic activation; 10 cycles, 30 sec decay between cycles.

TABLE A.7. Integral Counting Results

Atomic Number, Z	Element	Experimental Conditions				Experimental Sensitivity (counts/g · 10^6 n/(cm² · sec)) Uncorrected
		γ-Energy (MeV)	T_{act}	T_{dec}	T_{cnt}	
4	Be	0.1 –4.0	6 s[a]	2 s	6 s	$1.3 \cdot 10^3$
14	Si	0.05 –2.0	20 m	0.5 m	20 m	100
15	P	0.05 –2.0	20 m	0.5 m	20 m	$3 \cdot 10^3$
19	K	0.05 –2.0	20 m	0.5 m	20 m	70
20	Ca	0.05 –4.0	20 m	0.5 m	20 m	50
24	Cr	0.05 –1.0	20 m	0.5 m	20 m	100
26	Fe	0.05 –2.0	20 m	0.5 m	20 m	30
58	Ce	0.05 –2.0	20 m	0.5 m	20 m	400
59	Pr	0.05 –2.0	20 m	0.5 m	20 m	$1.3 \cdot 10^3$
60	Nd	0.05 –2.0	20 m	0.5 m	20 m	$1.6 \cdot 10^3$
63	Eu	0.025–1.65	20 m	0.5 m	20 m	$5.3 \cdot 10^4$
65	Tb	0.012–0.50	20 m	0.5 m	20 m	$2.7 \cdot 10^3$
67	Ho	0.05 –2.0	20 m	0.5 m	20 m	$1.1 \cdot 10^3$
69	Tm	0.012–0.50	20 m	0.5 m	20 m	$4.3 \cdot 10^3$
70	Yb	0.025–0.25	2 m	3 s	2 m	150
81	Tl	0.05 –2.0	10 m	0.5 m	10 m	670

[a] Cyclic activation; 10 cycles, 10 sec decay between cycles.

TABLE A.8. List of Elements with No Observed Activations[a]

Atomic Number, Z	Element
6	Carbon
7	Nitrogen
8	Oxygen
16	Sulfur
83	Bismuth

[a] Irradiation time 20 min, 0.5 min decay and 20 min counting time.

TABLE A.9. Microscopic Removal Cross-Section of the Elements for 3-MeV Neutrons

Element	Atomic Weight	σ (barn)	Element	Atomic Weight	σ (barn)
H	1.00797	2.25			
Li	6.939	1.50	Pd	106.4	2.69
Be	9.022	2.33	Ag	107.870	2.68
B	10.811	1.13	Cd	112.40	2.69
C	12.01115	1.35	In	114.82	2.81
N	14.0067	1.20	Sn	118.69	2.76
O	15.9994	0.98	Sb	121.75	2.85
F	18.9984	1.69	Te	127.60	3.18
Na	22.9898	1.65	I	126.9044	3.18
Mg	24.312	1.35	Cs	132.905	3.38
Al	27.9815	1.59	Ba	137.34	3.60
Si	28.086	1.26	La	138.91	3.57
P	30.9738	1.92	Ce	140.12	3.63
S	32.064	1.82	Pr	140.907	3.63
Cl	35.453	2.04	Nd	144.24	3.66
K	39.102	2.07	Sm	150.35	3.78
Ca	40.08	2.10	Eu	151.96	3.87
Sc	44.956	2.12	Gd	157.25	3.75
Ti	47.90	2.29	Tb	158.924	3.69
V	50.942	2.28	Dy	162.50	3.87
Cr	51.996	2.11	Ho	164.930	3.90
Mn	54.9380	2.18	Er	167.26	3.93
Fe	55.847	2.03	Tm	168.934	4.02
Co	58.9332	2.05	Yb	173.04	4.29
Ni	58.71	2.01	Lu	174.97	4.17
Cu	63.54	1.98	Hf	178.49	4.11
Zn	65.37	1.99	Ta	180.948	4.01
Ga	69.72	1.86	W	183.85	4.26
Ge	72.59	2.04	Re	186.2	4.29
As	74.9216	2.22	Os	190.2	4.17
Se	78.96	2.37	Ir	192.2	4.32
Br	79.909	2.27	Pt	195.09	4.35
Rb	85.47	2.33	Au	196.967	4.41
Sr	87.62	2.94	Hg	200.59	4.56
Y	88.905	2.41	Tl	204.37	4.77
Zr	91.22	2.46	Pb	207.19	4.98
Nb	92.906	2.46	Bi	208.980	4.77
Mo	95.94	2.48	Th	232.038	4.50
Ru	101.07	2.51	U	238.03	4.74
Rh	102.905	2.53			

DISCUSSION OF RESULTS AND CONCLUSIONS

A review of this study shows that 51 elements are sufficiently sensitive using 3 MeV neutron fluxes available from neutron generators to offer possible analytical applications, by photopeak integration, particularly when they are major constituents. In the basic design of our irradiation facility some sacrifice of sensitivity was made in order to obtain a high degree of precision and accuracy.[5] Because of this fact and using the best targets available, the maximum usable 3-MeV neutron flux that can be obtained is of the order of 10^6 n/(cm² sec). However, in most facilities, 3 MeV neutron fluxes of the order of 10^7 n/(cm² sec) can be obtained easily and would provide additional sensitivity information for those elements listed in Table A.7 where integral γ-counting was performed.

REFERENCES

1. F. A. Lundgren, S. S. Nargolwalla, *Anal. Chem.*, 40 (1968) 672.
2. P. D. LaFleur, Ed., NBS Technical Note 458, 1969.
3. A. F. Avery, D. E. Bendall, J. Butler, K. T. Spinney, AERE-R-3216, 1960.
4. E. Storm, H. I. Israel, LA-3753, 1967.
5. S. S. Nargolwalla, M. R. Crambes, J. R. DeVoe, *Anal. Chem.*, 40 (1968) 666.
6. J. R. Stehn, M. D. Goldberg, B. A. Magurno, R. Weiner-Chasman, Neutron Cross Sections, BNL-325, 2nd ed., Suppl. 2, 1964.
7. D. F. Covell, *Anal. Chem.*, 31 (1959) 1785.
8. M. C. Lederer, J. M. Hollander, I. Perlman, *Table of Isotopes*, 6th ed., John Wiley, New York, 1967.

INDEX

Absorption coefficient, total linear, 259, 261
Accelerating Tube, Einsel Lens, 82–83
　Van de Graff, 81–82
Activation Analysis, principles, 2
Activation Curves, 278
Aluminum, analytical methods for, 282–291
　interferences, 283
Analysis, activation on stream, 218, 221
Analysis conditions, prediction of, 277
Antimony, analytical methods, 291–292
Arsenic, analytical methods, 292–296
Attenuation effects, gamma attenuation, 167–170, 254
　neutron attenuation, 156–167

Barium, analytical methods, 296–304
Boron, analytical methods, 305–309
Bromine, analytical methods, 309–314

Cadmium, analytical methods, 314–315
Calcium, analytical methods, 315–322
Cerium, analytical methods, 323–330
　interferences, 330
Cesium, analytical methods, 330–334
Chlorine, analytical methods, 334–339
Chromium, analytical methods, 339–343
　interferences, 343
Cobalt, analytical methods, 343–354
　interferences, 354
Copper, analytical methods, 354–358
Counting integral factor, definition, 279
Cross-Sections, for 14.5 MeV neutrons, 626–631
　isotope effect on, 52
　removal, 158, 159, 259, 261

Decay curves, definition, 279
Detectors, neutron, 171–173
　shielding for, 196–197
Deuterium gas, inlet regulation, 78
Deuteron beam, characteristics, 28

Deuterons, as projectiles, 15
　beam current, 18
　D-D reaction, 19–23
　D-T reaction, 23–29
　nuclear reactions of, 15–16
Dysprosium, analytical methods, 359–363

Energy, binding, 44
　exoergic and endoergic, 11
　interaction, 10, 11
　kinetic, 13, 246
　matter equivalent, 10, 11
　nuclear mass, 11
　threshold, 13, 258
Erbium, analytical methods, 363–367
Errors, systematic due to sample attenuation, 252–269
Europium, analytical methods, 367–371
Evaluation of acceleration neutron sources, 128–219
Excitation functions, for (n,p) reactions, 50–51
　for (n,α) reactions, 53–54
　for (n,2n) reactions, 54–55

Fluorine, analytical methods, 218, 371–380
　interferences, 379
Flux, density, 135, 136
　distribution, 32–42
　fast and thermal, 30–31
　from tritium targets, 255
　pattern, 40
　systematic error, 243–252
　usable, 28

Gadolinium, analytical methods, 380–385
Gallium, analytical methods, 385–389
Gamma absorption law, 254
Gas leaks, gas, 78
　palladium, 79
Germanium, analytical methods, 388–393
Gold, analytical methods, 393–397

Hafnium, analytical methods, 398–402

INDEX

Indium, analytical methods, 402–406
Ion sources, 68
  duoplasmatron source, 74–76
  Penning source, 72–74
  RF source, 69–72
  sealed tube, 120–125
Instrumental error, 269–271
Iridium, analytical methods, 406–410
Iron, analytical methods, 410–419
  interferences, 415, 419

Lead, analytical methods, 419–423
Lutetium, analytical methods, 423–427

Magnesium, analytical methods, 428–436
  interferences, 432
Manganese, analytical methods, 436–444
  interferences, 440
Mercury, analytical methods, 444–448
Moderators for fast neutrons, 30
Molybdenum, analytical methods, 448–449

Neodymium, analytical methods, 449–460
  interferences, 460
Neutron activation analysis, advantages, 3ff
  comparison with other techniques, 274ff
  limitations, 6
Neutron generator, pulsed, dual pulsing, 88
    postacceleration pulsing, 85–86
    preacceleration pulsing, 86–87
    pulsed systems, 83–84
  pumped-type, accelerating tube, 80–83
    gas regulation, 78–80
    general operation, 65–67
    hazards, 142–178
    high voltage supply, 113–114
    ion sources, 68–78
    operational characteristics, 116–117
    pulsing system, 83–89
    safety considerations, 114–115
    schematic, 65
    shielding, 162–170
    targets, 101–113
    vacuum systems, 89–100
  sealed tube, design, 119–120
    ion source, 120–125
    performance, 126–128
    replenisher system, 125–126
    source, 118
    target, 126

Neutron generator facility, cost, 6
Neutrons, attenuation, 156–162
  biological effect, 137–142
  detectors, 171–173
  energies as function of angle, 21, 25
  fast, 19, 43
  from small accelerators, 15
  hazard, 156–173
  producing reactions, 19–29
  sealed tube neutron source, 118–128
  thermalization, 29–31
  yield, 16
Neutrons, 14.5 MeV, cross-sections for n,n',γ reactions, 49
  energies from D-D reactions, 19–23
  energies from D-T reactions, 19, 23–29
  flux distribution, 32–42
  $(n,^3He)$, 58
  $(n,n'p)$, 59
  $(n,2p)$, 58
  radiative capture cross-sections for, 45–47
  reactions of, 42–60
  thermalization of D-D produced neutrons, 31
  thermalization of D-T produced neutrons, 30
  unusual nuclear reactions $(n,n\alpha)$, $(n,\alpha n)$, 56
  yield from reactions, 16–19
Nickel, analytical methods, 460–464
Niobium, analytical methods, 464–465
Nitrogen, analytical methods, 465–473
Nuclear reactions, centrifugal barrier, 15
  coulomb barrier, 14–15
  model, 11–12

Oxygen, analytical methods, 210, 230, 473–480

Palladium, analytical methods, 480–484
Phosphorus, analytical methods, 484–495
  interferences, 489
Positron emitters, analytical methods, 604–617
Potassium, analytical methods, 495–499
Power supply, high voltage, 113–114
Praseodymium, analytical methods, 499–504
Protons and charged particles, 138–139
Pulse pick up, 270

INDEX

Pulsing system, frequency, 86
  plural, 88–89
  postacceleration, 85–86
  preacceleration, 86–88

Q-value, nuclear reaction, 10–11

Radiation, bremstrahlung, 173–176
  dose flux equivalence, 134–142
  expose, 133–142
  maximum (MPD), 134, 136, 138, 141
  prompt gamma, 176
  sky-shine, 170–171
Reactions, D-D, 19–23
  D-T, 23–29
  fast neutrons, 19
  induced by fast neutrons, 42–56
  (n,γ), 44–46, 51–52
  $(n,n'\gamma)$, 46–48
  (n,p), 48–51
  $(n,2n)$, 52–55
  notation, 10
  rare, 56–60
  recoil secondary, 57
Rubidium, analytical methods, 504–508
  interferences, 508
Ruthenium, analytical methods, 508–512

Safety considerations, 114–115
Samarium, analytical methods, 512–523
  interferences, 516
Sample containers, configuration, 204–207
  material, 203–204
  systematic error, 241–243
Sample encapsulation, techniques, for dense solids, 209
  for liquids, 208–209
  for powders, 207–208
  for reactive metals, 209–210
Sample handling, 210–211
Sample transport, auxiliary, 221–222
  biaxial rotating system, 215, 216
  error sources, 211
  systems, 212–221
Sample transport system, pneumatic, 212–218
Selenium, analytical methods, 523–524
Self-shielding, parameters, 253
Sensitivities, analytical, calculated for 14-MeV and thermal neutrons, 634–637
  experimental for, 14-MeV and thermal neutrons, 638–642
  3-MeV neutrons, 643–658
Shielding, biological, typical facilities, 179–196
Silicon, analytical methods, 524–540
  interferences, 528
Silver, analytical methods, 541–545
Sodium, analytical methods, 545–551
  interferences, 550
Stopping power, 16
Strontium, analytical methods, 551–558
Sulfur, analytical methods, 558–562
Systematic error, evaluation, 226–270
  flux, 243–252
  nuclear constants, 227–229
  reaction interferences, 229–231
  recoil effect, 231
  sample container, 241–243

Tantalum, analytical methods, 563–566
Targets, neutron generator, configuration, 112–113
  cooling, 106–107
  design considerations, 102–104
  deuterium, frozen, 104
  gas-on-metal, 105, 106, 107–112
  hydrogen isotope, solid, 105
  pumped type generators, 101–113
  sealed tube generators, 126
  thickness, 27
Terbium, analytical methods, 567
Tin, analytical methods, 567–575
Titanium, analytical methods, 575–579
Total linear absorption coefficient, 259, 261
Tritium, characteristics, 143
  contamination, 145, 150
  critical levels, 143
  handling, 150–153
  in neutron generators, 144–145
  measurement, 153–156
  penetration through materials, 151
  target flux, 255
Tungsten, analytical methods, 579–580

Vacuum systems, design criteria, 100
  gettering effect, 93
  ion pump, 92–96, 148
  materials of construction, 99–100
  oil diffusion pump, 90–92

roughing pump, 90
turbomolecular, 97
vacuum measuring devices, 97–99
Vanadium, analytical methods, 580–588
Voltage multiplier, 64

Working curves for activation analysis, 280

X- and gamma rays, 135–136

Yttrium, analytical methods, 588–595

Zinc, analytical methods, 596–600
Zirconium, analytical methods, 600–604
  interferences, 604